Undergraduate Texts in Physics

Series Editors

Kurt H. Becker, NYU Polytechnic School of Engineering, Brooklyn, NY, USA

Jean-Marc Di Meglio, Matière et Systèmes Complexes, Université Paris Diderot, Bâtiment Condorcet, Paris, France

Sadri D. Hassani, Department of Physics, Loomis Laboratory, University of Illinois at Urbana-Champaign, Urbana, IL, USA

Morten Hjorth-Jensen, Department of Physics, Blindern, University of Oslo, Oslo, Norway

Michael Inglis, Patchogue, NY, USA

Bill Munro, NTT Basic Research Laboratories, Optical Science Laboratories, Atsugi, Kanagawa, Japan

Susan Scott, Department of Quantum Science, Australian National University, Acton, ACT, Australia

Martin Stutzmann, Walter Schottky Institute, Technical University of Munich, Garching, Bayern, Germany

Undergraduate Texts in Physics (UTP) publishes authoritative texts covering topics encountered in a physics undergraduate syllabus. Each title in the series is suitable as an adopted text for undergraduate courses, typically containing practice problems, worked examples, chapter summaries, and suggestions for further reading. UTP titles should provide an exceptionally clear and concise treatment of a subject at undergraduate level, usually based on a successful lecture course. Core and elective subjects are considered for inclusion in UTP.

UTP books will be ideal candidates for course adoption, providing lecturers with a firm basis for development of lecture series, and students with an essential reference for their studies and beyond.

More information about this series at http://www.springer.com/series/15593

Claus Grupen

Astroparticle Physics

Second Edition

With Contribution by Dr. Tilo Stroh

 Springer

Claus Grupen
Universität Siegen FB Physik
Siegen, Germany

ISSN 2510-411X   ISSN 2510-4128   (electronic)
Undergraduate Texts in Physics
ISBN 978-3-030-27341-5   ISBN 978-3-030-27339-2   (eBook)
https://doi.org/10.1007/978-3-030-27339-2

Cover illustration: Interaction of an energetic cosmic ray muon in the ALPEH experiment

This Springer imprint is published by the registered company Springer Nature Switzerland AG
The registered company address is: Gewerbestrasse 11, 6330 Cham, Switzerland

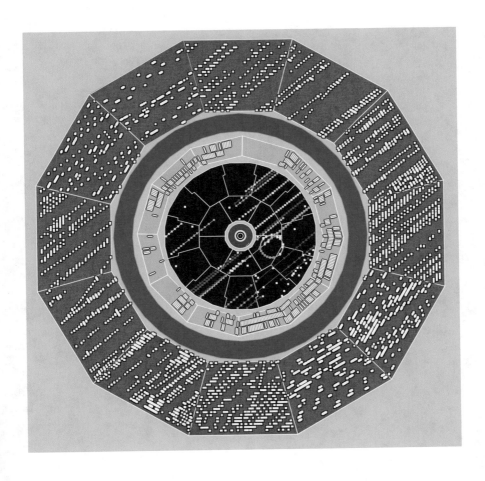

Muon shower in the ALEPH experiment at a depth of 125 m underground

Preface to the Present Second English Edition (2020)

Nothing lasts, nothing is finished, and nothing is perfect.

Alan Saporta

The first English edition of Astroparticle Physics was published in 2005. Even though the period of 14 years since the first edition of this book is short compared to the recently determined age of the universe by the Planck satellite, a lot has happened since 2005. In particular, there are new results in cosmology. The classical Big Bang model is able to explain the flatness, the horizon, and the monopole problems. However, what happened in the first fractions of a second is still unknown. What is bothering us is the still open problem of the dominance of dark matter and dark energy. CP-violating effects are known in the field of particle physics, but they are insufficient to explain the disappearance of antimatter. Results from the satellite experiment PAMELA and the AMS experiment on board the international space station ISS do find a surprisingly high positron excess at high energies. But there is also the possibility that these positrons originate from neutron stars, quasars, or active galactic nuclei.

In particle physics, the discovery of the long-searched-for Higgs boson at the Large Hadron Collider (LHC) at CERN in 2012 is certainly a highlight. With its mass of 125 GeV, this boson supports the Standard Model of particle physics. On the other hand, the non-observation of supersymmetric particles in the mass range of up to one TeV is somewhat disappointing. Supersymmetric particles were and still are considered as dark-matter candidates. With gravitational microlensing one can, however, 'see' the effect of dark matter in the Bullet Cluster, but this 'seeing' is a little indirect. So, the search for dark-matter particles still goes on.

A short flicker of hope for the detection of gravitational waves from the Big Bang came from the BICEP experiment at the South Pole, but it soon crumbled into dust. The expected gravitational waves should exhibit a fingerprint of the polarisation in the primordial cosmic background radiation, for which there seemed some evidence from BICEP. The polarisation of the cosmic blackbody radiation is, however, also influenced by cosmic dust, and the Planck satellite could not confirm BICEP's result. On the other hand, the LIGO detector with its two Michelson interferometers was the first to find excellent evidence for gravitational waves in 2015. This represented an unexpected real breakthrough. Up to the time of writing of this book 11 events were found, most of them dominated by mergers of two black holes, and one by a kilonova caused by a binary-neutron-star merger. The time structure of the gravitational-wave signals and the coincident measurement of the signals in two separate stations about 3000 kilometers apart provides excellent support for the correctness of these findings. This detection of gravitational waves opens up a new window for an astronomy in addition to the observation of celestial objects by electromagnetic radiation.

In cosmic rays the ICECUBE experiment found evidence for cosmogenic, high-energy neutrino events in the PeV range since 2013, at least one of them which possibly correlates with a radiation burst of a blazar residing in a galaxy at a distance of almost ten billion light-years. Therewith, also the neutrino astronomy enters the stage of astroparticle physics in the high-energy regime.

Also, new developments in detection techniques might gain additional scientific knowledge. For example, the radio measurements of high-energy cosmic rays provide a cost-effective technique for the measurement of primary cosmic rays in the range beyond 100 PeV in extensive air showers. The radio signal is believed to originate from geomagnetic synchrotron radiation. The pioneer experiments LOPES and LOFAR have opened up a new detection window by radio astronomy in the field of high-energy astroparticle physics. Radio experiments with their 100% duty time can observe the sky all day, in contrast to fluorescence and Cherenkov telescopes, which can only operate in cloudless and moonless nights. The Square Kilometre Array (SKA) presently under construction will be the most sensitive radio telescope in the future, and could contribute to uncover the origin and development of our universe.

The Planck satellite with its excellent angular resolution of five arc minutes (COBE: 7 degrees, WMAP: 13.5 arc minutes) and increased sensitivity has measured the cosmic background radiation in large detail, and has provided a set of new important cosmological parameters.

Compared to the first English edition we have added also new sections, like the one on extrasolar planets, which take into account new developments in astroparticle physics. Naturally, all chapters have been updated so that they present the most recent knowledge in this field.

The problem sections of the first edition of the book have also been updated and some problems to new sections have been added.

I thank Prof. Dr. Glen Cowan for his important ideas and suggestions, in particular, in the field of cosmology and the early universe, and for allowing me to use these ideas from the first edition of this book. I also acknowledge the numerous contributions of Dr. Tilo Stroh to the present edition, in particular, his successful efforts to produce a high-quality appearance of this book.

Siegen, Germany   Claus Grupen
September 2019

Preface to the First English Edition (2005)

This book on astroparticle physics is the translation of the book on 'Astroteilchenphysik' published in German by Vieweg, Wiesbaden, in the year 2000. It is not only a translation, however, but also an update. The young field of astroparticle physics is developing so rapidly, in particular with respect to 'new astronomies' such as neutrino astronomy and the detailed measurements of cosmic background radiation, that these new experimental results and also new theoretical insights need to be included.

The details of the creation of the universe are not fully understood yet and it is still not completely clear how the world will end, but recent results from supernovae observations and precise measurement of the primordial blackbody radiation seem to indicate with increasing reliability that we are living in a flat Euclidean universe which expands in an accelerated fashion.

In the past couple of years, cosmology has matured from a speculative science to a field of textbook knowledge with precision measurements at the percent level.

The updating process has been advanced mainly by my colleague Dr. Glen Cowan, who is lecturing on astroparticle physics at Royal Holloway College, London, and by myself. The chapter on 'Cosmology' has been rewritten, and chapters on 'The Early Universe', 'Big Bang Nucleosynthesis', 'The Cosmic Microwave Background', and 'Inflation' as well as a section on gravitational astronomy have been added. The old chapter on 'Unsolved Problems' chapter on 'Dark Matter', and part of it went into chapters on primary and secondary cosmic rays.

The book has been extended by a large number of problems related to astroparticle physics. Full solutions to all problems are given. To ease the understanding of theoretical aspects and the interpretation of physics data, a mathematical appendix is offered, where most of the formulae used are presented and/or derived. In addition, details on the thermodynamics of the early universe have been treated in a separate appendix.

Professor Dr. Simon Eidelman from the Budker Institute of Nuclear Physics in Novosibirsk and Dipl.Phys. Tilo Stroh have carefully checked the problems and proposed new ones. Dr. Ralph Kretschmer contributed some interesting and very intricate problems. I have also received many comments from my colleagues and students in Siegen.

The technical aspects of producing the English version lay in the hands of Ms. Ute Smolik, Lisa Hoppe, and Ms. Angelika Wied (text), Dipl.Phys. Stefan Armbrust (updating the figures), Dr. Glen Cowan and Ross Richardson (polished my own English translation), and M.Sc. Mehmet T. Kurt (helping with the editing). The final appearance of the book including many comments on the text, the figures, and the layout was accomplished by Dipl.Phys. Tilo Stroh and M.Sc. Nadir Omar Hashim.

Without the help of these people, it would have been impossible for me to complete the translation in any reasonable time, if at all. In particular, I would like to thank my colleague Prof. Dr. Torsten Fließbach, an expert on Einstein's theory of general relativity, for his critical assessment of the chapter on cosmology and for proposing significant improvements. Also, the contributions by Dr. Glen Cowan on the new insights into the evolution of the early universe and related subjects are highly appreciated. Dr. Cowan has really added essential ingredients to the last chapters of the book. Finally, Prof. Dr. Simon Eidelman, Dr. Armin Böhrer, and Dipl.Phys. Tilo Stroh read the manuscript with great care and made invaluable comments. I thank all my friends for their help in creating this English version of my book.

Siegen, February 2005   Claus Grupen

Preface to the German Edition

The field of astroparticle physics is not really a new one. Up until 1960, the physics of cosmic rays essentially represented this domain. Elementary particle physics in accelerators has evolved from the study of elementary particle processes in cosmic radiation. Among others, the first antiparticles (positrons) and the members of the second lepton generation (muons) were discovered in cosmic-ray experiments.

The close relationship between cosmology and particle physics was, however, recognized only relatively recently. Hubble's discovery of the expanding universe indicates that the cosmos originally must have had a very small size. At such primeval times, the universe was a microworld that can only be described by quantum-theoretical methods of elementary particle physics. Today, particle physicists try to recreate the conditions that existed in the early universe by using electron–positron and proton–antiproton collisions at high energies to simulate 'mini Big Bangs'.

The popular theories of elementary particle physics attempt to unify the various types of interactions in the Standard Model. The experimental confirmation of the existence of heavy vector bosons that mediate weak interactions (W^+, W^-, Z^0), and progress in the theoretical understanding of strong interactions seem to indicate that one may be able to understand the development of the universe just after the Big Bang. The high temperatures or energies that existed at the time of the Big Bang will, however, never be reached in earthbound laboratories. This is why a symbiosis of particle physics, astronomy, and cosmology is only too natural. Whether this new field is named astroparticle physics or particle astrophysics is more or less a matter of taste or the background of the author. This book will deal both with astrophysics and elementary particle physics aspects. We will equally discuss the concepts of astrophysics focusing on particles and particle physics using astrophysical methods. The guiding line is physics with astroparticles. This is why I preferred the term astroparticle physics over particle astrophysics.

After a relatively detailed historical introduction (Chap. 1) in which the milestones of astroparticle physics are mentioned, the basics of elementary particle physics (Chap. 2), particle interactions (Chap. 3), and measurement techniques

(Chap. 4) are presented. Astronomical aspects prevail in the discussion of acceleration mechanisms (Chap. 5) and primary cosmic rays (Chap. 6). In these fields, new disciplines such as neutrino and gamma-ray astronomy represent a close link to particle physics. This aspect is even more pronounced in the presentation of secondary cosmic rays (Chap. 7). On the one hand, secondary cosmic rays have been a gold mine for discoveries in elementary particle physics. On the other hand, however, they sometimes represent an annoying background in astroparticle observations.

The highlight of astroparticle physics is surely cosmology (Chap. 8) in which the theory of general relativity, which describes the macrocosm, is united with the successes of elementary particle physics. Naturally, not all questions have been answered; therefore a final chapter is devoted to open and unsolved problems in astroparticle physics (Chap. 9).

The book tries to bridge the gap between popular presentations of astroparticle physics and textbooks written for advanced students. The necessary basics from elementary particle physics, quantum physics, and special relativity are carefully introduced and applied, without rigorous derivation from appropriate mathematical treatments. It should be possible to understand the calculations presented with the knowledge of basic A-level mathematics. On top of that, the basic ideas discussed in this book can be followed without referring to special mathematical derivations.

I owe thanks to many people for their help during the writing of this book. Dr. Armin Böhrer read the manuscript with great care. Ms. Ute Bender and Ms. Angelika Wied wrote the text, and Ms. Claudia Hauke prepared the figures that were finalized by Dipl.Phys. Stefan Armbrust. I owe special thanks to Dr. Klaus Affholderbach and Dipl.Phys. Olaf Krasel who created the computer layout of the whole book in the LaTeX style. I am especially indebted to Dipl.Phys. Tilo Stroh for his constant help, not only as far as physics questions are concerned, but also in particular for applying the final touch to the manuscript with his inimitable, masterful eye for finding the remaining flaws in the text and the figures. Finally, I owe many thanks to the Vieweg editors, Ms. Christine Haite and Dipl.Math. Wolfgang Schwarz.

Geneva, July 2000

Claus Grupen

Contents

Chapter 1
Historical Introduction

Look into the past as guidance for the future.

Robert Jacob Goodkin

The field of astroparticle physics, or particle astrophysics, is relatively new. It is therefore not easy to describe the history of this branch of research. The selection of milestones in this book is necessarily subject to a certain arbitrariness and personal taste.

Historically, astroparticle physics is based on optical astronomy. As detector techniques improved, this observational science matured into astrophysics. This research topic involves many subfields of physics, like mechanics and electrodynamics, thermodynamics, plasma physics, nuclear physics, and elementary particle physics, as well as special and general relativity. Precise knowledge of particle physics is necessary to understand many astrophysical contexts, particularly since comparable experimental conditions cannot be prepared in the laboratory. The astrophysical environment therefore constitutes an important laboratory for high energy physicists.

The use of the term astroparticle physics is certainly justified, since astronomical objects have been observed in the 'light' of elementary particles. Of course, one could argue that X-ray or gamma-ray astronomy is more closely related to astronomy rather than to astroparticle physics. To be on the safe side, the new term astroparticle physics should be restricted to 'real' elementary particles. The observations of our Sun in the light of neutrinos, in the Homestake Mine (Davis experiment) in 1967, constitutes the birth of astroparticle physics, even though the first measurements of solar neutrinos by this radiochemical experiment were performed without directional correlation. It is only since the Kamiokande[1] experiment of 1987, and

---

[1] Kamiokande—Kamioka Nucleon Decay Experiment.

© Springer Nature Switzerland AG 2020
C. Grupen, *Astroparticle Physics*, Undergraduate Texts in Physics,
https://doi.org/10.1007/978-3-030-27339-2_1

later the Super-Kamiokande experiment, that one has been able to 'see' the Sun in real time by measuring the direction of the emitted neutrinos. Nature was also kind enough to explode a supernova in the Large Magellanic Cloud in 1987 (SN 1987A), whose neutrino burst could be recorded in the large water Cherenkov detectors of Kamiokande and IMB[2] and in the scintillator experiment at Baksan.

Presently, the fields of gamma and neutrino astronomy are expanding rapidly. The discovery of gravitational waves in 2015 has added a new domain to astronomy. Astronomy with charged particles, however, is a different matter. Irregular interstellar and intergalactic magnetic fields randomize the directions of charged cosmic rays. Only particles at very high energies travel along approximately straight lines through magnetic fields. This makes astronomy with charged particles possible, if the intensity of energetic primaries is sufficiently high.

Actually, there are tentative hints that the highest-energy cosmic rays ($> 10^{19}$ eV) have a non-uniform distribution and possibly originate from the supergalactic plane. This plane is an accumulation of galaxies in a disk-like fashion, in a similar way that stars form the Milky Way. Other possible sources, however, are individual galactic nuclei (like M87) at cosmological distances.

The milestones, which have contributed to the new discipline of astroparticle physics, shall be presented in chronological order. For that purpose, the relevant discoveries in astronomy, cosmic rays, and elementary particle physics will be considered in a well-balanced way. It is, of course, true that this selection is subject to personal bias.

1.1 Early Indications of Celestial Phenomena in the Sky

Only when creative people take ownership of cosmic discovery
will society accept science as the cultural activity that it is.

Neil deGrasse Tyson

It is interesting to point out the observations of the Vela supernova by the Sumerians 6000 years ago. This supernova exploded in the constellation Vela at a distance of 1500 light-years. Today the remnant of this explosion is visible, e.g., in the X-ray and gamma range. Vela X1 is a binary, one component of which is the Vela Pulsar. With a rotational period of 89 ms the Vela Pulsar is one of the 'slowest' pulsars so far observed in binaries. The naming scheme of X-ray sources is such that Vela X1 denotes the strongest ('the first') X-ray source in the constellation Vela.

The second spectacular supernova explosion was observed in China in 1054. The relic of this outburst is the Crab Nebula, whose remnant also emits X rays and gamma rays like Vela X1. Because of its time-independent brightness the Crab is often used as a 'standard candle' in gamma-ray astronomy (Fig. 1.1).

The observation of the northern lights (Gassendi 1621 and Halley 1716) as the aurora borealis ('northern dawn') lead Mairan, in 1733, to the idea that this

---

[2]IMB—Irvine–Michigan–Brookhaven collaboration.

Fig. 1.1 Crab Nebula [1, 2]

Fig. 1.2 Helical track of an electron in the Earth's magnetic field

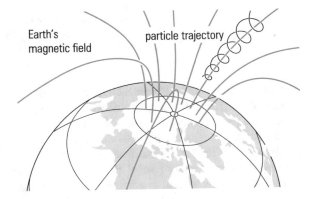

phenomenon might be of solar origin. Northern and southern lights are caused by solar electrons and protons incident in the polar regions traveling on helical trajectories along the Earth's magnetic field lines. At high latitudes, the charged particles essentially follow the magnetic field lines. This allows them to penetrate much deeper into the atmosphere, compared to equatorial latitudes, where they have to cross the field lines perpendicularly (Figs. 1.2 and 1.3).

It is also worth mentioning that the first correct interpretation of nebulae, as accumulations of stars that form galaxies, was given by a philosopher (Kant 1775) rather than by an astronomer.

Fig. 1.3 Aurora Borealis;
Photo credit: Dennis
Mammana [3]

1.2 Discoveries in the 20th Century

> *Astronomy is perhaps the science whose discoveries owe least to chance, in which human understanding appears in its whole magnitude, and through which man can best learn how small he is.*
>
> Georg Christoph Lichtenberg

The discovery of X rays (Röntgen 1895, Nobel Prize 1901), radioactivity (Becquerel 1896, Nobel Prize 1903), and the electron (Thomson 1897, Nobel Prize 1906) already indicated a particle physics aspect of astronomy. At the turn of the century Wilson (1900) and Elster and Geitel (1900) were concerned with measuring the remnant conductivity of air. Rutherford realized in 1903 that shielding an electroscope reduced the remnant conductivity (Nobel Prize 1908 for investigations on radioactive elements). It was only natural to assume that the radioactivity of certain ores present in the Earth's crust, as discovered by Becquerel, was responsible for this effect.

In 1910, Wulf measured a reduced intensity in an electrometer at the top of the Eiffel tower, apparently confirming the terrestrial origin of the ionizing radiation. Measurements by Hess (1911/1912, Nobel Prize 1936) with balloons at altitudes of up to 5 km showed that, in addition to the terrestrial component, there must also be a source of ionizing radiation, which becomes stronger with increasing altitude (Figs. 1.4 and 1.5).

This extraterrestrial component was confirmed by Kohlhörster two years later (1914). By developing the cloud chamber in 1912, Wilson made it possible to detect and follow the tracks left by ionizing particles (Nobel Prize 1927).

The extraterrestrial cosmic radiation that increases with altitude ('Höhenstrahlung') has numerous experimental possibilities (Fig. 1.6) and is of special importance to the development of astroparticle physics.

In parallel to these experimental observations, Einstein developed his theories of special and general relativity (1905 and 1916). The theory of *special relativity* is of paramount importance for particle physics, while the prevailing domain of

Fig. 1.4   Victor Hess at a balloon ascent for measuring cosmic radiation [4]

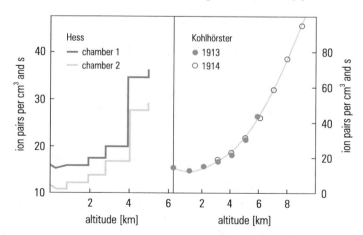

Fig. 1.5   Measurements of Hess (*left*) and Kohlhörster (*right*) showing the dependence of ionization on the altitude in the atmosphere [5, 6]

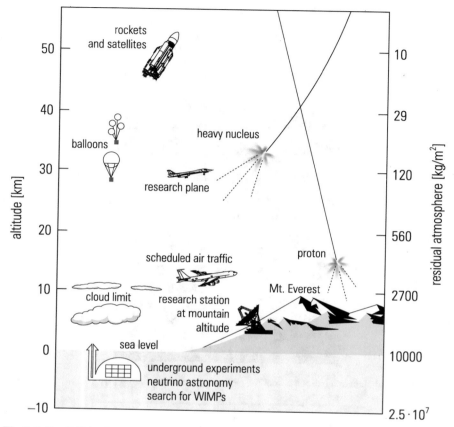

Fig. 1.6 Possibilities for experiments in the field of cosmic rays

general relativity is cosmology. Einstein received the Nobel Prize in 1921, however, not for his fundamental theories on relativity and gravitation, but for the correct quantum-mechanical interpretation of the photoelectric effect and the explanation of Brownian motion. Obviously the Nobel committee in Stockholm was not aware of the outstanding importance of the theories of relativity or possibly not even sure about the correctness of their predictions. This occurred even though Schwarzschild had already drawn correct conclusions for the existence of black holes as early as 1916, and Eddington had verified the predicted gravitational bending of light passing near the Sun during the solar eclipse in 1919. The experimental observation of the deflection of light in gravitational fields also constituted the discovery of *gravitational lensing*. This is when the image of a star appears to be displaced due to the gravitational lensing of light that passes near a massive object. This effect can also lead to double, multiple, or ring-shaped images of a distant star or galaxy if there is a massive object in the line of sight between the observer on Earth and the star

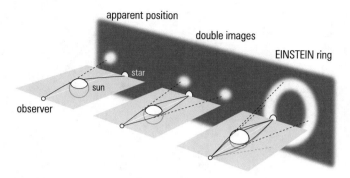

Fig. 1.7 Gravitational lensing. **a** light deflection, **b** double images, **c** Einstein ring

Fig. 1.8 Einstein cross [7]

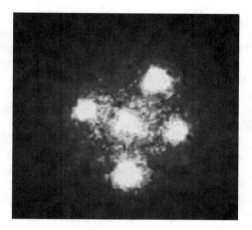

(Fig. 1.7). It was only in 1979 that multiple images of a quasar (double quasar) could be observed. This was followed in 1988 by an Einstein ring in a radio galaxy, as predicted by Einstein in 1936. Other configurations caused by gravitational lensing, like the Einstein cross were also seen (Figs. 1.7 and 1.8).

In the field of astronomy, stars are classified according to their brightness and colour of the spectrum (Hertzsprung–Russell diagram 1911). This scheme allowed a better understanding of the stellar evolution of main-sequence stars to red giants and white dwarves. In 1924 Hubble was able to confirm Kant's speculation that 'nebulae' are accumulations of stars in galaxies, by resolving individual stars in the Andromeda Nebula. Only a few years later (1929), he observed the redshift of the spectral lines of distant galaxies, thereby demonstrating experimentally that the universe is expanding.

In the meantime, a clearer picture about the nature of cosmic rays had emerged. Using new detector techniques in 1926, Hoffmann observed particle multiplication under absorbing layers ('Hoffmann's collisions'). In 1927, Clay demonstrated the dependence of the cosmic-ray intensity on the geomagnetic latitude. This was a

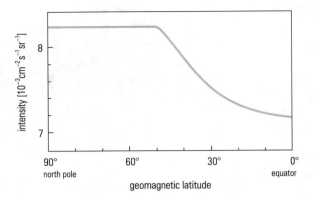

Fig. 1.9 Latitude effect: geomagnetic and atmospheric cutoff

clear indication of the charged-particle nature of cosmic rays, since photons would not have been influenced by the Earth's magnetic field.

Primary cosmic rays can penetrate deep into the atmosphere at the Earth's poles, by traveling parallel to the magnetic field lines. At the Equator they would feel the full component of the Lorentz force ($F = e(v \times B)$; F—Lorentz force, v—velocity of the cosmic-ray particle, B—Earth's magnetic field, e—elementary charge: at the poles $v \parallel B$ holds with the consequence of $F = 0$, while at the Equator one has $v \perp B$, which leads to $|F| = e \, v \, B$). This *latitude effect* was controversial at the time, because expeditions starting from medium latitudes ($\approx 50°$ north) to the Equator definitely showed this effect, whereas expeditions to the North Pole observed no further increase in cosmic-ray intensity. This result could be explained by the fact that charged cosmic-ray particles not only have to overcome the magnetic cutoff, but also suffer a certain ionization energy loss in the atmosphere. This atmospheric cutoff of about 2 GeV prevents a further increase in the cosmic-ray intensity towards the poles (Fig. 1.9). In 1929 Bothe and Kohlhörster could finally confirm the charged-particle character of cosmic rays at sea level by using coincidence techniques (Nobel Prize for Bothe 1954 for his discoveries in cosmic rays with his coincidence method).

As early as 1930, Störmer calculated trajectories of charged particles through the Earth's magnetic field to better understand the geomagnetic effects. In these calculations, he initially used positions far away from the Earth as starting points for the cosmic-ray particles. He soon realized, however, that most particles failed to reach sea level due to the action of the magnetic field. The low efficiency of this approach led him to the idea of releasing antiparticles from sea level to discover, where the Earth's magnetic field would guide them. In these studies, he observed that particles with certain momenta could be trapped by the magnetic field, which caused them to propagate back and forth from one magnetic pole to the other in a process called 'magnetic mirroring'. The accumulated particles form radiation belts, which were discovered in 1958 by Van Allen with experiments on board the Explorer I satellite (Fig. 1.10).

The final proof that primary cosmic rays consist predominantly of positively charged particles was established by the observation of the *east–west effect* (John-

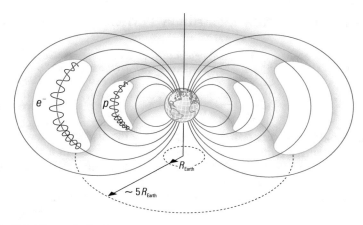

Fig. 1.10 Van Allen radiation belts

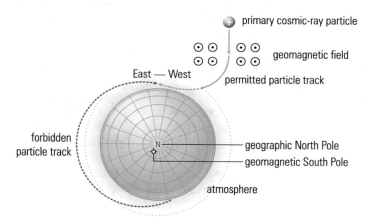

Fig. 1.11 East–west effect [8]

son and Alvarez and Compton, Nobel Prize Alvarez 1968, Nobel Prize Compton 1927). Considering the direction of incidence of cosmic-ray particles at the North Pole, one finds a higher intensity from the west compared to the east. The origin of this asymmetry relates to the fact that some possible trajectories of positively charged particles from easterly directions do not reach out into space (dashed tracks in Fig. 1.11). Therefore, the intensity from these directions is reduced.

In 1933, Rossi showed in a coincidence experiment that secondary cosmic rays at sea level initiate cascades in a lead absorber of variable thickness ('Rossi curve'). The absorption measurements in his apparatus also indicated that cosmic rays at sea level consist of a soft and a penetrating component.

1.3   Discoveries of New Elementary Particles in Cosmic Rays

> *The usual approach of science of constructing a mathematical*
> *model cannot answer the questions of why there should be a*
> *universe for the model to describe. Why does the universe go*
> *to all the bother of existing?*
>
> Stephen Hawking

Up to the thirties, only electrons, protons (as part of the nucleus), and photons were known as elementary particles. The positron was discovered in a cloud chamber by Anderson in 1932 (Nobel Prize 1936). This was the antiparticle of the electron, which was predicted by Dirac in 1928 (Nobel Prize 1933). This, and the discovery of the neutron by Chadwick in 1932 (Nobel Prize 1935), started a new chapter in elementary particle and astroparticle physics. Additionally in 1930, Pauli postulated the existence of a neutral, massless spin-$\frac{1}{2}$ particle to restore the validity of the energy, momentum, and angular-momentum conservation laws that appeared to be violated in nuclear beta decay (Nobel Prize 1945). This hypothetical enigmatic particle, the neutrino, could only be shown to exist in a reactor experiment in 1956 (Cowan and Reines, Nobel Prize to Reines in 1995). It eventually lead to a completely new branch of astronomy; *neutrino astronomy* is a classic example of a perfect interplay between elementary particle physics and astronomy.

It was reported that Landau (Nobel Prize 1962), within several hours of hearing about the discovery of the neutron, predicted the existence of cold, dense stars that consisted mainly of neutrons. In 1967, the existence of rotating neutron stars (pulsars) was confirmed by observing radio signals from these objects (Hewish and Bell, Nobel Prize for Hewish 1975).

Neutrons in a neutron star do not decay. This is because the Pauli exclusion principle (1925) forbids neutrons to decay into occupied electron states. The Fermi energy of remnant electrons in a neutron star is at around several 100 MeV, while the maximum energy transferred to electrons in neutron decay is 0.77 MeV. There are therefore no vacant electron levels available.

After discovering the neutron, the second building block of the nucleus, the question arose of how atomic nuclei could stick together. Although neutrons are electrically neutral, the protons would electrostatically repel each other. Based on the range of the nuclear force and Heisenberg's uncertainty principle (1927, Nobel Prize 1932), Yukawa conjectured in 1935 that unstable mesons of 200-fold electron mass could possibly mediate nuclear forces (Nobel Prize 1949). Initially it appeared that the muon, discovered by Anderson and Neddermeyer in a cloud chamber in 1937 (see Fig. 1.12), had the required properties of the hypothetical Yukawa particle. The muon, however, had no strong interactions with matter, and it soon became clear that the muon was a heavy counterpart of the electron. The fact that another electron-like particle existed caused Rabi (Nobel Prize 1944 for his resonance method for recording the magnetic properties of atomic nuclei) to remark: "Who ordered this?" Rabi's question remains unanswered to this day. The situation became even more critical

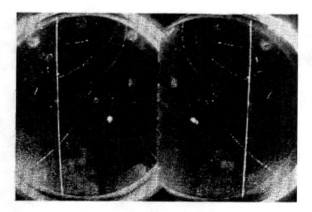

Fig. 1.12 Stereo view of a cosmic-ray muon in a cloud chamber (Anderson and Neddermeyer) [9]

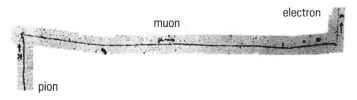

Fig. 1.13 Decay of a charged pion into a muon and subsequently into an electron in a nuclear emulsion [10]

when Perl (Nobel Prize 1995) discovered another, even heavier lepton, the tau, in 1975.

The discovery of the strongly interacting charged pions ($\pi^{\pm}$) in 1947 by Lattes, Occhialini, Powell, and Muirhead, using nuclear emulsions exposed to cosmic rays at mountain altitudes, solved the puzzle about the Yukawa particles (Nobel Prize 1950 to Cecil Powell for his development of the photographic method of studying nuclear processes and his discoveries regarding mesons made with this method; see Fig. 1.13). The pion family was supplemented in 1950 by the discovery of the neutral pion (π^0). Since 1949, pions can also be produced in particle accelerators.

Up to this time, elementary particles were predominantly discovered in cosmic rays. In addition to the muon ($\mu^{\pm}$) and the pions (π^+, π^-, π^0), tracks of charged and neutral kaons were observed in cloud-chamber events. Neutral kaons revealed themselves through their decay into two charged particles. This made the K^0 appear as an upside-down 'V', because only the ionization tracks of the charged decay products of the K^0 were visible in the cloud chamber (Rochester and Butler 1947, Fig. 1.14).

In 1951, some of the upside-down 'V's, which were thought to be neutral kaons, were in fact recognized as Lambda baryons, which also decayed relatively quickly into two charged secondaries ($\Lambda^0 \rightarrow p + \pi^-$). In addition, the Ξ and Σ hyperons

Fig. 1.14 Decay of a neutral
kaon in a cloud chamber.
The two tracks of the decay
pions are visible at the *lower
right* [11, 12]

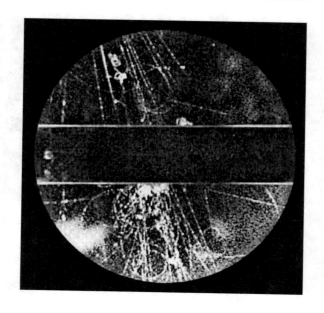

were discovered in cosmic rays (Ξ: Armenteros et al. 1952; Σ: Tomasini et al. 1953). In 1954 Yehuda Eisenberg exposed a stack of nuclear emulsions at an altitude of 30 km to cosmic rays, and he found tracks that could have originated from the decay of an Omega minus (Ω^-) [13].

Apart from studying local interactions of cosmic-ray particles, their global properties were also investigated. The showers observed under lead plates by Rossi were also found in the atmosphere (Pfotzer 1936). The interactions of primary cosmic rays in the atmosphere initiate *extensive air showers*, see Sect. 7.4, (Auger 1938). These showers lead to a maximum intensity of cosmic rays at altitudes of 15 km above sea level ('Pfotzer maximum', Fig. 1.15).

One year earlier (1937), Bethe and Heitler, and at the same time Carlson and Oppenheimer, developed the theory of electromagnetic cascades, which was successfully used to describe the extensive air showers.

In 1938, Bethe together with Weizsäcker, solved the long-standing mystery of the energy generation in stars. The fusion of protons leads to the production of helium nuclei, in which the binding energy of 6.6 MeV per nucleon is released, making the stars shine (Nobel Prize for Bethe in 1967).

In 1937, Forbush realized that a significant decrease of the cosmic-ray intensity correlated with an increased solar activity. The active Sun appears to create some sort of *solar wind* that consists of charged particles, whose flux generates a magnetic field in addition to the geomagnetic field. The solar activity thereby modulates the galactic component of cosmic rays (Figs. 1.16, 1.17).

The observation that the tails of comets always point away from the Sun led Biermann to conclude in 1951, that some kind of solar wind must exist. This more or less continuous particle flux was first directly observed by the Mariner 2 space probe in 1962. The solar wind consists predominantly of electrons and protons, with a small

Fig. 1.15 Intensity profile of cosmic-ray particles in the atmosphere

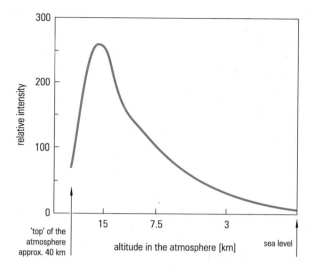

Fig. 1.16 Influence of the solar wind on the Earth's magnetic field

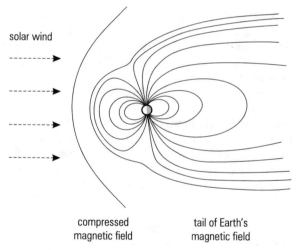

admixture of α particles. The particle intensities at a distance of one astronomical unit (the distance from Sun to Earth) are 2×10^8 ions/(cm^2 s). This propagating solar plasma carries part of the solar magnetic field with it, thereby preventing some primary cosmic-ray particles to reach the Earth.

In 1949 it became clear that primary cosmic rays consisted mainly of protons. Schein, Jesse, and Wollan used balloon experiments to identify protons as the carriers of cosmic radiation.

Fermi (Nobel Prize in 1938 for experiments on radioactivity and the theory of nuclear beta decay) investigated the interactions of cosmic-ray particles with atmospheric atomic nuclei and with the solar and terrestrial magnetic fields. By as early as

Fig. 1.17 Variation of the cosmic-ray intensity during a solar particle eruption (Forbush decrease). Such a solar wind of charged particles can present a serious radiation hazard for astronauts [14]

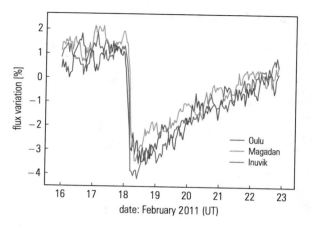

1949, he also had considered possible mechanisms, which might accelerate cosmic-ray particles to very high energies.

Meanwhile, it had been discovered that in addition to electrons, protons, and α particles, the whole spectrum of heavy nuclei existed in cosmic radiation (Freier, Bradt, Peters 1948). In 1950, ter Haar discussed supernova explosions as the possible origin of cosmic rays, an idea that was later confirmed by simulations and measurements.

After discovering the positron in 1932, the antiproton, the second known antiparticle, was found in an accelerator experiment by Chamberlain and Segrè in 1955 (Nobel Prize in 1959). Positrons (Meyer and Vogt, Earl 1961) and antiprotons (Golden 1979) were later observed in primary cosmic rays. It is, however, assumed that these cosmic-ray antiparticles do not originate from sources consisting of antimatter, but are produced in secondary interactions between primary cosmic rays and the interstellar gas or in the upper layers of the atmosphere.

1.4 Start of the Satellite Era

Having probes in space was like having a cataract removed.

Hannes Alfvén

The launch of the first artificial satellite (Sputnik, October 4th, 1957) paved the way for developments that provided completely new opportunities in astroparticle physics. The atmosphere represents an absorber with a thickness of ≈ 25 radiation lengths. The observation of primary X rays and gamma radiation was previously impossible due to their absorption in the upper layers of the atmosphere. This electromagnetic radiation can only be investigated—undisturbed by atmospheric absorption—at very high altitudes near the 'top' of the atmosphere. It still took some time until the first X-ray satellites (e.g., 1970 Uhuru, 1978 Einstein Observatory,

1983 Exosat; Nobel Prize for R. Giacconi 2002) and gamma satellites (e.g., 1967 Vela, 1969 OSO-3, 1972 SAS-2, 1975 COS-B)[3] were launched. They provided a wealth of new data in a hitherto unaccessible spectral range. The galactic center was found to be bright in X rays and gamma rays, and the first point sources of high-energy astroparticles could also be detected (Crab Nebula, Vela X1, Cygnus X3, ...).

With the discovery of *quasistellar radio sources* (quasars, 1960), mankind advanced as far as to the edge of the universe. Quasars appear to outshine whole galaxies if they are really located at cosmological distances. Their distance is determined from the redshift of their spectral lines. A very distant quasar was discovered in 1999 with a redshift of $z = \frac{\lambda - \lambda_0}{\lambda_0} = 7.085$. An object even farther away is the galaxy Abell 1835 IR 1916 with a redshift of $z = 10$. Its discovery was made possible through light amplification by a factor of about 50 resulting from strong gravitational lensing by a very massive galactic cluster in the line of sight to the distant galaxy [15]. As will be discussed in Chap. 8, this implies that the quasar is seen in a state when the universe was less than 5% of its present age. Consequently, this quasar resides at a distance of 13 billion light-years.[4] Initially, there was some controversy about whether the observed quasar redshifts were of gravitational or cosmological origin. Today, there is no doubt that the observed large redshifts are a consequence of the Hubble expansion of the universe. The Hubble telescope even found very distant galaxies with redshifts of up to $z = 11.9$. The largest redshift ever measured is actually that of the cosmic microwave background ($z \approx 1100$).

The expansion of the universe implies that it began in a giant explosion, some time in the past. Based on this Big Bang hypothesis, one arrives at the conclusion that this must have occurred about 14 billion years ago. The *Big Bang model* was in competition with the idea of a steady-state universe for quite some time. The steady-state model was based on the assumption that the universe as a whole was time independent with new stars being continuously created while old stars died out. On the other hand, Gamow had been speculating since the forties that there should be a residual radiation from the Big Bang. According to his estimate, the temperature of this radiation should be in the range of a few kelvin. Penzias and Wilson (Nobel Prize 1978) detected this echo of the Big Bang by chance in 1965, while they were trying to develop low-noise radio antennae (Fig. 1.18).[5] With this discovery, the Big Bang model finally gained general acceptance. The exact temperature of this blackbody radiation was measured by the COBE[6] satellite in 1992 as 2.726 ± 0.005 kelvin. The presently most accurate value of the blackbody temperature is

---

[3]OSO—Orbiting Solar Observatory; SAS—Small Astronomy Satellite.

[4]It has become common practice in the scientific literature that the number 10^9 is called a billion, while in other countries the billion is 10^{12}. Throughout this book the notation that a billion is equal to a thousand millions is used.

[5]The excrements of pigeons presented a severe problem during an attempt to reduce the noise of their horn antenna. When, after a thorough cleaning of the whole system, a residual noise still remained, Arno Penzias was reported to have said: "Either we have seen the birth of the universe, or we have seen another pile of pigeon shit."

[6]COBE—COsmic ray Background Explorer.

Fig. 1.18 Penzias and
Wilson in front of their horn
antenna used for the
measurement of the
blackbody radiation [16, 17]

2.725 $\pm$ 0.001 K, which was also confirmed by the satellites WMAP (Wilkinson
Microwave Anisotropy Probe) and Planck.

COBE also found spatial asymmetries of the 2.7-kelvin blackbody radiation at a
level of $\Delta T/T \approx 10^{-5}$. This implies that the early universe had a lumpy structure,
which can be considered as a seed for galaxy formation (Nobel Prize to J. C. Mather
and G. F. Smoot 2006). Results from WMAP and Planck confirmed these findings
with higher precision.

After the electron antineutrino had been directly measured by Cowan and Reines
in a reactor experiment, the famous two-neutrino experiment of Lederman, Schwartz,
and Steinberger in 1962 (Nobel Prize 1988) represented an important step for the
advancement of astroparticle physics. This experiment demonstrated that the neutrino emitted in nuclear beta decay is not identical with the neutrino occurring in
pion decay ($\nu_\mu \neq \nu_e$). At present, three generations of neutrinos are known (ν_e, ν_μ,
and ν_τ). Experiments at the Large Electron–Positron Collider (LEP) at CERN have
demonstrated in 1989 that there are only three neutrino generations with masses
below half of the Z mass. The direct observation of the tau neutrino was established
only relatively recently (July 2000) by the DONUT[7] experiment.

The observation of solar neutrinos by the Davis experiment in 1967 marked
the beginning of the discipline of neutrino astronomy (Nobel Prize for R. Davis
2002). In fact, Davis measured a deficit in the flux of solar neutrinos, which was

---

[7]DONUT—Direct Observation of NU Tau (ν_τ).

confirmed by subsequent experiments, GALLEX,[8] SAGE,[9] and Kamiokande (Nobel Prize for M. Koshiba 2002). It was considered unlikely that a lack of understanding of solar physics is responsible for the solar neutrino problem. In 1958 Pontecorvo highlighted the possibility of *neutrino oscillations*. Such oscillations ($\nu_e \rightarrow \nu_\mu$) are presently generally accepted as explanation of the solar neutrino deficit. This would imply that neutrinos have a very small non-vanishing mass. In the framework of the electroweak theory (Glashow, Salam, Weinberg 1967; Nobel Prize 1979) that unifies electromagnetic and weak interactions, a non-zero neutrino mass was not foreseen. The introduction of quarks as fundamental constituents of matter (Gell-Mann and Zweig 1964, Nobel Prize for Gell-Mann 1969), and their description by the theory of quantum chromodynamics, extended the electroweak theory to the *Standard Model of elementary particles* (Veltman, t'Hooft; Nobel Prize 1999).

In this model, the masses of elementary particles cannot be calculated a priori. Therefore, small non-zero neutrino masses should not represent a real problem for the standard model, especially since it contains 25 free parameters that have to be determined by experimental information. However, three neutrino generations with non-zero mass would add another 7 parameters (three for the masses and four mixing parameters). It is generally believed that the standard model will not be the final word of the theoreticians.

In 1998 the Super-Kamiokande experiment found evidence for a non-zero neutrino mass by studying the relative abundances of atmospheric electron and muon neutrinos. The observed deficit of atmospheric muon neutrinos is most readily and elegantly explained by the assumption that neutrinos oscillate from one lepton flavour to another ($\nu_\mu \rightarrow \nu_\tau$). This is only possible if neutrinos have mass.

The oscillation scenario for solar neutrinos was confirmed in 2001 by the SNO[10] experiment by showing that the total flavour-independent neutrino flux from the Sun arriving at Earth (ν_e, ν_μ, ν_τ) was consistent with solar-model expectations, demonstrating that some of the solar electron neutrinos had oscillated into a different neutrino flavour (Nobel Prize to T. Kajita and A. B. McDonald 2015). This observation finally solved the solar neutrino problem.

The idea of mixing was already known from the quark sector, where the d, s, and b couple as mixed states in weak interactions (Nobel Prize to M. Kobayashi, T. Maskawa, and Y. Nambu 2008).

The discovery of charmed mesons in cosmic rays (Niu et al. 1971; see Figs. 1.19 and 1.20) and the confirmation, by accelerator experiments, for the existence of a fourth quark (Richter and Ting 1974, Nobel Prize 1976) extended the standard model of Gell-Mann and Zweig to four quarks (up, down, strange, and charm).

The theory of general relativity and Schwarzschild's ideas on the formation of gravitational singularities were supported in 1970 by precise investigations of the strong X-ray source Cygnus X1. Optical observations of Cygnus X1 indicated that this compact X-ray source is ten times more massive than our Sun. The rapid variation

---

[8]GALLEX—German–Italian GALLium EXperiment.

[9]SAGE—Soviet–American Gallium Experiment.

[10]SNO—Sudbury Neutrino Observatory.

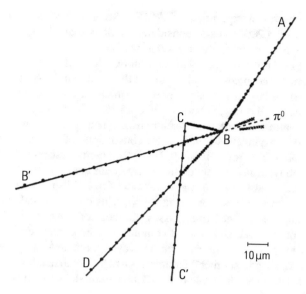

Fig. 1.19 z projection showing the pair production and decay of X particles (B and C) in an emulsion chamber seen by Niu et al. in 1971. The event was produced by a collision of a neutral cosmic-ray particle with a heavy element in the emulsion. The particle B decays at the point B into a charged particle B' and a neutral pion. The decay photons of the π^0 initiate an electron shower each, which is evident from the somewhat broader tracks. The particle C decays at C into a charged particle C' and further unseen neutral hadrons (which cannot be observed in a nuclear emulsion). The event is interpreted as pair production and decay of new particles with masses around $2\,\mathrm{GeV}/c^2$ and lifetimes consistent with those of charmed mesons [18]

in the intensity of X rays from this object leads to the conclusion that this source only has a diameter of about 10 km. A typical neutron star has a similar diameter to this, but is only three times as heavy as the Sun. An object that was as massive as Cygnus X1 would experience such a large gravitational contraction, which would overcome the Fermi pressure of degenerate neutrons. This leads to the conclusion that a black hole must reside at the center of Cygnus X1.

By 1974, Hawking had already managed to unify some aspects of the theory of general relativity and quantum physics. He was able to show that black holes could evaporate by producing fermion pairs from the gravitational energy outside the event horizon. If one of the fermions escaped from the black hole, its total energy and thereby its mass would be decreased (*Hawking radiation*). The time constants for the evaporation process of massive black holes, however, exceed the age of the universe by many orders of magnitude.

There were some hopes that gravitational waves, which would be measured on Earth, could resolve questions on the formation of black holes and other cosmic catastrophes. These hopes were boosted by gravitational-wave experiments by Weber in 1969. The positive signals of these early experiments have, so far, not been confirmed.

Fig. 1.20 x and y projection of the event in the previous figure [18]

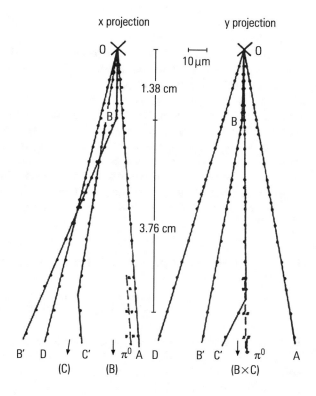

It is generally believed that the findings of Weber were due to mundane experimental backgrounds.

In contrast, Taylor and Hulse succeeded in providing indirect evidence for the emission of gravitational waves in 1974, by observing a binary star system that consisted of a pulsar and a neutron star (Nobel Prize 1993). They were able to precisely test the predictions of general relativity using this binary star system. The rotation of the orbital ellipse (periastron rotation) of this system is ten thousand times larger than the perihelion rotation of the planet Mercury. The decreasing orbital period of the binary is directly related to the energy loss by the emission of gravitational radiation. The observed speeding-up rate of the orbital velocities of the partners of the binary system and the slowing-down rate of the orbital period agree with the prediction based on the theory of general relativity to better than 1 per mill.

A breakthrough was achieved by the direct discovery of gravitational waves by the LIGO telescope in 2015. LIGO had observed three events of mergers of heavy black holes in two independent Michelson interferometers about 3000 km apart. LIGO (Advanced Laser Interferometer Gravitational Wave Observatory) had already searched for gravitational waves from 2002 to 2010, although initially without success. A substantially improved laser system allowed then 2015 the observation of mergers of black holes [19]. In the present setup LIGO can measure length variations of down to 10^{-19} m, corresponding to less than one thousandth of the proton diameter.

It is to be expected that there are processes occurring in the universe, which lack an immediate explanation. This was underlined by the discovery of *gamma-ray bursters* (GRB) in 1967. It came as a surprise when gamma-ray detectors on board military reconnaissance satellites, which were in orbit to check possible violations of the test-ban treaty on thermonuclear explosions, observed γ bursts. This discovery was withheld for a while due to military secrecy. However, when it became clear that the γ bursts did not originate from Earth but rather from outer space, the results were published. Gamma-ray bursters light up only once and are mostly short-lived, with burst durations lasting from 10 ms to a few seconds. There is also a component of bursts with relatively long duration. It is conceivable that γ bursts are caused by supernova explosions or by collisions between neutron stars.

1.5 Contributions of Accelerators to Cosmic Rays

> *Despite my resistance to hyperbole, the Large Hadron Collider*
> *belongs to a world that can only be described with superlatives.*
> Lisa Randall

It might appear that the elementary-particle aspect of astroparticle physics has been completed by the discovery of the b quark (Lederman 1977) and t quark (CDF collaboration 1995). There are now six known leptons (ν_e, e^-; ν_μ, μ^-; ν_τ, τ^-) along with their antiparticles ($\bar{\nu}_e$, e^+; $\bar{\nu}_\mu$, μ^+; $\bar{\nu}_\tau$, τ^+). These are accompanied by six quarks (up, down; charm, strange; top, bottom; [20]) and their corresponding six antiquarks (Fig. 1.21). These matter particles can be arranged in three families or 'generations'. Measurements of the primordial deuterium, helium, and lithium abundance in astrophysics had already given some indication that there may be only three families with light neutrinos. This astrophysical result was later confirmed beyond any doubt by experiments at the electron–positron collider LEP[11] in 1989 (see also Fig. 2.1). The standard model of elementary particles, with its three fermion generations, was also verified by the discovery of gluons, the carriers of the strong force (DESY,[12] 1979, Fig. 1.22, and the bosons of the weak interaction (W^+, W^-, Z) at CERN[13] in 1983; Nobel Prize for Rubbia and van der Meer 1984). The discovery of asymptotic freedom of quarks in the theory of the strong interaction by Gross, Politzer, and Wilczek was honored by the Nobel Prize in 2004.

In the standard model of electroweak and strong interactions, the mass generation is believed to come about by a *spontaneous symmetry breaking*, the so-called Higgs mechanism. This process favours the existence of at least one additional massive neutral boson. The missing Higgs particle providing masses for the fundamental

---

[11]LEP—Large Electron–Positron collider at CERN in Geneva.

[12]DESY—Deutsches Elektronen Synchrotron in Hamburg.

[13]CERN—Conseil Européen pour la Recherche Nucléaire.

Fig. 1.21 Periodic table of elementary particles [20]

fermions was found at the Large Hadron Collider (LHC) at CERN in 2012 (see Fig. 1.23; Nobel Prize for P. Higgs and F. Englert in 2013 for the theoretical prediction of the Higgs particle).

1.6 Renaissance of Cosmic Rays

A poem is a neutrino – mainly nothing – it has no mass and can pass through the earth undetected.

Mary Ruefle

The observation of the supernova explosion 1987A, along with the burst of extra-galactic neutrinos, represented the birth of real astroparticle physics. The measurement of only 25 neutrinos out of a possible 10^{58} emitted, allowed elementary particle physics investigations that were hitherto inaccessible in laboratory experiments. The dispersion of arrival times enabled physicists to derive an upper limit of the neutrino mass ($m_{\nu_e} < 10\,\mathrm{eV}$). The mere fact that the neutrino source was 170 000 light-years away in the Large Magellanic Cloud, allowed a lower limit on the neutrino lifetime to be estimated. The gamma line emission from SN 1987A gave confirmation that heavy elements up to iron, cobalt, and nickel were synthesized in the explosion, in

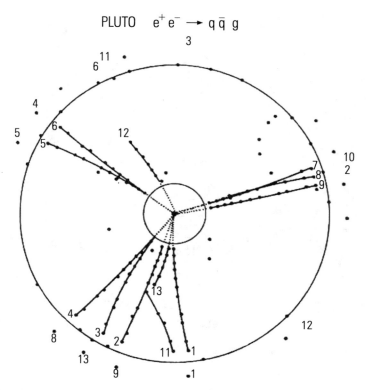

PLUTO $e^+ e^- \longrightarrow q \bar{q} \, g$

Fig. 1.22 The gluon as carrier of strong interactions was discovered at the electron–positron collider PETRA at DESY in Hamburg in 1979. All four PETRA experiments PLUTO, JADE, MARK J, and TASSO saw the gluon in the process of gluon bremsstrahlung. One year earlier PLUTO had already indirect evidence for the gluon when they could demonstrate that the Υ decay was best described by a three-gluon decay. The image shows a three-jet event in PLUTO [21]

Fig. 1.23 Invariant $\gamma\gamma$ mass distribution showing the production of the Higgs boson in the mass range around $125 \, \text{GeV}/c^2$. The background-subtracted results are from the ATLAS experiment at the LHC [22]

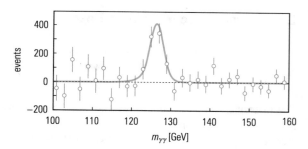

agreement with predictions of supernova models. As the first optically visible supernova since the discovery of the telescope, SN 1987A marked an ideal symbiosis of astronomy, astrophysics, and elementary particle physics (Fig. 1.24).

Fig. 1.24 Supernova
explosion SN 1987A in the
Tarantula Nebula [23]

The successful launch of the high-resolution X-ray satellite ROSAT[14] in 1990 paved the way for the discovery of numerous X-ray sources. The Hubble telescope, which was started in the same year, provided optical images of stars and galaxies in hitherto unprecedented quality, once the slightly defocusing mirror had been adjusted by a spectacular repair in space. The successful mission of ROSAT was followed by the X-ray satellites Chandra (named after Subrahmanyan Chandrasekhar, Nobel Prize 1983) and XMM (X-ray Multi-Mirror mission, renamed Newton Observatory in 2002), both launched in 1999. Presently, the launch of the INTEGRAL[15] satellite in 2002 and the start of the FERMI telescope (FGST—Fermi Gamma Ray Space Telescope) in 2008 provide also important contributions to the field of X-ray and gamma astronomy. NuSTAR (Nuclear Spectroscopic Telescope Array), launched in 2012, is a space-based X-ray telescope that uses Wolter telescopes to focus high-energy X rays from astrophysical sources. It covers the energy range from 3 to 79 keV, thereby extending the energy range of earlier space instruments.

The Compton Gamma-Ray Observatory (CGRO, launched in 1991) opened the door for GeV gamma astronomy. The earthbound atmospheric air Cherenkov telescopes (H.E.S.S., see also Fig. 1.25) and MAGIC (Major Atmospheric Gamma Imaging Cherenkov Telescopes) and other air-shower experiments were able to search for and identify TeV point sources in our Milky Way (e.g., Crab Nebula) and in extragalactic distances (1992, Markarian 501, Markarian 421). The active galactic nuclei of the Markarian galaxies are also considered excellent candidate sources of high-energy hadronic charged cosmic rays.

---

[14]ROSAT—ROentgen SATellite of the Max Planck Institute for Extraterrestrial Physics, Munich.

[15]International Gamma-Ray Astrophysics Laboratory.

Fig. 1.25 The H.E.S.S. Cherenkov telescope array in Namibia for the measurement of high-energy gamma rays (H.E.S.S. = High Energy Stereoscopic System) [24]

1.7 Open Questions

We will first understand how simple the universe is, when we realize, how strange it is.

Anonymous

Do you know the ordinances of the heavens? Can you establish their rule on the Earth?

Book Hiob 38:33

A still unsolved question of astroparticle physics is the problem of *dark matter* and *dark energy*. From the observation of orbital velocities of stars in our Milky Way and the velocities of galaxies in galactic clusters, it is clear that the energy density of the visible matter in the universe is insufficient to correctly describe its dynamics (Fig. 1.26).

Since the early nineties, the MACHO[16] and EROS[17] experiments have searched for compact, non-luminous, Jupiter-like objects in the halo of our Milky Way, using the technique of microlensing. Some candidates have been found, but their number is nowhere near sufficient to explain the missing dark matter in the universe. One can conjecture that exotic, currently unknown particles (supersymmetric particles, WIMPs,[18] ...), or massive neutrinos may contribute to solve the problem of the

---

[16]MACHO—search for MAssive Compact Halo Objects.

[17]EROS—Expérience pour la Recherche d'Objets Sombres.

[18]WIMP—Weakly Interacting Massive Particles.

Fig. 1.26 Sketch of typically measured rotational star velocities in galaxies in comparison to expectations based on Keplerian orbits

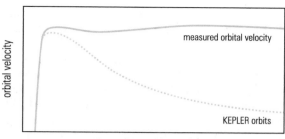

distance from galactic center

missing dark matter. A non-vanishing vacuum energy density of the universe is also known to play a decisive role in the dynamics and evolution of the universe. The idea of a vacuum energy had already been introduced in 1915 by Einstein in the form of a cosmological constant Λ.

In 1998 the Super-Kamiokande experiment found evidence for a non-zero neutrino mass by studying the relative abundances of atmospheric electron and muon neutrinos. The observed deficit of atmospheric muon neutrinos is most readily and elegantly explained by the assumption that neutrinos oscillate from one lepton flavour to another (e.g., $\nu_\mu \rightarrow \nu_\tau$). This is only possible if neutrinos have mass. The presently favoured mass of $0.05\,\text{eV}$ for ν_τ, however, is insufficient to explain the dynamics of the universe alone.

A very important unsolved problem in cosmology is the dominance of matter. It is true that *CP*-violating effects are known from particle physics. However, the size of the effect is insufficient to explain the disappearance of antimatter. Results from the satellite PAMELA and the AMS experiment on board the space station ISS do find an increase in the positron production at high energies, but these positrons could have also been produced in supernova explosions, neutron stars, quasars, or active galactic nuclei. It has been argued that the dominance of matter over antimatter may originate from the fact that neutrinos and antineutrinos behave possibly in a different way in neutrino oscillations [25]. The search for antiparticles in primary cosmic rays and the theoretical understanding of a possible strong matter–antimatter-asymmetric effect is still an important aspect in particle physics and astronomy.

An equally exciting discovery is the measurement of the acceleration parameter of the universe. Based on the ideas of the classical Big Bang, one would assume that the initial thrust of the explosion would be slowed down by gravitation. Observations on distant supernova explosions (1998), however, indicate that in early cosmological epochs the rate of expansion was smaller than today. The finding of an *accelerating universe*—which is now generally accepted—has important implications for cosmology (Nobel Prize to S. Perlmutter, B. P. Schmidt, and A. G. Riess 2013). It suggests that the largest part of the missing substance of the universe is stored as dark energy in a dynamical vacuum ('quintessence'?).

Finally, it should be highlighted that the discovery of *extrasolar planets* (Mayor and Queloz 1995) has led to the resumption of discussions on the existence of extraterrestrial intelligence. Until now about 4000 extrasolar planets have been discovered. Among those there are more than 500 systems with two up to seven planets (status 2018). These planets are usually not seen in the optical range, but are detected rather indirectly. These techniques include variations in their radial velocities, microlensing, or simple transits. Possibly the first direct evidence for an extrasolar planet was the observation of a planet orbiting the brown dwarf 2M1207 at a distance of 225 light-years. This finding was confirmed by the Hubble space telescope in 2006. It is estimated that there might be millions of planets with habitable zones in the Milky Way. The NASA planet searcher space telescope Kepler alone has found more than 2600 exoplanets orbiting other stars and a large number of planets with habitable zones. After running out of fuel at the end of 2018 the planet-hunting work will be passed on to the TESS[19] satellite, which is expected to find more than 10000 exoplanets.

The object Kepler-452b found in 2015 is an Earth-like planet in the habitable zone of its parent star. It looks as if it is a rocky planet like our Earth. The associated star is in the constellation Cygnus at a distance of 1400 light-years. One of the planets closest to Earth is orbiting the star Epsilon Eridani. It is also in the habitable zone of its star. Its distance from Earth is only 10 light-years. Even closer is Proxima Centauri b orbiting the 4.2-light-years-distant dwarf star Proxima Centauri in 11.2 days in a habitable zone. Possibly we are not the only intelligent beings in the universe pursuing astroparticle physics.

Summary

The hour of birth of astroparticle physics is certainly the historic balloon flight of Victor Hess in 1912. He discovered cosmic rays with an ionization chamber operated at different altitudes in the atmosphere. In the early days of cosmic rays—the name astroparticle physics was not familiar in those times—many discoveries of elementary particles were made: positrons, muons, and pions were the first new elementary particles. With the advent of accelerators the field of elementary particles was shifted to these new machines. Cosmic ray physics still continued, and in the seventies cosmic-ray accelerators experienced a renaissance. The measurement of solar neutrinos, the discovery of the supernova SN 1987A with the observation of neutrinos from this source, and the discovery of neutrino oscillations lead to a revival of cosmic rays. The discovery of gravitational waves adds to the comprehensive aspect of astroparticle physics. Today astroparticle physics is an active, interdisciplinary research field, which unifies astronomy, cosmic rays, and particle physics applying multi-messenger experiments for future discoveries.

---

[19]Transiting Exoplanet Survey Satellite.

1.8 Problems

1. Work out the

   (a) velocity of an Earth satellite in a low-altitude orbit,
   (b) the escape velocity from Earth,
   (c) the altitude of a geostationary satellite above ground level. Where can such a geostationary satellite be positioned?

2. What is the bending radius of a solar particle (proton, 1 MeV kinetic energy) in the Earth's magnetic field (0.5 gauss) for vertical incidence with respect to the field? Use the relation between the centrifugal force and the Lorentz force (6.1.1), and argue whether a classical calculation or a relativistic calculation is appropriate.

3. Estimate the average energy loss of a muon in the atmosphere (production altitude 20 km, muon energy $\approx 10\,\mathrm{GeV}$; check with Fig. 4.4).

4. What is the ratio of intensities of two stars, which differ by one unit in magnitude only (for the definition of the magnitude see the Glossary)?

5. Small astronomical objects like meteorites and asteroids are bound by solid-state effects while planets are bound by gravitation. Estimate the minimum mass, from where on gravitational binding starts to dominate as binding force. Gravitational binding dominates if the potential gravitational energy exceeds the total binding energy of the solid material, where the latter is taken to be proportional to the number of atoms in the object. The average atomic number is A, from which together with the Bohr radius r_B the average density can be estimated.

6. There is a statement in this chapter that a quasar at a redshift of $z = 7.085$ gives us information on the universe when it was less than about 5% of its present age. Can you convert the redshift into the age of the universe?

7. What is the distance of two features on the Sun that can still be resolved by the 200-inch Mt. Palomar telescope, when information in the green range (535 nm) is used?

   Hint: Find a formula for the resolution of a circular aperture by either an approximate treatment using an appropriate average considering an inscribed and a circumscribed square aperture or by looking up the formula for an Airy disk.

Chapter 2
The Standard Model of Elementary Particles

*Most basic ideas of science are essentially simple and can
usually be expressed in a language that everyone understands.*
Albert Einstein

Over the last years a coherent picture of elementary particles has emerged. Since the development of the atomic model, improvements in experimental resolution have allowed scientists to investigate smaller and smaller structures. Even the atomic nucleus, which contains practically the total mass of the atom, is a composite object. Protons and neutrons, the building blocks of the nucleus, have a granular structure that became obvious in electron–nucleon scattering experiments. In the naive quark parton model, a nucleon consists of three quarks. The onion-type phenomenon of ever smaller constituents of particles that were initially considered to be fundamental and elementary, may have come to an end with the discovery of quarks and their dynamics. While atoms, atomic nuclei, protons, and neutrons can be observed as free particles in experiments, quarks can never escape from their hadronic prison. In spite of an intensive search by numerous experiments, nobody has ever been able to find free quarks. Quantum chromodynamics, which describes the interaction of quarks, only allows the '*asymptotic freedom*' of quarks at high momenta. Bound quarks that are inside nucleons typically have low momenta and are subject to '*infrared slavery*'. This confinement does not allow the quarks to separate from each other.

Quarks are constituents of strongly interacting hadronic matter. The size of quarks is below 10^{-18} m. In addition to quarks, there are leptons that interact weakly and electromagnetically. With the resolution of the strongest microscopes (accelerators and storage rings), quarks and leptons appear to be point-like particles, having no internal structure. Three different types of leptons are known: electrons, muons, and taus. Each charged lepton has a separate neutrino: ν_e, ν_μ, ν_τ. Due to the precise investigations of the Z particle, which is the neutral carrier of weak interactions, it is known that there are exactly three particle families with light neutrinos (Fig. 2.1). This result was obtained from the measurement of the total Z decay width. According to Heisenberg's uncertainty principle, the resolution of complementary quantities is

© Springer Nature Switzerland AG 2020
C. Grupen, *Astroparticle Physics*, Undergraduate Texts in Physics,
https://doi.org/10.1007/978-3-030-27339-2_2

intrinsically limited by Planck's constant ($h = 6.626\,0693 \times 10^{-34}$ J s). The relation
between the complementary quantities of energy and time is

$$\Delta E\ \Delta t \geq \hbar/2 \ (\hbar = h/2\pi)\,. \tag{2.0.1}$$

If $\Delta t = \tau$ is the lifetime of the particle, relation (2.0.1) implies that the decay width
$\Delta E = \Gamma$ is larger when τ is shorter. If there are many generations of light neutrinos,
the Z particle can decay into all these neutrinos,

$$Z \rightarrow \nu_x + \bar{\nu}_x\,. \tag{2.0.2}$$

These decays can occur even if the charged leptons ℓ_x associated with the respective
generation are too heavy to be produced in Z decay. A large number of different
light neutrinos will consequently reduce the Z lifetime, thereby increasing its decay
width. The exact measurement of the Z decay width took place at the LEP storage ring
(Large Electron–Positron Collider) in 1989, enabling the total number of neutrino
generations to be determined: there are exactly three lepton generations with light
neutrinos (see Fig. 2.1).

The measurement of the primordial helium abundance had already allowed physicists to derive a limit for the number of neutrino generations. The nucleosynthesis
in the early universe was essentially determined by the number of relativistic particles, which were able to cool down the universe after the Big Bang. At temperatures of $\approx 10^{10}$ K that correspond to energies, where nucleons start to bind in nuclei
(≈ 1 MeV), these relativistic particles would have consisted of protons, neutrons,
electrons, and neutrinos. If many different neutrino flavours existed, a large amount

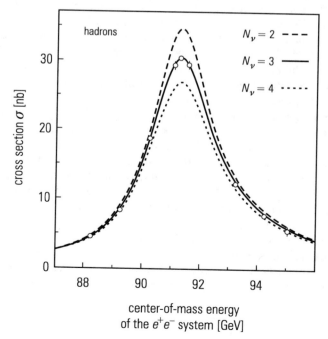

Fig. 2.1 Determination of the number of neutrino generations from Z decay

of energy would have escaped from the original fireball, owing to the low interaction probability of neutrinos. This has the consequence that the temperature would have decreased quickly. A rapidly falling temperature means that the time taken for neutrons to reach nuclear binding energies would have been very short, and consequently they would have had very little time to decay (lifetime $\tau_n = 885.7$ s). If there were many neutrons that did not decay, they would have been able to form helium together with stable protons. The primordial helium abundance is therefore an indicator of the number of neutrino generations. In 1990, the experimentally determined primordial helium abundance allowed physicists to conclude that the maximum number of different light neutrinos is at most four.

In addition, there are also three quark generations, which have a one-to-one correspondence with the three lepton generations:

$$\begin{pmatrix} \nu_e \\ e^- \end{pmatrix} \begin{pmatrix} \nu_\mu \\ \mu^- \end{pmatrix} \begin{pmatrix} \nu_\tau \\ \tau^- \end{pmatrix}$$
$$\begin{pmatrix} u \\ d \end{pmatrix} \begin{pmatrix} c \\ s \end{pmatrix} \begin{pmatrix} t \\ b \end{pmatrix}. \tag{2.0.3}$$

The properties of these fundamental matter particles are listed in Table 2.1.

Quarks have fractional electric charges (in units of the elementary charge). The different kinds of quarks ($u, d; c, s; t, b$) in the three respective generations (families) are characterized by a different flavour.

Up to the nineties of the last century one had assumed that neutrinos were massless. Their masses from direct measurements were compatible with being zero, and

Table 2.1 Periodic table of elementary particles: matter particles (fermions)

LEPTONs ℓ, spin $\frac{1}{2}\hbar$ (antileptons $\bar{\ell}$)

Electr. charge (e)	1st generation		2nd generation		3rd generation	
	Flavour	Mass (GeV/c^2)	Flavour	Mass (GeV/c^2)	Flavour	Mass (GeV/c^2)
0	ν_e electron neutrino	$<2 \times 10^{-9}$	ν_μ muon neutrino	$<1.7 \times 10^{-4}$	ν_τ tau neutrino	<0.018
-1	e electron	5.11×10^{-4}	μ muon	0.106	τ tau	1.777

QUARKs q, spin $\frac{1}{2}\hbar$ (antiquarks $\bar{q}$)

Electr. charge (e)	Flavour	$\simeq$ mass (GeV/c^2)	Flavour	$\simeq$ mass (GeV/c^2)	Flavour	$\simeq$ mass (GeV/c^2)
$+2/3$	u up	2×10^{-3} up to 8×10^{-3}	c charm	1.1 up to 1.7	t top	173
$-1/3$	d down	5×10^{-3} up to 15×10^{-3}	s strange	0.1 up to 0.3	b bottom	4.3

still are. Therefore, only upper limits on their masses could be given, and the limits obtained from these direct measurements are given in Table 2.1. Neutrino oscillations, however, have shown that neutrinos actually do have a small mass, as indicated by the measurements of solar neutrinos (Davis experiment), and by the Super-Kamiokande and SNO experiments, which can only be interpreted in terms of neutrino oscillations. Actually, in grand unified theories (GUTs) unifying electroweak and strong interactions, neutrinos are predicted to have small but non-zero masses. Neutrino oscillation measurements, however, provide only values for the difference of squared masses like $\delta m^2 = m_{\nu_1}^2 - m_{\nu_2}^2$ for two neutrinos, respectively, with the masses m_{ν_1} and m_{ν_2}. Under plausible assumptions and the known limit for the electron neutrino mass one can suppose $m_\nu < 2\,\mathrm{eV}/c^2$ to be valid for all neutrino flavours. If the mass hierarchy in the neutrino sector were the same as for the charged leptons, the neutrino masses could be around $50\,\mathrm{meV}/c^2$. Cosmological arguments based on the data from the Planck experiment suggest $\sum m_\nu < 0.23\,\mathrm{eV}/c^2$ for the sum of the three light neutrinos.

Only approximate values of masses for quarks can be given, because free quarks do not exist and the binding energies of quarks in hadrons can only be estimated roughly. For each particle listed in Table 2.1 there exists an antiparticle, which is in all cases different from the original particle. This means that there are actually 12 fundamental leptons and an equal number of quarks.

The interactions between elementary particles are governed by different forces. There are four forces in total, distinguished by strong, electromagnetic, weak, and gravitational interactions. In the 1960s, it was possible to unite the electromagnetic and weak interactions into the electroweak theory. The carriers of all the interactions are particles with integer spin (*bosons*), in contrast to the matter particles that all have half-integer spin (*fermions*). The properties of these bosons are compiled in Table 2.2.

Table 2.2 Periodic table of elementary particles: carriers of the forces (bosons)

Electroweak interaction	γ		W^-	W^+	Z
Spin ($\hbar$)	1		1	1	1
Electr. charge (e)	0		-1	$+1$	0
Mass (GeV/c^2)	0		80.3	80.3	91.2

Strong interaction	Gluon g
Spin ($\hbar$)	1
Electr. charge (e)	0
Mass (GeV/c^2)	0

Gravitational interaction	Graviton G
Spin ($\hbar$)	2
Electr. charge (e)	0
Mass (GeV/c^2)	0

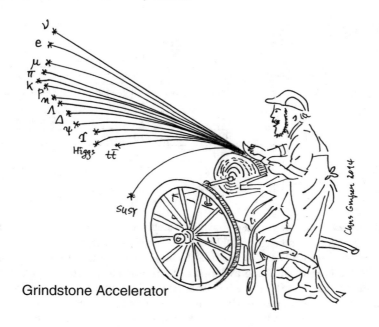

Grindstone Accelerator

Creation of elementary particles

The dawn of the quark model.

Too many particles

Table 2.3 Properties of interactions

| Interaction → | Gravitation | Electroweak interaction | | Strong |
Property ↓		Weak	Electromagnetic	
Acts on	Mass–energy	Flavour	Electric charge	Colour charge
Particles concerned	All	Quarks, leptons	All charged particles	Quarks, Gluons
Exchange particle	Graviton G	W^+, W^-, Z	γ	Gluons g
Relative strength	$\approx 10^{-40}$	10^{-5}	10^{-2}	1
Range	∞	$\approx 10^{-3}$ fm	∞	≈ 1 fm
Example	System Earth–Moon	β decay	Atomic binding	Nuclear binding

While the existence of the gauge bosons of electroweak interactions and the gluon of strong interactions are well established, the *graviton*, the carrier of the gravitational force, has not yet been discovered. The properties of interactions are compared in Table 2.3. It is apparent that gravitation can be completely neglected in the microscopic domain, because its strength in relation to strong interactions is only 10^{-40}.

In the primitive quark model, all strongly interacting particles (hadrons) are composed of valence quarks. A baryon is a three-quark system, whereas a meson consists of a quark and an antiquark. Examples of baryons include the proton, which is an *uud* system, and the neutron is an *udd* composite. Correspondingly, an example of a meson is the positively charged pion, which is an $u\bar{d}$ system. The existence of baryons consisting of three identical quarks with parallel spin ($\Omega^- = (sss)$, spin $\frac{3}{2}\hbar$) indicates that quarks must have a hidden quantum number, otherwise the Pauli exclusion principle would be violated. This hidden quantum number is called *colour*. Electron–positron interactions show that there are exactly three different colours. Each quark therefore comes in three colours, however, the observed hadrons are colour-neutral, i.e., they are colourless. If the three degrees of freedom in colour are denoted by red (r), green (g), and blue (b), the proton is a composite object made up from $u_{red}u_{green}d_{blue}$. In addition to valence quarks, there is also a sea of virtual quark–antiquark pairs in hadrons.

The quarks that form hadrons are held together by the exchange of gluons. Since gluons mediate the interactions between quarks, they must possess two colours: they carry a colour and an anticolour. Since there are three colours and anticolours each, one would expect that $3 \times 3 = 9$ gluons exist. Quantum chromodynamics, however, is only mediated by by eight gluons. A singlet consisting of a colour-neutral mixed state of *all* colours and anticolours, $\frac{1}{\sqrt{3}}(r\bar{r} + g\bar{g} + b\bar{b})$, does not carry colours and does not exist. The eight states of gluons are $r\bar{g}$, $r\bar{b}$, $g\bar{b}$, $g\bar{r}$, $b\bar{r}$, $b\bar{g}$, $\frac{1}{\sqrt{2}}(r\bar{r} - g\bar{g})$, $\frac{1}{\sqrt{6}}(r\bar{r} + g\bar{g} - 2b\bar{b})$. In a very simplified picture, the gluon radiation of a quark can be illustrated by the diagram shown in Fig. 2.2.

Nucleons in a nucleus are bound together by the residual interaction of gluons, in very much the same way as molecular binding is a result of the residual interactions of electric forces.

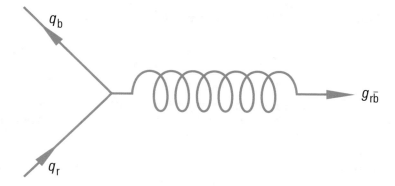

Fig. 2.2   Creation of coloured gluons by quarks

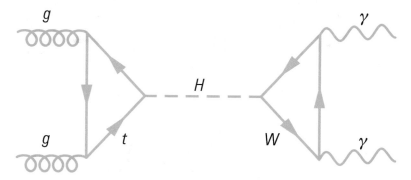

Fig. 2.3   Production of a Higgs particle by gluon fusion and subsequent decay into two photons. The gluons are provided by the colliding beam protons

The Standard Model of elementary particles has been supplemented by the discovery of the Higgs particle in 2012, possibly it is even complete now. The mass of the Higgs boson is around $125\,\mathrm{GeV}/c^2$. In the framework of the Brout–Englert–Higgs mechanism it is possible to assign masses to the fundamental fermions by spontaneous symmetry breaking. Figure 2.3 shows the production of a Higgs boson by gluon fusion and the decay of the Higgs particle into two energetic photons. This production and decay mechanism played a leading role in the discovery of the Higgs.

This success of experimental particle physics should not mislead us to assume that we have now completed the understanding of the Standard Model. This model still has 25 free parameters, which need to be adjusted by experiments. These parameters are: 12 values for the fermion masses (6 quarks, 6 leptons), 3 mixing angles and a phase of the Cabibbo–Kobayashi–Maskawa matrix, 3 mixing angles and a phase (if the neutrinos are Dirac particles) of the Pontecorvo–Maki–Nakagawa–Sakata matrix, 2 parameters for the Higgs mass and the vacuum expectation value of the Higgs field, and, finally, 3 couplings for the interactions: the fine-structure constant α, the coupling of the strong interaction α_s, and the coupling constant of the electroweak interaction.

2.1 Examples of Interaction Processes

It is possible in quantum mechanics to sneak quickly across a region which is illegal energetically.

Richard P. Feynman

Interactions of elementary particles can be graphically represented by Feynman diagrams,[1] which present a short-hand for the determination of cross sections. In the following, the underlying quark–lepton structure will be characterized for some interaction processes.

Rutherford scattering of electrons on protons is mediated by photons (Fig. 2.4).

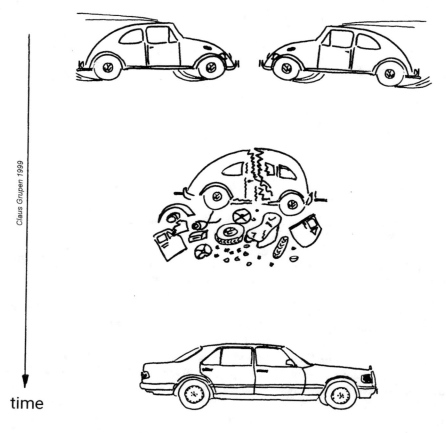

Claus Grupen 1999

time

production of a Higgs particle in
proton–proton interactions

Low cross sections

---

[1] See the Glossary.

Fig. 2.4 Rutherford
scattering of electrons on
protons

Fig. 2.5 Rutherford
scattering as photon–quark
scattering process

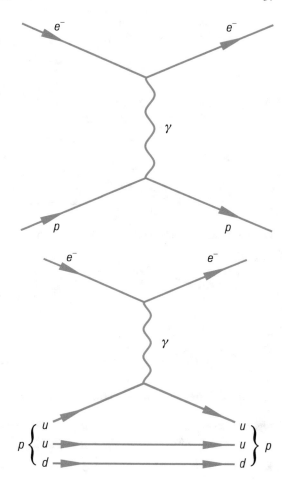

At high energies, however, the photon does not interact with the proton as a whole,
but rather only with one of its constituent quarks (Fig. 2.5). The other quarks of the
nucleon participate in the interaction only as spectators. As photons are electrically
neutral particles, they cannot change the nature of a target particle in an interaction. In
weak interactions, however, there are charged bosons that can cause an interchange
between particles within a family. As an example, Fig. 2.6 shows the scattering of an
electron neutrino on a neutron via a charged-current (W^+, W^- exchange) reaction.

   In a neutral-current interaction (Z exchange), the neutrino would not alter its
nature when scattered off the neutron. If electron neutrinos are scattered on electrons, charged and neutral currents can contribute. For the scattering of muon or tau
neutrinos on electrons only neutral currents contribute, because ν_μ and ν_τ are not
members of the electron family (Fig. 2.7).

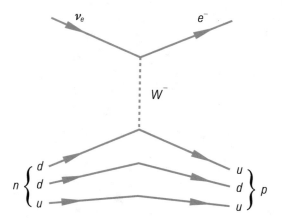

Fig. 2.6 Neutrino–nucleon scattering via a charged current

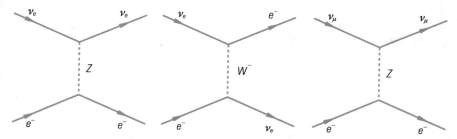

Fig. 2.7 Scattering possibilities of neutrinos on electrons

Fig. 2.8 Neutron decay

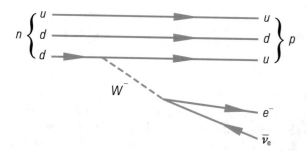

Decays of elementary particles can be described in a similar way. Nuclear beta decay of the neutron $n \rightarrow p + e^- + \bar{\nu}_e$ is mediated by a weak charged current (Fig. 2.8), where a d quark in the neutron is transformed into a u quark by the emission of a virtual W^-. The W^- immediately decays into members of the first lepton family ($W^- \rightarrow e^- \bar{\nu}_e$). In principle, the W^- could also decay according to $W^- \rightarrow \mu^- \bar{\nu}_\mu$ or $W^- \rightarrow \bar{u}d$, but this is not kinematically allowed. Muon decay can be described in a similar fashion (Fig. 2.9). The muon transfers its charge to a W^-, thereby transforming itself into the neutral lepton of the second family, the ν_μ. The W^- in turn decays again into in $e^- \bar{\nu}_e$.

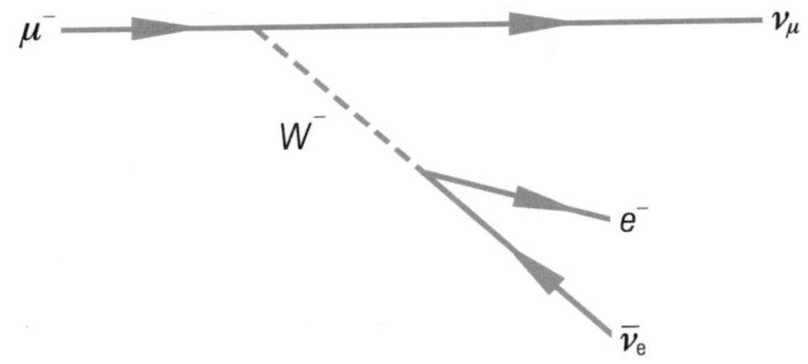

Fig. 2.9 Muon decay

Weak Interactions

Finally, pion decay will be discussed (Fig. 2.10). In principle, the W^+ can also decay in this case, into an $e^+ \nu_e$ state. Helicity reasons, however, strongly suppress this decay: as a spin-0 particle, the pion decays into two leptons that must have antiparallel spins due to angular-momentum conservation. The *helicity* is the projection of the spin onto the momentum vector, and it is fixed for the neutrino (for massless particles the spin is either parallel or antiparallel to the momentum). Particles normally carry negative helicity (spin $\parallel -p$, left-handed) so that the positron, as an antiparticle (spin $\parallel p$, right-handed), must take on an unnatural helicity (Fig. 2.11). The probability of carrying an abnormal helicity is proportional to $1 - \frac{v}{c}$ (where v is velocity of the charged lepton). Owing to the relatively high mass of the muon ($m_\mu \gg m_e$), it takes on a much smaller velocity compared to the electron in pion decay, i.e., $v(\mu) \ll v(e)$. The consequence of this is that the probability for the decay muon to take on an unnatural helicity is much larger compared to the positron. For this reason, the $\pi^+ \to e^+ \nu_e$ decay is strongly suppressed compared to the $\pi^+ \to \mu^+ \nu_\mu$ decay (the suppression factor is 1.23×10^{-4}).

Fig. 2.10   Pion decay

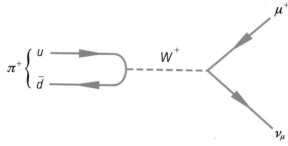

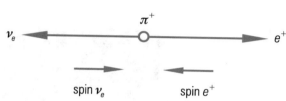

Fig. 2.11 Helicity conservation in π^+ decay

2.2   Quantum Numbers and Symmetries

I don't demand that a theory corresponds to reality because I don't know what it is.

Stephen Hawking

The various elementary particles are characterized by quantum numbers. In addition to the electric charge, the membership of a quark generation (quark flavour) or lepton generation (lepton number) is introduced as a quantum number. Leptons are assigned the lepton number $+1$ in their respective generation, whereas antileptons are given the lepton number -1. Lepton numbers for the different lepton families (L_e, L_μ, L_τ) are separately conserved, as is shown in the example of the muon decay:

$$
\begin{array}{ccccc}
\mu^- & \rightarrow & \nu_\mu & + e^- & + \bar{\nu}_e \\
L_\mu \quad 1 & & 1 & 0 & 0 \\
L_e \quad 0 & & 0 & 1 & -1
\end{array}
\qquad (2.2.1)
$$

The parity transformation P is the space inversion of a physical state. Parity is conserved in strong and electromagnetic interactions, however, in weak interactions it is maximally violated. This means that the mirror state of a weak process does not correspond to a physical reality. Nature distinguishes between right and left in weak interactions.

The operation of charge conjugation C applied to a physical state changes all the charges, meaning that particles and antiparticles are interchanged, whilst leaving quantities like momentum or spin untouched. Charge conjugation is also violated in weak interactions. In β decay, for example, left-handed electrons (negative helicity) and right-handed positrons (positive helicity) are favoured. Even though the symmetry operations P and C are not conserved individually, their combination CP, which is the application of space inversion (parity operation P) with subsequent interchange of particles and antiparticles (charge conjugation C) is a well-respected symmetry. This symmetry, however, is still broken in certain decays (K^0 and B^0 decays), but it is a common belief that the CPT symmetry (CP symmetry with additional time reversal) is conserved under all circumstances.

Some particles, like kaons, exhibit a very strange behaviour. They are produced copiously, but decay relatively slowly. These particles are produced in strong interactions, but they decay via weak interactions. This property is accounted for by introducing the quantum number *strangeness*, which is conserved in strong interactions, but violated in weak decays. Owing to the conservation of strangeness in strong interactions, only the associate production of strange particles, i.e., the combined production of hadrons, one of which contains a strange and the other an anti-strange quark, is possible, such as

$$\pi^- + p \to K^+ + \Sigma^- . \tag{2.2.2}$$

In this process, the $\bar{s}$ quark in the K^+ ($= u\bar{s}$) receives the strangeness $+1$, whilst the s quark in the Σ^- ($= dds$) is assigned the strangeness -1. In the weak decay of the $K^+ \to \pi^+\pi^0$, the strangeness is violated, since pions do not contain strange quarks (s).

Certain particles that behave in an identical way under strong interactions, but differ in their charge state, are integrated into isospin multiplets. Protons and neutrons are nucleons that form an isospin doublet of $I = 1/2$. When the nucleon isospin is projected onto the z axis, the state with $I_z = +1/2$ corresponds to a proton, whereas the $I_z = -1/2$ state relates to the neutron. The three pions (π^+, π^-, π^0) combine to form an isospin triplet with $I = 1$. In this case, $I_z = -1$ corresponds to the π^-, $I_z = +1$ is the π^+, whilst $I_z = 0$ relates to the π^0. The particle multiplicity m in an isospin multiplet is related to the isospin via the equation

$$m = 2I + 1 . \tag{2.2.3}$$

Finally, the *baryon number* should be mentioned. Quarks are assigned the baryon number $1/3$, and antiquarks are given $-1/3$. All baryons consisting of three quarks are therefore assigned the baryon number 1, whereas all other particles get the baryon number 0.

The properties of the conservation laws for the different interaction types in elementary particle physics are compiled in Table 2.4.

Unfortunately, there is a small but important complication in the quark sector. As can be seen from Table 2.1, there is a complete symmetry between leptons and quarks. Leptons, however, participate in interactions as free particles, whereas quarks do not. Due to quark confinement, spectator quarks always participate in the interactions in some way. For charged leptons, there is a strict law of lepton-number conservation: The members of different generations do not mix with each other. For the quarks, it

Table 2.4 Conservation laws of particle physics (conserved: +; violated: −)

Physics quantity	Interaction		
	Strong	Electromagnetic	Weak
Momentum	+	+	+
Energy (incl. mass)	+	+	+
Angular momentum	+	+	+
Electric charge	+	+	+
Quark flavour	+	+	−
Lepton number	./.	+	+
Parity	+	+	−
Charge conjugation	+	+	−
Strangeness	+	+	−
Isospin	+	−	−
Baryon number	+	+	+

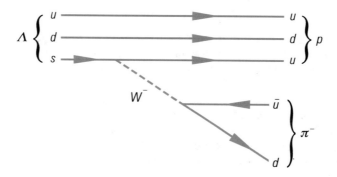

Fig. 2.12 Lambda decay: $\Lambda \to p + \pi^-$

was seen that weak processes can change the strangeness. In Λ decay, the s quark
belonging to the second generation can transform into a u quark of the first generation.
This would otherwise only be allowed to happen to a d quark (Fig. 2.12).

It appears as if the s quark can sometimes behave like the d quark. It is, in fact,
the d' and s' quarks that couple to weak interactions, rather than the d and s quarks.
The d' and s' quarks can be described as a rotation with respect to the d and s quarks.
This rotation is expressed by

$$d' = d \cos \theta_C + s \sin \theta_C ,$$
$$s' = -d \sin \theta_C + s \cos \theta_C , \tag{2.2.4}$$

where θ_C is the mixing angle (Cabibbo angle).

The reason that angles are used for weighting is based on the fact that the sum of the squares of the weighting factors, $\cos^2 \theta + \sin^2 \theta = 1$, automatically guarantees the correct normalization. θ_C has been experimentally obtained to be approximately $13°$ ($\sin \theta_C \approx 0.2255$). Since $\cos \theta_C \approx 0.9742$, the d' quark predominantly behaves like the d quark, albeit with a small admixture of the s quark.

The quark mixing originally introduced by Cabibbo was extended by Kobayashi and Maskawa to all three quark families, such that d', s', and b' are obtained from d, s, b by a rotation matrix. This matrix is called the Cabibbo–Kobayashi–Maskawa matrix (CKM matrix),

$$\begin{pmatrix} d' \\ s' \\ b' \end{pmatrix} = U \begin{pmatrix} d \\ s \\ b \end{pmatrix} . \tag{2.2.5}$$

The elements on the main diagonal of the (3×3) matrix U are very close to unity. The off-diagonal elements indicate the strength of the quark-flavour violation. A similar complication in the neutrino sector will be discussed later, where the eigenstates of the mass are not identical with the eigenstates of the weak interaction (see Sect. 6.3).

2.3 Unified Theory of Interactions

It is a kind of religion for intelligent atheists. Very simple was my explanation, and plausible enough—as most wrong theories are!
Herbert George Wells

The Standard Model of electroweak and strong interactions cannot be the final theory. The model contains too many free parameters, which have to be adjusted by hand. In addition, the masses of all fundamental fermions are initially zero. They only get their masses by a mechanism of spontaneous symmetry breaking (*Higgs mechanism*). Another very important point to note is that gravitation is not considered in this model at all, whereas it is the dominant force in the universe as a whole. There have been many attempts to formulate a Theory of Everything (TOE) that unites all interactions. A very promising candidate for such a global description is the *string theory*. String theory is based on the assumption that elementary particles are not point-like, but are one-dimensional strings. Different string excitations or oscillations correspond to different particles. In addition, certain string theories are supersymmetric. Supersymmetric particles (SUSY particles) have to be very heavy. Experiments at the LHC have so far failed to find such particles with masses below the TeV region. Preliminary analyses of LHC results from 2015/16 appeared to show a bump hinting at a boson decaying into photon pairs around a mass of $750\,\text{GeV}$. Subsequent measurements at higher luminosity, however, revealed this as statistical fluctuation.

Supersymmetric theories establish a symmetry between fermions and bosons. String theories and, in particular, superstring theories, are constructed in a higher-dimensional space. Out of the original 11 dimensions in the so-called M superstring theory, 7 must be compacted to a very small size, because they are not observed in nature.

Some theoretical theorists consider string theories presently as best candidates to unite quantum field theories and general relativity. They might even solve the problem of the three generations of elementary particles. In the framework of string theories in eleven dimensions the weakness of gravity might be related to the fact that part of the gravitational force is leaking into extra dimensions, while, e.g., electromagnetism, in contrast, is confined to the familiar four dimensions.

If gravity were really leaking into extra dimensions, the energy sitting there could give rise to dark energy influencing the structure of the universe (see Chap. 13 on dark energy and dark matter). Gravitational matter in extra dimensions would only be visible by its gravitational interactions. So, leaking gravity might explain cosmic puzzles.

An alternative to string theories is the approach of loop quantum gravity, which also tries to merge quantum mechanics and general relativity. In loop quantum gravity space and time are quantized. In contrast to string theories that do not make any testable predictions, loop quantum gravity predicts that the velocity of light has a small dependence on energy. Low-energy photons should travel slightly faster than high-energy photons. The granularity of space and time in quantum gravity is expected to slow down photons of the highest energies. This could in principle be tested by measuring possible delays in the arrival times of γ rays and visible light from cataclysmic events in cosmic-ray sources like in active galactic nuclei. This possible frequency-dependent effect should not only apply to electromagnetic waves but also to gravitational waves.

It is also conceivable that we live in a *holographic universe* in the sense that all information from a higher-dimensional space could be coded into a lower-dimensional space, just like a three-dimensional body can be represented by a two-dimensional hologram.

In Fig. 2.13, an overview of the historical successes of the unification of different theories is displayed with a projection into the future. One assumes that with increasing temperature ($\hat{=}$ energy), nature gets more and more symmetric. At very high temperatures, as existed at the time of the Big Bang, the symmetry was so perfect that all interactions could be described by one universal force. The reason that increasingly large accelerators with higher energies are being constructed is to track down this universal description of all forces.

According to present beliefs, the all-embracing theory of supergravity (SUGRA) is a field theory that combines the principles of supersymmetry and general relativity. There is no unique picture of what a theory of supergravity should look like. It can be formulated in 11 dimensions, and it is related to an 11-dimensional superstring theory embedded in a superordinate M-theory. The smallest constituents of such a theory are p-dimensional objects ('*branes*') of the size of the Planck length $L_P = \sqrt{\hbar G / c^3}$ (where G is the gravitational constant, $\hbar$ is Planck's constant, and c is the velocity of light). Seven of the ten spatial dimensions are compacted into a Calabi–

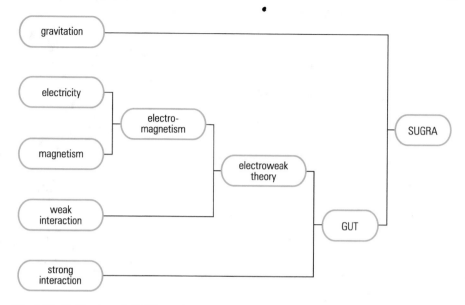

Fig. 2.13 Unification of all different interactions to obtain a Theory of Everything; GUT = Grand Unified Theory, SUGRA = Supergravitation

Yau space. According to taste, the 'M' in the M-theory stands for 'membrane', 'matrix', 'mystery', or 'mother (of all theories)'.

Summary

The Standard Model is one of the best confirmed theories in particle physics. Still, it is only a preliminary description of elementary particles and their interactions. However, there are far too many parameters, which have to be adjusted to experiments. In the early days of particle physics many of these parameters were provided by cosmic radiation, and even today astroparticle physics supplies important information, like neutrino properties and ideas on neutrino masses, which were originally not foreseen in the Standard Model. Even being incomplete, the Standard Model constitutes a fundamental milestone in approaching a final description of all particles and their interactions. Unfortunately, a fly in the ointment is still the missing breakthrough to include gravity into the common framework.

This is my idea of a unified theory!

2.4 Problems

1. Which of the following reactions or decays are allowed?

   (a) $\mu^- \rightarrow e^- + \gamma$,
   (b) $\mu^+ \rightarrow e^+ + \nu_e + \bar{\nu}_\mu + e^+ + e^-$,
   (c) $\pi^0 \rightarrow \gamma + e^+ + e^-$,
   (d) $\pi^+ \rightarrow \mu^+ + e^-$,
   (e) $\Lambda \rightarrow p + K^-$,
   (f) $\Sigma^+ \rightarrow n + \pi^+$,
   (g) $K^+ \rightarrow \pi^+ + \pi^- + \pi^+$,
   (h) $K^+ \rightarrow \pi^0 + \pi^0 + e^+ + \nu_e$.

2. What is the minimum kinetic energy of a cosmic-ray muon to survive to sea level from a production altitude of 20 km ($\tau_\mu = 2.197\,03\,\mu s$, $m_\mu = 105.658\,37\,\text{MeV}$)? For this problem one should assume that all muons have the given lifetime in their rest frame.

3. Work out the Coulomb force and the gravitational force between two singly charged particles of the Planck mass at a distance of $r = 1\,\text{fm}$!

4. In a fixed-target experiment positrons are fired at a target of electrons at rest. What positron energy is required to produce a Z ($m_Z = 91.188\,\text{GeV}$)?

Chapter 3
Kinematics and Cross Sections

The best way to escape a problem is to solve it.

Alan Saporta

In astroparticle physics the energies of participating particles are generally that high, that *relativistic kinematics* must be used. In this field of science it becomes obvious that mass and energy are only different facets of the same thing. Mass is a particularly compact form of energy, which is related to the total energy of a particle by the famous Einstein relation

$$E = mc^2 . \tag{3.0.1}$$

In this equation m is the mass of a particle, which moves with the velocity v, and c is the velocity of light in vacuum.

The experimental result that the velocity of light in vacuum is the maximum velocity in all inertial systems leads to the fact that particles with velocity near the velocity of light do not get much faster when accelerated, but mainly only become heavier,

$$m = \frac{m_0}{\sqrt{1 - \beta^2}} = \gamma m_0 . \tag{3.0.2}$$

In this equation m_0 is the rest mass, $\beta = v/c$ is the particle velocity, normalized to the velocity of light, and

$$\gamma = \frac{1}{\sqrt{1 - \beta^2}} \tag{3.0.3}$$

is the Lorentz factor. Using this result, (3.0.1) can also be written a

$$E = \gamma m_0 c^2 , \tag{3.0.4}$$

where $m_0 c^2$ is the rest energy of a particle. The momentum of a particle can be expressed as

© Springer Nature Switzerland AG 2020
C. Grupen, *Astroparticle Physics*, Undergraduate Texts in Physics,
https://doi.org/10.1007/978-3-030-27339-2_3

$$p = mv = \gamma m_0 \beta c .$$ (3.0.5)

Using (3.0.3), the difference

$$E^2 - p^2 c^2 = \gamma^2 m_0^2 c^4 - \gamma^2 m_0^2 \beta^2 c^4$$

can be written as

$$E^2 - p^2 c^2 = \frac{m_0^2 c^4}{1 - \beta^2}(1 - \beta^2) = m_0^2 c^4 .$$ (3.0.6)

This result shows that $E^2 - p^2 c^2$ is a Lorentz-invariant quantity called the invariant mass squared of the particle of mass m.[1] This quantity is the same in all systems and it equals the square of the rest energy. Consequently, the total energy of a relativistic particle can be expressed by

$$E = c\sqrt{p^2 + m_0^2 c^2} .$$ (3.0.7)

This equation holds for all particles. For massless particles or, more precisely, particles with rest mass zero, one obtains

$$E = cp .$$ (3.0.8)

Particles of total energy E without rest mass are also subject to gravitation, because they acquire a mass according to

---

[1] More precisely the invariant mass times the square of the speed of light squared.

$$m = E/c^2 . \tag{3.0.9}$$

The transition from relativistic kinematics to classical (Newtonian) mechanics ($p \ll m_0 c$) can also be derived from (3.0.7) by series expansion. The kinetic energy of a particle is obtained to

$$
\begin{aligned}
E^{\text{kin}} = E - m_0 c^2 &= c\sqrt{p^2 + m_0^2 c^2} - m_0 c^2 \\
&= m_0 c^2 \sqrt{1 + \left(\frac{p}{m_0 c}\right)^2} - m_0 c^2 \\
&\approx m_0 c^2 \left(1 + \frac{1}{2}\left(\frac{p}{m_0 c}\right)^2\right) - m_0 c^2 \\
&= \frac{p^2}{2m_0} = \frac{1}{2}m_0 v^2 ,
\end{aligned}
\tag{3.0.10}
$$

in accordance with classical mechanics. Using (3.0.4) and (3.0.5), the velocity can be expressed by

$$v = \frac{p}{\gamma m_0} = \frac{c^2 p}{E}$$

or

$$\beta = \frac{cp}{E} . \tag{3.0.11}$$

In relativistic kinematics it is usual to set $c = 1$. This simplifies all formulae. If, however, numerical quantities have to be calculated, the actual value of the velocity of light has to be considered.

3.1 Threshold Energies

Energy has mass and mass represents its energy.

Albert Einstein

In astroparticle physics frequently the problem occurs to determine the *threshold energy* for a certain process of particle production. This requires that in the center-of-mass system of the collision at least the masses of all particles in the final state of the reaction have to be provided. In storage rings the center-of-mass system is frequently identical with the laboratory system so that, for example, the creation of a particle of mass M in an electron–positron head-on collision (e^+ and e^- have the same total energy E) requires

$$2E \geq M . \tag{3.1.1}$$

If, on the other hand, a particle of energy E interacts with a target at rest as it is characteristic for processes in cosmic rays, the center-of-mass energy for such a process must first be calculated.

For the general case of a collision of two particles with total energy E_1 and E_2 and momenta p_1 and p_2 the Lorentz-invariant center-of-mass energy E_{CMS} can be determined using (3.0.7) and (3.0.11) and setting $c = 1$ in the following way:

$$
\begin{aligned}
E_{CMS} &= \sqrt{s} \\
&= \left\{ (E_1 + E_2)^2 - (p_1 + p_2)^2 \right\}^{1/2} \\
&= \left\{ E_1^2 - p_1^2 + E_2^2 - p_2^2 + 2E_1 E_2 - 2p_1 \cdot p_2 \right\}^{1/2} \\
&= \left\{ m_1^2 + m_2^2 + 2E_1 E_2 (1 - \beta_1 \beta_2 \cos\theta) \right\}^{1/2} .
\end{aligned}
\tag{3.1.2}
$$

In this equation θ is the angle between p_1 and p_2. For high energies ($\beta_1, \beta_2 \to 1$ and $m_1, m_2 \ll E_1, E_2$) and not too small angles θ (3.1.2) simplifies to

$$
E_{CMS} = \sqrt{s} \approx \left\{ 2E_1 E_2 (1 - \cos\theta) \right\}^{1/2} .
\tag{3.1.3}
$$

If one particle (for example, the particle of the mass m_2) is at rest (laboratory system $E_2 = m_2$, $p_2 = 0$), (3.1.2) leads to

$$
\sqrt{s} = \left\{ m_1^2 + m_2^2 + 2E_1 m_2 \right\}^{1/2} .
\tag{3.1.4}
$$

Using the relativistic approximation ($m_1^2, m_2^2 \ll 2E_1 m_2$), one gets

$$
\sqrt{s} \approx \sqrt{2E_1 m_2} .
\tag{3.1.5}
$$

In such a reaction only particles with masses $M \leq \sqrt{s}$ can be produced.

3.2   Examples for the Determination of Center-of-Mass Energies

The centre of the system of the world is immovable.

Sir Isaac Newton

Example 1:   Let us assume that a high-energy cosmic-ray proton (energy E_p, momentum p, rest mass m_p) produces a proton–antiproton pair on a target proton at rest:

$$
p + p \to p + p + p + \bar{p} .
\tag{3.2.1}
$$

According to (3.1.2) the center-of-mass energy can be calculated as follows:

$$\sqrt{s} = \left\{ (E_p + m_p)^2 - (\mathbf{p} - 0)^2 \right\}^{1/2}$$
$$= \left\{ E_p^2 + 2m_p E_p + m_p^2 - p^2 \right\}^{1/2}$$
$$= \left\{ 2m_p E_p + 2m_p^2 \right\}^{1/2} . \tag{3.2.2}$$

For the final state, consisting of three protons and one antiproton (the mass of the antiproton is equal to the mass of the proton), one has

$$\sqrt{s} \geq 4m_p . \tag{3.2.3}$$

From this the threshold energy of the incident proton can be derived to be

$$2m_p E_p + 2m_p^2 \geq 16\, m_p^2 ,$$
$$E_p \geq 7m_p \ (= 6.568\,\text{GeV}) , \tag{3.2.4}$$
$$E_p^{\text{kin}} = E_p - m_p \geq 6m_p .$$

Example 2: For the equivalent process of $e^+ e^-$ pair production by an energetic electron on an electron target at rest,

$$e^- + e^- \to e^- + e^- + e^+ + e^- , \tag{3.2.5}$$

one would expect the corresponding result $E_e^{\text{kin}} \geq 6m_e$. According to Eq. (3.1.4) one has

$$\sqrt{s} = \{ m_e^2 + m_e^2 + 2E_e m_e \}^{1/2} \quad \geq \quad 4m_e ,$$
$$E_e \quad \geq \quad 7m_e ,$$
$$E_e \quad \geq \quad 3.58\,\text{MeV} ,$$
$$E_e^{\text{kin}} = E_e - m_e \geq 3.07\,\text{MeV} . \tag{3.2.6}$$

Example 3: Let us consider the photoproduction of an electron–positron pair on a target electron at rest,

$$\gamma + e^- \to e^- + e^+ + e^- ; \tag{3.2.7}$$

$$\sqrt{s} = \{ m_e^2 + 2E_\gamma m_e \}^{1/2} \geq 3m_e ,$$
$$E_\gamma \geq 4m_e ,$$
$$E_\gamma \geq 2.04\,\text{MeV} . \tag{3.2.8}$$

Example 4: Consider the photoproduction of a neutral pion (mass $m_{\pi^0} \approx$ 135 MeV) on a target proton at rest (mass m_p):

$$\gamma + p \rightarrow p + \pi^0, \tag{3.2.9}$$

$$\sqrt{s} = \{m_0^2 + 2E_p m_0\}^{1/2} \geq m_0 + m_{\pi^0},$$
$$m_0^2 + 2E_p m_0 \geq m_0^2 + m_{\pi^0}^2 + 2m_0 m_{\pi^0},$$

$$E_p \geq \frac{2m_0 m_{\pi^0} + m_{\pi^0}^2}{2m_0} = m_{\pi^0} + \frac{m_{\pi^0}^2}{2m_0}, \tag{3.2.10}$$
$$E_p \geq m_{\pi^0} + 9.7\,\text{MeV} \approx 145\,\text{MeV}.$$

3.3 Four-Vectors

The physicist in preparing for his work needs three things, mathematics, mathematics, and mathematics.

Wilhelm Conrad Röntgen

For calculations of this kind it is practical to introduce Lorentz-invariant *four-vectors*. In the same way as time t and the position vector $s = (x, y, z)$ can be combined to form a four-vector, also a four-momentum vector

$$q = \begin{pmatrix} E \\ p \end{pmatrix} \text{ with } p = (p_x, p_y, p_z) \tag{3.3.1}$$

can be introduced. Because of

$$q^2 = \begin{pmatrix} E \\ p \end{pmatrix}^2 = E^2 - p^2 = m_0^2 \tag{3.3.2}$$

the square of the four-momentum is equal to the square of the rest mass. For photons one has

$$q^2 = E^2 - p^2 = 0. \tag{3.3.3}$$

Those particles, which fulfill (3.3.2) are said to lie on the *mass shell*. On-shell particles are also called real. Apart from that, particles can also borrow energy for a short time from the vacuum within the framework of Heisenberg's uncertainty principle. Such particles are called virtual. They are not on the mass shell. In interaction processes virtual particles can only occur as exchange particles.

Example 5:   Photoproduction of an electron–positron pair in the Coulomb field of a nucleus (see Fig. 3.1).

In this example the incoming photon γ is real, while the photon γ^* exchanged between the electron and the nucleus is virtual (Fig. 3.2)

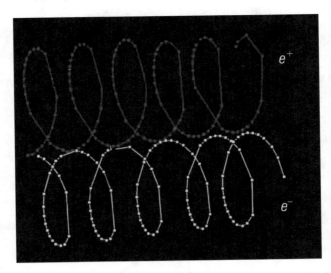

Fig. 3.1 Electron–positron pair production by a photon (coming from the left) in the time projection chamber of the ALEPH experiment at CERN. The electron and positron tracks are bent in the solenoidal magnetic field producing helical tracks each [26]

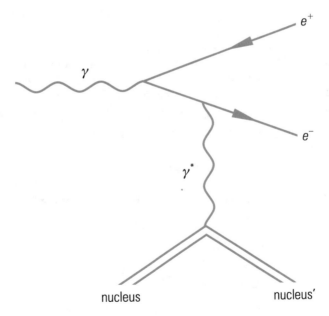

Fig. 3.2 The process $\gamma + \text{nucleus} \rightarrow e^+ + e^- + \text{nucleus}'$

$$q_\gamma + q_p \geq 2m_e + m_p \, ,$$
$$q_\gamma^2 + q_p^2 + 2q_\gamma q_p \geq 4m_e^2 + m_p^2 + 4m_e m_p \, .$$

Because of $q_\gamma^2 = 0$, and assuming further that the target nucleus is a proton, thus $q_p^2 = m_p^2$, one gets

$$2q_\gamma q_p \geq 4m_e^2 + 4m_e m_p \, .$$

The product of the four-momenta yields

$$2 \begin{pmatrix} E_\gamma \\ \boldsymbol{p}_\gamma \end{pmatrix} \begin{pmatrix} m_p \\ 0 \end{pmatrix} = 2m_p E_\gamma \geq 4m_e^2 + 4m_e m_p \, ,$$

because the target proton is at rest ($E = m_p$, $\boldsymbol{p} = 0$). Under these assumptions one obtains for the threshold energy

$$E_\gamma \geq 2m_e + \frac{2m_e^2}{m_p} \, .$$

The threshold energy is a little larger than $2m_e$, because the target proton receives some recoil. Without the proton as recoil partner this reaction would violate the momentum conservation.

Example 6: Electron–proton scattering (Fig. 3.3)

The virtuality of the exchanged photon γ^* can easily be determined from the kinematics based on the four-momentum vectors of the electron and proton. The four-momentum vectors are defined in the following way: incoming electron $q_e = \begin{pmatrix} E_e \\ \boldsymbol{p}_e \end{pmatrix}$, final-state electron $q'_e = \begin{pmatrix} E'_e \\ \boldsymbol{p}'_e \end{pmatrix}$, incoming proton $q_p = \begin{pmatrix} E_p \\ \boldsymbol{p}_p \end{pmatrix}$, final-state proton $q'_p = \begin{pmatrix} E'_p \\ \boldsymbol{p}'_p \end{pmatrix}$. Since energy and momentum are conserved, also four-momentum conservation holds:

$$q_e + q_p = q'_e + q'_p \, . \tag{3.3.4}$$

The four-momentum squared of the exchanged virtual photon $q_{\gamma^*}^2$ is determined to be

$$\begin{aligned} q_{\gamma^*}^2 &= (q_e - q'_e)^2 \\ &= \begin{pmatrix} E_e - E'_e \\ \boldsymbol{p}_e - \boldsymbol{p}'_e \end{pmatrix}^2 = (E_e - E'_e)^2 - (\boldsymbol{p}_e - \boldsymbol{p}'_e)^2 \\ &= E_e^2 - p_e^2 + E_e'^2 - p_e'^2 - 2E_e E'_e + 2\boldsymbol{p}_e \cdot \boldsymbol{p}'_e \\ &= 2m_e^2 - 2E_e E'_e (1 - \beta_e \beta'_e \cos\theta) \, , \end{aligned} \tag{3.3.5}$$

where β_e and β'_e are the velocities of the incoming and outgoing electron and θ is the angle between $\boldsymbol{p}_e$ and $\boldsymbol{p}'_e$. For high energies and not too small scattering angles (3.3.5) is simplified to

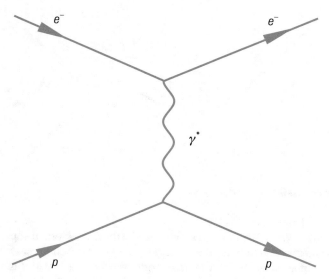

Fig. 3.3 The process $e^- + p \to e^- + p$

$$q_{\gamma^*}^2 = -2E_e E_e'(1 - \cos\theta)$$
$$= -4E_e E_e' \sin^2 \frac{\theta}{2} \,. \tag{3.3.6}$$

If $\sin\frac{\theta}{2}$ can be approximated by $\frac{\theta}{2}$, one gets for not too small angles

$$q_{\gamma^*}^2 = -E_e E_e' \theta^2 \,. \tag{3.3.7}$$

The mass squared of the exchanged photon in this case is negative! This means that the mass of γ^* is purely imaginary. Such photons are called *space-like*.

Example 7: Muon pair production in e^+e^- interactions (Fig. 3.4)
Assuming that electrons and positrons have the same total energy E and opposite momentum ($\boldsymbol{p}_{e^+} = -\boldsymbol{p}_{e^-}$), one has

$$q_{\gamma^*}^2 = (q_{e^+} + q_{e^-})^2 = \begin{pmatrix} E + E \\ \boldsymbol{p}_{e^+} + (-\boldsymbol{p}_{e^+}) \end{pmatrix}^2$$
$$= 4E^2 \,. \tag{3.3.8}$$

In this case the mass of the exchanged photon is $2E$, which is positive. Such a photon is called *time-like*. The muon pair in the final state can be created if $2E \geq 2m_\mu$.

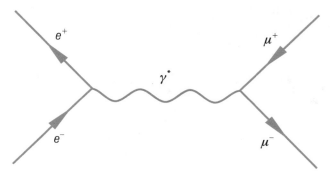

Fig. 3.4 The process $e^+ + e^- \to \mu^+ + \mu^-$

Example 8:   Detection of WIMPs

The search for WIMPs is of particular interest in the framework of supersymmetric theories and consequently for the hunting of weakly interacting massive particles. The determination of the maximum energy transfer of a WIMP (assumed mass $m_W = 100\,\text{GeV}$) to a target nucleus ($m_T = 12\,\text{GeV}$) at rest is for the search for supersymmetric and weakly interacting massive particles of particular interest. It is possible that WIMPs are gravitationally captured in the halo of our Milky Way. They will then also assume the halo velocity of about 300 km/s, corresponding to $\beta = 10^{-3}$. Because of this low velocity it is in order to use a classical approach. Let us assume that the velocity of the WIMP before the collision is u_1 and v_1 after the interaction. The target—initially assumed to be at rest—gets a velocity v_2 in the collision. The masses of the WIMP and target are m_W and m_T respectively.

Momentum conservation yields

$$m_W \cdot u_1 = m_W \cdot v_1 + m_T \cdot v_2 \qquad (3.3.9)$$

or

$$m_W \cdot v_1 = m_W \cdot u_1 - m_T \cdot v_2 . \qquad (3.3.10)$$

After squaring this equation, one gets

$$m_W^2 \cdot v_1^2 = m_W^2 \cdot u_1^2 + m_T^2 \cdot v_2^2 - 2 \cdot m_W m_T u_1 v_2 . \qquad (3.3.11)$$

Energy conservation (after multiplying with a factor of 2) gives:

$$m_W \cdot u_1^2 = m_W \cdot v_1^2 + m_T \cdot v_2^2 . \qquad (3.3.12)$$

This equation, multiplied by m_W, leads to

$$m_W^2 \cdot u_1^2 = m_W^2 \cdot v_1^2 + m_T m_W \cdot v_2^2 . \qquad (3.3.13)$$

Inserting this into the momentum conservation, one obtains

$$m_W^2 \cdot v_1^2 = m_W^2 \cdot v_1^2 + m_T m_W \cdot v_2^2 + m_T^2 \cdot v_2^2 - 2 \cdot m_W m_T u_1 v_2 , \qquad (3.3.14)$$

which leads to:

$$2 \cdot m_W m_T u_1 v_2 = m_T m_W \cdot v_2^2 + m_T^2 \cdot v_2^2 . \qquad (3.3.15)$$

After reducing and rearrangement one obtains

$$\frac{v_2}{u_1} = \frac{2 m_W}{m_W + m_T} . \qquad (3.3.16)$$

After squaring and multiplication with $\frac{m_T}{m_W}$ the ratio of the target to the WIMP energy is:

$$f = \frac{\frac{1}{2} m_T v_2^2}{\frac{1}{2} m_W u_1^2} = \frac{4 m_T m_W}{(m_W + m_T)^2} . \qquad (3.3.17)$$

Inserting numbers for the assumed masses, one finds $f = 0.383$. With $u_1 = 300\,km/s$ and $m_W = 100\,GeV$ the kinetic energy of the incident WIMP is 8×10^{-15} joule. This leads to a recoil energy of the target—using the factor f—of about 20 keV. This low energy of the recoil nucleus is very difficult to detect efficiently in the presence of background and other underground reactions.

3.4 Examples for the Treatment of Decays

Mathematics is the language in which God has written the universe.

Galileo Galilei

3.4.1 Two-Body Decays

All things are subject to decay.

John Dryden

The elegant formalism of four-momentum vectors for the calculation of kinematical relations can be also extended to decays of elementary particles. In a two-body decay of an elementary particle at rest the two decay particles get well-defined discrete energies because of momentum conservation.

Example 9:   The two-body decay $\pi^+ \rightarrow \mu^+ + \nu_\mu$
Four-momentum conservation yields

$$q_\pi^2 = (q_\mu + q_\nu)^2 = m_\pi^2 . \qquad (3.4.1)$$

In the rest frame of the pion the muon and neutrino are emitted in opposite directions, $p_\mu = -p_{\nu_\mu}$,

$$\left(\frac{E_\mu + E_\nu}{p_\mu + p_{\nu_\mu}} \right)^2 = (E_\mu + E_\nu)^2 = m_\pi^2 . \qquad (3.4.2)$$

Neglecting a possible non-zero neutrino mass for this consideration, one has

$$E_\nu = p_{\nu_\mu}$$

with the result

$$E_\mu + p_\mu = m_\pi .$$

Rearranging this equation and squaring it gives

$$E_\mu^2 + m_\pi^2 - 2E_\mu m_\pi = p_\mu^2 ,$$
$$2E_\mu m_\pi = m_\pi^2 + m_\mu^2 ,$$
$$E_\mu = \frac{m_\pi^2 + m_\mu^2}{2m_\pi} . \qquad (3.4.3)$$

For $m_\mu = 105.658\,369\,\text{MeV}$ and $m_{\pi^\pm} = 139.570\,18\,\text{MeV}$ one gets $E_\mu^{\text{kin}} = E_\mu - m_\mu = 4.09\,\text{MeV}$. For the two-body decay of the kaon, $K^+ \rightarrow \mu^+ + \nu_\mu$, (3.4.3) gives $E_\mu^{\text{kin}} = E_\mu - m_\mu = 152.49\,\text{MeV}$ ($m_{K^\pm} = 493.677\,\text{MeV}$).

Due to helicity conservation the decay $\pi^+ \to e^+ + \nu_e$ is strongly suppressed (see Fig. 2.11). Using (3.4.3) the positron would get in this decay a kinetic energy of $E_{e^+}^{\text{kin}} = E_{e^+} - m_e = \frac{m_\pi}{2} + \frac{m_e^2}{2m_\pi} - m_e = \frac{m_\pi}{2} \left(1 - \frac{m_e}{m_\pi}\right)^2 \approx 69.3\,\text{MeV}$, which is approximately half the pion mass. This is not a surprise, since the 'heavy' pion decays into two nearly massless particles.

Example 10: The decay $\pi^0 \to \gamma + \gamma$

The kinematics of the π^0 decay at rest is extremely simple. Each decay photon gets as energy one half of the pion rest mass. In this example also the decay of a π^0 in flight will be considered. If the photon is emitted in the direction of flight of the π^0, it will get a higher energy compared with the emission opposite to the flight direction. The decay of a π^0 in flight (Lorentz factor $\gamma = E_{\pi^0}/m_{\pi^0}$) yields a flat spectrum of photons between a maximum and minimum energy. Four-momentum conservation

$$q_{\pi^0} = q_{\gamma_1} + q_{\gamma_2}$$

leads to

$$q_{\pi^0}^2 = m_{\pi^0}^2 = q_{\gamma_1}^2 + q_{\gamma_2}^2 + 2q_{\gamma_1} q_{\gamma_2}. \tag{3.4.4}$$

Since the masses of real photons are zero, the kinematic limits are obtained from the relation

$$2q_{\gamma_1} q_{\gamma_2} = m_{\pi^0}^2. \tag{3.4.5}$$

In the limit of maximum or minimum energy transfer to the photons they are emitted parallel or antiparallel to the direction of flight of the π^0. This leads to

$$\boldsymbol{p}_{\gamma_1} \parallel -\boldsymbol{p}_{\gamma_2}. \tag{3.4.6}$$

Using this, (3.4.5) can be expressed as

$$2(E_{\gamma_1} E_{\gamma_2} - \boldsymbol{p}_{\gamma_1} \cdot \boldsymbol{p}_{\gamma_2}) = 4E_{\gamma_1} E_{\gamma_2} = m_{\pi^0}^2. \tag{3.4.7}$$

Because of $E_{\gamma_2} = E_{\pi^0} - E_{\gamma_1}$ (3.4.7) leads to the quadratic equation

$$E_{\gamma_1}^2 - E_{\gamma_1} E_{\pi^0} + \frac{m_{\pi^0}^2}{4} = 0 \tag{3.4.8}$$

with the symmetric solutions

$$E_{\gamma_1}^{\text{max}} = \frac{1}{2}(E_{\pi^0} + p_{\pi^0}),$$
$$E_{\gamma_1}^{\text{min}} = \frac{1}{2}(E_{\pi^0} - p_{\pi^0}). \tag{3.4.9}$$

Because of $E_{\pi^0} = \gamma m_{\pi^0}$ and $p_{\pi^0} = \gamma m_{\pi^0} \beta$ Eq. (3.4.9) can also be expressed as

$$E_{\gamma_1}^{\max} = \frac{1}{2}\gamma m_{\pi^0}(1 + \beta) = \frac{1}{2}m_{\pi^0}\sqrt{\frac{1+\beta}{1-\beta}},$$

$$E_{\gamma_1}^{\min} = \frac{1}{2}\gamma m_{\pi^0}(1 - \beta) = \frac{1}{2}m_{\pi^0}\sqrt{\frac{1-\beta}{1+\beta}}. \tag{3.4.10}$$

In the relativistic limit ($\gamma \gg 1$, $\beta \approx 1$) a photon emitted in the direction of flight of the π^0 gets the energy $E_\gamma^{\max} = E_{\pi^0} = \gamma m_{\pi^0}$ and the energy of the backward-emitted photon is zero. In a logarithmic scale the energy distribution of the photons is symmetric around half the π^0 mass. In case of a spectrum of neutral pions, the energy distributions of the decay photons are superimposed in such a way that the resulting spectrum has a maximum at half the π^0 mass in a logarithmic plot (see Fig. 3.5).

3.4.2 Three-Body Decays

Harmony makes small things grow. Lack of it makes big things decay.

Sallust

Much more difficult is the treatment of a three-body decay. Such a process is going to be explained for the example of the muon decay:

$$\mu^- \rightarrow e^- + \bar{\nu}_e + \nu_\mu. \tag{3.4.11}$$

Let us assume that the muon is originally at rest ($E_\mu = m_\mu$). Four-momentum conservation

$$q_\mu = q_e + q_{\bar{\nu}_e} + q_{\nu_\mu} \tag{3.4.12}$$

can be rephrased as

$$(q_\mu - q_e)^2 = (q_{\bar{\nu}_e} + q_{\nu_\mu})^2,$$

$$q_\mu^2 + q_e^2 - 2q_\mu q_e = m_\mu^2 + m_e^2 - 2\binom{m_\mu}{0}\binom{E_e}{p_e}$$

$$= (q_{\bar{\nu}_e} + q_{\nu_\mu})^2,$$

$$E_e = \frac{m_\mu^2 + m_e^2 - (q_{\bar{\nu}_e} + q_{\nu_\mu})^2}{2m_\mu}. \tag{3.4.13}$$

The electron energy is largest, if $(q_{\bar{\nu}_e} + q_{\nu_\mu})^2$ takes on a minimum value. For vanishing neutrino masses this means that the electron gets a maximum energy, if

$$q_{\bar{\nu}_e} q_{\nu_\mu} = E_{\bar{\nu}_e} E_{\nu_\mu} - p_{\bar{\nu}_e} \cdot p_{\nu_\mu} = 0. \tag{3.4.14}$$

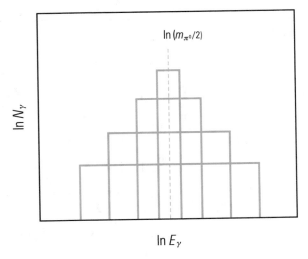

Fig. 3.5 Energy distribution of photons in π^0 decay for different pion energies in a logarithmic plot [27]

Equation (3.4.14) is satisfied for $\boldsymbol{p}_{\bar{\nu}_e} \parallel \boldsymbol{p}_{\nu_\mu}$. This yields

$$E_e^{\max} = \frac{m_\mu^2 + m_e^2}{2m_\mu} \approx \frac{m_\mu}{2} = 52.83 \,\text{MeV} . \qquad (3.4.15)$$

Fig. 3.6 Energy spectrum of
electrons from muon decay

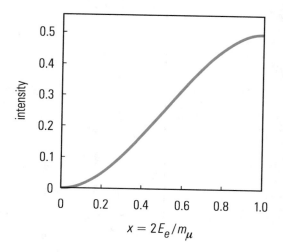

Fig. 3.7 Event display of
the decay of a charged kaon
into three pions in a nuclear
emulsion
$(K^{\pm} \to \pi^{\pm} + \pi^{-} + \pi^{+}$; in
a nuclear emulsion the
charge of a particle cannot be
determined) [28]

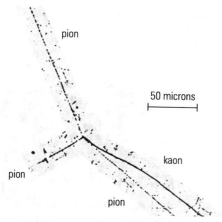

In this configuration the electron momentum $\boldsymbol{p}_e$ is antiparallel to both neutrino
momenta, which in turn are parallel to each other.

If the spins of all participating particles and the structure of weak interactions are
taken into consideration, one obtains for the electron spectrum, using the shorthand
$x = 2E_e/m_\mu \approx E_e/E_e^{\mathrm{max}}$,

$$N(x) = \mathrm{const}\, x^2(1.5 - x). \tag{3.4.16}$$

Just as in nuclear beta decay $(n \to p + e^- + \bar{\nu}_e)$ the available decay energy
in a three-body decay is distributed continuously among the final-state particles
(Fig. 3.6).

An example of a relatively rare decay mode of a charged kaon into three charged
pions (branching fraction 5.6%) is shown in Fig. 3.7. The slowing down kaon

increases its ionization towards the end of its range (see Chap. 4), thereby producing an intense, dark track, until it eventually stops and decays into three pions.

3.5 Lorentz Transformations

We have learned something about the laws of nature, their invariance with respect to the Lorentz transformation, and their validity for all inertial systems moving uniformly, relative to each other. We have the laws but do not know the frame to which to refer them.

Albert Einstein

If interaction or decay processes are treated, it is fully sufficient to consider the process in the center-of-mass system. In a different system (for example, the laboratory system) the energies and momenta are obtained by a *Lorentz transformation*. If E and p are energy and momentum in the center-of-mass system and if the laboratory system moves with the velocity β relative to $p_\parallel$, the transformed quantities E^* and $p_\parallel^*$ in this system are calculated to be (compare Fig. 3.8)

$$\begin{pmatrix} E^* \\ p_\parallel^* \end{pmatrix} = \begin{pmatrix} \gamma & -\gamma\beta \\ -\gamma\beta & \gamma \end{pmatrix} \begin{pmatrix} E \\ p_\parallel \end{pmatrix}, \quad p_\perp^* = p_\perp. \tag{3.5.1}$$

The transverse momentum component is not affected by this transformation. Instead of using the matrix notation, (3.5.1) can be written as

$$\begin{aligned} E^* &= \gamma E - \gamma\beta p_\parallel, \\ p_\parallel^* &= -\gamma\beta E + \gamma p_\parallel. \end{aligned} \tag{3.5.2}$$

For $\beta = 0$ and correspondingly $\gamma = 1$ one trivially obtains $E^* = E$ and $p_\parallel^* = p_\parallel$.

A particle of energy $E = \gamma_2 m_0$, seen from a system that moves with β_1 relative to the particle parallel to the momentum p, gets in this system the energy

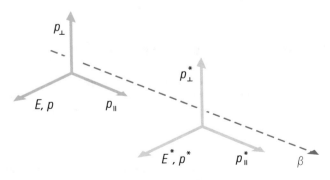

Fig. 3.8 Illustrating the Lorentz transformation

$$E^* = \gamma_1 E - \gamma_1 \beta_1 p_\parallel$$

$$= \gamma_1 \gamma_2 m_0 - \gamma_1 \frac{\sqrt{\gamma_1^2 - 1}}{\gamma_1} \sqrt{(\gamma_2 m_0)^2 - m_0^2}$$

$$= \gamma_1 \gamma_2 m_0 - m_0 \sqrt{\gamma_1^2 - 1} \sqrt{\gamma_2^2 - 1} \,. \tag{3.5.3}$$

If $\gamma_1 = \gamma_2 = \gamma$ (for a system that moves along with a particle), one naturally obtains

$$E^* = \gamma^2 m_0 - m_0(\gamma^2 - 1) = m_0 \,.$$

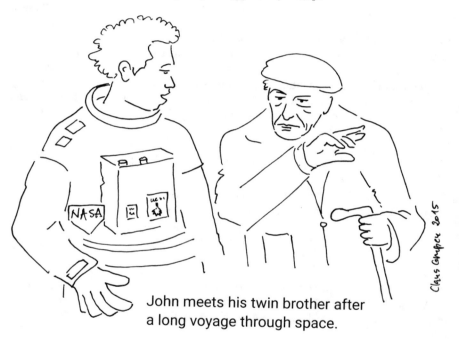

John meets his twin brother after a long voyage through space.

3.6 Determination of Cross Sections

> *Physicists are, as a general rule, highbrows. They think and talk in long, Latin words, and when they write anything down they usually include at least one partial differential and three Greek letters.*
>
> Stephen White

Apart from the kinematics of interaction processes the *cross section* for a reaction is of particular importance. In the most simple case the cross section can be considered as an effective area that the target particle represents for the collision with a projectile.

If the target has an area of πr_T^2 and the projectile size corresponds to πr_P^2, the geometrical cross section for a collision is obtained to be

$$\sigma = \pi (r_T + r_P)^2 . \tag{3.6.1}$$

In most cases the cross section also depends on other parameters, for example, on the energy of the particle. The atomic cross section σ_A, measured in cm^2, is related to the interaction length λ according to

$$\lambda \{cm\} = \frac{A}{N_A \{g^{-1}\} \varrho \{g/cm^3\} \sigma_A \{cm^2\}} \tag{3.6.2}$$

(N_A—Avogadro number, A—atomic mass of the target, ϱ—density). Frequently, the interaction length is expressed by $(\lambda \varrho) \{g/cm^2\}$. Correspondingly, the absorption coefficient is defined to be

$$\mu \{cm^{-1}\} = \frac{N_A \varrho \sigma_A}{A} = \frac{1}{\lambda} ; \tag{3.6.3}$$

equivalently, the absorption coefficient can also be expressed by $(\mu/\varrho) \{(g/cm^2)^{-1}\}$.
The absorption coefficient also provides a useful relation for the determination of interaction probabilities or rates,

$$\phi \{(g/cm^2)^{-1}\} = \frac{\mu}{\varrho} = \frac{N_A}{A} \sigma_A . \tag{3.6.4}$$

If σ_N is the cross section per nucleon, one has

$$\phi \{(g/cm^2)^{-1}\} = \sigma_N N_A . \tag{3.6.5}$$

If j is the particle flux per cm^2 and s, the number of particles dN scattered through an angle θ into the solid angle $d\Omega$ per unit time is

$$dN(\theta) = j \sigma(\theta) d\Omega , \tag{3.6.6}$$

where

$$\sigma(\theta) = \frac{d\sigma}{d\Omega}$$

is the differential scattering cross section, describing the probability of scattering of the projectile into the solid-angle element $d\Omega$, where

$$d\Omega = \sin \theta \, d\theta \, d\varphi \tag{3.6.7}$$

(φ—azimuthal angle, θ—polar angle).

For azimuthal symmetry one has

$$d\Omega = 2\pi \sin\theta \, d\theta = -2\pi \, d(\cos\theta) \, . \qquad (3.6.8)$$

Apart from the angular dependence the cross section can also depend on other quantities, so that a large number of differential cross sections are known, for example,

$$\frac{d\sigma}{dE}, \ \frac{d\sigma}{dp},$$

or even double differential cross sections such as

$$\frac{d^2\sigma}{dE \, d\theta} \, . \qquad (3.6.9)$$

To illustrate the idea of differential cross sections the transverse momentum distributions of pions and kaons in cosmic rays is compared to fixed-target data from accelerators as an example in Fig. 3.9 [29]. The double differential cross section for transverse momentum distributions of pions, kaons and protons obtained in proton–proton collisions is shown in Fig. 3.10 [31]. Here, as second variable the commonly used rapidity is introduced, which is interpreted as a measure for the relativistic velocity ($y = \frac{1}{2} \cdot \ln(\frac{E+p_z}{E-p_z})$).

Apart from the mentioned characteristic quantities there is quite a large number of other kinematical variables, which are used for the treatment of special processes and decays.

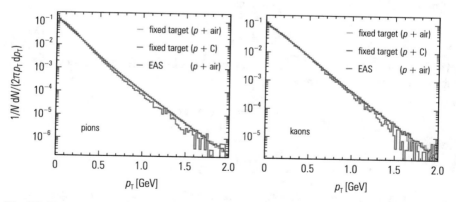

Fig. 3.9 Transverse momentum distributions of pions and kaons in cosmic rays from the KAS-CADE experiment (nucleons $\approx 160\,\text{GeV}$ on air) in comparison to fixed-target data from accelerators (protons $160\,\text{GeV}$ on air; protons $160\,\text{GeV}$ on carbon) [30]

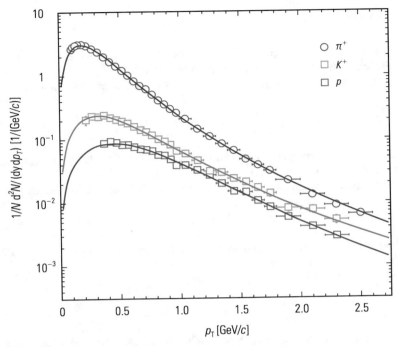

Fig. 3.10 Double differential transverse momentum distributions of pions, kaons, and protons obtained in proton–proton collisions at a center-of-mass energy of 900 GeV in ALICE at the LHC (p_T is the transverse momentum, y is the rapidity, which is defined as $y = \frac{1}{2} \cdot \ln(\frac{E+p_z}{E-p_z})$, and z is oriented along the beam direction, i.e., longitudinal momentum. Rapidity is commonly used as a measure for relativistic velocity.) [31]

Summary

In astroparticle physics the knowledge of threshold energies and cross sections is of fundamental importance. In the early days of cosmic rays it was conventional to use only the laboratory system. The reason was that cosmic ray physics was always done in fixed-target experiments. In particle physics, in particular, with storage rings, the description of physics quantities in the center-of-mass system is mandatory. The astrophysicist has to have extensive experience with both systems. The knowledge of energy dependent cross sections is of importance for the design and analysis of experiments, not only in astroparticle physics. In this chapter the transitions between different Lorentz systems are illustrated with typical processes of relevance for astroparticle physics.

3.7 Problems

1. What is the threshold energy for a photon, E_{γ_1}, to produce a $\mu^+\mu^-$ pair in a collision with a blackbody photon of energy 1 meV?

2. The mean free path λ (in g/cm^2) is related to the nuclear cross section σ_N (in cm^2) by

$$\lambda = \frac{1}{N_A \sigma_N},$$

where N_A is the Avogadro number, i.e., the number of nucleons per g, and σ_N is the cross section per nucleon.

The number of particles penetrating a target x unaffected by interactions is

$$N = N_0 \, e^{-x/\lambda}.$$

How many collisions happen in a thin target of thickness x ($N_A = 6.022 \times 10^{23}$ g^{-1}, $\sigma_N = 1$ b, $N_0 = 10^8$, $x = 0.1$ g/cm^2)?

3. The neutrino was discovered in the reaction

$$\bar{\nu}_e + p \rightarrow n + e^+,$$

where the target proton was at rest. What is the minimum neutrino energy to induce this reaction?

4. The scattering of a particle of charge z on a target of nuclear charge Z is mediated by the electromagnetic interaction. Work out the momentum transfer p_b, perpendicular to the momentum of the incoming particle for an impact parameter b (distance of closest approach)! For the calculation assume the particle track to be undisturbed, i.e., the scattering angle to be small.

5. The scattering of an electron of momentum p on a target nucleus of charge Z was treated in Problems 4 under the assumption that the scattering angle is small. Work out the general expression for the transverse momentum using the Rutherford scattering formula

$$\tan \frac{\vartheta}{2} = \frac{Z r_e}{b \beta^2}, \tag{3.7.1}$$

where ϑ is the scattering angle.

Chapter 4
Physics of Particle and Radiation Detection

Every physical effect can be used as a basis for a detector.

Anonymous

The measurement techniques relevant to astroparticle physics are rather diverse. The detection of astroparticles is usually a multistep process. In this field of research, particle detection is mostly indirect. It is important to identify the nature of the astroparticle in a suitable interaction process. The target for interactions is, in many cases, not identical with the detector that measures the interaction products. Cosmic-ray muon neutrinos, for example, interact via neutrino–nucleon interactions in the antarctic ice or in the ocean, subsequently producing charged muons. These muons suffer energy losses from electromagnetic interactions in the ice (water), in which, among others, Cherenkov radiation is produced. The Cherenkov light is recorded, via the photoelectric effect, by photomultipliers. This is then used to reconstruct the energy and the direction of incidence of the muon, which is approximately identical to the direction of incidence of the primary neutrino.

In this chapter, the primary interaction processes will first be described. The processes, which are responsible for the detection of the interaction products in the detector, will then be presented.

The cross sections for the various processes depend on the particle nature, the particle energy, and the target material. A useful relation to determine the interaction probability ϕ and the event rate is obtained from the atomic-(σ_A) or nuclear-interaction cross section (σ_N) according to

$$\phi\,\{(\mathrm{g/cm^2})^{-1}\} = \frac{N_A}{A}\sigma_A = N_A\,\{\mathrm{g}^{-1}\}\,\sigma_N\,\{\mathrm{cm^2}\}\,, \qquad (4.0.1)$$

© Springer Nature Switzerland AG 2020
C. Grupen, *Astroparticle Physics*, Undergraduate Texts in Physics,
https://doi.org/10.1007/978-3-030-27339-2_4

where N_A is Avogadro's number, A is the atomic mass of the target, and σ_A is the atomic cross section in $cm^2/atom$ (σ_N in $cm^2/nucleon$), see also (3.6.4) and (3.6.5). If the target represents an area density d {g/cm^2} and if the flux of primary particles is F {s^{-1}}, the event rate R is obtained as

$$R = \phi \{(g/cm^2)^{-1}\} \, d \, \{(g/cm^2)\} \, F \, \{(s^{-1})\} \,. \tag{4.0.2}$$

4.1 Interactions of Astroparticles

Observations are meaningless without a theory to interpret them.

Raymond A. Lyttleton

The primary particles carrying astrophysical information are nuclei (protons, helium nuclei, iron nuclei, . . .), electrons, photons, or neutrinos. These three categories of particles are characterized by completely different interactions. Protons and other nuclei will undergo strong interactions. They are also subject to electromagnetic and weak interactions, however, the corresponding cross sections are much smaller than those of strong interactions. Primary nuclei will therefore interact predominantly via processes of strong interactions. A typical interaction cross section for inelastic proton–proton scattering at energies of around $100\,GeV$ is $\sigma_N \approx 40\,mb$ ($1\,mb = 10^{-27}\,cm^2$). Since high-energy primary protons interact in the atmosphere via proton–air interactions, the cross section for proton–air collisions is of great interest. The dependence of this cross section on the proton energy is shown in Fig. 4.1.

For a typical interaction cross section of $250\,mb$, the *mean free path* of protons in the atmosphere (for nitrogen: $A = 14$) is, see Chap. 3, (3.6.2),

$$\lambda = \frac{A}{N_A \, \sigma_A} \approx 93\,g/cm^2 \,. \tag{4.1.1}$$

This means that the first interaction of protons occurs in the upper part of the atmosphere. If the primary particles are not protons but rather iron nuclei (atomic number $A_{Fe} = 56$), the first interaction will occur at even higher altitudes because the cross section for iron–air interactions is correspondingly larger.

Primary high-energy photons (energy $\gg 10\,MeV$) interact via the electromagnetic process of electron–positron pair production. The characteristic interaction length[1] ('radiation length') for electrons in air is $X_0 \approx 36\,g/cm^2$. For high-energy photons (energy $\geq 10\,GeV$), where pair production dominates, the cross section is 7/9 of the

---

[1]The radiation length for electrons is defined in (4.2.2). It describes the degrading of the electron energy by bremsstrahlung according to $E = E_0 \, e^{-x/X_0}$. This 'interaction length' X_0 is also characteristic for pair production by photons. The interaction length for hadrons (protons, pions, ...) is defined through (4.1.1), where σ_A is the total cross section. This length is sometimes also called collision length. If the total cross section in (4.1.1) is replaced by its inelastic part only, the resulting length is called absorption length.

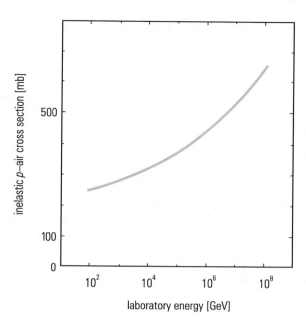

Fig. 4.1 Cross section for proton–air interactions

cross section for electrons ([32], Chap. 1), so the radiation length for photons is 9/7 of that for electrons, i.e., 47 g/cm². The first interaction of photon-induced electromagnetic cascades therefore also occurs in the uppermost layers of the atmosphere.

The detection of cosmic-ray neutrinos is completely different. They are only subject to weak interactions (apart from gravitational interactions). The cross section for neutrino–nucleon interactions is given by

$$\sigma_{\nu N} = 0.7 \times 10^{-38} \, E_\nu \, [\text{GeV}] \, \text{cm}^2/\text{nucleon} \,. \qquad (4.1.2)$$

Neutrinos of 100 GeV possess a tremendously large interaction length in the atmosphere:

$$\lambda \approx 2.4 \times 10^{12} \, \text{g/cm}^2 \,. \qquad (4.1.3)$$

This number has to be compared to the area density of the Earth of 7×10^9 g/cm² for the passage through its center. The vertex for possible neutrino–air interactions in the atmosphere should consequently be uniformly distributed along the neutrino track in the atmosphere.

Charged and/or neutral particles are created in the interactions, independent of the identity of the primary particle. These secondary particles will, in general, be recorded by the experiments or telescopes. To achieve this, a large variety of secondary processes can be used.

4.2   Interaction Processes Used for Particle Detection

> *I often say when you can measure what you are speaking about,*
> *and express it in numbers, you know something about it; but when*
> *you cannot measure it, when you cannot express it in numbers,*
> *your knowledge is of a meagre and unsatisfying kind.*
>
> Lord Kelvin (William Thomson)

Figures 4.2 and 4.3 show the main interaction processes of charged particles and photons, as they are typically used in experiments in astroparticle physics. In this overview, not only the interaction processes are listed, but also the typical detectors that utilize the corresponding interaction processes. The mechanism that dominates charged-particle interactions is the energy loss by ionization and excitation. This energy-loss process is described by the Bethe–Bloch formula:

$$-\left.\frac{dE}{dx}\right|_{\text{Ion.}} = K \cdot z^2 \frac{Z}{A} \cdot \frac{1}{\beta^2} \left\{ \frac{1}{2} \ln \frac{2m_e c^2 \beta^2 \gamma^2 T_{\max}}{I^2} - \beta^2 - \frac{\delta}{2} \right\}, \qquad (4.2.1)$$

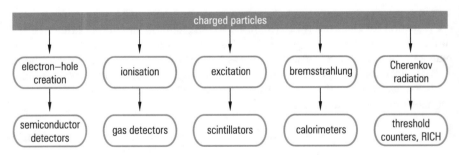

Fig. 4.2 Overview of interaction processes of charged particles

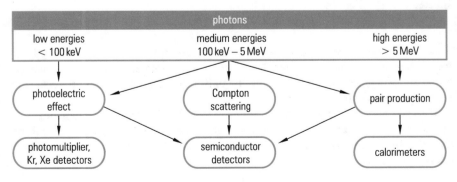

Fig. 4.3 Overview of interaction processes of photons

where

K – $4\pi N_A r_e^2 m_e c^2 = 0.307\,\mathrm{MeV}/(\mathrm{g/cm^2})$;
N_A – Avogadro number;
r_e – classical electron radius ($= 2.82\,\mathrm{fm}$);
$m_e c^2$ – electron rest energy ($= 511\,\mathrm{keV}$);
z – projectile charge;
Z, A – target charge and target mass;
β – projectile velocity ($= v/c$);
γ – $1/\sqrt{1 - \beta^2}$;
$T_{\max}$ – $\dfrac{2 m_e p^2}{m_0^2 + m_e^2 + 2 m_e E/c^2}$,
            maximum energy transfer to an electron,
            m_0—mass of the incident particle,
            p, E—momentum and total energy of the projectile;
I – average ionization energy of the target;
δ – density correction.

The energy loss of charged particles, according to the Bethe–Bloch relation, is illustrated in Fig. 4.4 and Fig. 4.5. It exhibits a $1/\beta^2$ increase at low energies. The

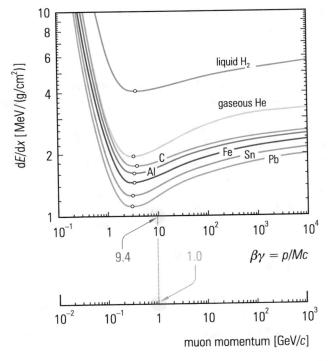

Fig. 4.4 Energy loss of charged particles in various targets. The curves are universal for all charged particles if the $\beta\gamma$ scale is used. The energy loss of different particles depends, of course, on their momenta, which are different for a particular particle type for a given $\beta\gamma$ value. For example, a muon of momentum of $1\,\mathrm{GeV}/c$ corresponds to $\beta\gamma = 9.4$. In general the momentum p is given by $p = M \cdot c \cdot \beta\gamma$

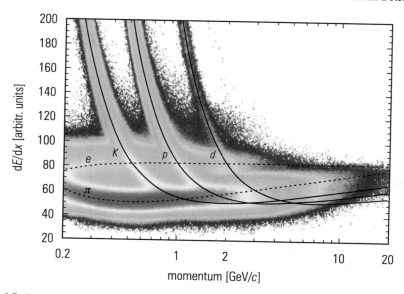

Fig. 4.5 Particle identification in the time projection chamber of the ALICE experiment at the LHC [33]

minimum ionization rate occurs at around $\beta\gamma \approx 3.5$. This feature is called the minimum of ionization, and particles with such $\beta\gamma$ values are said to be *minimum ionizing*. For high energies, the energy loss increases logarithmically ('relativistic rise') and reaches a plateau ('Fermi plateau') owing to the density effect. The energy loss of gases in the plateau region is typically 60% higher compared to the ionization minimum. The energy loss of singly charged minimum-ionizing particles by ionization and excitation in air is $1.8\,\text{MeV}/(\text{g}/\text{cm}^2)$ and $2.0\,\text{MeV}/(\text{g}/\text{cm}^2)$ in water (ice).

For the production of an electron–positron pair in air an average energy of 30 eV is required. In contrast, in semiconductor counters the average energy for the creation of an electron–hole pair is only 3 eV. The average energy needed to produce a scintillation photon in inorganic materials (e.g., NaI) is about 25 eV and ≈ 100 eV in organic materials. Semiconductor detectors can be made very small and allow as pixel detectors high spatial resolutions. In astroparticle physics they are mainly used in satellite experiments or on the International Space Station.

Equation (4.2.1) only describes the average energy loss of charged particles. The energy loss is distributed around the most probable value by an asymmetric Landau distribution. The average energy loss is about twice as large as the most probable energy loss. The ionization energy loss is the basis of a large number of particle detectors.

"This micropattern detector is so small, because we intend to detect tiny particles with it!"

The scintillation of gases is used in fluorescence telescopes for energies $\geq$EeV ($\geq 10^{18}$ eV) in the field of particle astronomy, like in the Auger experiment. In these experiments the atmosphere represents the target for the primary particle. The interaction products create scintillation light in the air, which is recorded in telescopes on the ground equipped with photomultipliers mounted in the focal plane of mirrors.

For high energies, the bremsstrahlung process becomes significant. The energy loss of electrons due to this process can be described by

$$-\left.\frac{\mathrm{d}E}{\mathrm{d}x}\right|_{\text{brems}} = 4\alpha N_A \frac{Z^2}{A} r_e^2 E \ln \frac{183}{Z^{1/3}} = \frac{E}{X_0}, \tag{4.2.2}$$

where α is the fine-structure constant ($\alpha^{-1} \approx 137$). The definition of the *radiation length* X_0 is evident from (4.2.2). The other quantities in (4.2.2) have the same meanings as in (4.2.1).

Energy loss due to bremsstrahlung is of particular importance for electrons. For heavy particles, the bremsstrahlung energy loss is suppressed by the factor $1/m^2$. The energy loss, however, increases linearly with energy, and is therefore important for all particles at high energies.

In addition to bremsstrahlung, charged particles can also lose some of their energy by direct electron–positron pair production, or by nuclear interactions. The energy loss due to these two interaction processes also varies linearly with energy. Muons as secondary particles in astroparticle physics play a dominant role in particle-detection

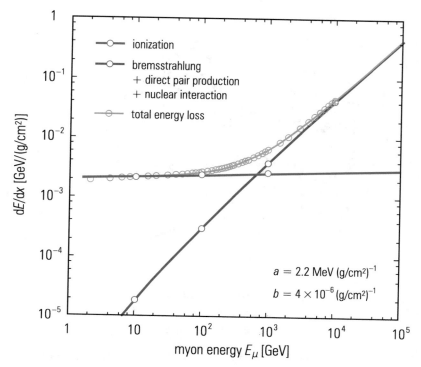

Fig. 4.6 Energy loss of muons in standard rock [34]

techniques, e.g., in neutrino astronomy. Muons are not subject to strong interactions and they can consequently travel relatively large distances. This makes them important for particle detection in astroparticle physics. The total energy loss of muons can be described by:

$$-\frac{dE}{dx}\bigg|_{\text{muon}} = a(E) + b(E)\,E\,, \tag{4.2.3}$$

where $a(E)$ describes the ionization energy loss, and $b(E)\,E$ summarizes the processes of muon bremsstrahlung, direct electron pair creation, and nuclear interaction. The energy loss of muons in standard rock depends on their energy. It is displayed in Fig. 4.6.

For particles with high energies, the total energy loss is dominated by bremsstrahlung and the processes that depend linearly on the particles' energies. These energy-loss mechanisms are therefore used as a basis for particle calorimetry. In calorimetric techniques, the total energy of a particle is dissipated in an active detector medium. The output signal of such a calorimeter is proportional to the absorbed energy. In this context, electrons and photons with energies exceeding 100 MeV can already be considered as high-energy particles because they initiate electromagnetic cascades. The mass of the muon is much larger than that of the electron, making *muon calorimetry* via energy-loss measurements only possible for energies beyond ≈ 1 TeV. This calorimetric technique is of particular importance in the field of TeV neutrino astronomy.

Another interaction mechanism, which can be sometimes rather annoying at accelerators and storage rings has recently been used to an advantage in experiments on extensive air showers. The large number of charged particles created in air showers are deflected in the Earth's magnetic field and produce synchrotron radiation, which is emitted in the radio domain. This geosynchrotron radiation allows to investigate air showers with a duty time of 100%, in contrast to the measurements of air showers using the fluorescence technique, which can only be done in the dark and even requires moonless nights.

4.3 Particle Identification

At least once per year, some group of scientists will become very excited and announce that:
• *The universe is even bigger than they thought!*
• *There are even more subatomic particles than they thought!*
Dave Barry

Identification means that the mass of the particle and its charge is determined. In elementary particle physics most particles have unit charge. But in the study, e.g., of the chemical composition of primary cosmic rays different charges must be distinguished.

Every effect of particles or radiation can be used as a working principle for a particle detector.

The deflection of a charged particle in a magnetic field determines its momentum p; the radius of curvature ρ is given by

$$\rho \propto \frac{p}{z} = \frac{\gamma m_0 \beta c}{z} , \qquad (4.3.1)$$

where z is the particle's charge, m_0 its rest mass, and $\beta = \frac{v}{c}$ its velocity. The quantity $\frac{p}{z}$ known in astroparticles physics is often called *magnetic rigidity* or, simply, *rigidity*. The particle velocity can be determined, e.g., by a time-of-flight method yielding

$$\beta \propto \frac{1}{\tau} , \qquad (4.3.2)$$

where τ is the flight time, which can be measured with a pair of scintillation counters or resistive plate chambers. Also the determination of the Cherenkov angle allows to obtain the particle velocity.

A calorimetric measurement provides a determination of the kinetic energy

$$E^{kin} = (\gamma - 1)m_0 c^2 \,, \tag{4.3.3}$$

where $\gamma = \frac{1}{\sqrt{1-\beta^2}}$ is the Lorentz factor.

From these measurements the ratio of m_0/z can be inferred, i.e., for singly charged particles we have already identified the particle. To determine the charge one needs another z-sensitive effect, e.g., the ionization energy loss

$$\frac{dE}{dx} \propto \frac{z^2}{\beta^2} \ln(a\beta\gamma) \tag{4.3.4}$$

(a is a material-dependent constant).

Now we know m_0 and z separately. In this way even different isotopes of elements can be distinguished. Neutral particles can only be identified via their interaction products. If all particles produced in such an interaction are measured the identity of the neutral particle can be determined. A particular case are neutrinos in a final state of an interaction. Neutrinos leave no tracks in the detector, but if the total energy of the final-state particles is known, e.g., as decay products of a heavy particle, neutrinos can be reconstructed from the missing energy and missing momentum and thereby also identified.

4.4 Principles of the Atmospheric Air Cherenkov Technique

> *A great pleasure in life is doing what people say you cannot do.*
> Walter Bagehot

The atmospheric Cherenkov technique is becoming increasingly popular for TeV γ astronomy since it allows to identify photon-induced electromagnetic showers, which develop in the atmosphere. A charged particle that moves in a medium with refractive index n, and has a velocity v that exceeds the velocity of light $c_n = c/n$, emits electromagnetic radiation known as *Cherenkov radiation*. There is a threshold effect for this kind of energy loss; Cherenkov radiation only occurs if

$$v \geq \frac{c}{n} \text{ or, equivalently, } \beta = \frac{v}{c} \geq \frac{1}{n} \,. \tag{4.4.1}$$

Cherenkov radiation is emitted under an angle of

$$\theta_C = \arccos \frac{1}{n\beta} \tag{4.4.2}$$

relative to the direction of the particle velocity. Due to this process, a particle of charge number z creates a certain number of photons in the visible spectral range ($\lambda_1 = 400$ nm up to $\lambda_2 = 700$ nm). The number of photons produced per centimeter is calculated from the following equation:

$$\frac{dN}{dx} = 2\pi\alpha z^2 \frac{\lambda_2 - \lambda_1}{\lambda_1 \lambda_2} \sin^2 \theta_C$$
$$\approx 490 \, z^2 \sin^2 \theta_C \, \text{cm}^{-1} .$$

(4.4.3)

These photons are emitted isotropically in azimuth (Fig. 4.7).

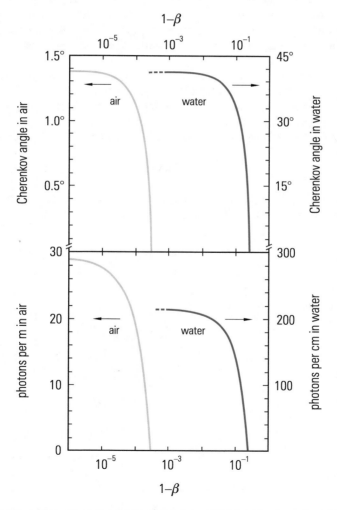

Fig. 4.7 Variation of the Cherenkov angle and photon yield of singly charged particles in water and air

For relativistic particles ($\beta \approx 1$), the Cherenkov angle is $42°$ in water and $1.4°$ in air (Fig. 4.7). In water, around 220 photons per centimeter are produced by a singly charged relativistic particle. The corresponding number in air is 30 photons per meter. Figure 4.7 shows the variation of the Cherenkov angle and the photon yield with the particle velocity for water and air. The atmospheric Cherenkov technique permits the identification of photon-induced electromagnetic showers that develop in the atmosphere and separates them from the more abundant hadronic cascades. Photons point back to their sources, while hadrons mainly produce only an isotropic background. The axis of the Cherenkov cone follows the direction of incidence of the primary photon. The Cherenkov cone for γ-induced cascades in air spans only $\pm 1.4°$, therefore the hadronic background in such a small angular range is relatively small. Apart from this background subtraction, also the different Cherenkov patterns in the photomultiplier matrix in the recording system, which is characteristically different for hadron and photon or electron induced showers, can be used for particle separation.

Apart from the atmospheric Cherenkov technique, the Cherenkov effect is also utilized in large water Cherenkov detectors for neutrino astronomy. The operation principle of a water Cherenkov counter is sketched in Fig. 4.8. Cherenkov radiation is emitted along a distance Δx. The Cherenkov cone projects an image on the detector surface, which is at a distance d from the source. The image is a ring with an average radius

$$r = d \tan \theta_C . \qquad (4.4.4)$$

This detection technique plays an important role in the huge experiments like Super-Kamiokande and in ICECUBE.

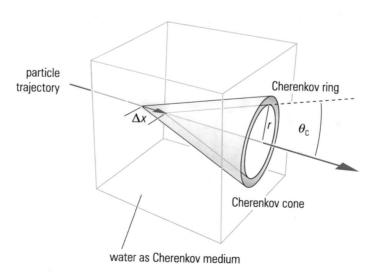

Fig. 4.8 Production of a Cherenkov ring in a water Cherenkov counter

Claus Grupen 2015

For very high energies charged particles can also be detected via transition radiation. They can even be identified with this technique because the intensity or energy, respectively, of the transition-radiation photons depends on the Lorentz factor of the incident particle. This allows, e.g., an electron–pion separation in transition radiation detectors, if the momentum of the particles is known.

4.5 Special Aspects of Photon Detection

Are not the rays of light very small bodies emitted from shining substances?

Sir Isaac Newton

The detection of photons is more indirect compared to charged particles. Photons first have to create charged particles in an interaction process. These charged particles will then be detected via the processes described, such as ionization, excitation, bremsstrahlung, and the production of Cherenkov radiation.

At comparatively low energies, as in X-ray astronomy, photons can be imaged by reflections at grazing incidence. Photons are detected in the focal plane of an X-ray telescope via the photoelectric effect. Semiconductor counters, X-ray CCDs,[2] or multiwire proportional chambers filled with a noble gas of high atomic number (e.g., krypton, xenon) can be used for focal detectors. These types of detectors provide spatial details, as well as energy information (see Fig. 4.9).

---

[2]CCD—Charge-Coupled Device (solid-state ionization chamber).

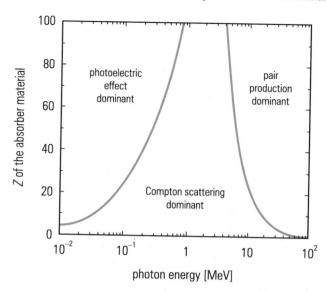

Fig. 4.9 Domains, in which various photon interactions dominate, shown in their dependence on the photon energy and the nuclear charge of the absorber

The Compton effect dominates for photons at MeV energies (see Fig. 4.9). In *Compton scattering*, a photon of energy E_γ transfers part of its energy ΔE to a target electron, thereby being redshifted. Based on the reaction kinematics, the ratio of the scattered photon energy E'_γ to the incident photon energy E_γ can be derived:

$$\frac{E'_\gamma}{E_\gamma} = \frac{1}{1 + \varepsilon(1 - \cos\theta_\gamma)} . \tag{4.5.1}$$

In this equation, $\varepsilon = E_\gamma/m_e c^2$ is the reduced photon energy and θ_γ is the scattering angle of the photon in the γ–electron interaction (Fig. 4.10). With a Compton telescope, not only the energy, but also the direction of incidence of the photons can be determined. In such a telescope, the energy loss of the Compton-scattered photon $\Delta E = E_\gamma - E'_\gamma$ is determined in the upper detector layer by measuring the energy of the Compton electron (see Fig. 4.10). The Compton-scattered photon of reduced energy will subsequently be detected in the lower detector plane, preferentially by the photoelectric effect. Based on the kinematics of the scattering process and using (4.5.1), the scattering angle θ_γ can be determined. As a consequence of the isotropic emission around the azimuth, the reconstructed photon direction does not point back to a unique position in the sky; it only defines a circle, respectively ellipse, in the sky. If, however, many photons are recorded from the source, the intercepts of these circles (or ellipses) define the position of the source. The detection of photons via the Compton effect in such Compton telescopes is usually performed using segmented large-area inorganic or organic scintillation counters that are read out via

Fig. 4.10 Principle of a
Compton telescope

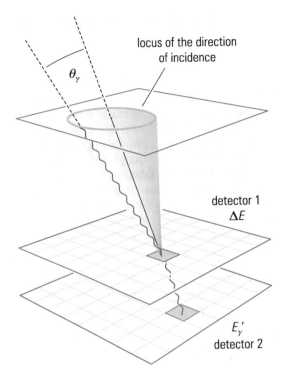

locus of the direction
of incidence

θ_γ

detector 1
ΔE

E_γ'
detector 2

photomultipliers. Alternatively, for high-resolution telescopes, semiconductor pixel detectors can also be used. This 'ordinary' Compton process is taken advantage of for photon detection. In astrophysical sources the inverse Compton scattering plays an important role. In such a process a low-energy photon might gain substantial energy in a collision with an energetic electron, and it can be shifted into the X-ray or γ-ray domain.

At high photon energies, the process of electron–positron pair creation dominates. Similarly to Compton telescopes, the electron and positron tracks enable the direction of the incident photon to be determined. The photon energy is obtained from the sum of the electron and positron energy. This is normally determined in electromagnetic calorimeters, in which electrons and positrons deposit their energy to the detector medium in alternating bremsstrahlung and pair-production processes. These electromagnetic calorimeters can be total-absorption crystal detectors such as NaI or CsI, or they can be constructed along the so-called sandwich principle. A sandwich calorimeter is a system, where absorber and detector layers alternate. Particle multiplication occurs preferentially in the passive absorber sheets, whilst the shower of particles produced is recorded in the active detector layers. Sandwich calorimeters can be compactly constructed and highly segmented, however, they are inferior to crystal calorimeters as far as the energy resolution is concerned.

4.6   Cryogenic Detection Techniques

> *Ice vendor to his son: 'Stick to it, there is a future in cryogenics.'*
>
> Anonymous

The detectors described so far can be used for the spectroscopy of particles from the keV range up to the highest energies. For many investigations the detection of particles of extremely low energy in the range between 1 and 1000 eV is of great interest. Calorimeters for such low-energy particles could be or are used for the detection of and search for low-energy cosmic neutrinos, weakly interacting massive particles (WIMPs) or other candidates of dark, non-luminous matter, for X-ray spectroscopy in astrophysics and material science [35, 36], single-optical-photon spectroscopy, and in other experiments. Since more than 20 years this field of experimental particle physics is developing intensively and by now it comprises dozens of projects.

Figure 4.11 demonstrates the superior performance of a cryogenic detector even when comparing it to an already excellent solid-state counter [36].

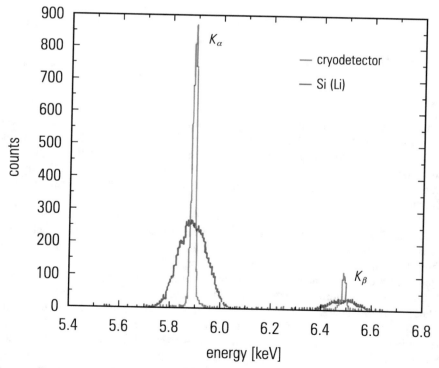

Fig. 4.11   X-ray spectrum of a ^{55}Fe (^{55}Mn) source showing the K_α and K_β lines. The narrow peaks were recorded with a superconducting tunnel diode with a resolution of 14.5 eV FWHM (0.24%). The curves labeled Si(Li) were measured at a twofold higher rate with a lithium-drifted silicon detector showing a resolution of 131 eV (2.2%) [35, 36]

To reduce the detection threshold and improve at the same time the energy resolution, it is only natural to replace the ionization or electron–hole pair production by *quantum transitions* requiring lower energies.

Phonons in solid-state materials have energies around 10^{-5} eV for temperatures around 100 mK. The other types of quasiparticles at low temperature are *Cooper pairs* in a superconductor, which are bound states of two electrons with opposite spin that behave like bosons and will form at sufficiently low temperatures a Bose condensate. Cooper pairs in superconductors have binding energies in the range between 4×10^{-5} eV (Ir) and 3×10^{-3} eV (Nb). Thus, even extremely low energy depositions would produce a large number of phonons or break up Cooper pairs. To avoid thermal excitations of these quantum processes, such detectors, however, would have to be operated at extremely low temperatures, typically in the milli-kelvin range. For this reason, such devices are called cryogenic detectors. *Cryogenic calorimeters* or *cryogenic detectors* can be subdivided in two main categories: firstly, detectors for quasiparticles in superconducting materials or suitable crystals, and secondly, phonon detectors in insulators.

One detection method is based on the fact that the superconductivity of a substance is destroyed by energy deposition if the detector element is sufficiently small. This is the working principle of superheated superconducting granules [37, 38]; see Fig. 4.12.

In this case the cryogenic calorimeter is made of a large number of superconducting spheres with diameters in the 100 μm range. If these granules are embedded in a magnetic field, and the energy deposition of a low-energy particle transfers one particular granule from the superconducting to the normal-conducting state, this

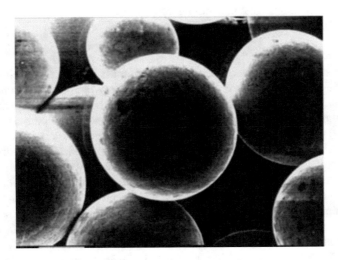

Fig. 4.12 Tin granules (130 μm diameter) as cryogenic calorimeter. A relatively small energy transfer can be sufficient to transfer a granule from the superconducting state into the normal-conducting state, thereby creating a detectable signal [39, 40]

transition can be detected by the suppression of the *Meissner effect*. This is, where the magnetic field, which does not enter the granule in the superconducting state, now again passes through the normal-conducting granule. The transition from the super-conducting to the normal-conducting state can be detected by pickup coils coupled to very sensitive preamplifiers or by SQUIDs (*Superconducting Quantum Interference Devices*). These *quantum interferometers* are extremely sensitive detection devices for magnetic effects. The operation principle of a SQUID is based on the *Josephson effect*, which represents a tunnel effect operating between two superconductors sep-arated by thin insulating layers. In contrast to the normal one-particle tunnel effect, known, e.g., from alpha decay, the Josephson effect involves the tunnelling of Cooper pairs. In Josephson junctions, interference effects of the tunnel current occur that can be influenced by magnetic fields. The structure of these interference effects is related to the size of the magnetic flux quanta.

The detection of transitions from the superconducting into the normal-conducting state is possible with appropriate materials and can yield signal amplitudes of about $100\,\mu V$ and recovery times of 10–50 ns. This already indicates that *superconducting strip counters* are possible candidates for microvertex detectors for future genera-tions of particle physics experiments or even astroparticle experiments on board the International Space Station in the future, if sufficient cooling can be provided.

An alternative method to detect *quasiparticles* is to let them directly tunnel through an insulating foil between two superconductors (SIS—Superconducting–Insulating–Superconducting transition). In this case the problem arises of keeping undesired leakage currents at an extremely low level.

In contrast to Cooper pairs, phonons, which can be excited by energy depositions in insulators, can be detected with methods of classical calorimetry. If ΔE is the absorbed energy, this results in a temperature rise of

$$\Delta T = \Delta E/mc,$$

where c is the specific heat capacity and m the mass of the calorimeter. If these calorimetric measurements are performed at very low temperatures, where c can be very small (the lattice contribution to the specific heat is proportional to T^3 at low temperatures), this method is also used to detect individual particles. In a real experiment, the temperature change is recorded with a thermistor, which is basically an NTC resistor (negative temperature coefficient), embedded into or fixed to an ultrapure crystal. The crystal represents the absorber, i.e., the detector for the radiation that is to be measured. Because of the discrete energy of phonons, one would expect discontinuous thermal energy fluctuations, which can be detected with electronic filter techniques. In Fig. 4.13 the principle of such a calorimeter is sketched [38].

In this way α particles and γ rays have been detected in a large TeO_2 crystal at 15 mK in a purely thermal detector with thermistor readout with an energy resolution of 4.2 keV FWHM for 5.4-MeV α particles [43]. Special bolometers have also been developed, in which heat and ionization signals are measured simultaneously.

Thermal detectors provide promise for improvements of energy resolutions. For example, a 1-mm cubic crystal of silicon kept at 20 mK would have a heat capacity

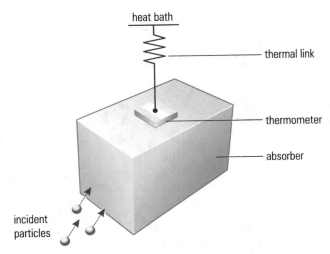

Fig. 4.13 Schematic view of a cryogenic calorimeter. The main components are the absorber for the incident particles, the thermometer for the measurement of the heat signal and the thermal coupling to the cryo bath [38, 41, 42]

of 5×10^{-15} J/K and a FWHM energy resolution of 0.1 eV (corresponding to $\sigma = 42$ meV) [44].

There are, however, still various important problems to be solved, before these values can be reached.

1. Non-uniformity of phonon collection, especially in large detectors.
2. Spatial non-uniformity of the recombination of electron–hole pairs trapped by various impurities.
3. Noise due to electromagnetic sources, especially microphonics (generation of noise due to mechanical, acoustic and electromagnetic excitation).
4. Problems with keeping the temperature of the bolometer constant, and consequently its gain.

Joint efforts in the fields of cryogenics, particle physics, and astrophysics are required, which may lead to exciting and unexpected results. One interesting goal would be to detect relic neutrinos of the Big Bang with energies around 200 μeV, or find evidence for dark-matter particles.

Cryogenic detectors allow to detect single photons over a wide spectral range. Therefore, these detectors have been used in astronomy already for some time. Energy-dispersive X-ray detectors benefit from the high energy resolution of cryodetectors, which allow to investigate the low-energy part of atomic spectra (below several keV) [45].

The required low temperatures for cryogenic detectors can be reached, e.g., by adiabatic demagnetisation.

As an example for an application of cryodetectors the search for *weakly interacting massive particles* (WIMPs) will be described in some more detail. The interaction

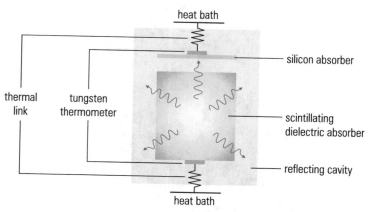

Fig. 4.14 Schematic of a cryogenic calorimeter with simultaneous measurement of the thermal due to phonons and the optical signal from photons [38, 41]

cross section for WIMP interactions is extremely small, so that possible backgrounds have to be reduced to a very low level. Unfortunately also the energy transfer of a WIMP to a target nucleus in a cryogenic detector is only in the range of ≈20 keV. An excellent method to discriminate a WIMP signal against the background caused, e.g., by local radioactivity, is to use scintillating crystals like $CaWO_4$, $CdWO_4$, or $ZnWO_4$. These scintillators allow to measure the light yield at low temperatures and the phonon production by WIMP interactions at the same time. *Nuclear recoils* due to WIMP–nucleon scattering produce mainly phonons and very little scintillation light, while in *electron recoils* also a substantial amount of scintillation light is created. A schematic view of such a cryogenic detector system is shown in Fig. 4.14 [38].

Particles are absorbed in a scintillating dielectric crystal. The scintillation light is detected in a silicon wafer while the phonons are measured in two tungsten thermometers, one of which can be coupled to the silicon detector to increase the sensitivity of the detector. The whole detector setup is enclosed in a reflecting cavity and operated at milli-kelvin temperatures. The response of a $CaWO_4$ cryogenic calorimeter to electron recoils and nuclear recoils is shown in Fig. 4.15 [38, 41, 46].

Electron recoils were created by irradiating the crystal with 122-keV and 136-keV photons from a ^{57}Co source and electrons from a ^{90}Sr β source (left panel). To simulate also WIMP interactions the detector was bombarded additionally with neutrons from an americium–beryllium source leading to phonon and scintillation-light yields as shown in the right-hand plot of Fig. 4.15. The light output due to electron recoils caused by photons or electrons (which constitute the main background for WIMP searches) is quite high, whereas nuclear recoils created by neutrons provide a strong phonon signal with only low light yield. It is conjectured that WIMP interactions will look similar to neutron scattering, thus allowing a substantial background rejection if appropriate cuts in the scatter diagram of light versus phonon yield are applied. However, the figure also shows that the suppression of electron recoils at energies below 20 keV becomes rather difficult.

Surprising capabilities of Squids (octopus)

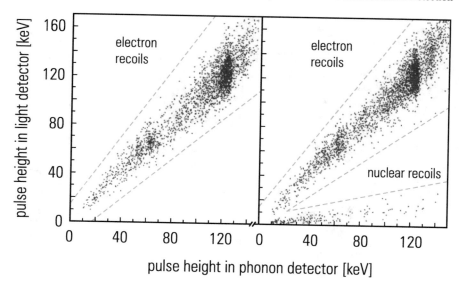

Fig. 4.15 Scatter diagram of the pulse height of the light signal versus the phonon signal in a CaWO$_4$ crystal. The *left-hand part* of the figure shows the response of the detector to photons and electrons only, while in the *right-hand part* also neutron interactions are included, which are supposed to simulate a WIMP signal [38, 41, 46]

4.7 Propagation and Interactions of Astroparticles in Galactic and Extragalactic Space

> *Man must rise above the Earth – to the top of the atmosphere and beyond – for only thus will he fully understand the world in which he lives.*
>
> Socrates

Now that the principles for the detection of primary and secondary particles have been described, the interactions of astroparticles traveling from their sources to Earth through galactic and extragalactic space shall be briefly discussed.

Neutrinos are only subject to weak interactions with matter, so their range is extremely large. The galactic or intergalactic space does not attenuate the neutrino flux, and magnetic fields do not affect their direction; therefore, neutrinos point directly back to their sources. Large hopes of neutrino astronomy rest on the ICE-CUBE detector at the South Pole and its planned extensions. These seem to be justified after their observation of high-energy neutrinos in the PeV range from extragalactic distances.

The matter density in our galaxy, and particularly in intergalactic space, is very low. This signifies that the ionization energy loss of primary protons traveling from their sources to Earth is extremely small. Protons can, however, interact with cosmic photons. Blackbody photons, in particular, represent a very-high-density target

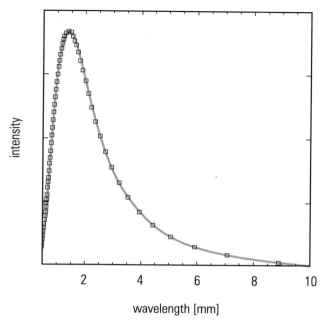

Fig. 4.16 Blackbody spectrum of cosmic microwave background photons

($\approx$400 photons/cm^3). The energy of these photons is very low, typically 250 µeV, and they follow a Planck distribution (Fig. 4.16), see also Chap. 11: 'The Cosmic Microwave Background'.

The process of pion production by blackbody photons interacting with high-energy protons requires the proton energy to exceed a certain threshold (Greisen–Zatsepin–Kuzmin cutoff). This threshold is reached if photo–pion production via the Δ resonance is kinematically possible in the photon–proton center-of-mass system ($p + \gamma \rightarrow p + \pi^0$). If protons exceed this threshold energy, they quickly lose their energy and fall below the threshold. The GZK cutoff limits the mean free path of the highest-energy cosmic rays (energy $> 6 \times 10^{19}$ eV) to less than about 50 Mpc, quite a small distance in comparison to typical extragalactic scales. Of course, energetic protons lose also energy by inverse Compton scattering on blackbody photons. In contrast to π^0 production via the Δ resonance this process has no threshold. Moreover, the cross section varies like $1/E_\gamma$, i.e., deceases rapidly with increasing photon energy. Compared to the resonant π^0 production the cross section for inverse Compton scattering of protons on blackbody photons is small and therefore has no significant influence on the shape of the primary proton spectrum. A further possible process, $p + \gamma \rightarrow p + e^+ + e^-$, even though it has a lower threshold than $p + \gamma \rightarrow p + \pi^0$, does not proceed through a resonance, and therefore its influence on the propagation of energetic protons in the dense photon field is of little importance. It does, however, modify the shape of the primary spectrum at the highest

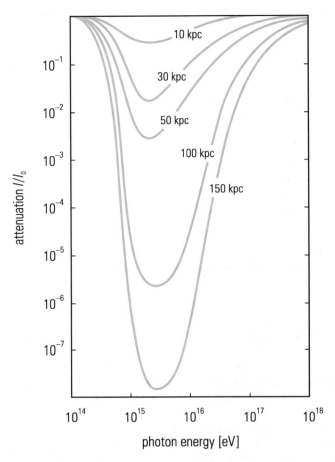

Fig. 4.17 Attenuation of the intensity of energetic primary photons by interactions with blackbody radiation

energies. In addition, primary protons (as charged particles) naturally interact with the galactic and extragalactic magnetic fields as well as the Earth's magnetic field. Only the most energetic protons (energy $\gg 10^{18}$ eV), which experience a sufficiently small magnetic deflection, can be used for particle astronomy.

Photons are not influenced by magnetic fields. They do, like protons, however, interact with blackbody photons to create electron–positron pairs via the $\gamma\gamma \rightarrow e^+e^-$ process. Owing to the low electron and positron masses, the threshold energy for this process is only about $\approx 10^{15}$ eV. The attenuation of primary photons (by interactions with blackbody photons), as a function of the primary photon energy, is shown in Fig. 4.17 for several distances to possible γ-ray sources. The process $\gamma\gamma \rightarrow e^+e^-$ limits the range of energetic photons ($> 10^{15}$ eV) effectively to about 50 kpc. For higher energies, $\gamma\gamma$ processes with different final states ($\mu^+\mu^-$, ...) also occur.

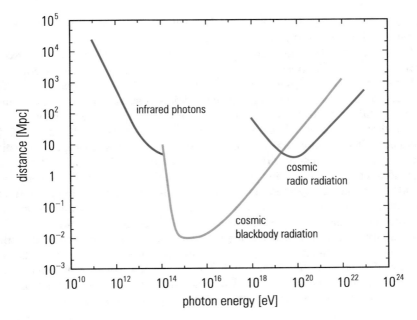

Fig. 4.18 Attenuation of high-energy photons by interactions with infrared photons, the cosmic blackbody radiation, and cosmic radio emission in its respective dependence on the photon energy. On the ordinate the distance is given that the photons of a given energy can pass without significant attenuation [47]

A potentially competing process, $\gamma\gamma \rightarrow \gamma\gamma$, is connected with a very small cross section (it is proportional to the fourth power of the fine-structure constant). In addition, the angular deflection of the photons due to this process is extremely small.

Apart from the attenuation of photons due to interactions with the cosmological background radiation, also processes at lower energies with starlight photons and infrared radiation become relevant. Even radiation from the radio domain attenuates photons at higher energies (Fig. 4.18). In addition, photons can also produce electron–positron pairs on cosmic protons, albeit with lower cross sections.

4.8 Characteristic Features of Detectors

> *Our eyes only see big dimensions, but beyond those there are others that escape detection because they are so small.*
>
> Brian Greene

The secondary interaction products of astroparticles are detected in an appropriate device, which can be a detector on board of a satellite, in a balloon, or at ground level, or even in an underground laboratory. The quality of the measurement depends on the energy and position resolution of the detector. In most cases the ionization energy loss is the relevant detection mechanism.

In gaseous detectors an average of typically 30 eV is required to produce an electron–ion pair. The liberated charges are collected in an external electric field and produce an electric signal that can be further processed. In contrast, in solid-state detectors, the average energy for the creation of an electron–hole pair is only ≈3 eV, resulting in an improved energy resolution. If, instead, excitation photons produced by the process of scintillation in a crystal detector are recorded, e.g., by photomultipliers, energy deposits of about 25 eV are necessary to yield a scintillation photon in inorganic materials (like NaI(Tl)), while in organic crystals ≈100 eV are required to create a scintillation photon. In cryogenic detectors much less energy is needed to produce charge carriers. This substantial advantage, which gives rise to excellent energy resolutions, is only obtained at the expense of operating the detectors at cryogenic temperatures, mostly in the milli-kelvin range.

Apart from these classical techniques also new methods of particle detection become popular. Different techniques are, for example, used for extensive air showers by measuring scintillation photons with fluorescence telescopes or photons created via geosynchrotron radiation. Also Cherenkov-generated radio emission or even acoustic detection of energetic events is used or is under investigation.

Summary

The detection of astroparticles is mostly indirect. It is easy to measure charged particles up to the several hundred GeV via their ionization in satellite experiments. For higher energies one runs out of intensity and has to resort to more indirect techniques. For electromagnetic radiation (radio, X rays or γ rays) different techniques are appropriate. All these techniques (except radio measurements) rest essentially on the transformation of photons into charged particles, which, in turn, can be measured via their ionization, scintillation or Cherenkov radiation. For the detection of processes with low energy transfers (like WIMP search or dark-matter search), cryogenic detectors are mandatory. They are operated at very low temperatures, in the milli-kelvin range or at even lower temperatures. A problem arises for the detection of charged particles and photons from sources at very large distances. The omnipresent blackbody radiation attenuates charged particles and photons significantly, so that these particles may not reach Earth at all. Also irregular galactic and intergalactic magnetic fields prevent particle astronomy from pointing to accelerators in the sky. The only exception are neutrinos, which reach out to the largest distances.

4.9 Problems

1. Show that (4.0.1) is dimensionally correct.
2. The average energy required for the production of

(a) a photon in a plastic scintillator is 100 eV,
(b) an electron–ion pair in air is 30 eV,
(c) an electron–hole pair in silicon is 3.65 eV,
(d) a quasiparticle (break-up of a Cooper pair in a superconductor) is 1 meV.

What is the relative energy resolution in these counters for a stopping 10-keV particle assuming Poisson statistics (neglecting the Fano effect[3])?

3. The simplified energy loss of a muon is parameterized by (4.2.3). Work out the range of a muon of energy E (=100 GeV) in rock ($\varrho_{rock} = 2.5\,g/cm^3$) under the assumption that a (=2 MeV/(g/cm^2)) and b (=4.4 $\times 10^{-6}$ cm^2/g for rock) are energy independent.

4. Show that the mass of a charged particle can be inferred from the Cherenkov angle θ_C and momentum p by

$$m_0 = \frac{p}{c}\sqrt{n^2\cos^2\theta_C - 1}\,,$$

where n is the index of refraction.

5. In a cryogenic argon calorimeter ($T = 1.1$ kelvin, mass 1 g) a WIMP (weakly interacting massive particle) deposits 10 keV. By how much does the temperature rise?
(The specific heat of argon at 1.1 K is $c_{sp} = 8 \times 10^{-5}\,J/(g\,K)$.)

6. Derive (4.5.1) using four-momenta.

7. Work out the maximum energy that can be transferred to an electron in a Compton process! As an example use the photon transition energy of 662 keV emitted by an excited ^{137}Ba nucleus after a beta decay from ^{137}Cs,

$$^{137}\text{Cs} \rightarrow\ ^{137}\text{Ba}^* + e^- + \bar{\nu}_e$$
$$\hookrightarrow\ ^{137}\text{Ba} + \gamma\ (662\,\text{keV})\ .$$

What kind of energy does the electron get for infinitely large photon energies? Is there, on the other hand, a minimum energy for the backscattered photon in this limit?

8. Figure 4.4 shows the energy loss of charged particles as given by the Bethe–Bloch formula. The abscissa is given as momentum and also as product of the normalized velocity β and the Lorentz factor γ. Show that $\beta\gamma = p/m_0c$ holds.

9. Equation (4.4.3) shows that the number of emitted Cherenkov photons N is proportional to $1/\lambda^2$. The wavelength for X-ray photons is shorter than that for the visible light region. Why then is Cherenkov light not emitted in the X-ray region?

---

[3]For any specific value of a particle energy the fluctuations of secondary particle production (like electron–ion pairs) are smaller than might be expected according to a Poissonian distribution. This is a simple consequence of the fact that the total energy loss is constrained by the fixed energy of the incident particle. This leads to a standard error of $\sigma = \sqrt{F\,N}$, where N is the number of produced secondaries and F, the Fano factor, is smaller than 1.

10. For non-relativistic particles of charge z the Bethe–Bloch formula can be approximated by

$$\frac{dE}{dx} = a \frac{z^2}{E} \ln(bE),$$

where a and b are material-dependent constants.

Work out the energy–range relation if $\ln(bE)$ can be approximated by $(bE)^{1/4}$.

Chapter 5
Acceleration Mechanisms

*Physics also solves puzzles. However, these puzzles are not
posed by mankind, but rather by nature.*

Maria Goeppert-Mayer

The origin of cosmic rays is one of the major unsolved astrophysical problems. The highest-energy cosmic rays possess macroscopic energies and their origin is likely to be associated with the most energetic processes in the universe. When discussing cosmic-ray origin, one must in principle distinguish between the power source and the acceleration mechanism. Cosmic rays can be produced by particle interactions at the sites of acceleration like in pulsars. The acceleration mechanism can, of course, also be based on conventional physics using electromagnetic or gravitational potentials such as in supernova remnants or active galactic nuclei. One generally assumes that in most cases cosmic-ray particles are not only produced in the sources but also accelerated to higher energies in or near the source. Candidate sites for cosmic-ray production and acceleration are supernova explosions, highly magnetized spinning neutron stars, i.e., pulsars, accreting black holes, and the centers of active galactic nuclei. However, it is also possible that cosmic-ray particles powered by some source experience acceleration during the propagation in the interstellar or intergalactic medium by interactions with extensive gas clouds. These gas clouds are created by magnetic-field irregularities and charged particles can gain energy while they scatter off the constituents of these 'magnetic clouds'.

In *top–down scenarios* energetic cosmic rays can also be produced by the decay of topological defects, domain walls, or cosmic strings, which could be relics of the Big Bang.

There is a large number of models for cosmic-ray acceleration [48]. This appears to indicate that the actual acceleration mechanisms are not completely understood and identified. On the other hand, it is also possible that different mechanisms are at work for different energies. In the following the most plausible ideas about cosmic-ray-particle acceleration will be presented.

© Springer Nature Switzerland AG 2020
C. Grupen, *Astroparticle Physics*, Undergraduate Texts in Physics,
https://doi.org/10.1007/978-3-030-27339-2_5

5.1 Cyclotron Mechanisms

> *The sun is an inexhaustible source of power—it is the continuously wound up clockwork, which keeps the mechanisms and activities on Earth going.*
>
> Robert Mayer

Even normal stars can accelerate charged particles up to the GeV range. This acceleration can occur in time-dependent magnetic fields. These magnetic sites appear as starspots or sunspots, respectively. The temperature of sunspots is slightly lower compared to the surrounding regions. They appear darker, because part of the thermal energy has been transformed into magnetic field energy. Sunspots in typical stars can be associated with magnetic field strengths of up to 1000 gauss (1 tesla $= 10^4$ gauss). The lifetime of such sunspots can exceed several rotation periods. The spatial extension of sunspots on the Sun can be as large as 10^9 cm. The observed Zeeman splitting of spectral lines has shown beyond any doubt that magnetic fields are responsible for the sunspots. Since the Zeeman splitting of spectral lines depends on the magnetic field strength, this fact can also be used to measure the strength of the magnetic fields on stars.

The magnetic fields in the Sun are generated by turbulent plasma motions, where the plasma consists essentially of protons and electrons. The motions of this plasma constitute currents, which produce magnetic fields. When these magnetic fields are generated and when they decay, electric fields are created, in which protons and electrons can be accelerated.

Figure 5.1 shows schematically a sunspot of extension $A = \pi R^2$ with a variable magnetic field $\boldsymbol{B}$.

The time-dependent change of the magnetic flux ϕ produces a potential U,

$$-\frac{\mathrm{d}\phi}{\mathrm{d}t} = \oint \boldsymbol{E} \cdot \mathrm{d}s = U \tag{5.1.1}$$

($\boldsymbol{E}$—electrical field strength, $\mathrm{d}s$—infinitesimal distance along the particle trajectory). The magnetic flux is given by

$$\phi = \int \boldsymbol{B} \cdot \mathrm{d}\boldsymbol{A} = B\pi R^2 , \tag{5.1.2}$$

where $\mathrm{d}\boldsymbol{A}$ is the infinitesimal area element. In this equation it is assumed that $\boldsymbol{B}$ is perpendicular to the area, i.e., $\boldsymbol{B} \parallel \boldsymbol{A}$, (the vector $\boldsymbol{A}$ is always perpendicular to the area). One turn of a charged particle around the time-dependent magnetic field leads to an energy gain of

$$E = eU = e\pi R^2 \frac{\mathrm{d}B}{\mathrm{d}t} . \tag{5.1.3}$$

A sunspot of an extension $R = 10^9$ cm and magnetic field $B = 2000$ gauss at a lifetime of one day ($\frac{\mathrm{d}B}{\mathrm{d}t} = 2000\,\text{gauss/day}$) leads to

Fig. 5.1 Principle of particle acceleration by variable sunspots

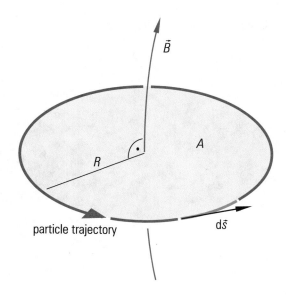

$$E = 1.6 \times 10^{-19}\,\text{A s}\,\pi\,10^{14}\,\text{m}^2\,\frac{0.2\,\text{V s}}{86\,400\,\text{s m}^2}$$
$$= 1.16 \times 10^{-10}\,\text{J} = 0.73\,\text{GeV}. \qquad (5.1.4)$$

Actually, particles from the Sun with energies beyond 10 GeV have been observed. This, however, might also represent the limit for the acceleration power of stars based on the cyclotron mechanism.

The cyclotron model can explain the correct energies, however, it does not explain why charged particles propagate in circular orbits around time-dependent magnetic fields. Circular orbits are only stable in the presence of guiding forces such as they are used in earthbound accelerators.

5.2 Acceleration by Sunspots

Living on Earth may be expensive, but it includes an annual free trip around the Sun.

Ashleigh Brilliant

Sunspots often come in pairs of opposite magnetic polarity (see Fig. 5.2).

The sunspots normally approach each other and merge at a later time. Let us assume that the left sunspot is at rest and the right one approaches the first sunspot with a velocity v. The moving magnetic dipole produces an electric field perpendicular to the direction of the dipole and perpendicular to its direction of motion v, i.e., parallel to $v \times B$. Typical solar magnetic sunspots can create electrical fields

Fig. 5.2 Sketch of a sunspot pair

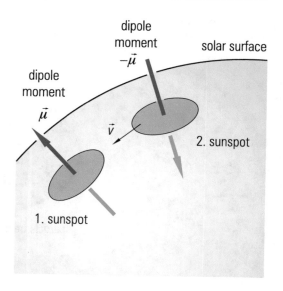

of 10 V/m. In spite of such a low field strength, protons can be accelerated since the collision energy loss is smaller than the energy gain in the low-density chromosphere. Under realistic assumptions (distance of sunspots 10^7 m, magnetic field strengths 2000 gauss, relative velocity $v = 10^7$ m/day) particle energies in the GeV range are obtained. This shows that the model of particle acceleration in approaching magnetic dipoles can only explain energies, which can also be provided by the cyclotron mechanism. The mechanism of approaching sunspots, however, sounds more plausible because in this case no guiding forces (like in the cyclotron model) are required.

In addition to merging sunspots, particle acceleration is also possible during strong solar flares (see Fig. 5.3), in which electric fields of several 100 V/m can be created. In these fields, which can extend over large distances, like 10^7 m, electrons and ions can be accelerated up to several GeV. Massive stars with strong stellar eruptions may reach even higher energies.

5.3 Shock Acceleration

> *Basic research is what I am doing when I don't know what I am doing.*
>
> Wernher von Braun

If a massive star has exhausted its hydrogen, the radiation pressure can no longer withstand the gravitational pressure and the star will collapse under its own gravity. The liberated gravitational energy increases the central temperature of a massive star to such an extent that helium burning can start. If the helium reservoir is used

Fig. 5.3 Post-eruptive loops in the wake of a solar flare recorded by the TRACE satellite [49]

up, the process of gravitational infall of matter repeats itself until the temperature is further increased so that the products of helium themselves can initiate fusion processes. These successive fusion processes can lead at most to elements of the iron group (Fe, Co, Ni). For higher nuclear charges the fusion reaction is endotherm, which means that without providing additional energy heavier elements cannot be synthesized. When the fusion process stops at iron, the massive star will implode. In this process part of its mass will be ejected into interstellar space. This material can be recycled for the production of a new star generation, which will contain—like the Sun—some heavy elements. As a result of the implosion a compact neutron star will be formed that has a density, which is comparable to the density of atomic nuclei. In the course of a supernova explosion some elements heavier than iron are produced if the copiously available neutrons are attached to the elements of the iron group, which—with successive β^- decays—allows elements with higher nuclear charge to be formed:

Fig. 5.4 Schematics of shock-wave acceleration

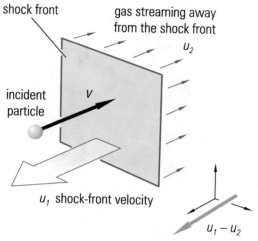

$$\begin{aligned}
{}^{56}_{26}\text{Fe} + n &\rightarrow {}^{57}_{26}\text{Fe}\,, \\
{}^{57}_{26}\text{Fe} + n &\rightarrow {}^{58}_{26}\text{Fe}\,, \\
{}^{58}_{26}\text{Fe} + n &\rightarrow {}^{59}_{26}\text{Fe}^* \\
&\qquad\hookrightarrow {}^{59}_{27}\text{Co} + e^- + \bar{\nu}_e\,, \\
{}^{59}_{27}\text{Co} + n &\rightarrow {}^{60}_{27}\text{Co}^* \\
&\qquad\hookrightarrow {}^{60}_{28}\text{Ni} + e^- + \bar{\nu}_e\,.
\end{aligned} \tag{5.3.1}$$

The ejected envelope of a supernova represents a *shock front* with respect to the interstellar medium. Let us assume that the shock front moves at a velocity u_1. Behind the shock front the gas recedes with a velocity u_2. This means that the gas has a velocity $u_1 - u_2$ in the laboratory system (see Fig. 5.4).

A particle of velocity v colliding with the shock front and being reflected gains the energy

$$\begin{aligned}
\Delta E &= \frac{1}{2}m(v + (u_1 - u_2))^2 - \frac{1}{2}mv^2 \\
&= \frac{1}{2}m(2v(u_1 - u_2) + (u_1 - u_2)^2)\,.
\end{aligned} \tag{5.3.2}$$

Since the linear term dominates ($v \gg u_1, u_2; u_1 > u_2$), this simple model provides a relative energy gain of

$$\frac{\Delta E}{E} \approx \frac{2(u_1 - u_2)}{v}. \tag{5.3.3}$$

A more general, relativistic treatment of shock acceleration including also variable scattering angles leads to

$$\frac{\Delta E}{E} = \frac{4}{3} \frac{u_1 - u_2}{c}, \tag{5.3.4}$$

where it has been assumed that the particle velocity v can be approximated by the speed of light c. Similar results are obtained if one assumes that particles are trapped between two shock fronts and are reflected back and forth from the fronts (see also Fig. 5.5).

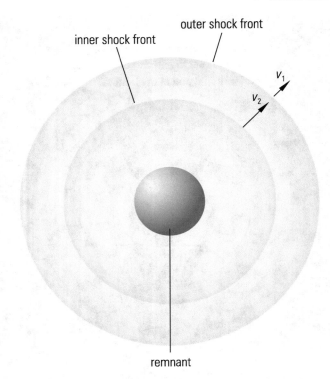

Fig. 5.5 Particle acceleration by multiple reflections between two shock fronts

Usually the inner front will have a much higher velocity (v_2) compared to the outer front (v_1), which is decelerated already in interactions with the interstellar material (Fig. 5.5). The inner shock front can provide velocities up to $20\,000\,\mathrm{km/s}$, as obtained from measurements of the Doppler shift of the ejected gas. The outer front spreads into the interstellar medium with velocities between some $100\,\mathrm{km/s}$ up to $1000\,\mathrm{km/s}$. For shock accelerations in active galactic nuclei even superfast shocks with $v_2 = 0.9\,c$ are discussed.

A particle of velocity v being reflected at the inner shock front gains the energy

$$\Delta E_1 = \frac{1}{2}m(v + v_2)^2 - \frac{1}{2}mv^2 = \frac{1}{2}m(v_2^2 + 2vv_2)\,. \qquad (5.3.5)$$

Reflection at the outer shock front leads to an energy loss

$$\Delta E_2 = \frac{1}{2}m(v - v_1)^2 - \frac{1}{2}mv^2 = \frac{1}{2}m(v_1^2 - 2vv_1)\,. \qquad (5.3.6)$$

On average, however, the particle gains an energy

$$\Delta E = \frac{1}{2}m(v_1^2 + v_2^2 + 2v(v_2 - v_1)).$$ (5.3.7)

Since the quadratic terms can be neglected and because of $v_2 > v_1$, one gets

$$\Delta E \approx mv\Delta v, \quad \frac{\Delta E}{E} \approx 2\frac{\Delta v}{v}.$$ (5.3.8)

This calculation followed similar arguments as in (5.3.2) and (5.3.3).

Both presented shock acceleration mechanisms are linear in the relative velocity. Sometimes this type of shock acceleration is called Fermi mechanism of first order. Under plausible conditions using the relativistic treatment, maximum energies of about 100 TeV can be explained in this way. It seems even possible that energies up to the galactic cosmic-ray knee at $\approx 10^{15}$ eV are attainable.

5.4   Fermi Mechanisms

> *Results! Why man, I have gotten a lot of results, I know several thousand things that don't work.*
>
> Thomas Edison

Fermi mechanism of second order (or more general Fermi mechanism) describes the interaction of cosmic-ray particles with magnetic clouds. At first sight it appears improbable that particles can gain energy in this way. Let us assume that a particle (with velocity v) is reflected from a gas cloud, which moves with a velocity u (Fig. 5.6).

If v and u are antiparallel, the particle gains the energy

$$\Delta E_1 = \frac{1}{2}m(v + u)^2 - \frac{1}{2}mv^2 = \frac{1}{2}m(2uv + u^2).$$ (5.4.1)

In case that v and u are parallel, the particle loses an energy

$$\Delta E_2 = \frac{1}{2}m(v - u)^2 - \frac{1}{2}mv^2 = \frac{1}{2}m(-2uv + u^2).$$ (5.4.2)

On average a net energy gain of

$$\Delta E = \Delta E_1 + \Delta E_2 = mu^2$$ (5.4.3)

results, leading to the relative energy gain of

$$\frac{\Delta E}{E} = 2\frac{u^2}{v^2}.$$ (5.4.4)

Fig. 5.6 Energy gain of a particle by a reflection from a magnetic gas cloud

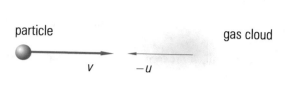

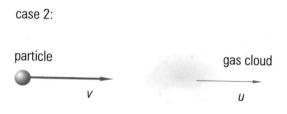

Since this acceleration mechanism is quadratic in the cloud velocity, this variant is often called Fermi mechanism of 2nd order. The result of (5.4.4) remains correct even under relativistic treatment. Since the cloud velocity is rather low compared to the particle velocities ($u \ll v \approx c$), the energy gain per collision ($\sim u^2$) is very small. Therefore, the acceleration of particles by the Fermi mechanism requires a very long time. In this acceleration type one assumes that magnetic clouds act as collision partners—and not normal gas clouds—because the gas density and thereby the interaction probability is larger in magnetic clouds.

Another important aspect is that cosmic-ray particles will lose some of their gained energy by interactions with the interstellar or intergalactic gas between two collisions. This is why this mechanism requires a minimum injection energy, above which particles can only be effectively accelerated. These injection energies could be provided by the Fermi mechanism of 1st order, that is, by shock acceleration (see also Fig. 5.6).

It appears possible that second-order Fermi acceleration is capable of accelerating cosmic rays up to ultra-high energies.

5.5 Pulsars

I think there should be a law of Nature, which prevents a star from behaving in such an absurd way.

Sir Arthur Stanley Eddington

Spinning magnetized neutron stars (pulsars) are remnants of supernova explosions. While stars typically have radii of 10^6 km, they shrink under a gravitational collapse to a size of just about 20 km. This process leads to densities of 6×10^{13} g/cm^3 comparable to nuclear densities. In this process electrons and protons are so closely packed that in processes of weak interactions neutrons are formed:

$$p + e^- \rightarrow n + \nu_e. \tag{5.5.1}$$

Since the Fermi energy of electrons in such a neutron star amounts to several hundred MeV, the formed neutrons cannot decay because of the Pauli principle, since the maximum energy of electrons in neutron beta decay is only 0.78 MeV and all energy levels in the Fermi gas of electrons up to this energy and even beyond are occupied.

The gravitational collapse of stars conserves the angular momentum. Therefore, because of their small size, rotating neutron stars possess extraordinarily short rotational periods.

Assuming orbital periods of a normal star of about one month like for the Sun, one obtains—if the mass loss during contraction can be neglected—pulsar frequencies ω_{Pulsar} of (Θ—moment of inertia)

$$\Theta_{\text{star}} \, \omega_{\text{star}} = \Theta_{\text{pulsar}} \, \omega_{\text{pulsar}},$$

$$\omega_{\text{pulsar}} = \frac{R_{\text{star}}^2}{R_{\text{pulsar}}^2} \, \omega_{\text{star}} \tag{5.5.2}$$

corresponding to pulsar periods of

$$T_{\text{pulsar}} = T_{\text{star}} \frac{R_{\text{pulsar}}^2}{R_{\text{star}}^2}. \tag{5.5.3}$$

For a stellar size $R_{\text{star}} = 10^6$ km, a pulsar radius $R_{\text{pulsar}} = 20$ km, and a rotation period of $T_{\text{star}} = 1$ month one obtains

$$T_{\text{pulsar}} \approx 1 \text{ ms}. \tag{5.5.4}$$

The gravitational collapse amplifies the original magnetic field extraordinarily. If one assumes that the magnetic flux, e.g., through the upper hemisphere of a star, is conserved during the contraction, the magnetic field lines will be tightly squeezed. One obtains (see Fig. 5.7)

Fig. 5.7 Increase of a magnetic field during the gravitational collapse of a star

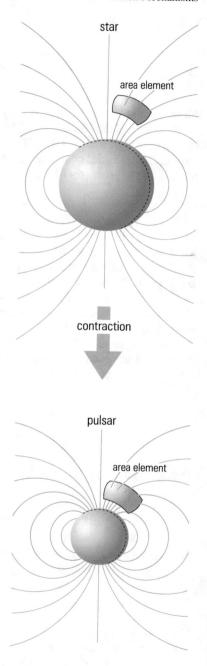

$$\int_{\text{star}} \boldsymbol{B}_{\text{star}} \cdot d\boldsymbol{A}_{\text{star}} = \int_{\text{pulsar}} \boldsymbol{B}_{\text{pulsar}} \cdot d\boldsymbol{A}_{\text{pulsar}},$$

$$\boldsymbol{B}_{\text{pulsar}} = \boldsymbol{B}_{\text{star}} \, \frac{R_{\text{star}}^2}{R_{\text{pulsar}}^2}. \tag{5.5.5}$$

For $B_{\text{star}} = 1000$ gauss magnetic pulsar fields of 2.5×10^{12} gauss $= 2.5 \times 10^8$ T are obtained! These theoretically expected extraordinarily high magnetic field strengths have been experimentally confirmed by measuring quantized energy levels of free electrons in strong magnetic fields ('Landau levels'). The rotational axis of pulsars usually does not coincide with the direction of the magnetic field. It is obvious that the vector of these high magnetic fields spinning around the non-aligned axis of rotation will produce strong electric fields, in which particles can be accelerated.

For a 30-ms pulsar with rotational velocities of

$$v = \frac{2\pi R_{\text{pulsar}}}{T_{\text{pulsar}}} = \frac{2\pi \times 20 \times 10^3 \text{ m}}{3 \times 10^{-2} \text{ s}} \approx 4 \times 10^6 \text{ m/s}$$

one obtains, using $\boldsymbol{E} = \boldsymbol{v} \times \boldsymbol{B}$ with $\boldsymbol{v} \perp \boldsymbol{B}$, electrical field strengths of

$$|\boldsymbol{E}| = v\,B \approx 10^{15} \text{ V/m}. \tag{5.5.6}$$

This implies that singly charged particles can gain $1\,\text{PeV} = 1000\,\text{TeV}$ per meter. However, it is not at all obvious how pulsars manage in detail to transform the rotational energy into the acceleration of particles. Pulsars possess a rotational energy of

$$E_{\text{rot}} = \frac{1}{2}\, \Theta_{\text{pulsar}}\, \omega_{\text{pulsar}}^2 = \frac{1}{2}\frac{2}{5}\, m\, R_{\text{pulsar}}^2\, \omega_{\text{pulsar}}^2 \tag{5.5.7}$$
$$\approx 7 \times 10^{42}\,\text{J} \approx 4.4 \times 10^{61}\,\text{eV}$$

($T_{\text{pulsar}} = 30\,\text{ms}$, $M_{\text{pulsar}} = 2 \times 10^{30}\,\text{kg}$, $R_{\text{pulsar}} = 20\,\text{km}$, $\omega = 2\pi/T$).

If the pulsars succeed to convert a fraction of only 1% of this enormous energy into the acceleration of cosmic-ray particles, one obtains an injection rate of

$$\frac{dE}{dt} \approx 1.4 \times 10^{42}\,\text{eV/s}, \tag{5.5.8}$$

if a pulsar lifetime of 10^{10} years is assumed.

If one considers that our galaxy contains 10^{11} stars and if the supernova explosion rate (pulsar creation rate) is assumed to be 1 per century, a total number of 10^8 pulsars have provided energy for the acceleration of cosmic-ray particles since the creation of our galaxy (age of the galaxy $\approx 10^{10}$ years). This leads to a total energy of 2.2×10^{67} eV for an average pulsar injection time of 5×10^9 years. For a total volume of our galaxy (radius 15 kpc, average effective thickness of the galactic disk 1 kpc) of 2×10^{67} cm^3 this corresponds to an energy density of cosmic rays of 1.1 eV/cm^3.

One has, of course, to consider that cosmic-ray particles stay only for a limited time in our galaxy and are furthermore subject to energy-loss processes. Still, the above presented crude estimate describes the actual energy density of cosmic rays of ≈ 1 eV/cm^3 rather well.

5.6 Binaries

The advantage of living near a binary is to get a tan in half the time.

Anonymous

Binaries consisting of a pulsar or neutron star and a normal star can also be considered as a site of cosmic-ray-particle acceleration. In such a binary system matter is permanently dragged from the normal star and whirled into an accretion disk around the compact companion. Due to these enormous plasma motions very strong electromagnetic fields are produced in the vicinity of the neutron star. In these fields charged particles can be accelerated to high energies (see Fig. 5.8).

The energy gain of infalling protons (mass m_p) in the gravitational potential of a pulsar (mass M_{pulsar}) is

normal star

matter
transfer

pulsar

Fig. 5.8 Formation of accretion disks in binaries

$$\Delta E = -\int_{\infty}^{R_{\text{pulsar}}} G\frac{m_p\, M_{\text{pulsar}}}{r^2}\, \mathrm{d}r = G\frac{m_p\, M_{\text{pulsar}}}{R_{\text{pulsar}}}$$
$$\approx 1.1 \times 10^{-11}\,\text{J} \approx 70\,\text{MeV} \tag{5.6.1}$$

$(m_p \approx 1.67 \times 10^{-27}\,\text{kg},\quad M_{\text{pulsar}} = 2 \times 10^{30}\,\text{kg},\quad R_{\text{pulsar}} = 20\,\text{km},\quad G \approx 6.67 \times 10^{-11}\,\text{m}^3\,\text{kg}^{-1}\,\text{s}^{-2}$ gravitational constant).

The matter falling into the accretion disk achieves velocities v, which are obtained under classical treatment from

$$\frac{1}{2}mv^2 = \Delta E = G\frac{m\, M_{\text{pulsar}}}{R_{\text{pulsar}}} \tag{5.6.2}$$

to provide values of

$$v = \sqrt{\frac{2G M_{\text{pulsar}}}{R_{\text{pulsar}}}} \approx 1.2 \times 10^8\,\text{m/s}. \tag{5.6.3}$$

The variable magnetic field of the neutron star, which is perpendicular to the accretion disk, will produce via the Lorentz force a strong electric field. Using

$$F = e(v \times B) = eE,\tag{5.6.4}$$

the particle energy E is obtained, using $v \perp B$, to

$$E = \int F \cdot \mathrm{d}s = evB\Delta s.\tag{5.6.5}$$

Under plausible assumptions ($v \approx c$, $B = 10^6$ T, $\Delta s = 10^5$ m) particle energies of 3×10^{19} eV are possible. Even more powerful are accretion disks, which form around black holes or the compact nuclei of active galaxies. One assumes that in these active galactic nuclei and in jets ejected from such nuclei, particles can be accelerated to the highest energies observed in primary cosmic rays.

The details of these acceleration processes are not yet fully understood. Sites in the vicinity of black holes—a billion times more massive than the Sun—could possibly provide the environment for the acceleration of the highest-energy cosmic rays. Confined highly relativistic jets are a common feature of such compact sources. It is assumed that the jets of particles accelerated near a black hole or the nucleus of a compact galaxy are injected into the radiation field of the source. Electrons and protons accelerated in the jets via shocks initiate electromagnetic and hadronic cascades. High-energy γ rays are produced by inverse Compton scattering off accelerated electrons. High-energy neutrinos are created in the decays of charged pions in the development of the hadronic cascade. It is assumed that one detects emission from these sources only if the jets are beamed into our line of sight. A possible scenario for the acceleration of particles in *beamed jets* from massive compact sources is sketched in Fig. 5.9.

Cosmic acceleration by roller coasters

Triggered by the fact that no generally accepted acceleration model for very high energies has been put forward, a number of exotic alternatives have been proposed.

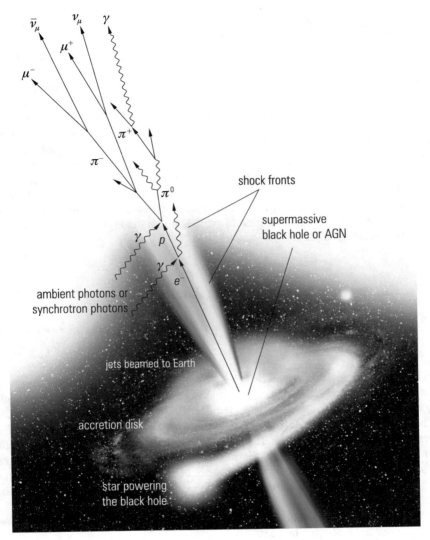

Fig. 5.9 Acceleration model for relativistic jets powered by a black hole or an active galactic nucleus. The various reactions are only sketched

A Cannon Ball model has been discussed, which describes mass ejections from compact sources in the form of cannon balls with relativistic velocities [50]. In a collapse of a massive object accretion disks are formed. The infall of matter onto the accretion disk can lead to the emission of discrete cannon balls, which are expected to be emitted back to back. Because strong, varying time dependent magnetic fields are involved, charged particles might be accelerated to the highest energies. The Cannon Ball model claims to be able to explain the luminosity, the slopes of the particle spectra and the position of the knee(s) and ankle(s) in terms of simple and standard physics.

Another possibility is the acceleration of energetic particles by resonant cyclotron emission. Relativistic shocks with strong, time-dependent magnetic fields, which in turn produce powerful electric fields, can accelerate electrons and positrons. Therefore the plasma wind in compact sources can be dominated by ions, where the magnetic field forces the ions to propagate along helical lines. These ions are expected to emit cyclotron waves (Alfvén waves) in the magnetosphere of the source, which can be absorbed in a resonant fashion by electrons and positrons. In this way the electrons and positrons might be accelerated to high Lorentz factors with good efficiency. Such a resonant cyclotron emission might be able to produce also high-energy particles.

The large number of proposed mechanisms for high-energy particle acceleration also connected with the jet formation from compact sources shows, that no generally accepted idea about the acceleration of particles with the highest energies of up to 10^{20} eV has emerged [51].

5.7   Energy Spectra for Primary Particles

> *Get first your facts right and then you can distort them as much as you please.*
>
> Mark Twain

At the present time it is not at all clear, which of the presented mechanisms contribute predominantly to the acceleration of cosmic-ray particles. There are good arguments to assume that the majority of galactic cosmic rays is produced by shock acceleration, where the particles emitted from the source are possibly further accelerated by the Fermi mechanism of 2nd order. In contrast, it is likely that the extremely energetic cosmic rays are predominantly accelerated in pulsars, binaries, or in jets emitted from black holes or active galactic nuclei. For shock acceleration in supernova explosions the shape of the energy spectrum of cosmic-ray particles can be derived from the acceleration mechanism.

Let E_0 be the initial energy of a particle and εE_0 the energy gain per acceleration cycle. After the first cycle one gets

$$E_1 = E_0 + \varepsilon E_0 = E_0(1 + \varepsilon) \tag{5.7.1}$$

while after the nth cycle (e.g., due to multiple reflection at shock fronts) one has

$$E_n = E_0(1 + \varepsilon)^n . \tag{5.7.2}$$

To obtain the final energy $E_n = E$, a number of

$$n = \frac{\ln(E/E_0)}{\ln(1 + \varepsilon)} \tag{5.7.3}$$

cycles is required. Let us assume that the escape probability per cycle is P. The probability that particles still take part in the acceleration mechanism after n cycles is $(1 - P)^n$. This leads to the following number of particles with energies in excess of E:

$$N(> E) \sim \sum_{m=n}^{\infty}(1 - P)^m . \tag{5.7.4}$$

Because of $\sum_{m=0}^{\infty} x^m = \dfrac{1}{1 - x}$ (for $x < 1$), (5.7.4) can be rewritten as

$$N(> E) \sim (1 - P)^n \sum_{m=n}^{\infty}(1 - P)^{m-n}$$

$$= (1 - P)^n \sum_{m=0}^{\infty}(1 - P)^m = \frac{(1 - P)^n}{P} , \tag{5.7.5}$$

where $m - n$ has been renamed m. Equations (5.7.3) and (5.7.5) can be combined to form the integral energy spectrum

$$N(> E) \sim \frac{1}{P}\left(\frac{E}{E_0}\right)^{-\gamma} \sim E^{-\gamma} , \tag{5.7.6}$$

where the spectral index γ is obtained from (5.7.5) and (5.7.6) with the help of (5.7.3) to

$$(1 - P)^n = \left(\tfrac{E}{E_0}\right)^{-\gamma} ,$$

$$n \ln(1 - P) = -\gamma \ln(E/E_0) ,$$

$$\gamma = -\tfrac{n \ln(1-P)}{\ln(E/E_0)} = \tfrac{\ln(1/(1-P))}{\ln(1+\varepsilon)} . \tag{5.7.7}$$

This simple consideration yields a power law of primary cosmic rays in agreement with observation.

The energy gain per cycle surely is rather small ($\varepsilon \ll 1$). If also the escape probability P is low (e.g., at reflections between two shock fronts), (5.7.7) is simplified to

$$\gamma \approx \frac{\ln(1 + P)}{\ln(1 + \varepsilon)} \approx \frac{P}{\varepsilon} . \tag{5.7.8}$$

Experimentally one finds that the index of the integral spectrum up to energies of 10^{15} eV is $\gamma = 1.7$. For higher energies the integral primary cosmic-ray-particle spectrum steepens with $\gamma = 2$.

This calculation can only motivate the shape of the emission spectrum from the sources. It cannot describe the details of the primary spectrum as measured at the edge of the atmosphere, in particular, it is asking too much that it should be able to explain the knee(s) and ankle(s). These features depend on the propagation of cosmic rays in the galaxy and on the energy-dependent question of the galatic or extragalactic origin.

Summary

An essential task of astroparticle physics is to understand the different acceleration mechanisms and to localize the accelerators in the sky. Obviously one thinks about acceleration methods known from earthbound accelerators, which certainly are also at work in astroparticle physics. In addition, there are also specific methods that only work in space, like shock acceleration in supernova explosions or collisions of magnetized clouds at galactic and extragalactic distances. These processes even allow an educated guess at the spectral slope of primary energy spectra. As far as the acceleration of the highest-energy particles that are emitted from active galactic nuclei is concerned, one has no real clue how that might work. There are models how this might be achieved, e.g., in jet acceleration, but the details of these processes are not really understood.

Victor Hess discovers cosmic rays by inflation.

5.8 Problems

1. Work out the kinetic energy of electrons accelerated in a betatron for the classical ($v \ll c$) and the relativistic case ($B = 1$ tesla, $R = 0.2$ m). See also Problem 1.2.
2. A star of 10 solar masses undergoes a supernova explosion. Assume that 50% of its mass is ejected and the other half ends up in a pulsar of 10 km radius. What is the Fermi energy of the electrons in the pulsar? What is the consequence of it?
3. It is assumed that active galactic nuclei are powered by black holes. What is the energy gain of a proton falling into a one-million-solar-mass black hole down to the event horizon?
4. If the Sun were to collapse to a neutron star ($R_{NS} = 50$ km), what would be the rotational energy of such a solar remnant ($M_\odot = 2 \times 10^{30}$ kg, $R_\odot = 7 \times 10^8$ m, $\omega_\odot = 3 \times 10^{-6}$ s^{-1})?
   Compare this rotational energy to the energy that a main-sequence star like the Sun can liberate through nuclear fusion!
5. In a betatron the change of the magnetic flux $\phi = \int B \, dA = \pi R^2 B$ induces an electric field,

$$\int E \, ds = -\dot{\phi} \,,$$

   in which particles can be accelerated,

$$E = -\frac{\dot{\phi}}{2\pi R} = -\frac{1}{2} R \dot{B} \,.$$

   The momentum increase is given by

$$\dot{p} = -eE = \frac{1}{2} e R \dot{B} \,. \tag{5.8.1}$$

   What kind of guiding field would be required to keep the charged particles on a stable orbit?
6. The lifetime of a star is essentially determined by its mass. Massive stars burn their hydrogen rapidly, which leads to a short lifespan. It has been reported that black holes have been observed that are almost 10 billion years old, that is, they must have been created in a period of less than about three billion years after the Big Bang, i.e., in the early period of the universe. Estimate the lifespan of a star of 100 solar masses if one assumes a very rough approximation for the mass–lifespan relation of stars like

$$T = T_\odot \cdot \left(\frac{M}{M_\odot} \right)^{-2} \,,$$

where $T_\odot$ is the lifespan of the Sun ($\approx$10 billion years) and $M_\odot$ is the solar mass. This relation is only a very crude estimate. The dependence of the stars' lifespan depends, of course, on the spectral type of the star and many other parameters.

7. The travel to extrasolar planets with standard rockets seems impractical, because the velocities that are reachable are insufficient. Photon propulsion seems to be an alternative. A photon has a certain momentum, which could be used for a photon drive. Let us assume a high-power laser with 1 petawatt power as propulsion mechanism (photon wavelength 400 nm). Work out the number of photons, the photon momentum, and the momentum transfer and the force exerted on the space vehicle.

8. Neutron stars are spinning very fast. There are even millisecond pulsars. If they were too fast, they might even disintegrate due to the increasing centrifugal forces. Let us assume that the star has a radius r and an angular velocity ω, and that its density is uniform. Estimate the minimum density of a star with a rotation period of one millisecond so that it does not disintegrate.

Chapter 6
Primary Cosmic Rays

It will be found that everything depends on the composition of the forces with which the particles of matter act upon one another, and from these forces, as a matter of fact, all phenomena of nature take their origin.

R. J. Boscovich

Cosmic rays provide important information about high-energy processes occurring in our galaxy and beyond. Cosmic radiation produced in the sources is usually called *primordial cosmic rays*. This radiation is modified during its propagation in galactic and extragalactic space. Particles of galactic origin pass on average through a column density of $6 \, \text{g/cm}^2$ before reaching the top of the Earth's atmosphere. Of course, the atmosphere does not really have a 'top' but it rather exhibits an exponential density distribution. It has become common practice to understand under the top of the atmosphere an altitude of approximately 40 km. This height corresponds to a residual column density of $5 \, \text{g/cm}^2$ corresponding to a pressure of 5 mbar due to the residual atmosphere above altitudes of 40 km. Cosmic rays arriving at the Earth's atmosphere are usually called *primary cosmic rays.*

Sources of cosmic rays accelerate predominantly charged particles such as protons and electrons. Since all elements of the periodic table are produced during element formation, nuclei as helium, lithium, and so on can be also accelerated. Cosmic rays represent an extraterrestrial or even extragalactic matter sample whose chemical composition exhibits certain features similar to the elemental abundance in our solar system.

Charged cosmic rays accelerated in sources can produce a number of secondary particles by interactions in the sources themselves.

These mostly unstable secondary particles, i.e., pions and kaons, produce stable particles in their decay, i.e., photons from $\pi^0 \to \gamma\gamma$ and neutrinos from $\pi^+ \to \mu^+ + \nu_\mu$ decays. Secondary particles also emerge from the sources and can reach Earth. Let us first discuss the originally accelerated charged component of primary cosmic rays.

© Springer Nature Switzerland AG 2020
C. Grupen, *Astroparticle Physics*, Undergraduate Texts in Physics,
https://doi.org/10.1007/978-3-030-27339-2_6

Fig. 6.1 Elemental abundance of primary cosmic rays for $1 \leq Z \leq 28$

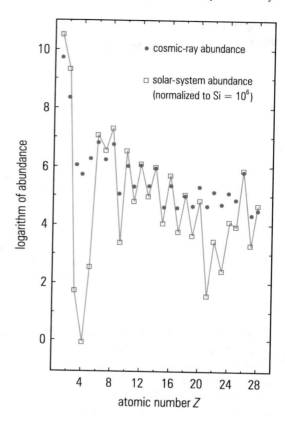

6.1   Charged Component of Primary Cosmic Rays

Coming out of space and incident on the high atmosphere, there is a thin rain of charged particles known as primary cosmic rays.
Michio Kaku

The elemental abundance of primary cosmic rays is shown in Figs. 6.1 and 6.2 in comparison to the chemical composition of the solar system. Protons are the dominant particle species ($\approx 85\%$) followed by α particles ($\approx 12\%$). Elements with a nuclear charge $Z \geq 3$ represent only a 3% fraction of charged primary cosmic rays. The chemical composition of the solar system, shown in Figs. 6.1 and 6.2, has many features in common with that of cosmic rays. However, remarkable differences are observed for lithium, beryllium, and boron ($Z = 3$–5), and for the elements below the iron group ($Z < 26$). The larger abundance of Li, Be, and B in cosmic rays can easily be understood by fragmentation of the heavier nuclei carbon ($Z = 6$) and, in particular, oxygen ($Z = 8$) in galactic matter on their way from the source to Earth.

In the same way the fragmentation or spallation of the relatively abundant element iron populates elements below the iron group. The general trend of the dependence

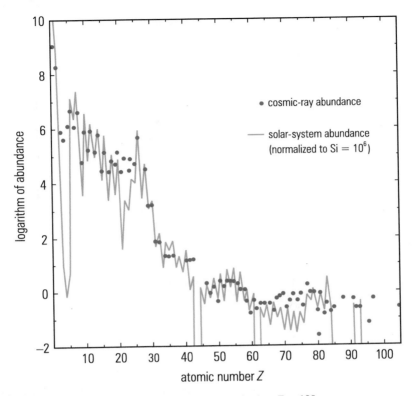

Fig. 6.2 Elemental abundance of primary cosmic rays for $1 \leq Z \leq 100$

of the chemical composition of primary cosmic rays on the atomic number can be understood by nuclear physics arguments. In the framework of the shell model it is easily explained that nuclear configurations with even proton and neutron numbers (even–even nuclei) are more abundant compared to nuclei with odd proton and neutron numbers (odd–odd nuclei). As far as stability is concerned, even–odd and odd–even nuclei are associated with abundances between ee and oo configurations. Extremely stable nuclei occur for filled shells ('magic nuclei'), where the magic numbers (2, 8, 20, 50, 82, 126) refer separately to protons and neutrons. As a consequence, doubly magic nuclei (like helium and oxygen) are particularly stable and correspondingly abundant. But nuclei with a large binding energy such as iron, which can be produced in fusion processes, are also relatively abundant in charged primary cosmic rays. The energy spectra of primary nuclei of hydrogen, helium, carbon, and iron are shown in Fig. 6.3 for a particular epoch of the solar cycle.

The low-energy part of the primary spectrum is modified by the Sun's and the Earth's magnetic field. The 11-year period of the sunspot cycle modulates the intensity of low-energy primary cosmic rays (<1 GeV/nucleon). The active Sun reduces

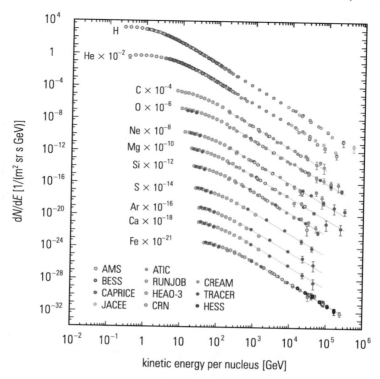

Fig. 6.3 Energy spectra of the main components of primary cosmic rays from direct measurements. The various data have been obtained from balloon measurements, from satellites, and an experiment at the International Space Station ISS [52]

the cosmic-ray intensity because a stronger magnetic field created by the Sun prevents galactic charged particles from reaching Earth.

In general, the intensity decreases with increasing energy so that a direct observation of the high-energy component of cosmic rays at the top of the atmosphere with balloons or satellites eventually runs out of statistics. Measurements of the charged component of primary cosmic rays at energies in excess of several hundred GeV must therefore resort to indirect methods. The atmospheric air Cherenkov technique (see Sect. 6.4: 'Gamma Astronomy') or the measurement of extensive air showers via air fluorescence or particle sampling (see Sect. 7.4: 'Extensive Air Showers') can in principle cover this part of the energy spectrum, however, a determination of the chemical composition of primary cosmic rays by this indirect technique is particularly difficult. Furthermore, the particle intensities at these high energies are extremely low. For particles with energies in excess of about 10^{19} eV the rate is only 1 particle per km^2 and year.

The all-particle spectrum of charged primary cosmic rays is relatively steep so that practically no details are observable. Only after multiplication of the intensity with

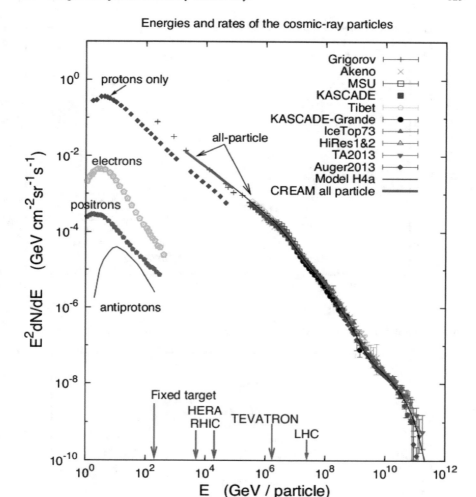

Fig. 6.4 The primary all-particle spectrum measured by different experiments scaled with the square of the primary energy. The contributions from electrons, positrons, and antiprotons as measured by the PAMELA experiment are shown [53]

a power of the primary energy, structures in the primary spectrum become visible (Figs. 6.4 and 6.5).

The results on the electron spectrum have been extended to very high energies (up to 20 TeV) by the H.E.S.S. experiment (Fig. 6.6). When cosmic-ray experiments without magnetic spectrometers measure electrons they cannot distinguish electrons form positrons. So, the electron spectrum is the combined spectrum of electromagnetically interacting charged particles, but, since positrons are suppressed compared to electrons, such spectra are called electron spectra. Around 1 TeV there is a struc-

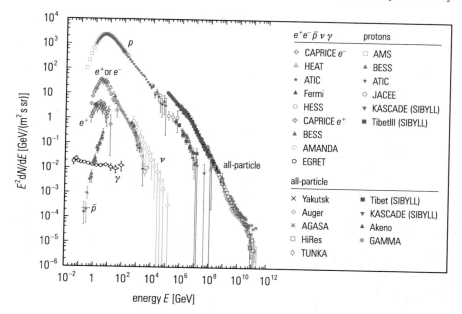

Fig. 6.5 Compilation of primary cosmic-ray spectra, including neutrinos and photons [54]

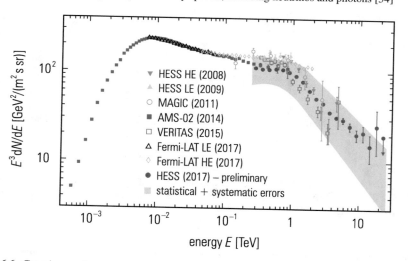

Fig. 6.6 Cosmic-ray-electron energy spectrum measured with H.E.S.S. (*red dots*) compared to previous measurements from various experiments. The shape of the electron spectrum and its steepening at energies around 1 TeV could give answers to various production and propagation models [56]

ture, which is also observed by the new CALET experiment on the ISS [55]. Insofar this structure should not be over-interpreted. Better statistics is required. This feature certainly requires further investigations.

The bulk of cosmic rays up to at least an energy of 10^{15} eV is believed to originate from within our galaxy. Above that energy, which is associated with the so-called '*knee*', the spectrum steepens. Above the so-called '*ankle*' at energies around 10^{19} eV the spectrum slightly flattens again for a short energy span and for energies in excess of 6×10^{19} eV the spectrum shows a strong cutoff (see also Figs. 6.7, 6.8, and 6.9).

For higher energies primary protons rapidly lose energy by interactions with the primordial background radiation. Around these energies also a crossover from the galactic component to a component of extragalactic origin might be conceivable. To compare the primary energies with energies from accelerators, one has to convert the center-of-mass energy, e.g., in storage rings to a laboratory energy. For the Large Hadron Collider (LHC) at CERN with a center-of-mass energy of 14 TeV the comparable energy in a laboratory system for the collision of a high-energy proton with a proton at rest using the relation $s = 2m E_{lab}$ (m is the proton mass) is obtained to be about 10^{17} eV.

Cosmic rays originate predominantly from within our galaxy. Galactic objects do not in general have such a combination of size and magnetic field strength to contain particles at very high energies.

Because of the equilibrium between the centrifugal and Lorentz force ($v \perp B$ assumed) one has

$$mv^2/\varrho = Z e v B , \qquad (6.1.1)$$

which yields for the momentum of singly charged particles

$$p = e \varrho B$$

(p is the particle momentum, B the magnetic field, v the particle velocity, m the particle mass, ϱ the bending radius or gyroradius). For a large-area galactic magnetic field of $B = 10^{-10}$ tesla in the galaxy (about 10^5 times weaker compared to the magnetic field on the surface of the Earth) and a gyroradius of 5 pc, from which particles start to leak from the galaxy, particles with momenta up to

$$p[\text{GeV}/c] = 0.3 \, B[\text{T}] \, \varrho[\text{m}] , \qquad (6.1.2)$$

$$p_{\text{max}} = 4.6 \times 10^6 \, \text{GeV}/c = 4.6 \times 10^{15} \, \text{eV}/c \qquad (6.1.3)$$

can be contained. 1 parsec (pc) is the popular unit of distance in astronomy (1 pc = 3.26 light-years = 3.0857×10^{16} m). Particles with energies exceeding 10^{15} eV start to leak from the galaxy. This causes the spectrum to get steeper to higher energies. Since the containment radius depends on the atomic number, see (6.1.1), the position of the knee should depend on the charge of primary cosmic rays in this scenario, i.e., the knee for iron would be expected at higher energies compared to the proton

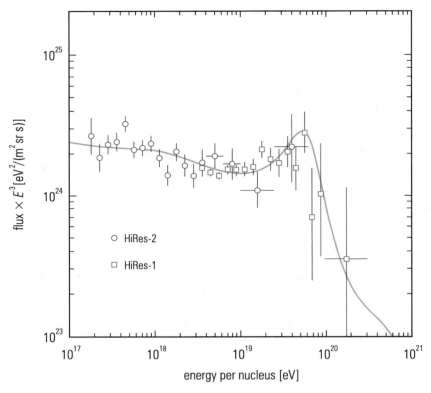

Fig. 6.7 Energy spectrum of primary cosmic rays scaled by a factor of E^3. The data are from the Utah high resolution experiment [57]

knee. And this was actually seen in the data of the KASCADE-Grande experiment at around 80 PeV (see Fig. 6.10).

Figure 6.10 shows some details on the measurement of the primary spectrum in the energy range from 100 TeV to 1 EeV. The air-shower data from KASCADE-Grande were separated into two mass groups, i.e., into groups of heavy and light primary masses according to the observed electron–muon ratio in the air showers. In this energy range, e.g., iron-induced showers tend to have more muons compared to proton-induced showers. Therefore, the sample of electron-poor showers is enriched in heavy primaries compared to the electron-rich sample, which is mainly caused by light primaries. The spectrum of heavy primaries (electron-poor) shows a clear knee-like feature at an energy of about 80 PeV (8×10^{16} eV), which is interpreted as 'iron knee'. The selection of heavy primaries enhances the knee-like feature that is already present in the all-particle spectrum. The first knee, the 'proton knee' is around 3 PeV. With the nuclear charge of iron ($Z = 26$), the observed steepening at 8×10^{17} eV fits well into the assumption that the containment of primaries in the Milky Way depends on the charge of the primaries. The light primaries (electron-

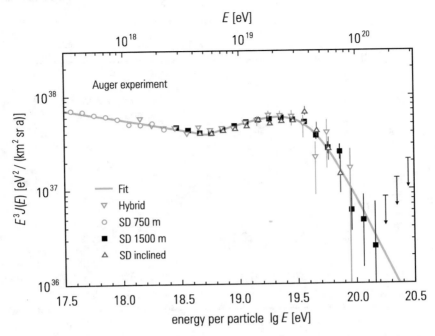

Fig. 6.8 Energy spectrum of primary cosmic rays scaled by E^3. The results from the Auger experiment clearly show the cutoff of the energy spectrum above 6×10^{19} eV (*SD* stands for the surface Cherenkov detectors, and *Hybrid* includes the surface detectors and results from the fluorescence telescopes) [58]

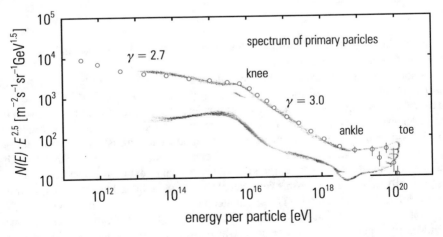

Fig. 6.9 Artist's view of the different structures of the spectrum of primary cosmic rays

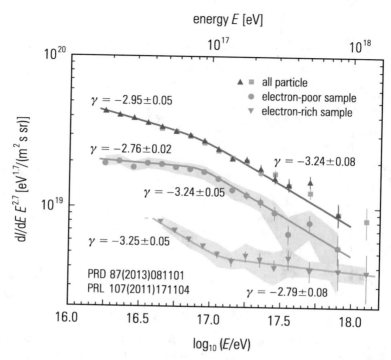

Fig. 6.10 Energy spectrum of primary particles in the energy range from 10 PeV to 1 EeV with data from KASCADE-Grande. The all-particle spectrum is separated into heavy primaries (electron-poor sample) and light primaries (electron-rich sample) [59]

rich) exhibit a flattening at somewhat higher energies, which could be understood as the onset of extragalactic protons.

Another possible reason for the knee in cosmic radiation could be related to the fact that 10^{15} eV is about the maximum energy that can be supplied by supernova explosions. For higher energies a different acceleration mechanism is required, which might possibly lead to a steeper energy spectrum. The knee could in principle also have its origin in a possible change of interaction characteristics of high-energy particles. It is in principle conceivable that the interaction cross section changes with energy giving rise to features at the knee of the primary cosmic-ray spectrum. There is presently no evidence for this from accelerator data at quite high energies, so this is considered rather unlikely. The slight steepening of the spectrum beyond 80 PeV is related to the fact that heavy nuclei like iron start to leak from the galaxy. One could argue that magnetic confinement is the reason for this interpretation. The events beyond several 10^{19} eV could be of extragalactic origin.

"Look! Here we go again, another supernova!"

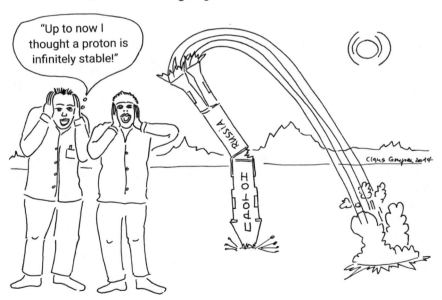

Proton rockets tend to decay

The small dip at energies around 10^{19} eV (see Fig. 6.7) could be the consequence of e^+e^- pair production of primary particles on photons of the blackbody radiation. By this process primary protons would lose some energy. The exact shape of the spectrum in this energy range depends on the variation of the energy-dependent cross section for e^+e^- pair production.

There are good reasons to expect a cutoff of the spectrum beyond 6×10^{19} eV. In 1966 it was realized by Greisen, Zatsepin, and Kuzmin (GZK) that cosmic rays above the energy of approximately 6×10^{19} eV would interact with the cosmic blackbody radiation. Protons of higher energies would rapidly lose energy by this interaction process causing the spectrum to be cut off at energies around 6×10^{19} eV. Primary protons with these energies produce pions on blackbody photons via the Δ resonance according to

$$\gamma + p \rightarrow p + \pi^0, \quad \gamma + p \rightarrow n + \pi^+, \tag{6.1.4}$$

thereby losing a large fraction of their energy.

The threshold energy for the photoproduction of pions can be determined from four-momentum conservation

$$(q_\gamma + q_p)^2 = (m_p + m_\pi)^2 \tag{6.1.5}$$

(q_γ, q_p are four-momenta of the photon or proton, respectively; m_p, m_π are proton and pion masses) yielding

$$E_p = (m_\pi^2 + 2m_p m_\pi)/4E_\gamma \tag{6.1.6}$$

for head-on collisions.

A typical value of the Planck distribution corresponding to the blackbody radiation of temperature 2.7 K is around 1.1 meV. With this photon energy the threshold energy for the photoproduction of pions is

$$E_p \approx 6 \times 10^{19} \text{ eV}. \tag{6.1.7}$$

It is, however, not guaranteed that the cutoff is due to this process of photoproduction. It is also conceivable that the possible sources of cosmic rays just run out of power to produce particles with energies beyond several 10^{19} eV. This scenario is actually supported by the recent Auger data.

The observation of several events in excess of 10^{20} eV (the 'toe' of primary cosmic rays), therefore, represents a certain mystery (see also Fig. 6.9). The Greisen–Zatsepin–Kuzmin cutoff limits the mean free path of high-energy protons to something like 50 Mpc. Photons as candidates for primary particles have even shorter mean free paths (≈ 50 kpc) because they produce electron pairs in gamma–gamma interactions with blackbody photons, infrared, starlight photons, or photons from the radio range ($\gamma\gamma \rightarrow e^+e^-$). This reasoning is supported by the Auger experiment, which has no evidence for photons beyond 10^{18} eV.

The hypothesis that primary neutrinos are responsible for the highest-energy events is rather unlikely. The interaction probability for neutrinos in the atmosphere is extremely small ($<10^{-4}$). Furthermore, the observed zenith-angle distribution of energetic events and the position of primary vertices of the cascade development in the atmosphere are inconsistent with the assumption that primary neutrinos are responsible for these events. Because of their low interaction probability one would expect that the primary vertices for neutrinos would be distributed uniformly in the atmosphere. In contrast, one observes that the first interaction takes place predominantly in the 100 mbar layer, which is characteristic of hadron or photon interactions.

One way out would be to assume that after all protons are responsible for the events with energies exceeding 6×10^{19} eV. This would support the idea that the sources of the highest-energy cosmic-ray events are relatively close. A candidate source is M87, an elliptic giant galaxy in the Virgo Cluster at a distance of 15 Mpc. From the center of M87 a jet of 1500 pc length is ejected that could be the source of energetic particles. M87, also known as Virgo A (3C274), is one of the strongest radio sources in the constellation Virgin.

An extreme alternative is the assumption that new, so far unknown elementary particles or unexpected phenomena or interaction processes are responsible for the extreme high-energy events.

Considering the enormous rigidity of these high-energy particles and the weakness of the intergalactic magnetic field, one would not expect substantial deflections of these particles over distances of 50 Mpc. This would imply that one can consider to do astronomy with these extremely high-energy cosmic rays. There is, however, no clear correlation of the arrival directions of these high-energy cosmic-ray events with known astronomical sources in the immediate neighbourhood of our galaxy or in the close local cluster of galaxies.

Antiparticles are extremely rare in primary cosmic rays. The measured primary antiprotons are presumably generated in interactions of primary charged cosmic rays with the interstellar gas. Antiprotons can be readily produced according to

$$p + p \rightarrow p + p + p + \bar{p}, \qquad (6.1.8)$$

while positrons are most easily formed in pair production by energetic photons. The flux of primary antiprotons for energies >10 GeV has been measured to be

$$\left. \frac{N(\bar{p})}{N(p)} \right|_{>10\,\text{GeV}} \approx 10^{-4}. \qquad (6.1.9)$$

The fraction of primary electrons in relation to primary protons is only 1%. Primary positrons constitute only 10% of the electrons at energies around 10 GeV. They are presumably also consistent with secondary origin. There is, however, an increase over the expected positron flux at energies around 100 GeV, as measured by the PAMELA and AMS experiment. The reason for this excess is so far unknown, but nearby supernova explosions or pulsars might have injected positrons into our galaxy.

To find out whether there are stars of antimatter in the universe, the existence of primary antinuclei (antihelium, anticarbon) must be established because secondary production of antinuclei with $Z \geq 2$ by cosmic rays is practically excluded. The non-observation of primary antimatter with $Z \geq 2$ is a strong hint that our universe is matter dominated. There are, however, some antihelium candidates found by the AMS experiment [60].

One might wonder whether the continuous bombardment of the Earth with predominantly positively charged particles (only 1% are negatively charged) would lead to a positive charge-up of our planet. This, however, is not true. When the rates of primary protons and electrons are compared, one normally considers only energetic particles. The spectra of protons and electrons are very different with electrons populating mainly low-energy regions. If *all* energies are considered, there are equal numbers of protons and electrons so that there is no charge-up of our planet.

The chemical composition of high-energy primary cosmic rays ($> 10^{15}$ eV) is to large extent an unknown territory. If the current models of nucleon–nucleon interactions are extrapolated into the range beyond 10^{17} eV (corresponding to a center-of-mass energy of $\gtrsim 14$ TeV in proton–proton collisions) and if the muon content and lateral distribution of muons in extensive air showers are taken as a criterion for the identity of the primary particle, then one would arrive at the conclusion that the chemical composition of primary cosmic rays beyond the knee ($> 10^{15}$ eV) changes towards to a higher fraction of heavy nuclei. The KASCADE-Grande experiment has clear indications of a steepening of the primary spectrum beyond 80 PeV, which is interpreted as *iron knee*. When in the sample of air showers the fraction of heavy primaries is enhanced by selecting muon-rich showers, there is a very pronounced steepening at 80 PeV in the position, where one would expect the iron knee (see Fig. 6.10). This means that heavy nuclei with energies beyond 80 PeV are leaking from our galaxy, in agreement with expectations from galactic containment.

Even though cosmic rays have been discovered more than 100 years ago, their origin is still an open question. It is generally assumed that active galactic nuclei, quasars, or supernova explosions are excellent source candidates for high-energy cosmic rays, but there is no direct evidence for this assumption. In the energy range up to 100 TeV individual sources have been identified by primary gamma rays. It is conceivable that gamma rays of these energies are decay products of elementary particles (π^0 decay, Centaurus A?), which have been produced by those particles that have been originally accelerated in the sources. Therefore, it would be interesting to see the sources of cosmic rays in the light of these originally accelerated particles.

This, however, presents a serious problem: photons and neutrinos travel on straight lines in galactic and intergalactic space, therefore pointing directly back to the sources. Charged particles, on the other hand, are subject to the influence of homogeneous or irregular magnetic fields. This causes the accelerated particles to travel along chaotic trajectories thereby losing all directional information before finally reaching Earth. Therefore, it is of very little surprise that the sky for charged particles with energies below 10^{14} eV appears completely isotropic. The level of observed

Fig. 6.11 Sketch of proton and iron trajectories in our Milky Way at energies of around 10^{18} eV

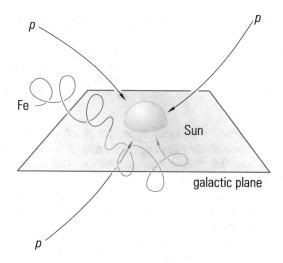

anisotropies lies below 0.5%. There is some hope that for energies exceeding 10^{18} eV a certain directionality could be found. It is true that also in this energy domain the galactic magnetic fields must be taken into account, however, the deflection radii are already rather large. The situation is even more complicated because of a rather uncertain topology of galactic magnetic fields. In addition, one must in principle know the time evolution of magnetic fields over the last $\approx$200 million years because the sources can easily reside at distances of >50 Mpc ($\widehat{=}163$ million light-years). For simultaneous observation of cosmic-ray sources in the light of charged particles and photons, one must take into account that charged particles are delayed with respect to photons because they travel on longer trajectories due to the bending by magnetic fields.

Since the magnetic deflection is proportional to the charge of a particle, proton astronomy is more promising than astronomy with heavy nuclei. This idea is outlined in Fig. 6.11, where the trajectories of protons and an iron nucleus ($Z = 26$) at an energy of 10^{18} eV are sketched for our galaxy. This figure clearly shows that one should only use—if experimentally possible—protons for *particle astronomy*. Actually, there are some hints that the origins of some of the events with energies $>10^{19}$ eV could lie in the supergalactic plane, a cluster of relatively close-by galaxies including our Milky Way ('local super galaxy'). It has also been discussed that the galactic center of our Milky Way and, in particular, the Cygnus region could be responsible for a certain anisotropy at 10^{18} eV. It must, however, be mentioned that claims for such a possible correlation are based on very low statistics and are therefore not unanimously supported. They certainly need further experimental confirmation.

6.2   Nature and Origin of the Highest-Energy Cosmic Rays

> *On what can we now place our hopes of solving the many riddles*
> *which still exist as to the origin and composition of cosmic rays?*
> *It must be emphasized here above all that to attain really decisive*
> *progress greater funds must be made available.*
>
> Victor Franz Hess

As already mentioned in earlier chapters, the highest-energy cosmic-ray particles are presumably of extragalactic origin. Some more detailed ideas will be mentioned in Sect. 7.7: 'Some Thoughts on the Highest Energies'. Here only some basic ideas will be presented.

The historic first event in the very-high-energy domain was the *Oh-My-God* event observed by John Linsley on the Dugway Proving Ground in Utah in October 1991. The energy of this spectacular event was 3×10^{20} eV.

We will discuss a few possibilities for sources for such rare events, sometimes also called Zevatrons, named in analogy to Lawrence Berkeley National Laboratory's Bevatron and Fermilab's Tevatron.

It has to be kept in mind that all such events are measured using the air-shower technique. The experimental error of the energy assignment is typically ±30%. A possible systematic uncertainty could arise from the Landau–Pomeranchuk–Migdal (LPM) effect, which may not have been correctly considered in the shower simulation. The LPM effect states that at high energies or high matter densities, the cross sections for bremsstrahlung and electron–positron pair production decrease. If not properly considered this might lead to a misassignment of the energy of a shower.

Using typical numbers of the magnetic field and the size of our galaxy one arrives at a maximum energy, which can probably be produced and stored, of

$$E_{max} = 10^5 \, \text{TeV} \, \frac{B}{3 \times 10^{-6} \, \text{G}} \, \frac{R}{50 \, \text{pc}} \, . \tag{6.2.1}$$

When $B = 3 \, \mu\text{G}$ and $R = 5 \, \text{kpc}$ are generously assumed, one gets

$$E_{max} = 10^7 \, \text{TeV} = 10^{19} \, \text{eV} \, . \tag{6.2.2}$$

This equation tells us that our Milky Way is unable to accelerate or store particles with higher energies, so that one has to assume that such particles must be extragalactic.

The threshold energy for the GZK cutoff via the photoproduction of pions on photons of the blackbody radiation was 6×10^{19} eV, leading to an attenuation length for protons of about 50 Mpc. Therefore, only nearby sources can be considered as candidates for the high-energy particles. Possibly the Markarian galaxies Mrk 421 and Mrk 501, standing at distances of about 100 Mpc or M87 (at 17 Mpc), are conceivable as candidates.

It has to be mentioned that the GZK cutoff can possibly be circumvented by assuming that the primaries are heavy nuclei. For iron primaries the GZK cutoff would be in that case at 3.4×10^{21} eV.

The chemical composition at high energies is subject of current research, and there is no general agreement about the outcome (see Fig. 6.12).

A somewhat extreme and drastic assumption would be to believe that the very energetic events were due to a violation of Lorentz invariance. If Lorentz transformations would depend not on the relative velocity differences rather on the absolute velocities, this would modify the threshold for the GZK cutoff. A more mundane idea for the cutoff would be to assume that cosmic accelerators just run out of power to produce higher-energy particles in sufficient numbers.

Photons as origin of high-energy showers are even more problematic. Photon–photon interactions with cosmic microwave photons or photons in the infrared or radio domain would prevent them to arrive from larger distances.

Neutrinos as candidates also have problems to explain high-energy events. Their interaction cross section is so small that one would need extreme neutrino fluxes to arrive at a significant rate of events. Also the distribution of vertices of air-shower events in the atmosphere is in conflict with a neutrino hypothesis.

Also WIMPs would only undergo rather weak or even superweak interactions, which would make their origin for the energetic events very unlikely.

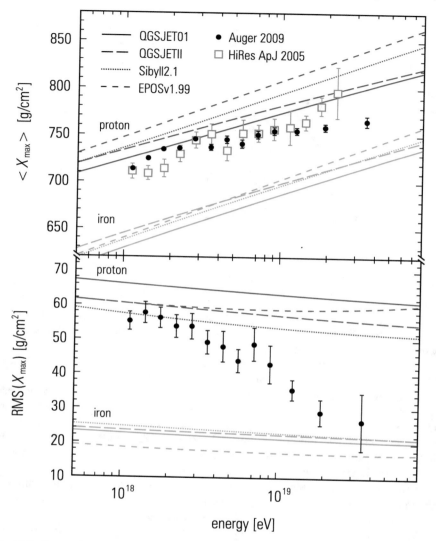

Fig. 6.12 Energy dependence of the position of the shower maximum X_{max} and its width for the Auger and HiRes experiments compared to various results of Monte Carlo simulations based on different hadronization models for protons and iron nuclei [61]

The fact that the arrival directions of the high-energy events do not convincingly cluster at some source candidate could be explained by the assumptions that the galactic or extragalactic fields are stronger than anticipated. There are in fact hints that the magnetic fields might lie in the μgauss rather than in the ngauss range.

Active galactic nuclei are frequently considered as potential candidates for the highest-energy particles. In particular, blazars with their powerful jets or even black holes are potential sources. The discovery of black-hole mergers in 2015 has shown that such cataclysmic events can convert masses effectively into radiation, so, why not also produce high-energy particles in these catastrophic events.

Particle jets from blazars or mergers of black holes are a popular scenario of the possible production of particles of extreme energy (see Fig. 6.13). If such reasoning were correct, these candidate sources should also be a rich source of high-energy neutrinos, and ICECUBE would have a chance to detect them.

Even though the origin of high-energy cosmic rays is an open question there are some tentative indications that at least some of them might come from the super-galactic plane (see also Fig. 7.36). Obviously more events are required to establish such a correlation. Apart from Auger, also ICECUBE has a chance to find possible point sources, in particular, since they have seen some extragalactic neutrinos in the PeV range. It is probably necessary—apart from collecting better statistics—to use larger detector systems or better detection techniques (e.g., JEM-EUSO).

Finally, it should be mentioned that some people try to explain an unknown by another unknown phenomenon. High-energy particles may not be the product of an acceleration. They could also be decay products of unstable, primordial objects. There are plenty of candidates for that:

- heavy SUSY particles from supersymmetric theories
- topological defects
- domain walls
- magnetic monopoles (if they exist)
- cosmic strings

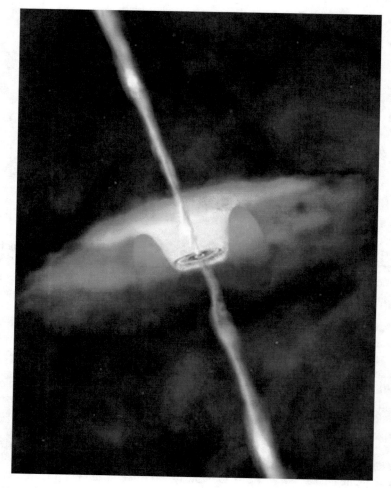

Fig. 6.13 Artist's view of a blazar ejecting jets, in which high-energy particles might be accelerated [62]

- cosmic loops of superconducting current
- necklaces
- massive metastable particles as remnants from the Big Bang or the inflation period
- particles from another universe?

As one can see there are many ideas for powerful hotspots to produce exotic events in the universe or beyond.

6.3 Neutrino Astronomy

Neutrino physics is largely an art of learning a great deal by observing nothing.

Haim Harari

The disadvantage of classical astronomies like observations in the radio, infrared, optical, ultraviolet, X-ray, or γ-ray band is related to the fact that electromagnetic radiation is quickly absorbed in matter. Therefore, with these astronomies one can only observe the surfaces of astronomical objects. In addition, energetic γ rays from distant sources are attenuated via $\gamma\gamma$ interactions with photons of the blackbody radiation by the process

$$\gamma + \gamma \rightarrow e^+ + e^- .$$

Energetic photons ($> 10^{15}$ eV), for example, from the Large Magellanic Cloud (LMC, 52 kpc distance) are significantly attenuated by this process (see Sect. 4.7).

Charged primaries can in principle also be used in astroparticle physics. However, the directional information is only conserved for very energetic protons ($> 10^{19}$ eV) because otherwise the irregular and partly not well-known galactic magnetic fields will randomize their original direction. For these high energies the Greisen–Zatsepin–Kuzmin cutoff also comes into play, whereby protons lose their energy via the photoproduction of pions off blackbody photons. For protons with energies exceeding 6×10^{19} eV the universe is no longer transparent (attenuation length $\lambda \approx 50$ Mpc). As a consequence of these facts, the requirement for an optimal astronomy can be defined in the following way:

1. The optimal astroparticles or radiation should not be influenced by magnetic fields.
2. The particles should not decay from source to Earth. This practically excludes neutrons as carriers unless neutrons have extremely high energy ($\tau_{neutron}^0 = 885.7$ s; at $E = 10^{19}$ eV one has $\gamma c \tau_{neutron}^0 \approx 300\,000$ light-years).
3. Particles and antiparticles should be different. This would in principle allow to find out whether particles originate from a matter or antimatter source. This requirement excludes photons because a photon is its own antiparticle, $\gamma = \bar{\gamma}$.
4. The particles must be penetrating so that one can look into the central part of the sources.
5. Particles should not be absorbed by interstellar or intergalactic dust or by infrared or blackbody photons.

These five requirements are fulfilled by neutrinos in an ideal way! One could ask oneself why *neutrino astronomy* has not been a major branch of astronomy all along. The fact that neutrinos can escape from the center of the sources is related to their low interaction cross section. This, unfortunately, goes along with an enormous difficulty to detect these neutrinos on Earth.

For solar neutrinos in the range of several 100 keV the cross section for neutrino–nucleon scattering is

$$\sigma(\nu_e N) \approx 10^{-45} \, \text{cm}^2/\text{nucleon} . \tag{6.3.1}$$

The interaction probability of these neutrinos with our planet Earth at central incidence is

$$\phi = \sigma \, N_A \, d \, \varrho \approx 4 \times 10^{-12} \tag{6.3.2}$$

(N_A is the Avogadro number, d the diameter of the Earth, ϱ the average density of the Earth). Out of the 7×10^{10} neutrinos per cm^2 and s radiated by the Sun and arriving at Earth only one or two at most are 'seen' by our planet.

As a consequence of this, neutrino telescopes must have an enormous target mass, and one has to envisage long exposure times. However, for high energies the interaction cross section rises with neutrino energy. Neutrinos in the energy range of several 100 keV can be detected by radiochemical methods. For energies exceeding 5 MeV large-volume water Cherenkov counters are an attractive possibility.

Neutrino astronomy is a very young branch of astroparticle physics. Up to now five different sources of neutrinos have been investigated. The physics results and implications of these measurements will be discussed in the following five sections.

6.3.1 Atmospheric Neutrinos

I have done a terrible thing: I have postulated a particle that cannot be detected.

Wolfgang Pauli

For real neutrino astronomy neutrinos from atmospheric sources are an annoying background. For the particle physics aspect of astroparticle physics atmospheric neutrinos have turned out to be a very interesting subject. Primary cosmic rays interact in the atmosphere with the atomic nuclei of nitrogen and oxygen. In these proton–air interactions nuclear fragments and predominantly charged and neutral pions are produced. The decay of charged pions (lifetime 26 ns) produces muon neutrinos:

$$\pi^+ \rightarrow \mu^+ + \nu_\mu , \; \pi^- \rightarrow \mu^- + \bar{\nu}_\mu . \tag{6.3.3}$$

Muons themselves are also unstable and decay with an average lifetime of 2.2 μs according to

$$\mu^+ \rightarrow e^+ + \nu_e + \bar{\nu}_\mu , \; \mu^- \rightarrow e^- + \bar{\nu}_e + \nu_\mu . \tag{6.3.4}$$

Therefore, the atmospheric neutrino beam contains electron and muon neutrinos and one would expect a ratio

$$\frac{N(\nu_\mu, \bar{\nu}_\mu)}{N(\nu_e, \bar{\nu}_e)} \equiv \frac{N_\mu}{N_e} \approx 2 , \tag{6.3.5}$$

as can be easily seen by counting the decay neutrinos in reactions (6.3.3) and (6.3.4).

Fig. 6.14 The Super-Kamiokande detector in the Kamioka mine in Japan. The cylindrical steel tank is about 40 m in height and contains 50 000 tons of ultrapure water [63]

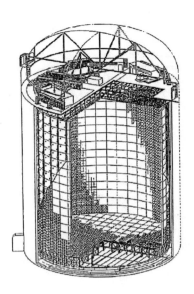

The presently largest experiments measuring atmospheric neutrinos are Super-Kamiokande (see Fig. 6.14) and ICECUBE (see Sect. 6.3.4). Neutrino interactions in the Super-Kamiokande detector are recorded in a tank of approximately 50 000 tons of ultrapure water. Electron neutrinos transfer part of their energy to electrons,

$$\nu_e + e^- \rightarrow \nu_e + e^- , \qquad (6.3.6)$$

or produce electrons in neutrino–nucleon interactions

$$\nu_e + N \rightarrow e^- + N' . \qquad (6.3.7)$$

Muon neutrinos are detected in neutrino–nucleon interactions according to

$$\nu_\mu + N \rightarrow \mu^- + N' . \qquad (6.3.8)$$

Electron antineutrinos and muon antineutrinos produce correspondingly positrons and positive muons. The charged leptons (e^+, e^-, μ^+, μ^-) can be detected via the Cherenkov effect in water. The produced Cherenkov light is measured in Super-Kamiokande with 13 000 photomultipliers of 50 cm cathode diameter. In the GeV range electrons initiate characteristic electromagnetic cascades of short range while muons produce long straight tracks. This presents a basis for distinguishing electron from muon neutrinos. On top of that, muons can be identified by their decay in the detector thereby giving additional evidence concerning the identity of the initiating neutrino species. Figures 6.15 and 6.16 show an electron and muon event in Super-Kamiokande. Muons have a well-defined range and produce a clear Cherenkov

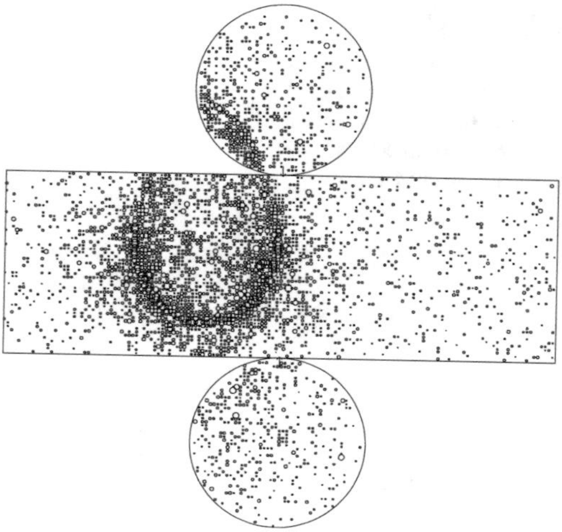

Fig. 6.15 Cherenkov pattern
of an electron in the
Super-Kamiokande detector.
The contour of an electron is
somewhat fuzzy because of
the shower development [63]

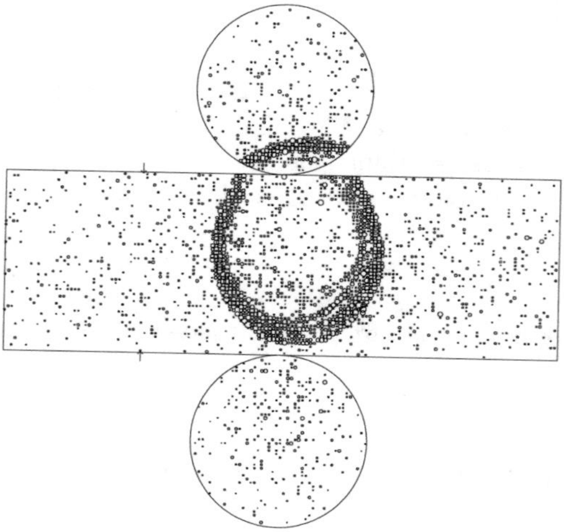

Fig. 6.16 Cherenkov pattern
of a muon in the
Super-Kamiokande detector.
The contour of the muon is
sharply bounded compared
to the electron pattern [63]

pattern with sharp edges while electrons initiate electromagnetic cascades thereby
creating a fuzzy ring pattern.

The result of the Super-Kamiokande experiment is that the number of electron-
neutrino events corresponds to the theoretical expectation while there is a clear deficit
of events initiated by muon neutrinos.

Because of the different acceptance for electrons and muons in the water
Cherenkov detector, the ratio of neutrino-induced muons to electrons is compared
to a Monte Carlo simulation. For the double ratio

$$R = \frac{(N_\mu/N_e)_{\text{data}}}{(N_\mu/N_e)_{\text{Monte Carlo}}} \qquad (6.3.9)$$

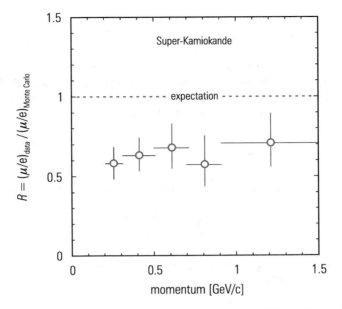

Fig. 6.17 Double ratio of the electron–muon rate comparing data and Monte Carlo [63]

one would expect the value $R = 1$ in agreement with the standard interaction and propagation models. However, the Super-Kamiokande experiment obtains

$$R = 0.69 \pm 0.06, \qquad (6.3.10)$$

which represents a clear deviation from expectation (see Fig. 6.17).

After careful checks of the experimental results and investigations of possible systematic effects the general opinion prevails that the deficit of muon neutrinos can only be explained by *neutrino oscillations*.

Mixed particle states are known from the quark sector (see Sect. 2.2). Similarly, it is conceivable that in the lepton sector the eigenstates of weak interactions ν_e, ν_μ, and ν_τ are superpositions of mass eigenstates ν_1, ν_2, and ν_3. A muon neutrino ν_μ born in a pion decay could be transformed during the propagation from the source to the observation in the detector into a different neutrino flavour. If the muon neutrino in reality was a mixture of two different mass eigenstates ν_1 and ν_2, these two states would propagate at different velocities if their masses were not identical and so the mass components get out of phase with each other. This could possibly result in a different neutrino flavour at the detector. If, however, all neutrinos were massless, they would all propagate precisely at the velocity of light, and the mass eigenstates can never get out of phase with each other.

For an assumed two-neutrino mixing of ν_e and ν_μ the weak eigenstates could be related to the mass eigenstates by the following two equations:

$$\nu_e = \nu_1 \cos \theta + \nu_2 \sin \theta \, ,$$
$$\nu_\mu = -\nu_1 \sin \theta + \nu_2 \cos \theta \, . \tag{6.3.11}$$

The *mixing angle* θ determines the degree of mixing. This assumption requires that the neutrinos have non-zero mass and, in addition, $m_1 \neq m_2$ must hold.

In the framework of this oscillation model the probability that an electron neutrino stays an electron neutrino, can be calculated to be:

$$P_{\nu_e \to \nu_e}(x) = 1 - \sin^2 2\theta \, \sin^2 \left(\pi \, \frac{x}{L_\nu} \right) , \tag{6.3.12}$$

where x is the distance from the source to the detector and L_ν is the oscillation length

$$L_\nu = \frac{2.48 \, E_\nu \text{[MeV]}}{(m_1^2 - m_2^2) \, \text{[eV}^2/c^4\text{]}} \, \text{m} \, . \tag{6.3.13}$$

The expression $m_1^2 - m_2^2$ is usually abbreviated as δm^2. Equations (6.3.12) and (6.3.13) can be combined to give

$$P_{\nu_e \to \nu_e}(x) = 1 - \sin^2 2\theta \, \sin^2 \left(1.27 \, \delta m^2 \frac{x}{E_\nu} \right) , \tag{6.3.14}$$

where δm^2 is measured in eV^2, x in km, and E_ν in GeV. The idea of a two-neutrino mixing is graphically presented in Fig. 6.18.

For the general case of mixing of all three neutrino flavours one obtains as generalization of (6.3.11)

$$\begin{pmatrix} \nu_e \\ \nu_\mu \\ \nu_\tau \end{pmatrix} = U_N \begin{pmatrix} \nu_1 \\ \nu_2 \\ \nu_3 \end{pmatrix} , \tag{6.3.15}$$

where U_N is the (3×3) neutrino mixing matrix.

This matrix is constructed analogously to the Cabibbo–Kobayashi–Maskawa mixing matrix, just as in the quark sector (s. Chap. 2). The idea of neutrino mixing originated from the works of Pontecorvo, Maki, Nakagawa, and Sakata, consequently this matrix is named PMNS matrix [64, 65].

The deficit of muon neutrinos can now be explained by the assumption that some of the muon neutrinos transform themselves during propagation from the point of production to the detector into a different neutrino flavour, e.g., into tau neutrinos. The sketch shown in Fig. 6.18 demonstrated that for an assumed mixing angle of 45° all neutrinos of a certain type have transformed themselves into a different neutrino flavour after propagating half the oscillation length. If, however, muon neutrinos have oscillated into tau neutrinos, a deficit of muon neutrinos will be observed in the detector because tau neutrinos would only produce taus in the water Cherenkov counter, but not muons. Since, however, the mass of the tau is rather high (1.77 GeV/c^2), tau neutrinos normally would not meet the requirement to provide

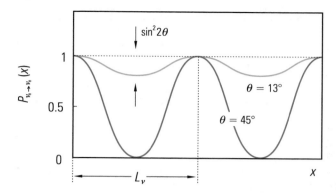

Fig. 6.18 Oscillation model for ($\nu_e \leftrightarrow \nu_\mu$) mixing for two different mixing angles; shown is the probability $P_{\nu_e \to \nu_e}(x)$

the necessary center-of-mass energy for tau production. Consequently, they would escape from the detector without interaction. If the deficit of muon neutrinos would be interpreted by ($\nu_\mu \to \nu_\tau$) oscillations, the mixing angle and the difference of mass squares δm^2 can be determined from the experimental data. The measured value of the double ratio $R = 0.69$ leads to

$$\delta m^2 \approx 2 \times 10^{-3} \, \text{eV}^2 \qquad (6.3.16)$$

at maximal mixing ($\sin^2 2\theta = 1$, corresponding to $\theta = 45°$).[1] If one assumes that in the neutrino sector a similar mass hierarchy as in the sector of charged leptons exists ($m_e \ll m_\mu \ll m_\tau$), then the mass of the heaviest neutrino can be estimated from (6.3.16),

$$m_{\nu_\tau} \approx \sqrt{\delta m^2} \approx 0.045 \, \text{eV} . \qquad (6.3.17)$$

The validity of this conclusion relies on the correctly measured absolute fluxes of electron and muon neutrinos. Because of the different Cherenkov pattern of electrons and muons in the water Cherenkov detector the efficiencies for electron neutrino and muon neutrino detections might be different. To support the oscillation hypothesis one would therefore prefer to have an additional independent experimental result. This is provided in an impressive manner by the ratio of upward- to downward-going muons. Upward-coming atmospheric neutrinos have traversed the whole Earth ($\approx 12\,800$ km). They would have a much larger probability to oscillate into tau neutrinos compared to downward-going neutrinos, which have traveled typically only 20 km. Actually, according to the experimental result of the Super-Kamiokande

---

[1]The 90% confidence limit for δm^2 given by the Super-Kamiokande experiment is $1.3 \times 10^{-3} \, \text{eV}^2 \leq \delta m^2 \leq 3 \times 10^{-3} \, \text{eV}^2$. The accelerator experiment K2K sending muon neutrinos to the Kamioka mine gets $\delta m^2 = 2.8 \times 10^{-3} \, \text{eV}^2$ [66]. K2K—from KEK to Kamioka, long-baseline neutrino oscillation experiment.

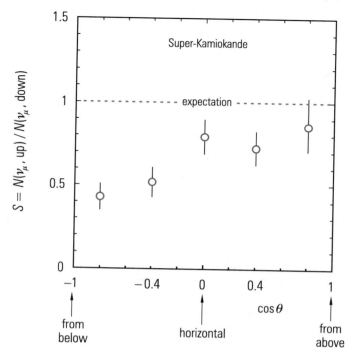

Fig. 6.19 Ratio of measured ν_μ fluxes in Super-Kamiokande as function of the zenith angle [63]

collaboration, the upward-going muon neutrinos, which have traveled through the whole Earth, are suppressed by a factor of two compared to the downward-going muons. This is taken as a strong indication for the existence of oscillations (see Fig. 6.19). For the ratio of upward- to downward-going muon neutrinos one obtains

$$S = \frac{N(\nu_\mu, \text{up})}{N(\nu_\mu, \text{down})} = 0.54 \pm 0.06, \qquad (6.3.18)$$

which presents a clear effect in favour of oscillation.

Details of the observed zenith-angle dependence of atmospheric ν_e and ν_μ fluxes also represent a particularly strong support for the oscillation model.

Since the production altitude L and energy E_ν of atmospheric neutrinos are known (≈ 20 km for vertically downward-going neutrinos), the observed zenith-angle dependence of electron and muon neutrinos can also be converted into a dependence of the rate versus the reconstructed ratio of L/E_ν. Figure 6.20 shows the ratio data/Monte Carlo for fully contained events as measured in the Super-Kamiokande experiment. The data exhibit a zenith-angle- (i.e., distance-) dependent deficit of muon neutrinos, while the electron neutrinos follow the expectation for no oscillations. The observed behaviour is consistent with ($\nu_\mu \leftrightarrow \nu_\tau$) oscillations, where a best fit is obtained for $\delta m^2 = 2.2 \times 10^{-3}$ eV^2 for maximal mixing ($\sin 2\theta = 1$).

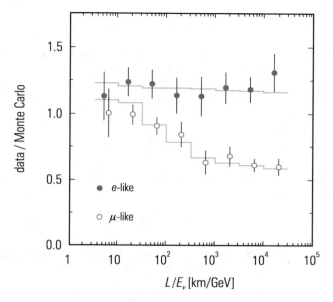

Fig. 6.20 Ratio of the fully contained events in Super-Kamiokande as a function of $\frac{L}{E_\nu}$, where L is the reconstructed production altitude of the neutrinos for electron and muon events. The *lower histogram* for muon-like events corresponds to the expectation for ν_μ oscillations into ν_τ with the parameters $\delta m^2 = 2.2 \times 10^{-3}\,\mathrm{eV}^2$ and $\sin 2\theta = 1$ [63]

If all results of the Super-Kamiokande experiment are put together, one gets under the assumption of $(\nu_\mu \leftrightarrow \nu_\tau)$ oscillations of atmospheric neutrinos the parameters $\delta m^2 = 2.4 \times 10^{-3}\,\mathrm{eV}^2$ and $\sin^2(2\theta) > 0.95$. It is, of course, also conceivable that the muon neutrinos oscillate into electron neutrinos. For this possibility one would get $\delta m^2 = 7.5 \times 10^{-5}\,\mathrm{eV}^2$ resp. $\sin^2(2\theta) = 0.85$. Assuming a mass hierarchy as in the sector of charged leptons and assuming $(\nu_\mu \leftrightarrow \nu_\tau)$ oscillations one would get for the mass of the heaviest neutrino (ν_τ) $m_{\nu_\tau} \approx \sqrt{\delta m^2} \approx 50\,\mathrm{meV}$.

In the Standard Model of elementary particles neutrinos have zero mass. Therefore, neutrino oscillations represent an important extension of the physics of elementary particles. In this example of neutrino oscillations the synthesis between astrophysics and particle physics becomes particularly evident.

6.3.2 Solar Neutrinos

> *Three things cannot be long hidden: the sun, the moon, and the truth.*
>
> Buddha

The Sun is a nuclear fusion reactor. In its interior hydrogen is burned to helium. The longevity of the Sun is related to the fact that the initial reaction

$$p + p \rightarrow d + e^+ + \nu_e \tag{6.3.19}$$

proceeds via the weak interaction. 86% of solar neutrinos are produced in this proton–proton reaction. Deuterium made according to (6.3.19) fuses with a further proton to produce helium 3,

$$d + p \rightarrow {}^3\text{He} + \gamma . \tag{6.3.20}$$

In ${}^3\text{He}$–${}^3\text{He}$ interactions

$$^3\text{He} + {}^3\text{He} \rightarrow {}^4\text{He} + 2p \tag{6.3.21}$$

the isotope helium 4 can be formed. On the other hand, the isotopes ${}^3\text{He}$ and ${}^4\text{He}$ could also produce beryllium,

$$^3\text{He} + {}^4\text{He} \rightarrow {}^7\text{Be} + \gamma . \tag{6.3.22}$$

${}^7\text{Be}$ is made of four protons and three neutrons. Light elements prefer symmetry between the number of protons and neutrons. ${}^7\text{Be}$ can capture an electron yielding ${}^7\text{Li}$,

$$^7\text{Be} + e^- \rightarrow {}^7\text{Li} + \nu_e , \tag{6.3.23}$$

where a proton has been transformed into a neutron. On the other hand, ${}^7\text{Be}$ can react with one of the abundant protons to produce ${}^8\text{B}$,

$$^7\text{Be} + p \rightarrow {}^8\text{B} + \gamma . \tag{6.3.24}$$

${}^7\text{Li}$ produced according to (6.3.23) will usually interact with protons forming helium,

$$^7\text{Li} + p \rightarrow {}^4\text{He} + {}^4\text{He} , \tag{6.3.25}$$

while the boron isotope ${}^8\text{B}$ will reduce its proton excess by β^+ decay,

$$^8\text{B} \rightarrow {}^8\text{Be} + e^+ + \nu_e , \tag{6.3.26}$$

and the resulting ${}^8\text{Be}$ will disintegrate into two helium nuclei. Apart from the dominant pp neutrinos (reaction (6.3.19)), further 14% are generated in the electron-capture reaction (6.3.23), while the ${}^8\text{B}$ decay contributes only at the level of 0.02% albeit yielding high-energy neutrinos. In total, the solar neutrino flux at Earth amounts to about 7×10^{10} particles per cm^2 and second.

The energy spectra of different reactions, which proceed in the solar interior at a temperature of 15 million kelvin, are shown in Fig. 6.21. The Sun is a pure electron-neutrino source. It does not produce electron antineutrinos and, in particular, no other neutrino flavours (ν_μ, ν_τ).

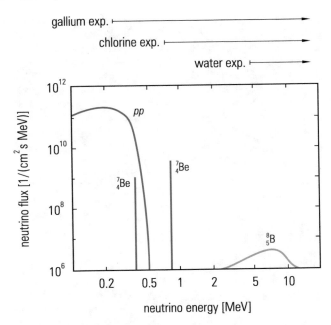

Fig. 6.21 Neutrino spectra from solar fusion processes. The reaction thresholds for the gallium, chlorine, and water Cherenkov experiments are indicated. The threshold for the SNO experiment is around 5 MeV. The line fluxes of the beryllium isotopes are given in $cm^{-2} s^{-1}$

Three radiochemical experiments and two water Cherenkov experiments have been or are trying to measure the flux of solar neutrinos.

The historically first experiment for the search of solar neutrinos is based on the reaction

$$\nu_e + {}^{37}Cl \rightarrow {}^{37}Ar + e^- , \qquad (6.3.27)$$

where the produced ^{37}Ar has to be extracted from a huge tank filled with 380 000 liters of perchloroethylene (C_2Cl_4). Because of the low capture rate of less than one neutrino per day the experiment must be shielded against atmospheric cosmic rays. Therefore, it is operated in a gold mine at about 1500 m depth under the Earth's surface (see Fig. 6.22). After a run of typically one month the tank is flushed with a noble gas and the few produced ^{37}Ar atoms are extracted from the detector and subsequently counted. Counting is done by means of the electron-capture reaction of ^{37}Ar, where again ^{37}Cl is produced. Since the electron capture occurs predominantly from the K shell, the produced ^{37}Cl atom is now missing one electron in the innermost shell (in the K shell). The atomic electrons of the ^{37}Cl atom are rearranged under emission of either characteristic X rays or by the emission of Auger electrons. These Auger electrons and, in particular, the characteristic X rays are the basis for counting ^{37}Ar atoms produced by solar neutrinos.

Fig. 6.22 The detector of the chlorine experiment of R. Davis for the measurement of solar neutrinos. The detector is installed at a depth of 1480 m in the Homestake Mine in South Dakota. It is filled with 380 000 liters of perchloroethylene [67]. With kind permission of the Brookhaven National Laboratory

In the course of 30 years of operation a deficit of solar neutrinos has become more and more evident. The experiment led by Davis only finds 27% of the expected solar neutrino flux. To solve this neutrino puzzle, two further neutrino experiments were started. The gallium experiment GALLEX in a tunnel through the Gran Sasso mountains in Italy and the Soviet–American gallium experiment (SAGE) in the Caucasus measure the flux of solar neutrinos also in radiochemical experiments. Solar neutrinos react with gallium according to

$$\nu_e + {}^{71}\text{Ga} \rightarrow {}^{71}\text{Ge} + e^- . \tag{6.3.28}$$

In this reaction ${}^{71}\text{Ge}$ is produced and extracted like in the Davis experiment and counted. The gallium experiments have the big advantage that the reaction threshold for the reaction (6.3.28) is as low as 233 keV so that these experiments are sensitive to neutrinos from the proton–proton fusion while the Davis experiment with a threshold of 810 keV essentially only measures neutrinos from the ^{8}B decay. GALLEX and SAGE have also measured a deficit of solar neutrinos. They only find 52% of the expected rate, which presents a clear discrepancy to the prediction on the basis of the standard solar model. However, the discrepancy is not so pronounced as in the Davis experiment. A strong point for the gallium experiments is that the neutrino capture rate and the extraction technique have been checked with neutrinos of an artificial ^{51}Cr source. It could be convincingly shown that the produced ^{71}Ge atoms could be successfully extracted in the expected quantities.

The Kamiokande and Super-Kamiokande experiment, respectively, measure solar neutrinos via the reaction

$$\nu_e + e^- \rightarrow \nu_e + e^- \tag{6.3.29}$$

at a threshold of 5 MeV in a water Cherenkov counter. Since the emission of the knock-on electron follows essentially the direction of the incident neutrinos, the detector can really 'see' the Sun. This directionality gives the water Cherenkov counter a superiority over the radiochemical experiments. Figure 6.23 shows the neutrino counting rate of the Super-Kamiokande experiment as a function of the angle with respect to the Sun. The Super-Kamiokande experiment also measures a low flux of solar neutrinos representing only 40% of the expectation.

A reconstructed image of the Sun in the light of neutrinos is shown in Fig. 6.24.

Many proposals have been made to solve the solar neutrino problem. The obvious idea for elementary particle physicists was to doubt the correctness of the standard solar model. The flux of ^{8}B neutrinos varies with the central temperature of the Sun like $\sim T^{18}$. A reduction by only 5% of the central solar temperature would bring the Kamiokande experiment already in agreement with the now reduced expectation. However, solar astrophysicists consider even a somewhat lower central temperature of the Sun rather unlikely.

The theoretical calculation of the solar neutrino flux uses the cross sections for the reactions (6.3.19)–(6.3.26). An overestimate of the reaction cross sections would also lead to a too high expectation for the neutrino flux. A variation of these cross sections in a range, which is considered realistic by nuclear physicists, was insufficient to explain the discrepancy between the experimental data and expectation.

There have been further ideas proposed to solve the solar neutrino problem. If neutrinos had a mass, they could also possess a magnetic moment. If their spin is rotated while propagating from the solar interior to the detector at Earth, one would not be able to measure these neutrinos because the detectors are insensitive to neutrinos of wrong helicity.

Finally, solar neutrinos could decay on their way from Sun to Earth into particles, which might be invisible to the neutrino detectors.

A drastic assumption would be that the solar fire has gone out. In the light of neutrinos this would become practically immediately evident (more precisely: in 8 min). The energy transport from the solar interior to the surface, however, requires a time of several 100 000 years so that the Sun would continue to shine for this period even though the nuclear fusion at its center has come to an end.

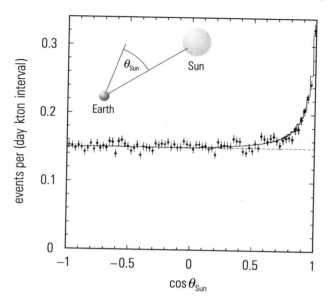

Fig. 6.23 Arrival directions of neutrinos measured in the Super-Kamiokande experiment [63]

Fig. 6.24 Reconstructed image of the Sun in the light of solar neutrinos. Due to the limited spatial and angular resolution of the Super-Kamiokande experiment and the small scattering angle of the electron with respect to the incoming neutrino the image of the Sun appears larger as it actually is. About 50 days of exposure were needed to detect these solar neutrinos in the 50 000-ton water detector. Photo credit: Kamioka Observatory, ICRR (Institute for Cosmic Ray Research), The University of Tokyo [63]

Fig. 6.25 The large SNO detector in a nickel mine in Ontario, Canada [68]

The Sudbury Neutrino Observatory (SNO) has finally demonstrated that the solar model for neutrino generation is correct. The SNO detector is installed in a nickel mine in Ontario in Canada at a depth of 2000 m. It consists of a 1000-ton heavy-water target (D_2O), which is mounted in an acrylic vessel of 12 m diameter (see Fig. 6.25).

The interaction target is viewed by 9600 photomultipliers. The detector cavity outside the vessel contains 7000 tons of normal, light water. The purpose of this shield is to reduce the radiation background from cosmic rays and the environmental radiation from the rock and the dust in the mine. The detection threshold is with 5 MeV in this experiment rather high. To break up the deuterons in the heavy water at least the binding energy of 2.2 MeV must be provided.

The big advantage of the SNO experiment is that it can tell the difference between charged and neutral currents. The reaction

(a) $\nu_e + d \rightarrow p + p + e^-$

can only proceed via charged currents with electron neutrinos, while neutral currents, such as

(b) $\nu_x + d \rightarrow p + n + \nu'_x$ $(x = e, \mu, \tau)$,

are possible for all neutrino flavours. The neutrons produced in this reaction are captured by deuterons giving rise to the emission of 6.25-MeV photons, which signal the NC interaction. While the ν_e flux as obtained by the CC reaction is only 1/3 of the predicted solar neutrino flux, the total neutrino flux measured by the NC reaction is in agreement with the expectation of solar models, thereby providing evidence for a non-ν_e component.

This result solves the long-standing neutrino problem. It does not, however, resolve the underlying mechanism of the oscillation process. It is not at all clear, whether the ν_e oscillate into ν_μ or ν_τ. It is considered very likely that matter oscillations via the MSW effect in the Sun play a role for the transmutation of the solar electron neutrinos into other neutrino flavours, which unfortunately cannot directly be measured in a light-water Cherenkov counter.

The oscillation mechanism suggested by the different solar experiments ($\nu_e \rightarrow \nu_\mu$) was confirmed at the end of 2002 by the KamLAND[2] reactor-neutrino detector, which removed all doubts about possible uncertainties of the standard solar model predictions.

In addition to the vacuum oscillations described by (6.3.15), solar neutrinos can also be transformed by so-called matter oscillations. The flux of electron neutrinos and its oscillation property can be modified by neutrino–electron scattering when the solar neutrino flux from the interior of the Sun encounters collisions with the abundant number of solar electrons. This matter effect is particularly relevant for high-energy solar electron neutrinos. Flavour oscillations can even be magnified in a resonance-like fashion by matter effects so that certain energy ranges of the solar ν_e spectrum are depleted. The possibility of matter oscillations has first been proposed by Mikheyev, Smirnov, and Wolfenstein. The oscillation property of the MSW effect is different from that of vacuum oscillations. It relates to the fact that $\nu_e e^-$ scattering contributes a term to the mixing matrix that is not present in vacuum. Due to this charged-current interaction (Fig. 6.26), which is not possible for ν_μ and ν_τ in the Sun, the interaction Hamiltonian for ν_e is modified compared to the other neutrino flavours. This leads to alterations for the energy difference of the neutrino eigenvalues in matter compared to vacuum. Therefore, electron neutrinos are singled out by this additional interaction process in matter.

---

[2] KamLAND—Kamioka Liquid-scintillator Anti-Neutrino Detector.

Fig. 6.26 Feynman diagram for matter oscillations via the MSW effect. Given the energy of solar neutrinos and the fact that there are only target electrons in the Sun, this process can only occur for ν_e, but not for ν_μ and ν_τ

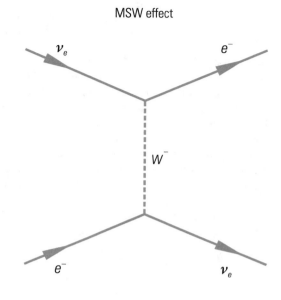

"Illustration of neutrino oscillations"

Depending on the electron density in the Sun the originally dominant mass eigenstate of ν_e can propagate into a different mass eigenstate, for which the neutrino detectors are not sensitive. One might wonder, how such matter oscillations work in the Sun. The probability for a neutrino to interact in matter is extremely small. The way the solar electron density affects the propagation of solar neutrinos, however, depends on amplitudes, which are square roots of probabilities. Therefore, even though the probabilities of interactions are small, the neutrino flavours can be significantly altered because of the amplitude dependence of the oscillation mechanism.

If the three neutrino flavours ν_e, ν_μ, ν_τ would completely mix, only $1/3$ of the original electron neutrinos would arrive at Earth. Since the neutrino detectors, however, are blind for MeV neutrinos of ν_μ and ν_τ type, the experimental results could be understood in a framework of oscillations. Obviously, the solar neutrino problem cannot be solved that easily. The results of the four so far described experiments that measure solar neutrinos do not permit a unique solution in the parameter space $\sin^2 2\theta$ and δm^2, compare (6.3.12) and (6.3.13). If $(\nu_e \rightarrow \nu_\mu)$ or $(\nu_e \rightarrow \nu_\tau)$ oscillations are assumed and if it is considered that the MSW effect is responsible for the oscillations, a δm^2 on the order of 4×10^{-4}–$2 \times 10^{-5}\,\mathrm{eV}^2$ and a large-mixing-angle solution, although disfavouring maximal mixing, is presently favoured. Assuming a mass hierarchy also in the neutrino sector, this would lead to a ν_μ or ν_τ mass of 0.02–0.004 eV. This is not necessarily in contradiction to the results from atmospheric neutrinos since solar neutrinos could oscillate into muon neutrinos and atmospheric muon neutrinos into tau neutrinos (or into so far undiscovered sterile neutrinos, which are not even subject to weak interactions).

For the numerous low-energy solar neutrinos, as they are measured in the Homestake and the gallium experiments, one can neglect the MSW effect, and one can apply the formalism of vacuum oscillations. This is related to the fact that the solar core, where the hydrogen fusion proceeds, is much larger than the oscillation length. Therefore, one has to average over the oscillation factor, which leads to the standard behaviour of vacuum oscillations.

With these oscillation scenarios one can explain consistently the different experiments (Homestake, Gallium, Borexino, and SNO experiment), which have provided slightly different results on the solar neutrino fluxes [69].

Under the assumption that solar electron neutrinos oscillate into ν_μ one would get for the oscillation parameters $\sin^2 2\theta \approx 0.09$ and $\delta m^2 = 2.4 \times 10^{-3}\,\mathrm{eV}^2$. Of course, one has to aim at determining all parameters of the neutrino mixing matrix. Presently one favours that electron neutrinos oscillate into muon neutrinos and muon neutrinos convert into tau neutrinos. This is also supported by the size of the mixing angles $(\nu_e \rightarrow \nu_\mu$: $\sin^2 2\theta \approx 0.85$; resp. $\nu_\mu \rightarrow \nu_\tau$: $\sin^2 2\theta > 0.95)$.

The oscillations scenario, which is favoured for the solar neutrinos $(\nu_e \rightarrow \nu_\mu)$, was also supported by the KamLAND experiment (Kamioka Liquid-scintillator Anti-Neutrino Detector) in 2002 using reactor neutrinos. This finally eliminated all doubts about the oscillation hypothesis. Therefore, it was only natural to award the Nobel Prize for the discovery of neutrino mass to Kajita (Kamiokande and Super-Kamiokande) and McDonald (SNO) in 2015.

Still, details of neutrino oscillations and the determination of the various mixing parameters in the Pontecorvo–Maki–Nakagawa–Sakata matrix is a matter of current research. The possible effect of Majorana-type neutrinos and hypothetical sterile neutrinos complicates the phenomenon of neutrino oscillations considerably [70].

A follow-up project on Super-Kamiokande will be the very large water Cherenkov detector Hyper-Kamiokande to be installed in Kamioka at 650 m underground. Hyper-Kamiokande will use 260 000 tons of pure water viewed at by 40 000 photomultipliers to investigate neutrino oscillations and search for *CP* violation in the neutrino sector. Understanding the neutrino is not only important for particle physics but it is also connected to deep questions on the origin of matter and the understanding of cosmology.

"Animalistic oscillations"

The recent Borexino experiment aims at a very special goal by measuring low-energy solar neutrinos. The experiment is installed in the Gran Sasso laboratory in Italy. Borexino is a liquid scintillator detector with a sensitive volume of $315\,\mathrm{m}^3$. The main goal is to measure the low-energy neutrinos from the $^7\mathrm{Be}$ capture. The energy threshold for this process is rather low ($250\,\mathrm{keV}$), which requires to carefully shield the experiment against all kinds of cosmic and local environmental background. Apart from that this detector is also sensitive to geoneutrinos, e.g., from the uranium–thorium decay chain.

6.3.3 Supernova Neutrinos

> *All humans are brothers. We came from the same supernova.*
> Allan Sandage

The brightest supernova since the observation of Kepler in the year 1604 was discovered by Ian Shelton at the Las Campanas observatory in Chile on February 23, 1987 (see Fig. 6.27). The region of the sky in the Tarantula Nebula in the Large Magellanic Cloud (distance 170 000 light-years), in which the supernova exploded, was routinely photographed by Robert McNaught in Australia already 20 hours earlier. However, McNaught developed and analyzed the photographic plate only the following day. Ian Shelton was struck by the brightness of the supernova that was visible to the naked eye. For the first time a progenitor star of the supernova explosion could be located. Using earlier exposures of the Tarantula Nebula, a bright blue supergiant,

Fig. 6.27 Supernova in the tarantula nebula; ©: Australian Astronomical Observatory, Photo by David Malin, based on CCD exposures of the Anglo Australian Telescope [71]

Sanduleak, was found to have exploded. Sanduleak was an inconspicuous star of 10-fold solar mass with a surface temperature of 15 000 K. During the hydrogen burning period Sanduleak increased its brightness reaching a luminosity 70 000 higher than the solar luminosity. After the hydrogen supply was exhausted, the star expanded to become a red supergiant. In this process its central temperature and pressure rose to such values that He burning became possible. In a relatively short time (600 000 years) the helium supply was also exhausted. Helium burning was followed by a gravitational contraction, in which the nucleus of the star reached a temperature of 740 million kelvin and a central density of 240 kg/cm^3. These conditions enabled carbon to ignite. In a similar fashion contraction and fusion phases occurred leading via oxygen, neon, silicon, and sulphur finally to iron, the element with the highest binding energy per nucleon.

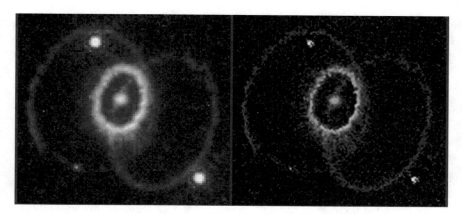

Fig. 6.28 Image of the SN-1987A explosion. The *left image* was taken by the Wide Field Planetary Camera 2 (WFPC2) of the Hubble telescope. The *right-hand image* was sharped by a deconvolution of the camera imperfections [72]

The pace of these successive contraction and fusion phases got faster and faster until finally iron was reached. Once the star has reached such a state, there is no way to gain further energy by fusion processes. Therefore, the stability of Sanduleak could no longer be maintained. Finally, in the last period of its life ($\approx 100\,000$ years) Sanduleak shrank to become a blue supergiant and the star collapsed under its own gravity.

Figure 6.28 shows the result of the supernova explosion some time later. As a consequence of the various burning and collapse phases a number of shock waves were generated, which emitted spherical matter jets that formed different rings. The detailed course of events of the supernova explosion and the formation of ring-like structures was tried to reconstruct with Monte Carlo simulations, without so far leading to conclusive results.

During this process the electrons of the star were forced into the protons and a neutron star of approximately 20 km diameter was produced. In the course of this deleptonization a *neutrino burst* of immense intensity was created,

$$e^- + p \rightarrow n + \nu_e. \tag{6.3.30}$$

In the hot phase of the collapse corresponding to a temperature of 10 MeV ($\approx 10^{11}$ K), the thermal photons produced electron–positron pairs that, however, were immediately absorbed because of the high density of the surrounding matter. Only the weak-interaction process via a virtual Z,

$$e^+ + e^- \rightarrow Z \rightarrow \nu_\alpha + \bar{\nu}_\alpha, \tag{6.3.31}$$

allowed energy to escape from the hot stellar nucleus in the form of neutrinos. In this reaction all three neutrino flavours ν_e, ν_μ, and ν_τ were produced 'democratically' in equal numbers. The total neutrino burst comprised 10^{58} neutrinos and even at Earth the neutrino flux from the supernova was comparable to that of solar neutrinos for a short period.

Actually, the neutrino burst of the supernova was the first signal to be registered on Earth. The large water Cherenkov counters of Kamiokande and IMB (Irvine–Michigan–Brookhaven) and the Baksan experiment recorded a total of 25 out of the emitted 10^{58} neutrinos (see Fig. 6.29). The energy threshold of the Kamiokande experiment was as low as 5 MeV. In contrast, the IMB collaboration could only measure neutrinos with energies exceeding 19 MeV. The Baksan liquid scintillator was lucky to record—even though their fiducial mass was only 200 t—five coincident events with energies between 10 and 25 MeV.

Since the neutrino energies in the range of 10 MeV are insufficient to produce muons or taus, only electron-type neutrinos were recorded via the reactions

$$
\begin{aligned}
\bar{\nu}_e + p &\rightarrow e^+ + n, \\
\bar{\nu}_e + e^- &\rightarrow \bar{\nu}_e + e^-, \\
\nu_e + e^- &\rightarrow \nu_e + e^-.
\end{aligned} \tag{6.3.32}
$$

In spite of the low number of measured neutrinos on Earth some interesting astrophysical conclusions can be drawn from this supernova explosion. If E_ν^i is the energy of individual neutrinos measured in the detector, ε_1 the probability for the interaction of a neutrino in the detector, and ε_2 the probability to also see this reaction, then the total energy emitted in form of neutrinos can be estimated to be

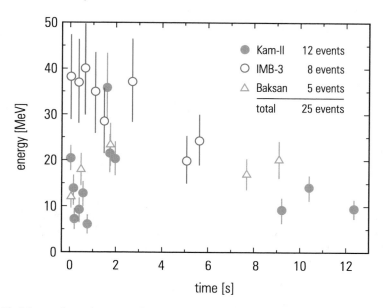

Fig. 6.29 Measured neutrino events from the supernova SN 1987A [73]

$$E_{\text{total}} = \sum_{i=1}^{20} \frac{E_\nu^i}{\varepsilon_1(E_\nu^i)\,\varepsilon_2(E_\nu^i)}\, 4\pi r^2 \, f(\nu_\alpha, \bar{\nu}_\alpha)\,, \qquad (6.3.33)$$

where the correction factor f takes into account that the water Cherenkov counters are not sensitive to all neutrino flavours. Based on the 25 recorded neutrino events in Super-Kamiokande, the IMB experiment, and at Baksan a total energy of

$$E_{\text{total}} = (6 \pm 2) \times 10^{46} \text{ J} \qquad (6.3.34)$$

is obtained. It is hard to comprehend this enormous energy. (The world energy consumption is 10^{21} joule per year.) During the 10 s lasting neutrino burst Sanduleak radiated more energy than the rest of the universe and hundred times more than the Sun in its total lifetime of about 10 billion years.

Measurements over the last 40 years have ever tightened the limits for neutrino masses. At the time of the supernova explosion the mass limit for the electron neutrino from measurements of the tritium beta decay (^{3}H $\rightarrow$ ^{3}He $+ e^- + \bar{\nu}_e$) was about 10 eV. Under the assumption that all supernova neutrinos are emitted practically at the same time, one would expect that their arrival times at Earth would be subject to a certain spread if the neutrinos had mass. Neutrinos of non-zero mass have different velocities depending on their energy. The expected difference of arrival times Δt of two neutrinos with velocities v_1 and v_2 emitted at the same time from the supernova is

$$\Delta t = \frac{r}{v_1} - \frac{r}{v_2} = \frac{r}{c}\left(\frac{1}{\beta_1} - \frac{1}{\beta_2}\right) = \frac{r}{c}\frac{\beta_2 - \beta_1}{\beta_1 \beta_2}. \tag{6.3.35}$$

If the recorded electron neutrinos had a rest mass m_0, their energy would be

$$E = mc^2 = \gamma m_0 c^2 = \frac{m_0 c^2}{\sqrt{1-\beta^2}}, \tag{6.3.36}$$

and their velocity

$$\beta = \left(1 - \frac{m_0^2 c^4}{E^2}\right)^{1/2} \approx 1 - \frac{1}{2}\frac{m_0^2 c^4}{E^2}, \tag{6.3.37}$$

since one can safely assume that $m_0 c^2 \ll E$. This means that the neutrino velocities are very close to the velocity of light. Obviously, the arrival-time difference Δt depends on the velocity difference of the neutrinos. Using (6.3.35) and (6.3.37) one gets

$$\Delta t \approx \frac{r}{c}\frac{\frac{1}{2}\frac{m_0^2 c^4}{E_1^2} - \frac{1}{2}\frac{m_0^2 c^4}{E_2^2}}{\beta_1 \beta_2} \approx \frac{1}{2}m_0^2 c^4 \frac{r}{c}\frac{E_2^2 - E_1^2}{E_1^2 E_2^2}. \tag{6.3.38}$$

The experimentally measured arrival-time differences and individual neutrino energies allow in principle to work out the electron neutrino rest mass

$$m_0 = \left\{\frac{2\Delta t}{r\,c^3}\frac{E_1^2 E_2^2}{E_2^2 - E_1^2}\right\}^{1/2}. \tag{6.3.39}$$

Since, however, not all neutrinos are really emitted simultaneously, (6.3.39) only allows to derive an upper limit for the neutrino mass using pairs of particles of known energy and known arrival-time difference. Using the results of the Kamiokande and IMB experiments a mass limit of the electron neutrino of

$$m_{\nu_e} \leq 10\,\text{eV} \tag{6.3.40}$$

could be established. This result was obtained in a measurement time of approximately 10 s. It demonstrates the potential superiority of astrophysical investigations over laboratory experiments.

Similarly, a possible explanation for the deficit of solar neutrinos by assuming neutrino decay was falsified by the mere observation of electron neutrinos from a distance of 170 000 light-years. For an assumed neutrino mass of $m_0 = 10\,\text{meV}$ the Lorentz factor of 10-MeV neutrinos would be

$$\gamma = \frac{E}{m_0 c^2} \approx 10^9. \tag{6.3.41}$$

This would allow to derive a lower limit for the neutrino lifetime from $\tau_\nu^0 = \tau_\nu/\gamma$ to

$$\tau_\nu^0 = 170\,000\,\text{a}\,\frac{1}{\gamma} \approx 5000\,\text{s}\,. \tag{6.3.42}$$

The supernova 1987A has turned out to be a rich astrophysical laboratory. It has shown that the available supernova models can describe the spectacular death of massive stars on the whole correctly. Given the agreement of the measured neutrinos fluxes with expectation, the supernova neutrinos do not seem to require oscillations. On the other hand, the precision of simulations and the statistical errors of measurements are insufficient to draw a firm conclusion about such a subtle effect for supernova neutrinos. The probability that such a spectacle of a similarly bright supernova in our immediate vicinity will happen again in the near future to clarify whether supernova neutrinos oscillate or not is extremely small. However, if it would happen in our Milky Way, one might expect to record tens of thousands neutrinos with the present and future much larger detectors. This might then give a further input to the effect of long-distance neutrino oscillations. It is not a surprise that the decision on the oscillation scenario has come from observations of solar and atmospheric neutrinos and accelerator experiments with well-defined, flavour-selected neutrino beams. With the experimental evidence of cosmic-ray-neutrino experiments (Davis, GALLEX, SAGE, Super-Kamiokande, SNO) and accelerator and reactor experiments (K2K, KamLAND) there is now unanimous agreement that oscillations in the neutrino sector are an established fact.

6.3.4 High-Energy Galactic and Extragalactic Neutrinos

Discover the force of the heavens O Men: Once recognised it can be put to use: No use could be seen in unknown things.
Johannes Kepler

The measurement of high-energy neutrinos ($\geq$TeV range) represents a big experimental challenge. The arrival direction of such neutrinos, however, would directly point back to the sources of cosmic rays. Therefore, a substantial amount of work is devoted to studies for neutrino detectors in the TeV range and the development of experimental setups for the measurement of galactic and extragalactic high-energy neutrinos. The reason to restrict oneself to high-energy neutrinos is obvious from the inspection of Fig. 6.30. The neutrino echo of the Big Bang has produced energies below the meV range. About a second after the Big Bang weak interactions have transformed protons into neutrons and neutrons into protons thereby producing neutrinos ($p + e^- \rightarrow n + \nu_e$, $n \rightarrow p + e^- + \bar{\nu}_e$). The temperature of these primordial neutrinos should be at 1.9 K at present time. The detection of these blackbody neutrinos is a real challenge, and nobody presently has an idea how to measure them. However, the measurement of neutrinos of higher energy is by now standard. The observation of solar ($\approx$MeV range) and supernova neutrinos ($\approx$10 MeV) is experimentally established. Atmospheric neutrinos represent a background for neutrinos

Fig. 6.30 Comparison of
different neutrino fluxes in
different energy domains
[74]

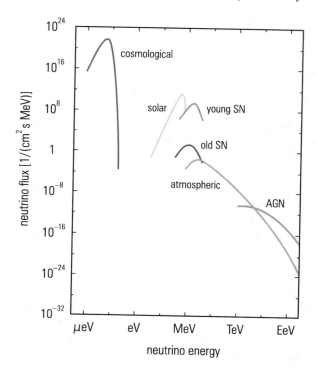

from astrophysical sources. Atmospheric neutrinos originate essentially from pion and muon decays. Their production spectra can be inferred from the measured atmospheric muon spectra. However, they have also been measured directly, but their intensity is only known to an accuracy of about 30%. The different sources of cosmic neutrinos are sketched in Fig. 6.30, where the flux of very-high-energy neutrinos is only a rough estimate.

In this chapter we will deal with these very-high-energy neutrinos. Pioneer work was started with large-volume water and ice Cherenkov detectors. The Baikal Deep Underwater Neutrino Telescope below the surface of Lake Baikal started early (2003). First attempts to measure high-energy neutrinos in the ocean with DUMAND (Deep Underwater Muon And Neutrino Detector) failed because of difficulties to deploy long strings of photomultipliers in the Pacific Ocean near Hawaii. As a consequence of this, the Hawaii team moved to the Antarctic and installed AMANDA (Antarctic Muon And Neutrino Detector Array) in the antarctic ice. AMANDA was quite successful, but is was too small to collect the rare interactions of very-high-energy cosmic neutrinos. Consequently it was enlarged to a detection volume of 1 km^3 (ICECUBE). There are even plans to upgrade ICECUBE by a factor of 10 in volume. In the Mediterranean smaller detectors are installed (NESTOR,[3] ANTARES,[4]

---

[3]NESTOR—Neutrino Extended Submarine Telescope with Oceanographic Research Project.

[4]ANTARES—Astronomy with a Neutrino Telescope and Abyss environmental RESearch.

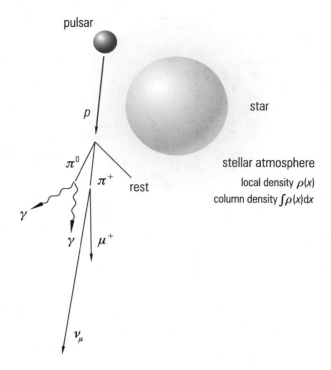

Fig. 6.31 Production mechanism of high-energy neutrinos in a binary

NEMO[5]) and a common large neutrino detector (KM3NeT) with a volume of 1 km^3 is being prepared.

It is generally assumed that binaries are good candidates for the production of energetic neutrinos. A binary consisting of a pulsar and a normal star could represent a strong neutrino source (Fig. 6.31).

The pulsar and the star rotate around their common center of mass. If the stellar mass is large compared to the pulsar mass, one can assume for illustration purposes of the neutrino production mechanism that the pulsar orbits the companion star on a circle. There are models, which suggest that the pulsar can manage to accelerate protons to very high energies. These accelerated protons collide with the gas of the atmosphere of the companion star and produce predominantly secondary pions in the interactions. The neutral pions decay relatively fast ($\tau_{\pi^0} = 8.4 \times 10^{-17}$ s) into two energetic γ rays, which would allow to locate the astronomical object in the light of γ rays. The charged pions produce energetic neutrinos by their ($\pi \to \mu\nu$) decay. Whether such a source radiates high-energy γ quanta or neutrinos depends crucially on subtle parameters of the stellar atmosphere. If pions are produced in a proton interaction such as

---

[5]NEMO—Neutrino Ettore Majorana Observatory.

Fig. 6.32 Competition between production and absorption of photons and neutrinos in a binary system [75]

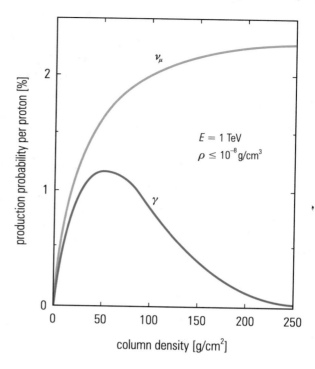

$$p + \text{nucleus} \rightarrow \pi^+ + \pi^- + \pi^0 + \text{anything} \,, \qquad (6.3.43)$$

equal amounts of neutrinos and photons would be produced by the decays of charged and neutral pions ($\pi^+ \rightarrow \mu^+ + \nu_\mu, \pi^- \rightarrow \mu^- + \bar{\nu}_\mu, \pi^0 \rightarrow \gamma + \gamma$). With increasing column density of the stellar atmosphere, however, photons would be re-absorbed, and for densities of stellar atmospheres of $\varrho \leq 10^{-8}$ g/cm^3 and column densities of more than 250 g/cm^2 this source would only be visible in the light of neutrinos (Fig. 6.32).

The source would shine predominantly in muon neutrinos (ν_μ or $\bar{\nu}_\mu$). These neutrinos can be recorded in a detector via the weak charged current, in which they produce muons (Fig. 6.33).

Muons created in these interactions follow essentially the direction of the incident neutrinos. The energy of the muon is measured by its energy loss in the detector. For energies exceeding the TeV range, muon bremsstrahlung and direct electron pair production by muons dominate. The energy loss by these two processes is proportional to the muon energy and therefore allows a calorimetric determination of the muon energy (compare Sect. 7.3, Fig. 7.17).

Because of the low interaction probability of neutrinos and the small neutrino fluxes, neutrino detectors must be very large and massive. Since the whole detector volume has to be instrumented to be able to record the interactions of neutrinos and the energy loss of muons, it is necessary to construct a simple, cost-effective detector.

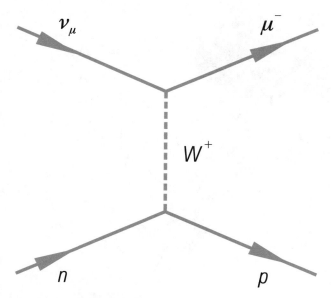

Fig. 6.33 Reaction for muon-neutrino detection

The only practicable candidates, which meet this condition, are huge water or ice Cherenkov counters. Because of the extremely high transparency of ice at large depths in Antarctica and the relatively simple instrumentation of the ice, ice Cherenkov counters are presently the most favourable choice for a realistic neutrino telescope. To protect the detector against the relatively high flux of atmospheric particles, it has become common practice to use the Earth as an absorber and concentrate on neutrinos, which enter the detector 'from below'. The principle of such a setup is sketched in Fig. 6.34.

Protons from cosmic-ray sources produce pions on a target (e.g., stellar atmosphere, galactic medium), which provide neutrinos and γ quanta in their decay. Photons are frequently absorbed in the galactic medium or disappear in $\gamma\gamma$ interactions with blackbody photons, infrared radiation, or starlight photons. The remaining neutrinos traverse the Earth and are detected in an underground detector. The neutrino detector itself consists of a large array of photomultipliers that record the Cherenkov light of muons produced in ice (or in water). In such neutrino detectors the photomultipliers have to be mounted in a suitable distance on strings and many of such strings will be deployed in ice (or water). The mutual distance of the photomultipliers on the strings and the string spacing depends on the absorption and scattering length of Cherenkov light in the detector medium. The installation of photomultiplier strings at a depth of 1000 m in AMANDA had shown that the ice was not free from bubbles. Only at depth of more than 1500 m the pressure (≥ 150 bar) is sufficient to make the bubbles disappear, thereby providing excellent transparency with absorption lengths of 300 m.

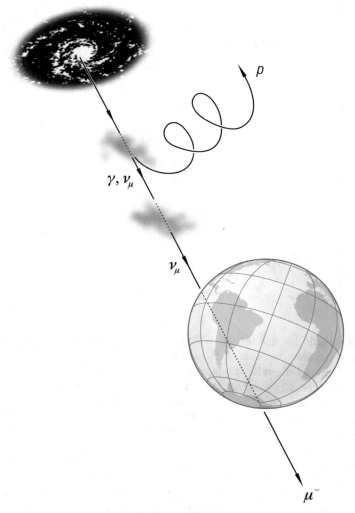

Fig. 6.34 Neutrino production, propagation in intergalactic space, and detection at Earth

The direction of incidence of neutrinos can be inferred from the arrival times of the Cherenkov light at the photomultipliers. In a water Cherenkov counter in the ocean bioluminescence and potassium-40 activity presents an annoying background, which is not present in ice. In practical applications it became obvious that the installation of photomultiplier strings in the antarctic ice is much less problematic compared to the deployment in the ocean.

Figure 6.35 shows the ICECUBE detector at the South Pole.

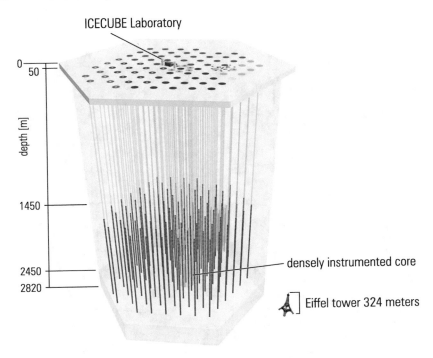

ICECUBE Laboratory

densely instrumented core

Eiffel tower 324 meters

Fig. 6.35 Setup of the ICECUBE experiments at the South Pole [76]

Presently, the ICECUBE detector is the largest neutrino telescope. It has an instrumented volume of one cubic kilometer. It extends to a depth of 2820 m under the antarctic ice shield. The experiment is complemented by a surface detector IceTop, a radio array, and a denser instrumented DeepCore. ICECUBE has 86 strings with approximately 5500 digital optical modules. Figure 6.36 shows an energetic muon in ICECUBE.

To measure the low fluxes of extragalactic neutrinos an instrumented volume of at least 1 km^3 is needed. A short estimation is in order to show this.

It is considered realistic that a point source in our galaxy produces a neutrino spectrum according to

$$\frac{\mathrm{d}N}{\mathrm{d}E_\nu} = 2 \times 10^{-11} \frac{100}{E_\nu^2 \, [\mathrm{TeV}^2]} \, \mathrm{cm}^{-2} \, \mathrm{s}^{-1} \, \mathrm{TeV}^{-1} . \tag{6.3.44}$$

This leads to an integral flux of neutrinos of

$$\Phi_\nu(E_\nu > 100 \, \mathrm{TeV}) = 2 \times 10^{-11} \, \mathrm{cm}^{-2} \, \mathrm{s}^{-1} \tag{6.3.45}$$

(see also Fig. 6.30 for extragalactic sources).

The interaction cross section of high-energy neutrinos was measured at accelerators to be

$$\sigma(\nu_\mu N) = 6.7 \times 10^{-39} \, E_\nu \, [\text{GeV}] \, \text{cm}^2/\text{nucleon} \, . \qquad (6.3.46)$$

If this linear dependence is valid up to high energies, one would arrive at a cross section of $6.7 \times 10^{-34} \, \text{cm}^2/\text{nucleon}$ for 100-TeV neutrinos. For a target thickness of one kilometer an interaction probability W per neutrino of

$$W = N_A \, \sigma \, d \, \varrho = 4 \times 10^{-5} \qquad (6.3.47)$$

is obtained ($d = 1 \, \text{km} = 10^5 \, \text{cm}$, $\varrho(\text{ice}) \approx 1 \, \text{g/cm}^3$).

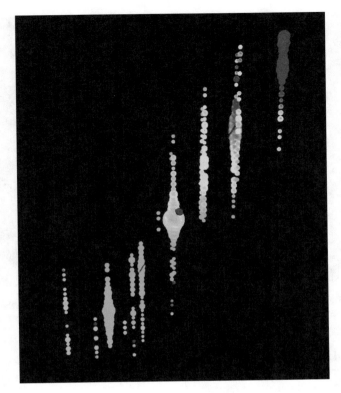

Fig. 6.36 Track of a muon, which has been produced in ICECUBE by a high-energy cosmic muon neutrino [76]

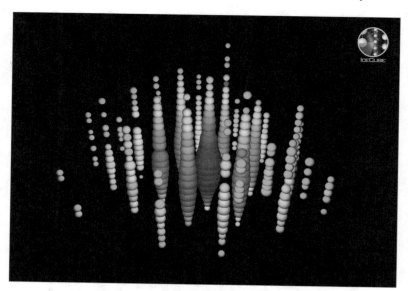

Fig. 6.37 High-energy event in the ICECUBE detector, presumably induced by an electron neutrino. The energy of this event is 1.14 PeV. The event pattern also shows the difficulty to determine the direction of incidence of the original neutrino [76]

The total interaction rate R is obtained from the integral neutrino flux Φ_ν, the interaction probability W, the effective collection area $A_{\mathrm{eff}} = 1\,\mathrm{km}^2$, and a measurement time t. This leads to an event rate of

$$R = \Phi_\nu \, W \, A_{\mathrm{eff}} \,, \tag{6.3.48}$$

corresponding to 250 events per year. For large absorption lengths of the produced Cherenkov light the effective collection area of the detector is even larger than the cross section of the instrumented volume. Assuming that there are about half a dozen sources in our galaxy, the preceding estimate would lead to a counting rate of about four events per day. In addition to this rate from point sources one would also expect to observe events from the diffuse neutrino background that, however, carries little astrophysical information.

Excellent candidates within our galaxy are the supernova remnants of the Crab Nebula and Vela, the galactic center, and Cygnus X3. Extragalactic candidates could be represented by the Markarian galaxies Mrk 421 and Mrk 501, by M87, or by quasars (e.g., 3C273).

ICECUBE has already measured a large number of neutrino events of astrophysical interest. Figure 6.37 shows a high-energy event, which was presumably initiated by an electron neutrino. Figure 6.38 is a sky map showing the arrival directions of cosmic neutrinos. There is no clear clustering of events, even though there are some events pointing back to the galactic center. However, one has to consider that the angular resolution, i.e., the pointing accuracy for events generated by electron neutrinos, which produce electromagnetic showers in the detector, is only moderate.

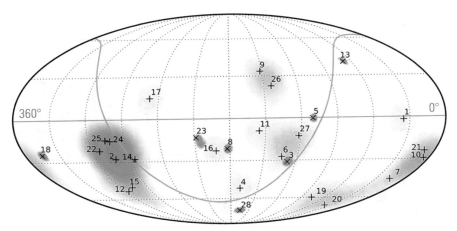

Fig. 6.38 Sky map of neutrino events in ICECUBE in equatorial coordinates. The *blue line* is the galactic plane. The galactic center is close to the event with the number 14. ν_μ-like events (with a detected muon) are flagged with '$\times$' and ν_e-like events (with an electron shower) with '$+$' [76]

However, in September 2017 ICECUBE recorded an energetic neutrino of 290 TeV, which was coincident in time and direction with an energetic gamma-ray flare observed by the Fermi satellite, where the Fermi signal pointed to a known blazar (TXS 0506+056) in the northern sky with high accuracy (see Fig. 6.39). The distance of the source was estimated to be about four billion light-years. Gamma rays with energies up to 400 GeV from this source were also observed by the Major Atmospheric Gamma Imaging Cherenkov (MAGIC) Telescopes. If confirmed by similar events this neutrino event from ICECUBE would imply the first observation of a hadronic cosmic accelerator.

These successes demonstrate that ICECUBE can provide excellent results to neutrino astrophysics. To obtain better statistics an extension of ICECUBE to an instrumented volume of 10 km³ is planned (ICECUBE-Gen2).

6.3.5 Geoneutrinos

> *Radioactive decay is key ingredient behind Earth's heat.*
> Glenn Horton-Smith

Geoneutrinos are not directly a topic on astroparticle physics. However, the availability of experiments looking for cosmic neutrinos has opened up this new type of research as a byproduct. Nearly half of the Earth's heat comes from the decay of radioactive isotopes inside. To better understand the sources of the Earth's heat, the measurement of antineutrinos from the decay of radioactive elements is a relatively new tool. The elemental composition of Earth has only been explored down to relatively shallow depths. Geoneutrinos provide a technique to probe directly the Earth's

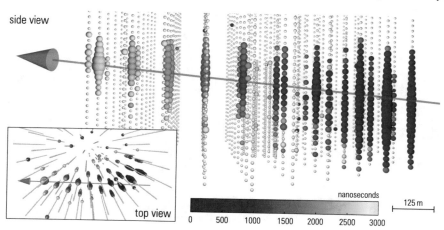

Fig. 6.39 Event display for neutrino event IceCube-170922A coincident in time and direction with an energetic gamma-ray flare observed by the Fermi satellite and the MAGIC telescope from the blazar TXS 0506+056 [77]. The high-energy neutrino (≈ 300 TeV) enters from below from the right-hand side and produces a muon that initiates a large shower

interior beyond the depth of 12 km, which so far has been achieved by drilling, but to the center of the Earth there is quite a long way to go.

The key elements responsible for a major fraction of the heat production are the elements ^{238}U, ^{232}Th, and ^{40}K, which lead to a calculated neutrino luminosity of the Earth of 10^6 cm^{-2} s^{-1} corresponding to a total neutrino flux of 32×10^{24} s^{-1}. According to various estimates the decay of these elements generates a radiogenic heating of about 20 TW, which constitutes about 50% of the total heat power of the Earth. The uranium (^{238}U $\rightarrow$ ^{206}Pb + 8 ^{4}He + 6e^- + 6$\bar{\nu}_e$) and thorium (^{232}Th $\rightarrow$ ^{208}Pb + 6 ^{4}He + 4e^- + 4$\bar{\nu}_e$) decay chains provide electron antineutrinos of about 3 MeV maximum energy. The potassium decay (^{40}K $\rightarrow$ ^{40}Ca + e^- + $\bar{\nu}_e$) with a branching ratio of 89% only leads to $\bar{\nu}_e$ of maximum energy of 1.3 MeV. The flux of antineutrinos from the ^{235}U decay chain is relatively low [78].

To measure the antineutrinos one uses the inverse-beta-decay reaction $\bar{\nu}_e + p \rightarrow e^+ + n$. For this reaction there is a threshold energy of 1.8 MeV corresponding to the difference between the rest-mass energies of the neutron plus positron and the proton. Due to this threshold antineutrinos from the potassium decay cannot be recorded in this reaction. However, these neutrinos can be measured by scattering on electrons. Antineutrinos from ^{238}U and ^{232}Th decay are detected by light signals from positron annihilation and photons from deuteron formation after $n + p \rightarrow d + \gamma$. These two signals are coincident in time and space and provide a powerful tool to reject, e.g., cosmic rays, which would only cause single signals. It does not, however, veto reactor antineutrinos, because they would exhibit the same signature as geoneutrinos.

The first measurement of geoneutrinos was accomplished by Kamiokande (2003) and KamLAND (2005) (Kamioka Liquid Scintillator Antineutrino Detector) and later by Borexino (see Figs. 6.40 and 6.41). KamLAND consists of a 1000-ton liquid

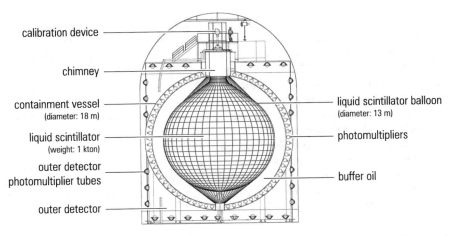

calibration device

chimney

containment vessel
(diameter: 18 m)

liquid scintillator
(weight: 1 kton)

outer detector
photomultiplier tubes

outer detector

liquid scintillator balloon
(diameter: 13 m)

photomultipliers

buffer oil

Fig. 6.40 The KamLAND antineutrino detector at the Kamioka observatory at 1000 m underground [79]

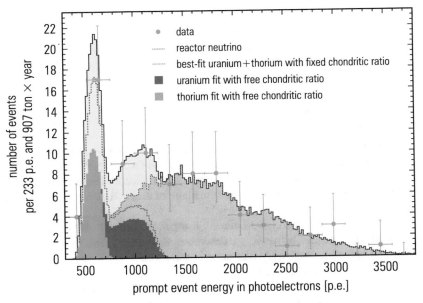

Fig. 6.41 Energy spectrum of 46 prompt $\bar{\nu}_e$ candidates from the Borexino experiment. The maximum energy of $\bar{\nu}_e$ from the ^{238}U decay is 3.26 MeV corresponding to about 1400 photoelectrons measured, and the maximum energy from ^{232}Th is 2.25 MeV (signal of ^{238}U and ^{232}Th *in yellow*). For the contribution of the uranium (*in blue*) and thorium (*in turquoise*) antineutrinos a free parameter was used. For the best fit for the sum of uranium and thorium antineutrinos a fixed chondritic ratio (typical for terrestrial upper-mantle rocks) was assumed. The *full line* is a fit to the data. The background from reactor antineutrinos is substantial (*ocher area*). However, the expected geogenic antineutrino signal (*dashed blue line*) stands clearly out at low energies [80]

scintillator detector surrounded by 1845 photomultipliers. It is set up in the Kamioka observatory 1000 m underground to shield against cosmic rays. Borexino is a high-purity liquid scintillator calorimeter with extremely low intrinsic radioactivity. The scintillation counter is placed in a stainless-steel container. It is shielded by a water tank to protect it against external radiation and cosmic-ray muons. The experiment is installed in the Italian Gran Sasso.

Geoneutrinos have to be detected against the strong background of antineutrinos from some 450 man-made reactors. In the past there have been also antineutrinos from natural reactors, such as the Oklo reactor in Gabon.

Although the measurement of geoneutrinos has started as byproduct of neutrino astrophysics, it has acquired a life of its own, and many larger experiments are being prepared or proposed to improve the knowledge about the Earth's interior, which cannot by studied at the moment by other means.

6.4 Gamma Astronomy

Let there be light.

Bible, Genesis

6.4.1 Introduction

The observation of stars in the optical spectral range belongs to the field of classical astronomy. Already the Chinese, Egyptians, and Greeks performed numerous systematic observations and learned a lot about the motion of heavenly bodies. The optical range, however, covers only a minute range of the total electromagnetic spectrum (Fig. 6.42).

All parts of this spectrum have been used for astronomical observations. From large wavelengths (radio astronomy), the sub-optical range (infrared astronomy), the classical optical astronomy, the ultraviolet astronomy, and X-ray astronomy one arrives finally at the *gamma-ray astronomy*.

Gamma-ray astronomers are used to characterize gamma quanta not by their wavelength λ or frequency ν, but rather by their energy,

$$E = h\nu.\tag{6.4.1}$$

Planck's constant in practical units is

$$h = 4.136 \times 10^{-21}\,\text{MeV s}.\tag{6.4.2}$$

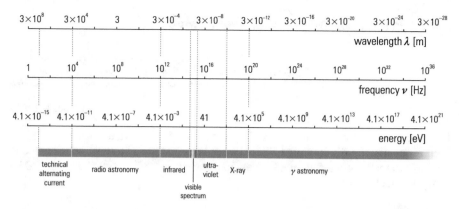

Fig. 6.42 Spectral ranges of electromagnetic radiation

The frequency ν is measured in Hz = 1/s. The wavelength λ is obtained to be

$$\lambda = c/\nu \,, \tag{6.4.3}$$

where c is the speed of light in vacuum ($c = 299\,792\,458$ m/s).

In atomic and nuclear physics one distinguishes gamma rays from X rays by the production mechanism. X rays are emitted in transitions of electrons in the atomic shell while gamma rays are produced in transformations of the atomic nucleus. This distinction also results naturally in a classification of X rays and gamma rays according to their energy. X rays typically have energies below 100 keV. Electromagnetic radiation with energies in excess of 100 keV is called γ rays. There is no upper limit for the energy of γ rays. Even cosmic γ rays with energies of 10^{15} eV = 1 PeV have been observed.

An important, so far unsolved problem of astroparticle physics is the origin of cosmic rays (see also Sects. 6.1 and 6.3). Investigations of charged primary cosmic rays are essentially unable to answer this question because charged particles have to pass through extended irregular magnetic fields on their way from the source to Earth. This causes them to be deflected in an uncontrolled fashion thereby 'forgetting' their origin. Therefore, particle astronomy with charged particles is only possible at extremely high energies when the particles are no longer significantly affected by cosmic magnetic fields. This would require to go to energies in excess of 10^{19} eV that, however, creates another problem because the flux of primary particles at these energies is extremely low. Whatever the sources of cosmic rays are, they will also be able to emit energetic penetrating γ rays, which are not deflected by intergalactic or stellar magnetic fields and therefore point back to the sources. It must, however, be kept in mind that also X and γ rays from distant sources might be subject to time dispersions. Astronomical objects in the line of sight of these sources can distort their trajectory by gravitational lensing thus making them look blurred and causing time-of-flight dispersions in the arrival time also for electromagnetic radiation.

6.4.2 *Production Mechanisms for γ Rays*

> *In general, the objects in the universe that are very high-energy*
> *objects, or the processes that are high-energy processes, will*
> *radiate more in the short wavelength range towards the gamma*
> *rays or the X rays.*
>
> Claude Nicollier

Possible sources for cosmic rays and thereby also for γ rays are supernovae and their remnants, rapidly rotating objects like pulsars and neutron stars, active galactic nuclei, and matter-accreting black holes. In these sources γ rays can be produced by different mechanisms.

(a) Synchrotron radiation:

The deflection of charged particles in a magnetic field gives rise to an accelerated motion. An accelerated electrical charge radiates electromagnetic waves (Fig. 6.43). This 'bremsstrahlung' of charged particles in magnetic fields is called *synchrotron radiation*. In circular earthbound accelerators the production of synchrotron radiation is generally considered as an undesired energy-loss mechanism. On the other hand, synchrotron radiation from accelerators is widely used for structure investigations in atomic and solid state physics as well as in biology and medicine.

Synchrotron radiation produced in cosmic magnetic fields is predominantly emitted by the lightweight electrons. The energy spectrum of synchrotron photons is continuous. The power P radiated by an electron of energy E in a magnetic field of strength B is [81, 82]

$$P \sim E^2 \, B^2 \, . \tag{6.4.4}$$

(b) Bremsstrahlung:

A charged particle, which is deflected in the Coulomb field of a charge (atomic nucleus or electron), emits bremsstrahlung photons (Fig. 6.44). This mechanism is to a certain extent similar to synchrotron radiation, only that in this case the deflection of the particle occurs in the Coulomb field of a charge rather than in a magnetic field.

The probability for *bremsstrahlung* ϕ varies with the square of the projectile charge z and also with the square of the target charge Z, see also (4.2.2). ϕ is proportional to the particle energy E and it is inversely proportional to the mass squared of the deflected particle:

$$\phi \sim \frac{z^2 Z^2 E}{m^2} \, . \tag{6.4.5}$$

Because of the smallness of the electron mass bremsstrahlung is predominantly created by electrons. The energy spectrum of bremsstrahlung photons is continuous and decreases like $1/E_\gamma$ to high energies.

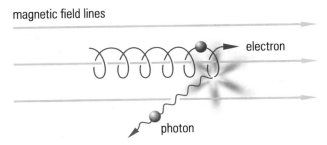

Fig. 6.43 Production of synchrotron radiation by deflection of charged particles in a magnetic field

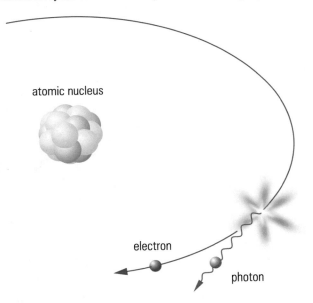

Fig. 6.44 Production of bremsstrahlung by deflection of charged particles in the Coulomb field of a nucleus

(c) Inverse Compton Scattering:

In the twenties of the last century Compton discovered that energetic photons can transfer part of their energy to free electrons in a collision, thereby losing a certain amount of energy. In astrophysics the *inverse Compton effect* plays an important role. Electrons accelerated to high energies in the source collide with the numerous photons of the blackbody radiation ($E_\gamma \approx 250\,\mu\text{eV}$, photon density $N_\gamma \approx 400/\text{cm}^3$) or starlight photons ($E_\gamma \approx 1\,\text{eV}$, $N_\gamma \approx 1/\text{cm}^3$) and transfer part of their energy to the photons, which are 'blueshifted' (Fig. 6.45).

(d) π^0 Decay:

Protons accelerated in the sources can produce charged and neutral pions in proton–proton or proton–nucleus interactions (Fig. 6.46). A possible process is

Fig. 6.45 Collision of an energetic electron with a low-energy photon. The electron transfers part of its energy to the photon and is consequently slowed down

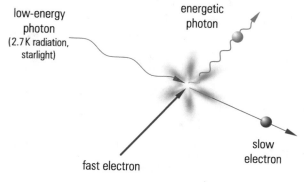

Fig. 6.46 π^0 production in proton interactions and π^0 decay into two photons

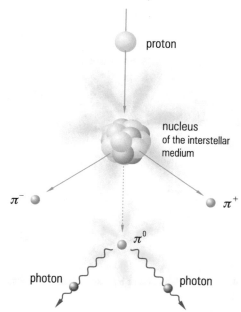

$$p + \text{nucleus} \rightarrow p' + \text{nucleus}' + \pi^+ + \pi^- + \pi^0 . \qquad (6.4.6)$$

Charged pions decay with a lifetime of 26 ns into muons and neutrinos, while neutral pions decay rapidly ($\tau = 8.4 \times 10^{-17}$ s) into two γ quanta,

$$\pi^0 \rightarrow \gamma + \gamma . \qquad (6.4.7)$$

If the neutral pion decays at rest, both photons are emitted back to back. In this decay they get each half of the π^0 rest mass ($m_{\pi^0} = 135$ MeV). In the π^0 decay in flight the photons get different energies depending on their direction

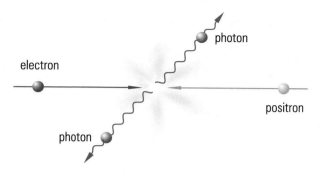

Fig. 6.47 e^+e^- pair production into two photons

of emission with respect to the direction of flight of the π^0 (see Example 10, Sect. 3.4). Since most pions are produced at low energies, photons from this particular source have energies of typically 70 MeV.

(e) Photons from Matter–Antimatter Annihilation:

In the same way as photons can produce particle pairs (pair production), charged particles can annihilate with their antiparticles into energy. The dominant sources for this production mechanism are electron–positron and proton–antiproton annihilations,

$$e^+ + e^- \rightarrow \gamma + \gamma .$$ (6.4.8)

Momentum conservation requires that at least two photons are produced. In e^+e^- annihilation at rest the photons get 511 keV each corresponding to the rest mass of the electron or positron, respectively (Fig. 6.47). An example for a proton–antiproton annihilation reaction is

$$p + \bar{p} \rightarrow \pi^+ + \pi^- + \pi^0 ,$$ (6.4.9)

where the neutral pion decays into two photons.

(f) Photons from Nuclear Transformations:

Heavy elements are 'cooked' in supernova explosions. In these processes not only stable but also radioactive isotopes are produced. These radioisotopes will emit, mostly as a consequence of a beta decay, photons in the MeV range like, e.g.,

$$
\begin{aligned}
{}^{60}\text{Co} \quad\rightarrow\quad & {}^{60}\text{Ni}^{**} + e^- + \bar{\nu}_e \\
& \hookrightarrow {}^{60}\text{Ni}^* + \gamma(1.17\,\text{MeV}) \\
& \qquad \hookrightarrow {}^{60}\text{Ni} + \gamma(1.33\,\text{MeV}) .
\end{aligned}
$$ (6.4.10)

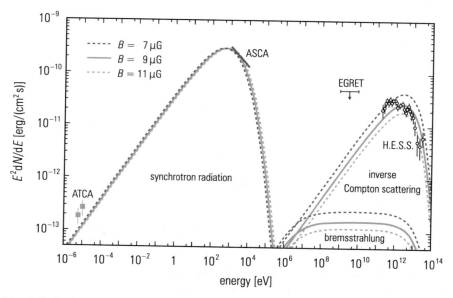

Fig. 6.48 Typical contributions from γ sources in various spectral ranges (also in the range of blackbody radiation (ATCA)) to the γ spectrum. Experimental results from ATCA (Australia telescope compact array), ASCA (Advanced satellite for cosmology and astrophysics), H.E.S.S. (High energy stereoscopic system), and an upper limit from EGRET (Energetic gamma ray experiment telescope) are shown [83]

(g) Cherenkov Radiation:

Another production mechanism, already discussed in Sect. 4.4, is Cherenkov radiation, where particles, which are faster than light in a medium of refractive index n, create electromagnetic radiation preferentially in the blue energy domain.

(h) Annihilation of Neutralinos:

In somewhat more exotic scenarios, energetic γ rays could also originate from annihilation of neutralinos, the neutral supersymmetric partners of ordinary particles, according to

$$\chi + \bar{\chi} \rightarrow \gamma + \gamma \,. \tag{6.4.11}$$

Figure 6.48 shows the various production mechanisms to the gamma-ray spectrum of a typical source. At low energies up to about $100\,\mathrm{keV}$ the synchrotron mechanism dominates. At higher energies, also beyond the TeV range, inverse Compton scattering and bremsstrahlung are the main production mechanisms. In this energy range also photons from π^0 decays could contribute, however, no convincing candidate for a hadronic accelerator has been found so far, which would be needed to produce neutral pions. The electron–positron annihilation would produce a monoenergetic line at $511\,\mathrm{keV}$, which is also occasionally found in some sources. Naturally, the details of the γ spectra depend on the properties of the source, such as the magnetic field and the gas and dust density.

6.4.3 Detection of γ Rays

Gamma rays are the sort of radiation you should avoid.
Neil deGrasse Tyson

In principle the inverse production mechanisms of γ rays can be used for their detection (see also Chap. 4). For γ rays with energies below several hundred keV the photoelectric effect dominates,

$$\gamma + \text{atom} \rightarrow \text{atom}^+ + e^- . \tag{6.4.12}$$

The photoelectron can be recorded, e.g., in a scintillation counter. For energies in the MeV range as it is typical for nuclear decays, Compton scattering has the largest cross section,

$$\gamma + e^-_{\text{at rest}} \rightarrow \gamma' + e^-_{\text{fast}} . \tag{6.4.13}$$

In this case the material of a scintillation counter can also act as an electron target, which records at the same time the scattered electron. For higher energies ($\gg 1\,\text{MeV}$) electron–positron pair creation dominates,

$$\gamma + \text{nucleus} \rightarrow e^+ + e^- + \text{nucleus}' . \tag{6.4.14}$$

Figure 6.49 shows the dependence of the mass attenuation coefficient μ for the three mentioned processes in a NaI scintillation counter.

This coefficient is defined through the photon intensity attenuation in matter according to

$$I(x) = I_0\, e^{-\mu x} \tag{6.4.15}$$

(I_0—initial intensity, $I(x)$—photon intensity after attenuation by an absorber of thickness x).

Since pair production dominates at high energies, this process is used for photon detection in the GeV range. Figure 6.50 shows a typical setup of a satellite experiment for the measurement of γ rays in the GeV range.

Energetic photons are converted into e^+e^- pairs in a modular tracking-chamber system (e.g., in a stack of semiconductor silicon counters). The energies E_{e^+} and E_{e^-} are measured in an electromagnetic calorimeter (mostly a crystal-scintillator calorimeter, NaI(Tl) or CsI(Tl)) so that the energy of the original photon is

$$E_\gamma = E_{e^+} + E_{e^-} . \tag{6.4.16}$$

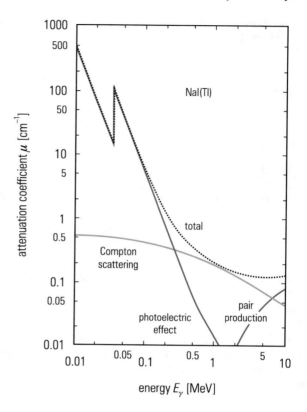

Fig. 6.49 Mass attenuation coefficient for photons in a sodium-iodide (NaI(Tl)) scintillation counter

The direction of incidence of the photon is derived from the electron and positron momenta, where the photon momentum is determined to be $\boldsymbol{p}_\gamma = \boldsymbol{p}_{e^+} + \boldsymbol{p}_{e^-}$. For high energies ($E \gg m_e c^2$) the approximations $|\boldsymbol{p}_{e^+}| = E_{e^+}/c$ and $|\boldsymbol{p}_{e^-}| = E_{e^-}/c$ are well satisfied.

The detection of electrons and positrons in the crystal calorimeter proceeds via electromagnetic cascades. In these showers the produced electrons initially radiate bremsstrahlung photons that convert into $e^+ e^-$ pairs. In alternating processes of bremsstrahlung and pair production the initial electrons and photons decrease their energy until absorptive processes like photoelectric effect and Compton scattering for photons on the one hand and ionization loss for electrons and positrons on the other hand halt further particle multiplication (Fig. 6.51; see also Fig. 6.52).

The anticoincidence counter in Fig. 6.50 serves the purpose of identifying incident charged particles and rejecting them from the analysis.

The Compton Gamma-Ray Observatory (CGRO) was an example for this kind of technique for the measurement of γ-ray sources and γ-ray spectra. The CGRO satellite took data from 1991 to 2000 and provided important experimental results for the understanding of the γ emission of different astrophysical objects. On board the CGRO were the following experiments: BATSE (Burst and Transient Source

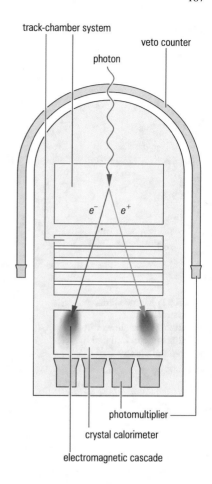

Fig. 6.50 Sketch of a satellite experiment for the measurement of γ rays in the GeV range

Experiment), OSSE (Oriented Scintillation Spectrometer Experiment), COMPTEL (Imaging Compton Telescope), and EGRET (Energetic Gamma Ray Experiment Telescope). BATSE has searched for γ bursts in the energy range from 20 to 600 keV. OSSE has analyzed sources in the MeV range. COMPTEL used the Compton effect to localize sources with relatively high angular resolution in the MeV region. Finally, EGRET covered the high-energy domain (20 MeV to 30 GeV) to obtain an image of the γ sky. EGRET has also discovered many γ-ray sources and has measured their spectra.

More recently the Fermi Gamma-Ray Space Telescope (FGST, in the preparation phase this satellite was named Gamma-Ray Large Area Space Telescope, GLAST) provided important γ-ray results in the energy range up to 300 GeV. FGST was started in 2008 and is still in orbit yielding high-quality results for γ astronomy. While charting the γ sky, FGST has discovered giant bubbles emanating from above and below the galactic plane with X-ray and γ-ray emission. The X-ray and γ-ray

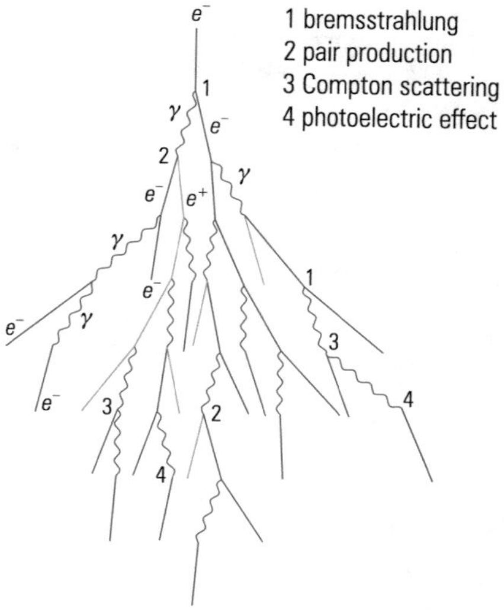

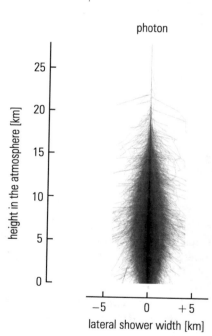

1 bremsstrahlung
2 pair production
3 Compton scattering
4 photoelectric effect

Fig. 6.51 Schematic representation of an electromagnetic cascade

photon
height in the atmosphere [km]
lateral shower width [km]

Fig. 6.52 Monte Carlo simulation of an electromagnetic shower in the atmosphere, initiated by a photon of energy 10^{14} eV [84]

emissions extend over huge regions of about 10 kpc above and below the galactic plane. The origin of these bubbles is not yet understood. A possible candidate for these bubbles is the black hole at the center of our galaxy. Figure 6.53 shows an impression of these γ-ray bubbles.

For energies in excess of several 100 GeV the photon intensities from cosmic-ray sources are so small that other techniques for their detection must be applied, since sufficiently large setups cannot be installed on board of satellites. In this context the detection of photons via the *atmospheric Cherenkov technique* plays a special role (see Figs. 6.54 and 6.55).

This is the domain of the imaging Cherenkov telescopes like H.E.S.S. (High Energy Stereoscopic System) in Namibia, MAGIC (Major Atmospheric Gamma Imaging Cherenkov Telescopes, see Fig. 6.56) on La Palma, and VERITAS (Very Energetic Radiation Imaging Telescope Array System) on the Mount Hopkins in Arizona. These experiments provide data for the high-energy γ range beyond TeV. The Imaging Cherenkov telescopes have charted the northern and southern sky and found many sources emitting in the TeV range. H.E.S.S. has scanned the southern sky including the galactic center (see Fig. 1.25 in Sect. 1.7), and it is not a surprise that this experiment has discovered a large number of high-energy γ sources.

The HAWC experiment (High-Altitude Water Cherenkov Observatory [85]) in the Sierra Negra near Puebla in Mexico at an altitude of 4100 m is a new ground-based experiment in the field of γ-ray astronomy providing interesting data in the energy range from 100 GeV to 100 TeV since 2016.

Presently, a giant Cherenkov Telescope Array (CTA) is in preparation, which should scan the southern sky (in Chile) and as well the northern sky (on La Palma) for γ-ray sources in the TeV range and beyond. In principle, the Auger experiment in Argentina should also be able to look for γ events with energies $> 10^{18}$ eV. So far, no γ-induced events were found at these very high energies.

To explain and illustrate the principle of imaging Cherenkov telescopes, one has to consider the following: When γ rays enter the atmosphere they produce—like already described for the crystal calorimeter—a cascade of electrons, positrons, and photons, which are generally of low energy. This shower does not only propagate longitudinally but it also spreads somewhat laterally in the atmosphere (see also Fig. 6.52). For initial photon energies below 10^{13} eV ($= 10$ TeV) the shower particles, however, do not reach sea level. Relativistic electrons and positrons of the cascade, which follow essentially the direction of the original incident photon, emit blue light in the atmosphere that is known as Cherenkov light. Charged particles, whose velocities exceed the speed of light, emit this characteristic electromagnetic radiation (see Chap. 4). Since the speed of light in atmospheric air is

$$c_n = c/n \tag{6.4.17}$$

(n is the index of refraction of air; $n = 1.000\,273$ at 20 °C and 1 atm), electrons with velocities

$$v \geq c/n \tag{6.4.18}$$

Fig. 6.53 Image of the γ-ray bubbles discovered by the Fermi satellite near the galactic center. Already the ROSAT satellite reported indications for the γ-ray bubbles. The bubbles extend over a region of about 50 000 light-years (diameter about 25 000 light-years) and start from the center of the Milky Way. It is speculated that the bubbles originate from the supermassive black hole at the center of our galaxy [86]

will emit Cherenkov light. This threshold velocity of

$$v = c_n = 299\,710\,637\,\text{m/s} \tag{6.4.19}$$

corresponds to a kinetic electron energy of

$$
\begin{aligned}
E_{\text{kin}} &= E_{\text{total}} - m_0 c^2 = \gamma m_0 c^2 - m_0 c^2 \\
&= (\gamma - 1)m_0 c^2 = \left(\frac{1}{\sqrt{1 - v^2/c^2}} - 1 \right) m_0 c^2 \\
&= \left(\frac{1}{\sqrt{1 - 1/n^2}} - 1 \right) m_0 c^2 \\
&= \left(\frac{n}{\sqrt{n^2 - 1}} - 1 \right) m_0 c^2 \approx 21.36\,\text{MeV}.
\end{aligned}
\tag{6.4.20}
$$

The production of Cherenkov radiation in an optical shock wave (Fig. 6.54) is the optical analogue to sound shock waves, which are created when aeroplanes exceed the velocity of sound.

In this way energetic primary γ quanta can be recorded at ground level via the produced Cherenkov light even though the electromagnetic shower does not reach sea level. The Cherenkov light is emitted under a characteristic angle of

Fig. 6.54 Emission of Cherenkov radiation in an optical shock wave for particles traversing a medium of refractive index n with a velocity exceeding the velocity of light in that medium ($v > c_0/n$)

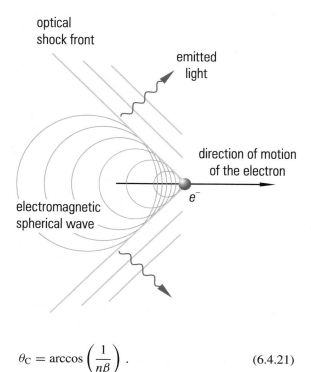

$$\theta_C = \arccos\left(\frac{1}{n\beta}\right). \tag{6.4.21}$$

For electrons in the multi-GeV range the opening angle of the Cherenkov cone is only 1.4°. Actually the Cherenkov angle is somewhat smaller ($\approx 1°$), since the shower particles are produced at large altitudes, where the density of air and thereby also the index of refraction is smaller.

A simple Cherenkov detector, therefore, consists of a parabolic mirror, which collects the Cherenkov light and a set of photomultipliers that record the light collected at the focal point of the mirror. Figure 6.55 shows the principle of photon measurements via the atmospheric Cherenkov technique. Large Cherenkov telescopes with mirror diameters ≥ 10 m allow to measure comparatively low-energy photons (<100 GeV) with correspondingly small shower size even in the presence of light from the night sky (see Fig. 6.56). For even higher energies ($>10^{15}$ eV) the electromagnetic cascades initiated by the photons reach sea level and can be recorded with techniques like those, which are used for the investigations and measurements of extensive air showers (particle sampling, air scintillation, cf. Sect. 7.4). At these energies it is anyhow impossible to explore larger regions of the universe in the light of γ rays. The intensity of energetic primary photons is attenuated by photon–photon interactions predominantly with the numerous ambient photons of the 2.7-K blackbody radiation. For the process

$$\gamma + \gamma \rightarrow e^+ + e^- \tag{6.4.22}$$

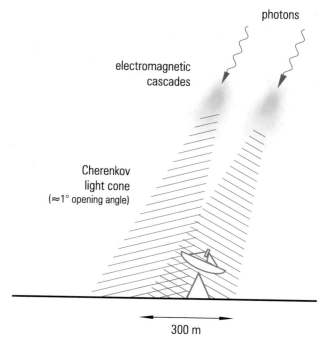

Fig. 6.55 Measurement of the Cherenkov light of photon-induced electromagnetic cascades in the atmosphere

twice the electron mass must be provided in the $\gamma\gamma$ center-of-mass system. For a primary photon of energy E colliding with a target photon of energy ε at an angle θ the threshold energy is

$$E_{\text{threshold}} = \frac{2m_e^2}{\varepsilon(1 - \cos\theta)}. \tag{6.4.23}$$

For a central collision ($\theta = 180°$) and a typical blackbody photon energy of $\varepsilon \approx 250\,\mu\text{eV}$ the threshold is

$$E_{\text{threshold}} \approx 10^{15}\,\text{eV}. \tag{6.4.24}$$

The cross section rises rapidly above threshold, reaches a maximum of 200 mb at twice the threshold energy, and decreases thereafter. For even higher energies further absorptive processes with infrared or starlight photons occur ($\gamma\gamma \rightarrow \mu^+\mu^-$) so that distant regions of the universe ($> 100\,\text{kpc}$) are practically inaccessible for energetic photons ($\gg 100\,\text{TeV}$). Photon–photon interactions, therefore, cause a horizon for γ astronomy, which allows us to explore the nearest neighbours of our local group of galaxies in the light of high-energy γ quanta, but they attenuate the γ intensity for larger distances so strongly that a meaningful observation becomes impossible (cf. Chap. 4, Figs. 4.17 and 4.18).

Fig. 6.56 MAGIC Cherenkov telescopes on La Palma, Spain; the mirror diameter of the largest telescope is 17 m (MAGIC = Major atmospheric gamma imaging Cherenkov telescopes); with kind permission of Razmik Mirzoyan, MAGIC [87]

Distant γ-ray sources, whose high-energy photons are absorbed by blackbody, infrared, or starlight photons, can still be observed in the energy range <1 TeV.

6.4.4 Observation of γ-Ray Point Sources

Keep your eyes on the stars, and your feet on the ground.
Theodore Roosevelt

First measurements of galactic γ rays were performed in the seventies with satellite experiments. The results of these investigations (Fig. 6.57) clearly show the galactic center, the Crab Nebula, the Vela X1 pulsar, Cygnus X3, and Geminga as γ-ray point sources. Recent satellite measurements with the Compton gamma-ray observatory (CGRO) show a large number of further γ-ray sources. There were four experiments on board the CGRO satellite (BATSE, OSSE, EGRET, COMPTEL).[6] These four telescopes cover an energy range from 30 keV up to 30 GeV. Apart from γ-ray bursts (see Sect. 6.4.5) numerous galactic pulsars and a large number of extragalactic

---

[6]OSSE—Oriented Scintillation Spectrometer Experiment
COMPTEL—imaging COMpton TELescope
EGRET—Energetic Gamma Ray Telescope Experiment
BATSE—Burst And Transient Source Experiment.

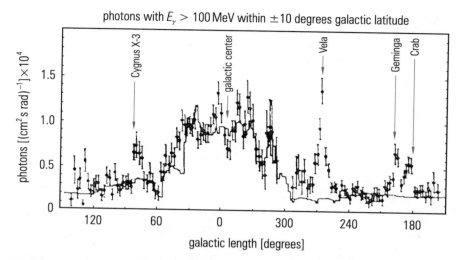

Fig. 6.57 One of the first measurements of the intensity of galactic γ rays for photon energies >100 MeV (data from SAS-2, 1972–1973) [88]

sources (AGN)[7] have been discovered. It was found out that old pulsars can transform their rotational energy more efficiently into γ rays compared to young pulsars. One assumes that the observed γ radiation is produced by synchrotron radiation of energetic electrons in the strong magnetic fields of the pulsars.

Among the discovered active galaxies highly variable blazars (extremely variable objects on short time scales with strong radio emission) were found, which had their maximum of emission in the gamma range. In addition, gamma quasars at high redshifts ($z > 2$) were observed. In this case the gamma radiation could have been produced by inverse Compton scattering of energetic electrons off photons.

Figure 6.58 shows a complete *all-sky survey* in the light of γ rays in galactic coordinates. Apart from the galactic center and several further point sources, the galactic disk is clearly visible. Candidates for point sources of galactic γ rays are pulsars, binary pulsar systems, and supernovae. Extragalactic sources are believed to be compact active galactic nuclei (AGN), quasi-stellar radio sources (quasars), blazars, and accreting black holes. According to common belief black holes could be the 'powerhouses' of quasars. Black holes are found at the center of galaxies, where the matter density is highest thereby providing sufficient material for the formation of accretion disks around black holes. According to the definition of a black hole no radiation can escape from it, however, the infalling matter heats up already before reaching the event horizon so that intensive energetic γ rays can be emitted. In this context the Hawking radiation, which has no importance for γ-ray astronomy, will not be considered, since black holes have a very low 'temperature'.

---

[7]AGN—Active Galactic Nuclei.

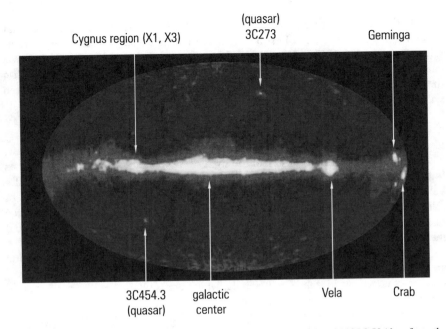

Fig. 6.58 All-sky survey in the light of γ rays with energies in excess of 100 MeV (data from the EGRET detector on board the CGRO). One clearly recognizes the correspondence of the sources (Cygnus X3, Vela, Geminga, Crab, and the galactic center) with the structures seen in Fig. 6.57. In addition, some extragalactic sources outside the galactic plane are visible [89]

There is an increasing number of space-based gamma-ray telescopes, like the NASA spacecraft Swift ('Swift Gamma Ray Explorer' or 'Neil Gehrels Swift Observatory'), carrying an instrument for gamma-ray-burst observations. BeppoSAX[8] and HETE[9]-2 have observed numerous X-ray and optical counterparts to bursts allowing distance determinations of γ-ray sources. INTEGRAL (INTErnational Gamma-Ray Astrophysics Laboratory; 15 keV–10 MeV) is an ESA mission, and AGILE[10] is a small all-Italian mission by the INFN collaboration and other institutes. Fermi was launched by NASA in 2008, which also includes a burst monitor to study gamma-ray bursts.

In the TeV range γ-ray point sources have been discovered with the air Cherenkov technique. Recently, a supernova remnant as a source of high-energy photons has been observed by the H.E.S.S.[11] experiment. The γ-ray spectrum of this object, SNR

---

[8] BeppoSAX—X-ray satellite named after Guiseppe 'Beppo' Occhialini.

[9] HETE—High Energy Transient Explorer.

[10] AGILE—Astro-rivelatore Gamma a Immagini ultra LEggero, Italian γ-ray satellite.

[11] H.E.S.S.—High Energy Stereoscopic System.

RX J1713.7-3946 in the galactic plane 7 kpc off the galactic center, near the solar system (distance 1 kpc), can best be explained by the assumption that the photons with energies near 10 TeV are produced by π^0 decays, i.e., this source is a good candidate for a hadron accelerator. The imaging air Cherenkov telescope MAGIC on the Canary Island La Palma with its 17-m-diameter mirror will compete in this field with the H.E.S.S. telescope in Namibia.

Apart from galactic sources (Crab Nebula) extragalactic objects emitting TeV photons (Markarian 421, Markarian 501, and M87) have also been unambiguously identified. Markarian 421 is an elliptic galaxy with a highly variable galactic nucleus. The luminosity of Markarian 421 in the light of TeV photons would be 10^{10} times higher than that of the Crab Nebula if isotropic emission were assumed. One generally believes that this galaxy is powered by a massive black hole, which emits jets of relativistic particles from its poles. It is conceivable that this galaxy at a distance of approximately 400 million light-years beams the high-energy particle jets—and thereby also the photon beam—exactly into the direction of Earth.

The highest γ energies from cosmic sources have been recorded by earthbound air-shower experiments, but also by air Cherenkov telescopes. It has become common practice to consider the Crab Nebula, which emits photons with energies up to 100 TeV, as a standard candle. γ-ray sources found in this energy regime are mostly characterized by an extreme variability. In this context the X-ray source Cygnus X3 plays a special role. In the eighties γ rays from this source with energies up to 10^{16} eV (10 000 TeV) were claimed to be seen. These high-energy γ rays appeared to show the same variability (period 4.8 h) as the X rays coming from this object. It has to be noted, however, that a high-energy gamma outburst from this source has never been seen again [90].

Apart from the investigation of cosmic sources in the light of high-energy γ rays, the sky is also searched for γ quanta of certain fixed energy. This γ-ray line emission hints at radioactive isotopes, which are formed in the process of nucleosynthesis in supernova explosions. It could be shown beyond any doubt that the positron emitter ^{56}Ni was produced in the supernova explosion 1987A in the Large Magellanic Cloud. This radioisotope decays into ^{56}Co with a half-life of 6.1 days. The light curve of this source showed a luminosity maximum followed by an exponential brightness decay. This could be traced back to the radioactive decay of the daughter ^{56}Co to the stable isotope ^{56}Fe with a half-life of 77.1 days (see Fig. 6.59).

Interesting results are also expected from an all-sky survey in the light of the 511-keV line from e^+e^- annihilation. This γ-ray line emission could indicate the presence of antimatter in our galaxy. The observation of the distribution of cosmic antimatter could throw some light on the problem why our universe seems to be matter dominated.

6.4.5 γ-Ray Bursters

We are all of us stars, and we deserve to twinkle.

Marilyn Monroe

Cosmic objects, which emit sudden single short outbursts of γ rays, have been discovered in the early seventies by American reconnaissance satellites. The purpose of these satellites was to check the agreement on the stop of nuclear weapon tests in the atmosphere. The recorded γ rays, however, did not come from the surface of the Earth or the atmosphere, but rather from outside sources and, therefore, were not related to explosions of nuclear weapons that also are a source of γ rays.

γ-*ray bursts* occur suddenly and unpredictably with a rate of approximately one burst per day. The durations of the γ-ray bursts are mostly short ranging from fractions of a second up to 100 s. There appear to be two distinct classes of γ-ray bursts, one with short (≈ 0.3 s), the other with longer durations (≈ 30 s), indicating the existence of two different populations of γ-ray bursters. Figure 6.60 shows the γ-ray light curve of a typical short burst. Within only one second the γ-ray intensity increases by a factor of nearly 10. The γ-ray bursters appear to be uniformly distributed over the whole sky. Because of the short burst duration it is very difficult to identify a γ-ray burster with a known object. At the beginning of 1997 researchers succeeded for the first time to associate a γ-ray burst with a rapidly fading object in the optical regime. From the spectral analysis of the optical partner one could conclude that the distance of this γ-ray burster was about several billion light-years.

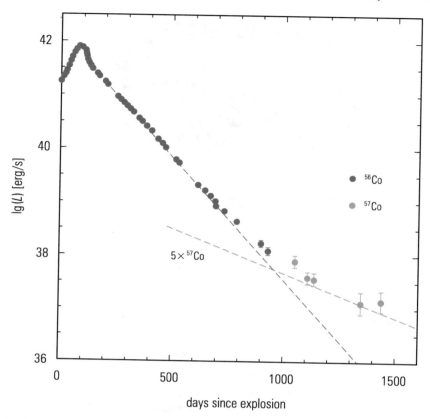

Fig. 6.59 The bolometrically measured light curve of SN 1987A. The *dashed lines* correspond to a conversion of ^{56}Co and—with lower intensity—of ^{57}Co into the optical, infrared, and ultraviolet spectral range [91]

The angular distribution of 2704 γ-ray bursters recorded up to the end of the BATSE measurement period is shown in Fig. 6.61 in galactic coordinates. From this diagram it is obvious that there is no clustering of γ-ray bursters along the galactic plane. Therefore, the most simple assumption is that these exotic objects are at cosmological distances, which means that they are extragalactic. Measurements of the intensity distributions of bursts show that weak bursts are relatively rare. This could imply that the weak (i.e., distant) bursts exhibit a lower spatial density compared to the strong (near) bursts.

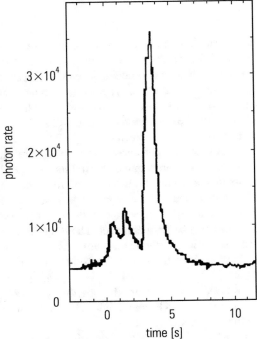

Fig. 6.60 Light curve of a typical γ burst, recorded on April 21, 1991 [92]

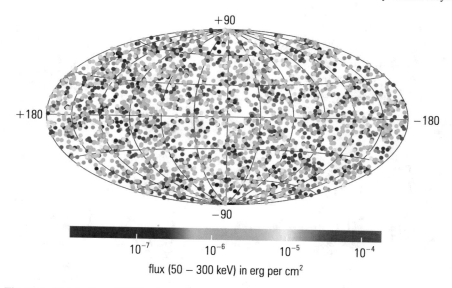

Fig. 6.61 Distribution of 2704 γ bursts in galactic coordinates from the BATSE detector on board the CGRO satellite [93]

Even though violent supernova explosions are considered to be excellent candidates for γ-ray bursts, it is not obvious whether also other astrophysical objects are responsible for this enigmatic phenomenon. The observed spatial distribution of γ-ray bursters suggests that they are at extragalactic distances. In this context the deficit of weak bursts in the intensity distribution could be explained by the redshift of spectral lines associated with the expansion of the universe. This would also explain why weaker bursts have softer energy spectra.

As already pointed out, these light curves of γ bursts are distinctly different. There are at least two classes of bursts that differ in their burst duration. One has to distinguish short bursts with a typical duration of 0.3 s from the long-duration bursts ($\approx$30 s), s. Fig. 6.62. It is assumed that the short bursts are created as mergers of two neutron stars or a neutron star with a black hole ('*kilonova*'). The short duration also indicates that the sources must be very compact. The long-duration bursts are more likely to originate from supernova collapses or with the final phases of massive stars. The rare super-long bursts with a duration of more than 10 000 s may be members of a third burst category [95]. The detection of a 90-min-long burst, which included the detection of an 18-GeV photon, offers the possibility for telescopes operating at other wavelengths to detect a GRB source while it is still active. This could also throw some light on the source's identity.

As already indicated a large fraction of long-duration γ-ray bursts is believed to be caused by violent supernova explosions (e.g., hypernova explosions with a collapse into a rotating black hole). This seems to have been confirmed by the association of the γ-ray burst GRB030329 with the supernova explosion SN2003dh. The observation

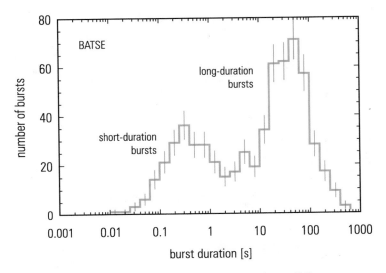

Fig. 6.62 Distribution of the frequency of short and long γ-ray bursts [94]

of the optical afterglow of this burst allowed to measure its distance to be at 800 Mpc. The afterglow of the burst, i.e., the optical luminosity of the associated supernova, reached a magnitude of 12 in the first observations after the γ-ray burst. Such a bright supernova might have been visible to the naked eye in the first minutes after the explosion. Such hypernova explosions are considered to be rare events, which are probably caused by stars of the 'Wolf–Rayet' type. Wolf–Rayet stars are massive objects ($M > 20\,M_\odot$) that initially consist mainly of hydrogen. During their burning phase they strip themselves off their outer layers thus consisting mainly of helium, oxygen, and heavy elements. When they run out of fuel, the core collapses and forms a black hole surrounded by an accretion disk. It is believed that in this moment a jet of matter is ejected from the black hole that represents the γ-ray burst ('collapsar model').

As candidates for short-duration γ-ray bursts, in addition to collisions of neutron stars and collisions of neutron stars with black holes, coalescences of two neutron stars forming a black hole, asteroid impacts on neutron stars, or exploding primordial mini black holes are discussed. From the short burst durations one can firmly conclude that the spatial extension of γ-ray bursters must be very small. However, only the exact localization and detailed observation of afterglow partners of γ-ray bursts will allow to clarify the problem of their origin. One important input to this question is the observation of a γ-ray burster also in the optical regime by the 10-m telescope on Mauna Kea in Hawaii. Its distance could be determined to be 9 billion light-years.

A spectacular event occurred when the merger of two neutron stars, event GW170817, was observed by the Advanced LIGO and Virgo detectors in 2017. At the same time a gamma-ray burst GRB 170817A was observed independently by the Fermi Gamma-ray Burst Monitor and the International Gamma-Ray Astro-

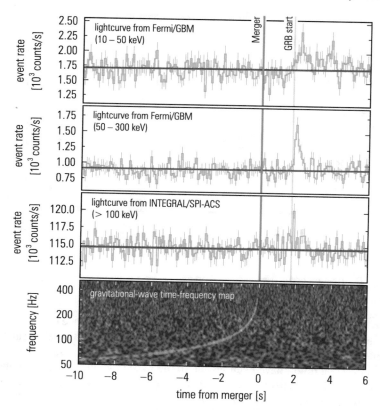

Fig. 6.63 Multi-messenger detection of GW170817 and GRB 170817A. The time evolution of the γ-ray, resp. X-ray signals from Fermi and INTEGRAL in the energy range from 10 keV to over 100 keV are compared with the gravitational time–frequency map from LIGO [96]

physics Laboratory (INTEGRAL). The time difference between the arrival of the gravitational wave and the GRB was about 2 s (see Fig. 6.63). This delay could naively be interpreted as being due to different velocities for the gravitational radiation and the light signal. A more orthodox explanation is that due to the gravitational deflection the light signal might not have traveled on a straight line in contrast to the gravitational-wave signal. The delay of the gamma-ray-burst signal could also be due to the fact that it was emitted at a time somewhat later than the merger event, which was spread over some time? The time delay could also be used to place new bounds on the violation of Lorentz invariance or enable a test of the equivalence principle.

In the year 1986 a variant of a family of γ-ray bursters was found. These objects emit sporadically γ bursts from the same source. The few so far known quasi-periodic γ-ray bursters all reside in our galaxy or in the nearby Magellanic Clouds. Most of these objects could be identified with young supernova remnants. These 'soft gamma-ray repeaters' appear to be associated with enormous magnetic fields. If such a magnetar rearranges its magnetic field to reach a more favourable energy state, a star quake might occasionally occur, in the course of which γ bursts are emitted. The γ-ray burst of the magnetar SGR-1900+14 was recorded by seven research satellites on August 27, 1998. From the observed slowing down of the rotational period of this magnetar one concludes that this object possesses a superstrong magnetic field of 10^{11} tesla exceeding the magnetic fields of normal neutron stars by a factor of 1000.

With these properties γ bursters are also excellent candidates as sources of cosmic rays. It is frequently discussed that the birth or the collapse of neutron stars could be associated with the emission of narrowly collimated particle jets. If this were true, we would only be able to see a small fraction of γ bursters. The total number of γ bursters would then be sufficiently large to explain the observed particle fluxes of cosmic rays. The enormous time-dependent magnetic fields would also produce strong electric fields, in which cosmic-ray particles could be accelerated up to the highest energies.

A spectacular event has also been recorded on June 16 in 2018. A very bright optical transient on an astonishingly short time scale occurred in the Hercules constellation 200 million light-years away. As an exclamation of surprise by astronomers ('Holy Cow!') this event was named 'the Cow'. It was also observed in different spectral ranges by X-ray and gamma-ray telescopes (NuSTAR, XMM, INTEGRAL, and Swift) and also by earthbound radio telescopes. Even ICECUBE has seen two neutrino events that might have come from the Cow. The astronomers are puzzled by this event and its origin is not clear. As source a magnetar, a very special supernova, or a compact star devoured by a black hole are discussed. This occurrence demonstrates that multi-messenger observations in astrophysics may provide ample information to understand what is going on in exceptional situations [97].

6.5 X-Ray Astronomy

The universe is popping all over the place.

Riccardo Giacconi

6.5.1 Introduction

X rays differ from γ rays by their production mechanism and their energy. X rays are produced if electrons are decelerated in the Coulomb field of atomic nuclei or in transitions between atomic electron levels. Their energy ranges between approximately 1–100 keV. In contrast, γ rays are usually emitted in transitions between nuclear levels, nuclear transformations, or in elementary-particle processes.

After the discovery of X rays in 1895 by Wilhelm Conrad Röntgen, X rays were mainly used in medical applications because of their high penetration power. X rays with energies exceeding 50 keV can easily pass through 30 cm of tissue (absorption probability $\approx 50\%$). The column density of the Earth's atmosphere, however, is too large to allow extraterrestrial X rays to reach sea level. In the keV energy region, corresponding to the brightness maximum of most X-ray sources, the range of X rays in air is 10 cm only. To be able to observe X rays from astronomical objects one therefore has to operate detectors at the top of the atmosphere or in space. This would imply balloon experiments, rocket flights, or satellite missions.

Balloon experiments can reach a flight altitude of 35–40 km. Their flight duration amounts to typically between 20 and 40 h. At these altitudes, however, a substantial fraction of X rays is already absorbed. Balloons, therefore, can only observe X-ray sources at energies exceeding 50 keV without appreciable absorption losses. In contrast, rockets normally reach large altitudes. Consequently, they can measure X-ray sources unbiased by absorption effects. However, their flight time of typically several minutes, before they fall back to Earth, is extremely short. Satellites have the big advantage that their orbit is permanently outside the Earth's atmosphere allowing observation times of several years.

In 1962 X-ray sources were discovered by chance when an American Aerobee rocket with a detector consisting of three Geiger counters searched for X rays from the Moon. No lunar X radiation was found, but instead extrasolar X rays from the constellations Scorpio and Sagittarius were observed. This was a big surprise because it was known that our Sun radiates a small fraction of its energy in the X-ray range and—because of solid-angle arguments—one did not expect X-ray radiation from other stellar objects. This is because the distance of the nearest stars is more than 100 000 times larger than the distance of our Sun. The brightness of such distant sources must have been enormous compared to the solar X-ray luminosity to be able to detect them with detectors that were in use in the sixties. The mechanism what made the sources Scorpio and Sagittarius shine so bright in the X-ray range, therefore, was an interesting astrophysical question.

6.5.2 Production Mechanisms for X Rays

It seemed at first a new kind of invisible light. It was clearly something new something unrecorded.

Wilhelm Conrad Röntgen

The sources of X rays are similar to those of gamma rays. Since the energy spectrum of electromagnetic radiation usually decreases steeply with increasing energy, X-ray sources outnumber gamma-ray sources. In addition to the processes already discussed in Sect. 6.4 (gamma-ray astronomy) like synchrotron radiation, bremsstrahlung, and inverse Compton scattering, a further production mechanism for X rays like thermal radiation from hot cosmic sources has to be considered. The Sun with its effective surface temperature of about 6000 K emits predominantly in the eV range. Sources with a temperature of several million kelvin would also emit X rays as blackbody radiation.

The measured spectra of many X-ray sources exhibit a steep intensity drop to very small energies, which can be attributed to absorption by cold material in the line of sight. At higher energies a continuum follows, which can be described either by a power law ($\sim E^{-\gamma}$) or by an exponential, depending on the type of the dominant production mechanism for X rays. Sources, in which relativistic electrons produce X rays by synchrotron radiation or inverse Compton scattering, can be characterized by a power-law spectrum like $E^{-\gamma}$. On the other hand, one obtains an exponential decrease to high energies if thermal processes dominate. A bremsstrahlung spectrum is usually relatively flat at low energies. In most cases more than one production process contributes to the generation of X rays. According to the present understanding, the X rays of most X-ray sources appear to be of thermal origin. For thermal X rays one has to distinguish two cases.

1. In a hot gas ($\approx 10^7$ K) the atoms are ionized. Electrons of the thermal gas produce in an optically thin medium (practically no self absorption) X rays by bremsstrahlung and by atomic-level transitions. The second mechanism requires the existence of atoms that still have at least one bound electron. At temperatures exceeding $\approx 10^7$ K, however, the most abundant atoms like hydrogen and helium are completely ionized so that in this case bremsstrahlung is the dominant source. In this context one understands under bremsstrahlung the emission of X rays, which are produced by interactions of electrons in the Coulomb field of positive ions of the plasma in continuum transitions (*thermal bremsstrahlung*). For energies $h\nu > kT$ the spectrum decreases exponentially like $e^{-h\nu/kT}$ (k: Boltzmann constant). On the other hand, if $h\nu \ll kT$, the spectrum is nearly flat. The assumption of low optical density of the source leads to the fact that the emission spectrum and production spectrum are practically identical.

2. A hot optically dense body produces a *blackbody spectrum* independently of the underlying production process because both emission and absorption processes are involved. Therefore, an optically dense bremsstrahlung source, which absorbs its own radiation, would also produce a blackbody spectrum.

The emission P of a blackbody is given by Planck's law,

$$P \sim \frac{\nu^3}{e^{h\nu/kT} - 1} \, . \tag{6.5.1}$$

For high energies ($h\nu \gg kT$) P can be described by an exponential,

$$P \sim e^{-h\nu/kT} \, , \tag{6.5.2}$$

while at low energies ($h\nu \ll kT$), because of

$$e^{h\nu/kT} = 1 + \frac{h\nu}{kT} + \cdots , \tag{6.5.3}$$

the spectrum decreases to low frequencies like

Fig. 6.64 Standard X-ray spectra by various production processes

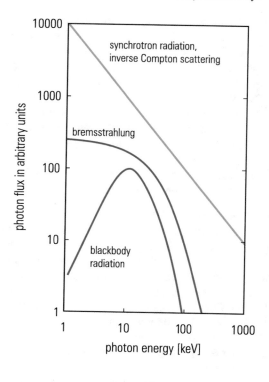

$$P \sim \nu^2 . \tag{6.5.4}$$

The total radiation S of a hot body is described by the Stefan–Boltzmann law

$$S = \sigma \, T^4 , \tag{6.5.5}$$

where σ is the Stefan–Boltzmann constant.

Typical energy spectra for various production mechanisms are sketched in Fig. 6.64.

Figure 6.65 shows early X-ray spectra of three sources in the constellation Cygnus. The strongest X-ray source, meaningfully denoted by Cyg X-1, has a somewhat higher X-ray luminosity (please note the downscaling factors) compared to Cygnus X3. The data cover the X-ray range from 2 to 10 keV. The deficiency of low-energy X rays is consistent with either self-absorption in the source or interstellar attenuation.

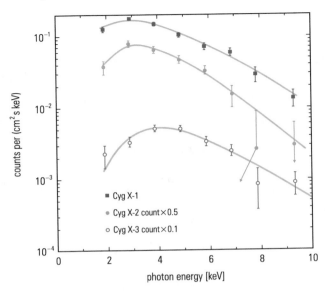

Fig. 6.65 Early observation of X-ray spectra of the sources in the constellation Cygnus during a rocket flight in 1966 [98]

6.5.3 Detection of X Rays

> *Great discoveries are made accidentally less often than the populace likes to think.*
>
> William Cecil Dampier

The observation of X-ray sources is more demanding compared to optical astronomy. X rays cannot be imaged with lenses since the index of refraction in the keV range is very close to unity. If X rays are incident on a mirror, they will be absorbed rather than reflected. Therefore, the direction of incidence of X rays has to be measured by different techniques. The most simple method for directional observation is based on the use of slit or wire collimators, which are mounted in front of an X-ray detector. In this case the observational direction is given by the alignment of the space probe. Such a geometrical system achieves resolutions on the order of 0.5°. By combining various types of collimators angular resolutions of one arc minute can be obtained.

In 1952 Wolter had already proposed how to build X-ray telescopes based on total reflection. To get reflection at grazing incidence rather than absorption or scattering, the imaging surfaces have to be polished to better than a fraction of 10^{-3} of the optical wavelength. Typically, systems of stacked assemblies of paraboloids or combinations of *parabolic* and *hyperbolic mirrors* are used (see Fig. 6.66).

To be able to image X rays in the range between 0.5 and 10 nm with this technique, the angles of incidence must be smaller than 1.5° (Fig. 6.67). The wavelength λ[nm] is obtained from the relation

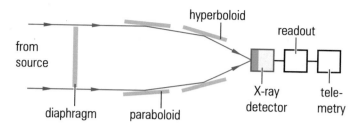

Fig. 6.66   Cross section through an X-ray telescope with parabolic and hyperbolic mirrors

Fig. 6.67 Angular-
dependent reflection power
of metal mirrors

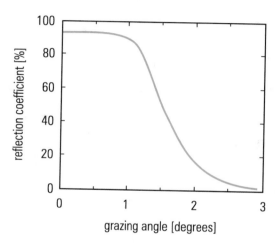

$$\lambda = \frac{c}{\nu} = \frac{hc}{h\nu} = \frac{1240}{E[\text{eV}]} \, \text{nm} \, . \tag{6.5.6}$$

The mirror system images the incident X rays onto the common focal point. In X-ray satellites usually several X-ray devices are installed as focal-point detectors. They are usually mounted on a remotely controllable device. Depending on the particular application an appropriate detector can be moved into the focal point. In multimirror systems angular resolutions of one arc second are obtained. The requirement of grazing incidence, however, considerably limits the acceptance of X-ray telescopes.

As detectors for X rays crystal spectrometers (Bragg reflection), proportional counters, photomultipliers, single-channel electron multipliers (channeltrons), semiconductor counters, or X-ray CCDs (charge-coupled devices) are in use.

In proportional counters the incident photon first creates an electron via the photoelectric effect, which then produces an avalanche in the strong electrical field (Fig. 6.68).

In the proportional domain gas amplifications of 10^3–10^5 are achieved. Since the absorption cross section for the photoelectric effect varies proportional to Z^5, a heavy noble gas (Xe, $Z = 54$) with a quencher should be used as counting gas.

Fig. 6.68 Working principle
of a proportional counter for
the measurement of X rays

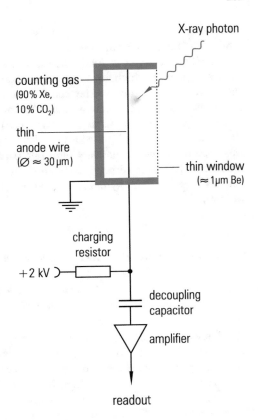

Thin foils ($\approx 1\,\mu$m) made from beryllium ($Z = 4$) or carbon ($Z = 6$) are used as entrance windows. The incident photon transfers its total energy to the photoelectron. This energy is now amplified during avalanche formation in a proportional fashion. Therefore, this technique not only allows to determine the direction of the incidence of the X-ray photon but also its energy.

With photomultipliers or channeltrons the incident photon is also converted via the photoelectric effect into an electron. This electron is then amplified by ionizing collisions in the discrete or continuous electrode system. The amplified signal can be picked up at the anode and further processed by electronic amplifiers.

The energy measurement of X-ray detectors is based on the number of charge carriers, which are produced by the photoelectron. In gas proportional chambers typically 30 eV are required to produce an electron–ion pair. Semiconductor counters possess the attractive feature that only approximately 3 eV are needed to produce an electron–hole pair. Therefore, the energy resolution of semiconductor counters is better by a factor of approximately $\sqrt{10}$ compared to proportional chambers. As solid-state materials silicon, germanium, or gallium arsenide can be considered.

Because of the easy availability and the favourable noise properties mostly silicon semiconductor counters are used.

If a silicon counter is subdivided in a matrix-like fashion into many quadratic elements (pixels), which are shielded against each other by potential wells, the produced energy depositions can be read out line by line. Because of the charge coupling of the pixels this type of silicon image sensor is also called charge-coupled device. Commercial CCDs with areas of 1 cm × 1 cm at a thickness of 300 μm have about 10^5 pixels. Even though the shifting of the charge in the CCD is a serial process, these counters have a relatively high rate capability. Presently, time resolutions of 1 ms up to 100 μs have been obtained. This allows rate measurements in the kHz range, which is extremely interesting for the observation of X-ray sources with high variability.

6.5.4   Observation of X-Ray Sources

The most remarkable discovery in all of astronomy is that the stars are made of atoms of the same kind as those on the earth.
                                                                                Richard P. Feynman

The Sun was the first star, of which X rays were recorded (Friedmann et al. 1951). In the range of X rays the Sun is characterized by a strong variability. In strong flares its intensity can exceed the X-ray brightness of the quiet Sun by a factor of 10 000.

In 1959 the first X-ray telescope was built (R. Giacconi, Nobel Prize 2002) and flown on an Aerobee rocket in 1962. During its six minutes flight time it discovered the first extrasolar X-ray sources in the constellation Scorpio. The observation time could be extended with the first X-ray satellite Uhuru (meaning 'freedom' in Swahili), which was launched from a base in Kenya 1970. Every week in orbit it produced more results than all previous experiments combined.

In the course of time a large number of X-ray satellites has provided more and more precision information about the X-ray sky. The satellite with the highest resolution up to 1999 was a common German–British–American project: the ROentgen SATellite ROSAT (see Fig. 6.69). ROSAT measured X rays in the range of 0.1–2.5 keV with a Wolter telescope of 83 cm diameter. As X-ray detectors, multiwire proportional chambers (PSPC)[12] with 25 arc seconds resolution and a channel-plate multiplier (HRI)[13] with 5 arc seconds resolution were in use. One of the PSPCs was permanently blinded by looking by mistake into the Sun. The second PSPC stopped operation after a data-taking period of 4 years because its gas supply was exhausted. Since then only the channel-plate multiplier was available as an X-ray detector. Compared to earlier X-ray satellites, ROSAT had a much larger geometrical acceptance, better angular and energy resolution, and a considerably increased signal-to-noise ratio: per angular pixel element the background rate was only one event per day.

---

[12]PSPC—Position Sensitive Proportional Chamber.

[13]HRI—High Resolution Imager.

Fig. 6.69 Image of the first X-ray satellite ROSAT; start in 1990, and active until 1999; re-entry into the atmosphere 2011 [99]

In a sky survey ROSAT has discovered about 130 000 X-ray sources. For comparison: the earlier flown Einstein Observatory HEAO[14] had only found 840 sources. The most frequent type of X-ray sources are nuclei of active galaxies ($\approx$65 000) and normal stars ($\approx$50 000). About 13 000 galactic clusters and 500 normal galaxies were found to emit X rays. The smallest class of X-ray sources are supernova remnants with approximately 300 identified objects.

In 1999 the X-ray satellite AXAF (Advanced X-ray Astrophysics Facility) was successfully started. In honor of the Indian–American astrophysicist Subrahmanyan Chandrasekhar it was renamed as Chandra. With Chandra (s. Fig. 6.70) and the 1999 started X-ray satellite XMM (X-ray Multi-Mirror Mission; renamed in 2000 in XMM-Newton resp. Newton Observatory; s. Fig. 6.71) improvements in angular resolution over ROSAT were achieved. The NuSTAR[15] dual X-ray telescope was launched in 2012 and is already in operation for seven years. NuSTAR extends the energy range compared to Chandra and XMM substantially up to 78 keV using a Wolter telescope (s. Fig. 6.72). Chandra, the Newton Observatory, and NuSTAR have good angular and energy resolutions and provide an excellent understanding of X-ray sources and the non-luminous matter in the universe. Instead of silicon devices, NuSTAR uses cadmium–zinc–tellurium (CdZnTe) state-of-the-art room-temperature semiconductors, which have a very high photon conversion efficiency. From the experimental point of view one has to take extreme care that the sensitive focal pixel detectors do not suffer radiation damage from low-energy solar particles (p, α, e) emitted during solar eruptions.

---

[14]HEAO—High Energy Astronomy Observatory.

[15]NuSTAR—Nuclear Spectroscopic Telescope Array.

"Our first X-ray image from the constellation Virgo!"

On the other hand, the X-ray mission ASTRO-E, started early in the year 2000, had to be abandoned because the booster rocket did not carry the satellite into an altitude required for the intended orbit. The satellite presumably burned up during re-entry into the atmosphere.

Supernova remnants (SNR) represent the most beautiful X-ray sources in the sky. ROSAT found that the Vela Pulsar also emits X rays with a period of 89 ms known from its optical emission. It appears that the X-ray emission of Vela X1 is partially of thermal origin. The supernova remnant SNR 1572 that was observed by the Danish astronomer Tycho Brahe shows a nearly spherically expanding shell in the X-ray range (Fig. 6.73). The shell expands into the interstellar medium with a velocity of about 50 km/s and it heats up in the course of this process to several million degrees.

The topology of the X-ray emission from the Crab Pulsar allows to identify different components: the pulsar itself is very bright in X rays compared to the otherwise

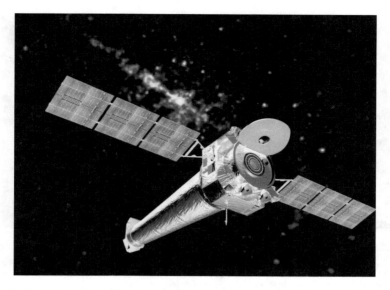

Fig. 6.70   The X-Ray satellite Chandra; start 1999, angular resolution 0.5 arc seconds [100]

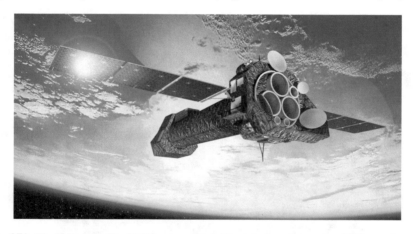

Fig. 6.71   The X-Ray satellite XMM, renamed into Newton observatory; start 1999, angular resolution 6 arc seconds [101]

more diffuse emission. The main component consists of a toroidal configuration, which is caused by synchrotron radiation of energetic electrons and positrons in the magnetic field of the pulsar. In addition, electrons and positrons escape along the magnetic field lines at the poles, where they produce X rays in a helical wind (see Fig. 6.74).

A large number of X-ray sources are binaries. In these binaries mostly a compact object—a white dwarf, a neutron star, or a black hole—accretes matter from a nearby

Fig. 6.72 The X-Ray satellite NuSTAR; start 2012, angular resolution 9.5 arc seconds. NuSTAR is smaller (about two meters only) compared to Chandra or XMM. It had to fit into the relatively small Pegasus launch vehicle. To achieve the required 10 m focal length needed for its dual telescopes, NuSTAR was mounted on a deployable mast, which was extended after the instrument was in orbit [102]

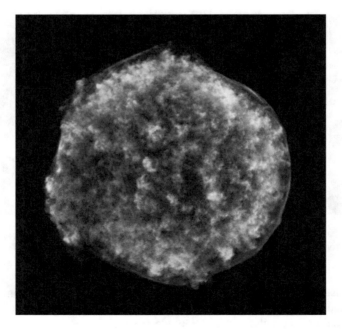

Fig. 6.73 X-ray image of the type Ia Supernova SN 1572 (Tycho), recorded with the Chandra telescope [100]

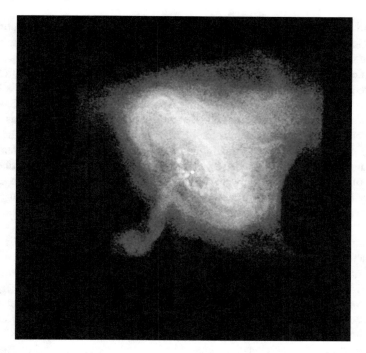

Fig. 6.74 X-ray image of the Crab Nebula, recorded with the Chandra telescope [100]

companion. The matter flowing to the compact object frequently forms an accretion disk (see, for example, Fig. 5.8), however, the matter can also be transported along the magnetic field lines landing directly on the neutron star. In such cataclysmic variables a mass transfer from the companion, e.g., to a white dwarf, can be sufficient to maintain a permanent hydrogen burning. If the ionized hydrogen lands on a neutron star also thermonuclear X-ray flashes can occur. Initially, the incident hydrogen fuses in a thin layer at the surface of the neutron star to helium. If a sufficient amount of matter is accreted, the helium produced by fusion can achieve such high densities and temperatures that it can be ignited in a thermonuclear explosion forming carbon.

The observation of thermal X rays from galactic clusters allows a mass determination of the hot plasma and the total gravitational mass of the cluster. This method is based on the fact that the temperature is a measure for the gravitational attraction of the cluster. A high gas temperature—characterized by the energy of the emitted X rays—represents via the gas pressure the counterforce to gravitation and prevents the gas from falling into the center of the cluster. Measurements of X rays from galactic clusters have established that the hot plasma between the galaxies is five times more massive than the galaxies themselves. The discovery of X-ray-emitting massive hot plasmas between the galaxies is a very important input for the understanding of the dynamics of the universe.

In the present understanding of the evolution of the universe all structures are hierarchically formed from objects of the respective earlier stages: stellar clusters combine to galaxies, galaxies form groups of galaxies, which grow to galactic clusters, which in turn produce superclusters. Distant, i.e., younger galactic clusters, get more massive while close-by galactic clusters hardly grow at all. This allows to conclude that nearby galactic clusters have essentially collected all available matter gravitationally. The mass of these clusters appears to be dominated by gas clouds, into which stellar systems are embedded like raisins in a cake. Therefore, the X-ray-emitting gas clouds allow to estimate the matter density in the universe. The X-ray observations from ROSAT suggest a value of approximately 30% of the critical mass density of the universe. If this were all, this would mean that the universe expands eternally (see Chap. 8). This figure of 30%—if correct—would imply that ROSAT has already seen dark matter, because baryons constitute only about 5% of the critical density.

Among others, Chandra has provided details of the Crab Nebula (Fig. 6.74). Chandra discovered a ring around the central pulsar, which was never seen before. Also X-ray emission from the supermassive black hole Sagittarius A* in the center of our Milky Way was observed. The earliest images of the shock waves from SN 1987A were seen by Chandra in X rays. Chandra also found evidence that sources originally assumed to be pulsars were supernova remnants, or possibly even other very compact objects like quark stars. Furthermore, Chandra has measured X-ray data on the collisions of superclusters, which might provide evidence for dark matter.

XMM-Newton has presented detailed results on the X-ray spectroscopy concerning the corona of stars. A very precise X-ray sky survey allowed to understand the evolution of active galactic nuclei in the early universe. Furthermore, XMM-Newton has determined the rotational velocities of black holes. Just as Chandra, XMM-Newton has also charted many supernova explosions in great detail.

Figure 6.75 shows a detailed XMM image of the galaxy NGC 7314. NGC 7314 is a spiral galaxy in the constellation southern fish (Piscis Austrinus, distance 50 million light-years). This galaxy was discovered by Herschel as early as 1834. XMM-Newton was lucky to find in the X-ray image of this galaxy another, very distant X-ray cluster, which only appeared in the projection as part of the foreground galaxy.

Among others, NuSTAR has obtained an excellent high-energy X-ray view of a portion of our Andromeda Galaxy at a distance of 2.5 million light-years (see Fig. 6.76). In addition, many X-ray binaries, also those consisting of black holes or neutron stars have been observed.

Chandra, XMM Newton, and NuSTAR have discovered a large number of new X-ray sources. Of particular interest are active galactic nuclei, which are supposed to be powered by black holes. Many or most of the black holes residing at the centers of distant galaxies may be difficult to find since they are hidden deep inside vast amounts of absorbing dust so that only energetic X rays or γ rays can escape. Already now these new X-ray satellites have observed galaxies at high redshifts emitting huge amounts of energy in the form of X rays, far more than can ever expected to be produced by star formation. Therefore, it is conjectured that these galaxies must contain actively accreting supermassive black holes.

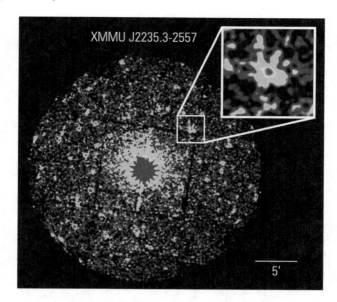

Fig. 6.75 Image of the galaxy NGC 7314 by XMM-Newton. By lucky coincidence the cluster XMMU J2235.3-2557 at a distance of 9000 million parsec (shown in the *white box in the upper right*) could be identified in this image of the galaxy [103]

Fig. 6.76 Image of the Andromeda Nebula featuring a part of the galaxy observed by NuSTAR in X rays. The central part of Andromeda, which cannot be resolved in the optical image, is found to consist of many different X-ray sources. The background image of Andromeda was taken by NASA's Galaxy evolution explorer in ultraviolet light [104]

Fig. 6.77   ROSAT X-ray of the Moon [99]

Even though presently known classes of sources will probably dominate the statistics of Chandra, XMM-Newton, and NuSTAR discoveries, the possibility of finding completely new and exciting populations of X-ray sources also exists.

A very recent result is the observation of the collision of two galactic clusters in Abell 754 at a distance of about 9 million light-years in the light of X rays. These clusters with millions of galaxies merge in a catastrophic collision into one single very large cluster.

The diffuse X-ray background, which was discovered relatively early, consists to a large extent (75%) of resolved extragalactic sources. It could easily be that the remaining diffuse part of X rays consists of so far non-resolved distant X-ray sources.

In spectra of many X-ray sources the iron line of 5.9 keV is observed. This is evidence either for the formation of iron in supernova explosions or originates from earlier sources, whose material was already processed in older star generations.

A surprising result of the recent investigations was that practically all stars emit X rays. A spectacular observation was also the detection of X rays from the Moon. However, the Moon does not emit these X rays itself. It is rather reflected corona radiation from our Sun in the same way as the Moon also does not shine in the optical range but rather reflects the sunlight (Fig. 6.77).

6.6 Gravitational-Wave Astronomy

What would physics look like without gravitation?
Albert Einstein

What is crooked cannot be straightened.
Ecclesiastes 1:15, The Bible

The discovery of gravitational waves, predicted by Einstein, by the LIGO telescopes in 2015 came as a surprise. It opened up a new window to the universe. LIGO measured for the first time a gravitational-wave signal from the merger of two black holes with their two independent Michelson interferometers. LIGO (Advanced Laser Interferometer Gravitational Wave Observatory) with its interferometer arms' length of 4 km had searched already for gravitational waves from 2002 until 2010, however, without success in this period. LIGO was then substantially improved and was actually successful in the first measurement period after the upgrade by measuring a convincing signal from a merger of two black holes (s. Figs. 6.78 and 6.79 [19]). This gravitational-wave signal produced a relative elongation of the interferometer arms of 10^{-21}. In the present setup LIGO can measure length variations of 10^{-19} m. This corresponds to less than one thousandth of the proton diameter.

For successful operation of LIGO the quality of the LASER beams and the mirrors in the interferometer is of outstanding importance. The achievable resolution of an

Fig. 6.78 Merging of two black holes, measured by LIGO, shown in a computer simulation. Photo credit: Caltech/MIT/LIGO Laboratory [19]

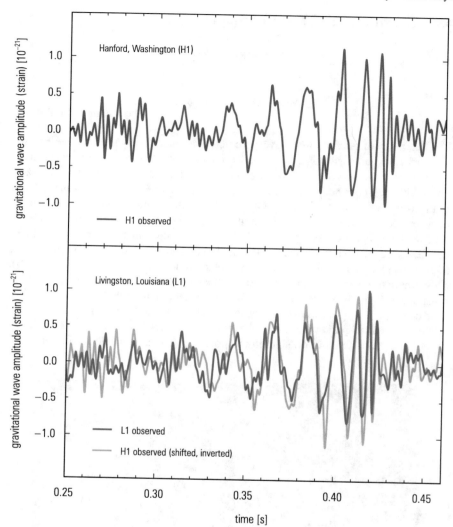

Fig. 6.79 Signals as measured by the two gravitational-wave detectors in the two detector stations in Hanford and Livingston. Shown are the light signals of the photodiodes at the point of interference of the two detector arms, which represent a compression and elongation of space-time (*top, in red*: Hanford (observed); *below, in blue*: Livingston (observed) compared with the Hanford signal, *in red* (observed, shifted, and inverted)). Photo credit: Caltech/MIT/LIGO Laboratory [19]

interferometer depends on the photon statistics of the LASER beam, i.e., on the quantum noise. The quantum nature of light leads to small phase and amplitude variations, which is called the shot noise. These fluctuations are symmetrical for classical light for phase and amplitude. To reduce these inherent quantum effects,

LIGO used the technique of squeezed light. Squeezed light reduces either the phase or the amplitude of light at the expense of the other component. Coherent squeezed states of light can be produced using birefringent non-linear optics. The squeezed light is then detected with homodyne detectors. This type of detection represents a method of extracting information from the LASER signal by measuring the strength of amplitude and phase fluctuations, i.e., the modulation of the light field, rather than aiming for a photon-number measurement. In addition, LIGO installed active and passive vibration isolation by using an elaborate setup of test masses and heavy quadrupole pendulum systems. These significant improvements enabled to reduce the noise in the antenna system substantially.

LIGO consists of two stations in Hanford, Washington and Livingston, Louisiana, at a distance of about 3000 km. With a time delay of 6.9 ms both stations recorded a textbook-like gravitational signal (s. Fig. 6.79). The signal originated from the merger of a binary consisting of two black holes with masses of $36\,M_\odot$ and $29\,M_\odot$, which circled each other with increasing frequency. Due to the energy loss in this process they approached each other and finally merged. Due to angular-momentum conservation their rotational frequency around the common center of mass increased, leading to the time-dependent gravitational-wave signal. The binary is in the southern sky at a distance of about 400 Mpc. The energy emitted in gravitational waves corresponds to the mass loss of the two merging black holes and is estimated to be about $3\,M_\odot \cdot c^2$.

The localization of the source of the gravitational signal is a difficult matter. The time delay of 6.9 ms between the two detector stations allows to determine a ring in the sky, from which the signal originated. The variability of the signal in the two interferometers enables further refinements. This lead to an estimation of the origin of the signal source of 1.4 billion light-years (440 Mpc). This corresponds to a distance far beyond the supergalactic plane, which is at a distance of about 200 million light-years.

LIGO has meanwhile been complemented by VIRGO in Pisa, Italy and will be further extended by a station in India. Having several stations allows to localize the position of possible sources of gravitational waves with higher accuracy.

Only 0.4 s after the gravitational-wave signal from LIGO, the space telescope Fermi (FGST) has obtained evidence for a weak source of gamma rays above 50 keV from the same region of the sky. The chance coincidence of the LIGO signal with the X-ray signal from Fermi is only 0.2%. The position of this Fermi X-ray source is also not correlated with other astrophysical, solar, or terrestrial events.

Only a little later (also in 2015) a second gravitational signal from LIGO was measured. This was also a very robust signal of two colliding and coalescing black holes, which eventually merged into a common black hole. The masses of these two merging black holes were determined to be 14 and 8 solar masses. Meanwhile a third gravitational signal from LIGO (31 and 19 solar masses) and further candidates were recorded.

In 2017 LIGO has recorded gravitational waves from the collisions of two neutron stars, known as a kilonova. The masses of the two colliding objects were too small to originate from a black-hole merger. This spectacular event was also seen with electromagnetic radiation in various other spectral ranges (see also Sect. 6.4.5 on gamma-ray bursters) and by the VIRGO gravitational interferometer. With these outstanding results gravitational-wave astronomy has opened up a new window to the universe.

However, gravitational-wave astronomy is still in its infancy. It is a new, young branch in astronomy promising important results about the early universe. Gravitational waves had been predicted by Einstein as early as 1916. Apart from the observation of Taylor and Hulse (see below) concerning the energy loss of a binary pulsar (PSR 1913+16) due to the emission of gravitational waves over a period starting from 1974 (Nobel Prize 1993) there was no direct evidence of the existence of gravitational waves. Nobody doubted the correctness of Einstein's prediction, especially since the results of Taylor and Hulse on the energy loss of the binary pulsar system by emission of gravitational radiation agree with the theoretical expectation of general relativity impressively well (to better than 0.1%).

Taylor and Hulse had observed the binary system PSR 1913+16 consisting of a pulsar and a neutron star over a period of more than 30 years. The two massive objects rotate around their common center of mass on elliptical orbits. The radio emission from the pulsar can be used as precise clock signal. When the pulsar and neutron star are closest together (periastron), the orbital velocities are largest and the gravitational field is strongest. For high velocities and in a strong gravitational field time is slowed down. This relativistic effect can be checked by looking for changes in the arrival time of the pulsar signal. In this massive and compact pulsar system the periastron time changes in a single day by the same amount, for which the planet Mercury needs a century in our solar system. Space-time in the vicinity of the binary is greatly warped.

The theory of relativity predicts that the binary system will lose energy with time as the orbital rotation energy is converted into gravitational radiation. Figure 6.80 shows the prediction based on Einstein's theory of general relativity in comparison to the experimental data. The excellent agreement between theory and experiment presented at that time the best—albeit indirect—evidence for gravitational waves.

The technique of pulsar timing (PTA = Pulsar Timing Array) allows to give evidence on a possible gravitational-wave background. This requires an accurate investigation of the arrival times of signals from the pulsars. If a gravitational wave would cross the passage of a photon from the pulsar, the curvature of space-time induced by the gravitational wave would modify the timing of the pulsar signal. With such a pulsar timing array the signals of 24 different pulsars were monitored over a period of 11 years. No effect of arrival-time modification was observed. If a gravitational-wave background would exist, the Parkes Pulsar Timing Array should have seen something [105].

The direct observation of gravitational waves has opened now a new window of astronomy to investigate the most violent and turbulent astrophysical processes in the universe. It might give information on processes, where dark matter or dark energy is involved. The observation of the gravitational waves from the early universe and the Big Bang is a real challenge.

However, as far as the direct observation of gravitational waves is concerned, the situation is to a certain extent similar to neutrino physics around 1950. At that

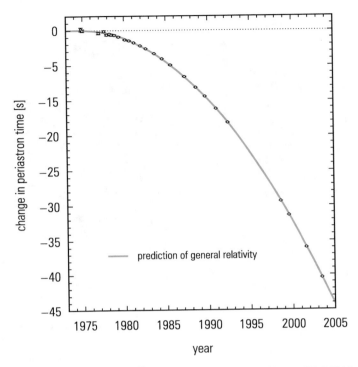

Fig. 6.80 Observed variation of the periastron time in the binary system PSR B1913+16 over a period of 30 years compared to the expectation based on Einstein's theory of general relativity. The agreement between theory and observation is better than 0.1% [106]

time nobody really doubted the existence of the neutrino, but there was no strong neutrino source available to test the prediction. Only the oncoming nuclear reactors were sufficiently powerful to provide a large enough neutrino flux to be observed. The problem was related to the low interaction cross section of neutrinos with matter.

Compared to gravitational waves the neutrino interaction with matter can be called 'strong'. The extremely low interaction probability of gravitational waves ensures that they provide a new window to cataclysmic processes in the universe at the expense of a very difficult detection. For neutrinos—depending on the energy—most astrophysical sources are almost transparent. Only at very high energies, where the cross section increases significantly, stellar objects become opaque. Because of the feeble interaction of gravitational waves they can propagate out of the most violent cosmological sources even more freely compared to neutrinos.

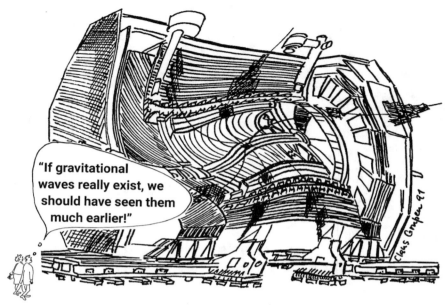

With electromagnetic radiation in various spectral ranges astronomical objects can be imaged. This is because the wavelength of electromagnetic radiation is generally very small compared to the size of astrophysical objects. Gravitational waves are transversal waves and exist in two polarization modes. Their frequency is in the range of 10^{-4} and 10^4 Hz, and their wavelengths are much larger compared to electromagnetic radiation so that imaging with this radiation is almost out of question. In electromagnetic radiation time-dependent electric and magnetic fields propagate through space-time. For gravitational waves one is dealing with oscillations of space-time itself.

Electromagnetic radiation is emitted when electric charges are accelerated or decelerated. In a similar way gravitational waves are created whenever there is a non-spherical acceleration of mass–energy distributions. There is, however, one important difference. Electromagnetic radiation is dipole radiation, while gravitational waves present quadrupole radiation. This is equivalent to saying that the quantum of gravitational waves, the graviton, has spin $2\hbar$.

Electromagnetic waves are created by time-varying dipole moments, where a dipole consists of one positive and one negative charge. In contrast, gravity has no charge, there is only positive mass. Negative mass does not exist. Even antimatter has the same positive mass just as ordinary matter. The antiproton has exactly the same mass as the proton (experimental limit $|m_p - m_{\bar{p}}| < 0.66$ eV). Therefore, it is not possible to produce an oscillating mass dipole. In a two-body system one mass accelerated to the left creates, because of momentum conservation, an equal and opposite action on the second mass that moves it to the right. For two equal masses the spacing may change but the center of mass remains unaltered. Consequently, there is no monopole or dipole moment. Therefore, the lowest order of oscillation

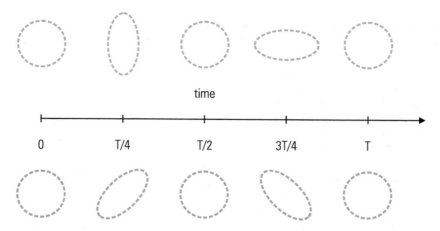

Fig. 6.81 Oscillation modes of a spherical antenna for gravitational radiation. Upon incidence of a gravitational wave the antenna will be excited to quadrupole modes. Depending on the polarization of the gravitational wave the deformation of the antenna results along a vertical axis (*on top*, the so-called + polarisation) or along an axis rotated by 45 degrees (*lower part*, the so-called × polarisation) [107]

generating gravitational waves is due to a time-varying quadrupole moment. In the same way as a distortion of test masses creates gravitational waves, where, e.g., the simplest non-spherical motion is one, in which horizontal masses move inside and vertical masses move apart, a gravitational wave will distort an antenna analogously by compression in one direction and elongation in the other (see Fig. 6.81).

The *quadrupole character of gravitational radiation* therefore leads to an action like a tidal force: it squeezes the antenna along one axis while stretching it along the other. Due to the weakness of the gravitational force the relative elongation of an antenna will be at most on the order of $h \approx 10^{-21}$ even for the most violent cosmic catastrophes. This corresponds to a distortion of space-time, which would decrease the distance from Sun to Earth by the diameter of the hydrogen atom.

There is, however, one advantage of gravitational waves compared to the measurement of electromagnetic radiation: Electromagnetic observables like the energy flux from astrophysical sources are characterized by a $1/r^2$ dependence due to solid-angle reasons. By contrast, the direct observable of gravitational radiation (h) decreases with distance only like $1/r$ (the energy flux, however, shows the same radial dependence as for electromagnetic waves). h depends linearly on the second derivative of the quadrupole moment of the astrophysical object and it is inversely proportional to the distance r,

$$h \sim \frac{G}{c^4} \frac{\ddot{Q}}{r} \quad (G: \text{Newton's constant}). \tag{6.6.1}$$

Consequently, an improvement of the sensitivity of a gravitational detector by a factor of 2 increases the measurable volume, where sources of gravitational waves may reside, by a factor of 8. The disadvantage of limited ability to image with gravitational radiation goes along with the advantage that gravitational-wave detectors have a nearly 4π steradian sensitivity over the sky.

The most promising candidates as sources for gravitational radiation are mergers of binary systems, accreting black holes, collisions of neutron stars, or special binaries consisting of two black holes orbiting around their common center of mass (like in the radio galaxy 3C66B, which appears to be the result of a merger of two galaxies).

The suppression of noise in these antennae is the most difficult problem. There have been stand-alone gravitational-wave detectors, mostly in the form of metal cylinders. Pioneering work in this field was done by J. Weber in 1969. Weber used an aluminum cylinder of 1.5 tons as antenna for gravitational waves as resonator. Piezo sensors were mounted on the surface of the cylinder to signal its deformations upon the impact of a gravitational wave. Because of the background noise only coincident signals from at least two antennas would be required. Weber claimed to have seen such coincident signals over a distance of 1000 km, but his results could never be reproduced.

The current gravitational-wave detectors are all based on the principle of Michelson interferometry. The interferometer GEO600 in Hannover, Germany, with an arm length of 600 m, has contributed essential ideas on the improvements on the sensitivity of these interferometers, which went into the LIGO upgrade.

VIRGO in Pisa, Italy with an arm length of 3 km, which is effectively increased by multiple reflections to 120 km, is in operation since 2017. Virgo is sensitive in the frequency range 10–1000 Hz. There are also plans to set up gravitational-wave detectors on the ground or in space in Japan (KAGRA[16] and DECIGO[17]) and Australia (AIGO[18]).

LISA (Laser Interferometer Space Antenna) is a very ambitious project to install a Michelson interferometer in space [108]. LISA is planned to consist of three satellites arranged in an equilateral triangle with sides 2.5 million km long. The three satellites will fly on Earth-like heliocentric orbits. The distance between the satellites has to be precisely monitored to enable the detection of a passing gravitational wave. A space distortion of $h = 10^{-21}$ would lead to a variation of the arm length of about 5 pm in LISA, corresponding to a tenfold diameter of an iron *nucleus*. This requires to install the satellites to a precision of better than a few *picometer* in space, which is a real technical challenge. LISA is mainly meant for the detection of low-frequency gravitational waves, which are supposed to originate from collisions of galaxies, mergers of extremely massive black holes, or of heavy binaries in our galaxy. To measure low-frequency gravitational waves of around 0.1–1 Hz with their minute elongations and compressions, background noise from Earth-like or thermal noise and influences of gravitational gradients at Earth must be excluded. Therefore, a Michelson interferometer in space is favoured or even necessary.

The essential features of LISA have been successfully tested with the LISA pathfinder. The goals of LISA pathfinder have been reached and its performance was even better than expected.

Another more mundane problem is the financing of such a multi-lateral project. The United States left the project because of budget cuts, and LISA has been replaced by the European project eLISA (Evolved Laser Interferometer Space Antenna). The start of eLISA is foreseen for 2034.

On April 10, 2019 a network of eight antennas, the Event Horizon Telescope Collaboration, has published a photo of the black hole from the center of M87. Most images of black holes in the past were just illustrations. The black hole itself is, of course, invisible, but outside the event horizon large quantities of matter swirl around the black hole and emit radiation in various spectral ranges. The Event Horizon Telescope team has combined the information from eight telescopes distributed over the whole world to form an interferometric image of the hot matter whirling around the black hole like in a maelstrom. Since the interferometric information of the various telescopes essentially comes from the radio domain, the reconstructed image must be shown in false colour to make it visible. The photo then shows in principle the hot matter in the accretion disk rotating around the black hole sitting in the center of M87, see Fig. 6.82.

Gravitational-wave detectors have opened up an exciting new window into the universe. With gravitational waves one has a chance to directly look back into the Big Bang, because the universe is transparent for this type of radiation. For electromagnetic waves the early phase of the universe is barred. Using gravitational

---

[16]KAGRA—Kamioka Gravitational Wave Detector.

[17]DECIGO—DECi-hertz Interferometer Gravitational wave Observatory (future Japanese space gravitational-wave antenna).

[18]AIGO—Australian International Gravitational Observatory.

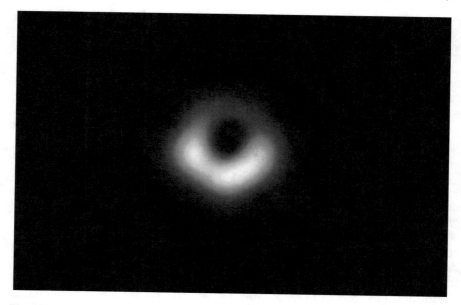

Fig. 6.82 Reconstructed radio image of the black hole in M87 based on information from worldwide distributed radio antennas by the event horizon telescope collaboration [109]

radiation, electromagnetic waves, neutrinos, and other astronomical techniques calls for a multi-messenger astronomy for the exploration of our universe.

Summary

Particles from space are either charged particles (nuclei and electrons), neutral particles (neutrinos), electromagnetic radiation in various spectral ranges (radio, optical, X rays, or gamma radiation), or gravitational radiation. All these particles, which are messengers from our Milky Way or beyond, provide specific information on the universe. Charged particles tell us about the chemical composition of primary cosmic rays, gamma rays and X rays allow to identify sources in the high-energy domain. However, because of absorption of electromagnetic radiation, one can only study the surfaces of astronomical objects. Neutrinos, on the other hand, provide directly a look into the interior of the sources. This goes along with difficulties to detect them. Even more so, gravitational waves allow to explore the most violent processes in the sky. A dream of astronomers is to combine all these detection techniques in a multi-messenger effort to study the universe and its evolution in the course of time.

6.7 Problems

Problems for Sect. 6.1

1. What is the reason that primary cosmic-ray nuclei like carbon, oxygen, and neon are more abundant than their neighbours in the periodic table of elements (nitrogen, fluorine, sodium)?
2. In Sect. 6.1 it is stated that primary cosmic rays consist of protons, α particles, and heavy nuclei. Only 1% of the primary particles are electrons. Does this mean that the planet Earth will be electrically charged in the course of time because of the continuous bombardment by predominantly positively charged primary cosmic rays? What kind of positive excess charge could have been accumulated during the period of existence of our planet if this were true?
3. The chemical abundance of the sub-iron elements in primary cosmic rays amounts to about 10% of the iron flux.

    (a) Estimate the fragmentation cross section for collisions of primary iron nuclei with interstellar/intergalactic nuclei.
    (b) What is the chance that a primary iron nucleus will survive to sea level?

Problems for Sect. 6.2

1. Neutrons as candidates for the highest-energy cosmic rays have not been discussed so far. What are the problems with neutrons?
2. The Oh-My-God event observed by John Linsley at Dugway Proving Ground had an energy of 3×10^{20} eV. Assuming that the particle was initiated by a proton, work out the velocity of this extremely-high-energy event.
3. Let us assume that the Oh-My-God event (see Problem 2 for this section) originated from the Andromeda Galaxy (distance 2.2 million light-years). How would a massless alien traveling along with the energetic proton perceive the flight time when he were sitting on top of this proton?
4. There are many estimates for cross sections for the detection of dark-matter particles. Many predictions for a nuclear cross section of a 1-TeV SUSY dark-matter particle range in the region of around 10^{-9} pb. If this were correct, would ICE-CUBE have a chance to see such particles?
5. Cryptons are hypothetical superheavy particles, which are thought to exist in string theories. Superheavy cryptons have been proposed as possible dark-matter particles. They could also be responsible for the extremely-high-energy cosmic-ray events beyond the Greisen–Zatsepin–Kuzmin cutoff. They are supposed to be sufficiently long-lived to be still around.
    Work out the crypton flux if they were responsible for the 11 cosmic-ray events with energies $\geq 10^{20}$ observed in the Japanese AGASA experiment. The AGASA air-shower experiment covers an area of about $100\,\text{km}^2$ and was operated over 14 years. Estimate the crypton flux assuming a cross section of $\approx 10^{-8}$ pb for their interaction in the atmosphere.

Problems for Sect. 6.3

1. The Sun converts protons into helium according to the reaction

$$4p \rightarrow {}^4\text{He} + 2e^+ + 2\nu_e \,.$$

The solar constant describing the power of the Sun at Earth is $P \approx 1400\,\text{W/m}^2$. The energy gain per reaction corresponds to the binding energy of helium ($E_B({}^4\text{He}) = 28.3\,\text{MeV}$). How many solar neutrinos arrive at Earth?

2. If solar electron neutrinos oscillate into muon or tau neutrinos they could in principle be detected via the reactions

$$\nu_\mu + e^- \rightarrow \mu^- + \nu_e \,, \quad \nu_\tau + e^- \rightarrow \tau^- + \nu_e \,.$$

Work out the threshold energy for these reactions to occur.

3. Radiation exposure due to solar neutrinos.

   (a) Use (6.3.1) to work out the number of interactions of solar neutrinos in the human body (tissue density $\varrho \approx 1\,\text{g cm}^{-3}$).

   (b) Neutrinos interact in the human body by

   $$\nu_e + N \rightarrow e^- + N' \,,$$

   where the radiation damage is caused by the electrons. Estimate the annual dose for a human under the assumption that on average 50% of the neutrino energy is transferred to the electron.

   (c) The equivalent dose is defined as

   $$H = (\Delta E/m)\, w_R \tag{6.7.1}$$

   (m is the mass of the human body, w_R the radiation weighting factor ($= 1$ for electrons), $[H] = 1\,\text{Sv} = 1 w_R\,\text{J kg}^{-1}$, and ΔE the energy deposit in the human body). Work out the annual equivalent dose due to solar neutrinos and compare it with the normal natural dose of $H_0 \approx 2\,\text{mSv/a}$.

4. Neutrino oscillations.[19]

   In the most simple case neutrino oscillations can be described in the following way (as usual $\hbar$ and c will be set to unity): in this scenario the lepton flavour eigenstates are superpositions

   $$|\nu_e\rangle = \cos\theta |\nu_1\rangle + \sin\theta |\nu_2\rangle \,,$$
   $$|\nu_\mu\rangle = -\sin\theta |\nu_1\rangle + \cos\theta |\nu_2\rangle$$

---

[19]This problem is difficult and its solution is mathematically demanding.

of mass eigenstates $|\nu_1\rangle$ and $|\nu_2\rangle$. All these states are considered as wave packets with well-defined momentum. In an interaction, e.g., a ν_e is assumed to be generated with momentum p, which then propagates as free particle, $|\nu_e; t\rangle = e^{-iHt}|\nu_e\rangle$. For the mass eigenstates one has $e^{-iHt}|\nu_i\rangle = e^{-iE_{\nu_i}t}|\nu_i\rangle$ with $E_{\nu_i} = \sqrt{p^2 + m_i^2}$, $i = 1, 2$. The probability to find a muon neutrino after a time t is

$$P_{\nu_e \to \nu_\mu}(t) = |\langle \nu_\mu | \nu_e; t\rangle|^2.$$

(a) Work out $P_{\nu_e \to \nu_\mu}(t)$. After a time t the particle is at $x \approx vt = pt/E_{\nu_i}$. Show that—under the assumption of small neutrino masses—the oscillation probability as given in (6.3.14) can be derived.

(b) Estimate for $(\nu_\mu \leftrightarrow \nu_\tau)$ oscillations with maximum mixing ($\sin 2\theta = 1$), $E_\nu = 1\,\text{GeV}$ and under the assumption of $m_{\nu_\mu} \ll m_{\nu_\tau}$ (that is, $m_{\nu_\tau} \approx \sqrt{\delta m^2}$) the mass of the τ neutrino, which results if the ratio of upward-to-downward-going atmospheric muon neutrinos is assumed to be 0.54.

(c) Does the assumption $m_{\nu_\mu} \ll m_{\nu_\tau}$ make sense here?

5. Work out the radiation exposure by geoneutrinos for humans. Here only the effect of the charged lepton in the neutrino reaction shall be considered.
6. Work out the heat power generated by geoneutrinos in the Earth.
7. The neutrino–nucleon cross section is very small but rises with energy, compare Eq. (6.3.46). How many high-energy neutrinos can make it through the Moon without interaction, e.g., what is the transparency of the Moon for 100-TeV neutrinos for central incidence?

Problems for Sect. 6.4

1. Estimate the detection efficiency for 1-MeV photons in a NaI(Tl) scintillation counter of 3 cm thickness.
   (Hint: Use the information on the mass attenuation coefficient from Fig. 6.49.)
2. Estimate the size of the cosmological object that has given rise to the γ-ray light curve shown in Fig. 6.60.
3. What is the energy threshold for muons to produce Cherenkov light in air ($n = 1.000\,273$) and water ($n = 1.33$)?
4. The Crab Pulsar emits high-energy γ rays with a power of $P = 3 \times 10^{27}\,\text{W}$ at 100 GeV. How many photons of this energy will be recorded by the FGST experiment (Fermi Gamma-ray Space Telescope; formerly GLAST: Gamma-ray Large Area Space Telescope) per year if isotropic emission is assumed? What is the minimum flux from the Crab that FGST will be able to detect (in $\text{J}/(\text{cm}^2\,\text{s})$)? (Collecting area of FGST: $A = 8000\,\text{cm}^2$, distance of the Crab: $R = 3400$ light-years.)
5. The solar constant describing the solar power arriving at Earth is $P_S \approx 1400\,\text{W}/\text{m}^2$.

   (a) What is the total power radiated by the Sun?
   (b) Which mass fraction of the Sun is emitted in 10^6 years?
   (c) What is the daily mass transport from the Sun delivered to Earth?

Problems for Sect. 6.5

1. The radiation power emitted by a blackbody in its dependence on the frequency is given by (6.5.1). Convert this radiation formula into a function depending on the wavelength.

2. The total energy emitted per second by a star is called its luminosity. The luminosity depends both on the radius of the star and its temperature. What would be the luminosity of a star ten times larger than our Sun ($R = 10\,R_\odot$) but at the same temperature? What would be the luminosity of a star of the size of our Sun but with ten times higher temperature?

3. The power radiated by a relativistic electron of energy E in a transverse magnetic field B through synchrotron radiation can be worked out from classical electrodynamics to be

$$P = \frac{e^2 c^3}{2\pi} C_\gamma E^2 B^2 \,,$$

where

$$C_\gamma = \frac{4}{3}\pi \frac{r_e}{(m_e c^2)^3} \approx 8.85 \times 10^{-5}\,\mathrm{m\,GeV^{-3}}\,.$$

Work out the energy loss due to synchrotron emission per turn of a 1-TeV electron in a circular orbit around a pulsar at a distance of 1000 km. What kind of magnetic field had the pulsar at this distance?

4. Consider the energy loss by radiation of a particle moving in a transverse homogeneous time-independent magnetic field, see also Problem 3 for this section. The radiated power is given by

$$P = \frac{2}{3}\frac{e^2}{m_0^2 c^3}\gamma^2 |\dot{\boldsymbol{p}}|^2 \,.$$

Calculate the time dependence of the particle energy and of the bending radius of the trajectory for the

(a) ultrarelativistic case,
(b) general case.[20]

5. An X-ray detector on board of a satellite with collection area of $1\,\mathrm{m}^2$ counts 10-keV photons from a source in the Large Magellanic Cloud with a rate of $1/h$. How many 10-keV photons are emitted from the source if isotropic emission is assumed? The detector is a Xe-filled proportional counter of 1 cm thickness (the attenuation coefficient for 10-keV photons is $\mu = 125\,(\mathrm{g/cm^2})^{-1}$, density $\varrho_{Xe} = 5.8 \times 10^{-3}\,\mathrm{g/cm^3}$).

6. X rays can be produced by inverse Compton scattering of energetic electrons (E_i) off blackbody photons (energy ω_i). Show that the energy of the scattered photon ω_f is related to the scattering angles φ_i and φ_f by

---

[20]The solution to this problem is mathematically demanding.

$$\omega_f \approx \omega_i \, \frac{1 - \cos \varphi_i}{1 - \cos \varphi_f} \, , \qquad (6.7.2)$$

where φ_i and φ_f are the angles between the incoming electron and the incoming and outgoing photon. The above approximation holds if $E_i \gg m_e \gg \omega_i$.

7. What is the temperature of a cosmic object if its maximum blackbody emission occurs at an energy of $E = 50\,\text{keV}$?

   (Hint: The solution of this problem leads to a transcendental equation, which needs to be solved numerically.)

Problems for Sect. 6.6

1. A photon propagating to a celestial object of mass M will gain momentum and will be shifted towards the blue. Work out the relative gain of a photon approaching the Sun's surface from a height of $H = 1\,\text{km}$. Analogously, a photon escaping from a massive object will be gravitationally redshifted.

   (Radius of the Sun $R_\odot = 6.9635 \times 10^8$ m, Mass of the Sun $M_\odot = 1.993 \times 10^{30}$ kg, acceleration due to Sun's gravity $g_\odot = 2.7398 \times 10^2$ m/s^2.)

2. Accelerated masses radiate gravitational waves. The emitted energy per unit time is worked out to be

$$P = \frac{G}{5c^5} \, \dddot{Q}^2 \, ,$$

where Q is the quadrupole moment of a certain mass configuration (e.g., the system Sun–Earth). For a rotating system with periodic time dependence ($\sim \sin \omega t$) each time derivative contributes a factor ω, hence

$$P \approx \frac{G}{5c^5} \, \omega^6 \, Q^2 \, .$$

For a system consisting of a heavy-mass object like the Sun (M) and a low-mass object, like Earth (m), the quadrupole moment is on the order of mr^2. Neglecting numerical factors of order unity, one gets

$$P \approx \frac{G}{c^5} \, \omega^6 \, m^2 \, r^4 \, .$$

Work out the power radiated from the system Sun–Earth and compare it with the gravitational power emitted from typical fast-rotating laboratory equipments.

Chapter 7
Secondary Cosmic Rays

There are more things in heaven and earth, Horatio, than are
dreamt of in your philosophy.

Shakespeare, Hamlet

For the purpose of astroparticle physics the influence of the Sun's and the Earth's
magnetic field is a perturbation, which complicates a search for the sources of cosmic
rays. The solar activity produces an additional magnetic field, which prevents part
of galactic cosmic rays from reaching Earth. Figure 7.1, however, shows that the
influence of the Sun is limited to primary particles with energies below 10 GeV.
The flux of low-energy primary cosmic-ray particles is anti-correlated to the solar
activity.

On the other hand, the solar wind, whose magnetic field modulates primary cosmic rays, is a particle stream in itself, which can be measured at Earth. The particles
constituting the solar wind (predominantly protons and electrons) are of low energy
(MeV region). These particles are captured to a large extent by the Earth's magnetic
field in the Van Allen belts or they are absorbed in the upper layers of the Earth's
atmosphere (see Fig. 1.10). Figure 7.2 shows the flux densities of protons and electrons in the Van Allen belts. The proton belt extends over altitudes from 2000 to
15 000 km. It contains particles with intensities up to $10^8/(\text{cm}^2\,\text{s})$ and energies up to
1 GeV. The electron belt consists of two parts. The inner electron belt with flux densities of up to 10^9 particles per cm^2 and s is at an altitude of approximately 3000 km,
while the outer belt extends from about 15 000 to 25 000 km. The inner part of the
radiation belts is symmetrically distributed around the Earth while the outer part is
subject to the influence of the solar wind and consequently deformed by it.

© Springer Nature Switzerland AG 2020
C. Grupen, *Astroparticle Physics*, Undergraduate Texts in Physics,
https://doi.org/10.1007/978-3-030-27339-2_7

Fig. 7.1 Modulation of the
primary spectrum by the
11-year cycle of the Sun

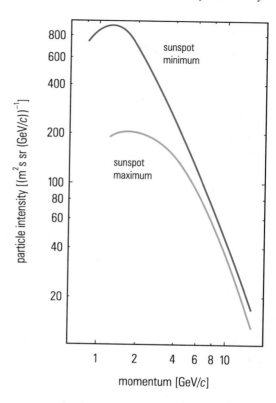

7.1   Propagation in the Atmosphere

Astroparticles are messengers from different worlds.

Anonymous

Primary cosmic rays are strongly modified by interactions with atomic nuclei in the
atmospheric air. The column density of the atmosphere amounts to approximately
$1000\,\text{g/cm}^2$, corresponding to an atmospheric pressure of about 1000 hPa. The residual atmosphere for flight altitudes of scientific balloons ($\approx$35–40 km) corresponds to
approximately several g/cm^2. For inclined directions the thickness of the atmosphere
increases strongly (approximately like $1/\cos\theta$, with θ—zenith angle).

For the interaction behaviour of primary cosmic rays the thickness of the atmosphere in units of the characteristic interaction lengths for the relevant particles
species in question is important. The *radiation length* for photons and electrons in
air is $X_0 = 36.66\,\text{g/cm}^2$. The atmosphere therefore corresponds to a depth of 27 radiation lengths. The relevant *interaction length* for hadrons in air is $\lambda = 90.0\,\text{g/cm}^2$,
corresponding to 11 interaction lengths per atmosphere. This means that practically
not a single particle of original primary cosmic rays arrives at sea level. Already
at altitudes of 15–20 km primary cosmic rays interact with atomic nuclei of the

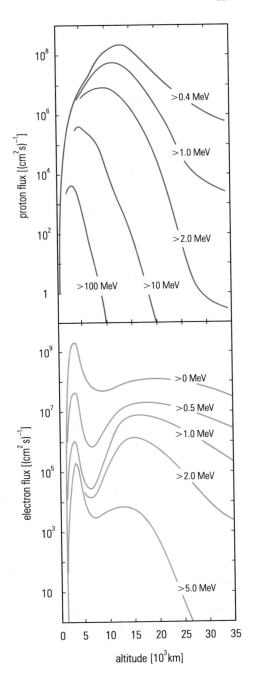

Fig. 7.2 Flux densities of protons and electrons in the radiation belts of the Earth

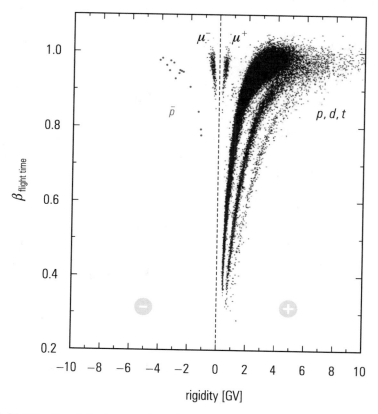

Fig. 7.3 Identification of singly charged particles in primary cosmic rays, measured at flight altitudes of a balloon at a residual atmosphere of $5\,\text{g/cm}^2$ [110]

air and initiate—depending on energy and particle species—electromagnetic and/or hadronic cascades.

The momentum spectrum of the singly charged component of primary cosmic rays at the top of the atmosphere is shown in Fig. 7.3. In this diagram the particle velocity $\beta = v/c$ is shown as a function of momentum (more precisely the rigidity is plotted, which is the momentum divided by the charge of the primary particle p/Z; since in this case $Z = 1$, momentum and rigidity are identical). Clearly visible are the bands of hydrogen isotopes as well as the low flux of primary antiprotons. Even at these altitudes several muons have been produced via pion decays. Since muon and pion mass are very close, it is impossible to separate them out in this scatter diagram. Also relativistic electrons and positrons would populate the bands labeled μ^+ and μ^-. One generally assumes that the measured antiprotons are not of primordial origin, but are rather produced by interactions in interstellar or interplanetary space or even in the residual atmosphere above the balloon.

The transformation of primary cosmic rays in the atmosphere is presented in Fig. 7.4. Protons with approximately 85% probability constitute the largest fraction of primary cosmic rays. Since the interaction length for hadrons is $90 \, \mathrm{g/cm^2}$, primary protons initiate a hadron cascade already in their first interaction approximately at an altitude corresponding to the 100-mbar layer. The secondary particles most copiously produced are pions. Kaons, on the other hand, are only produced with a probability of 10–15% compared to pions. Figure 7.5 shows the K/π ratio as it is measured in strong interactions. For center-of-mass energies $\sqrt{s}$ in the range from 20 GeV to 300 GeV the K/π ratio is more or less constant. To translate center-of-mass energies to laboratory energies (see Chap. 3 on kinematics) one finds that 300 GeV in the center of mass corresponds to 45 TeV in the laboratory system. Neutral pions initiate via their decay ($\pi^0 \rightarrow \gamma + \gamma$) electromagnetic cascades, whose development is characterized

Fig. 7.4 Transformation of
primary cosmic rays in the
atmosphere

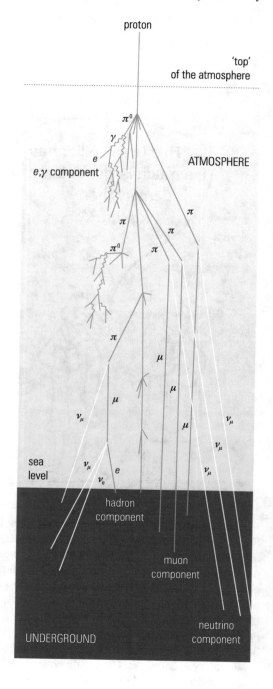

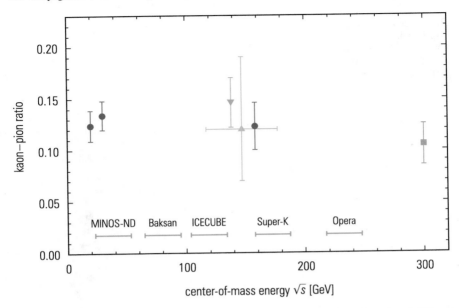

Fig. 7.5 Kaon–pion ratio as obtained in strong interactions. Compilation of results of different experiments for the measurement of the K/π ratio. (In the legend 'Super-K' stands for Super-Kamiokande.) The data are from proton–proton interactions and collisions of heavy ions. The typical center-of-mass energies of the various experiments are converted to laboratory energies as indicated above the horizontal axis [111]

by the shorter radiation length ($X_0 \approx \frac{1}{3}\lambda$ in air). This shower component is absorbed relatively easily and is therefore also named a soft component. Charged pions and kaons can either initiate further interactions or decay.

The competition between decay and interaction probability is a function of energy. For the same Lorentz factor charged pions (lifetime 26 ns) have a smaller decay probability compared to charged kaons (lifetime 12.4 ns). The leptonic decays of pions and kaons produce the penetrating muon and neutrino components ($\pi^+ \rightarrow \mu^+ + \nu_\mu$, $\pi^- \rightarrow \mu^- + \bar{\nu}_\mu$; $K^+ \rightarrow \mu^+ + \nu_\mu$, $K^- \rightarrow \mu^- + \bar{\nu}_\mu$). Muons can also decay and contribute via their decay electrons to the soft component and neutrinos to the neutrino component ($\mu^+ \rightarrow e^+ + \nu_e + \bar{\nu}_\mu$, $\mu^- \rightarrow e^- + \bar{\nu}_e + \nu_\mu$).

The energy loss of relativistic muons not decaying in the atmosphere is low ($\approx 1.8\,\text{GeV}$). They constitute with 80% of all charged particles the largest fraction of secondary particles at sea level.

Some secondary mesons and baryons can also survive down to sea level. Most of the low-energy charged hadrons observed at sea level are locally produced. The total fraction of hadrons at ground level, however, is very small.

Apart from their longitudinal development electromagnetic and hadronic cascades also spread out laterally in the atmosphere. The lateral size of an electromagnetic cascade is caused by *multiple scattering* of electrons and positrons, while in hadronic cascades the *transverse momenta* at production of secondary particles are responsible for the lateral width of the cascade. Figures 7.6, 7.7, and 7.8 show a comparison of

Fig. 7.6 Simulation of a
1-TeV photon in the
atmosphere. Scale: vertical
about 30 km, lateral 10 km
[84]

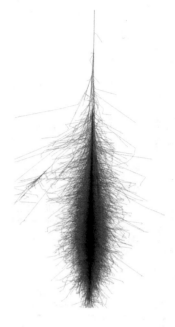

Fig. 7.7 Simulation of a
1-TeV proton in the
atmosphere. Scale: vertical
about 30 km, lateral 10 km
[84]

Fig. 7.8 Simulation of an iron nucleus of 1 TeV in the atmosphere. Scale: vertical about 30 km, lateral 10 km [84]

Fig. 7.9 Particle composition in the atmosphere as a function of atmospheric depth [112]

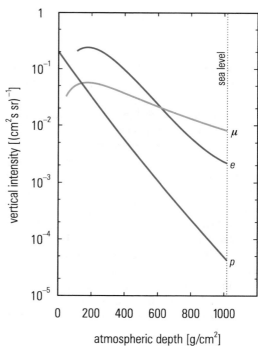

Fig. 7.10 Particle composition in the atmosphere as function of the atmospheric depth for particles with energies >1 GeV [112]

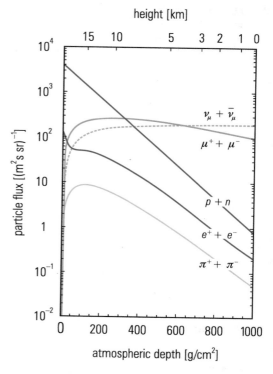

the shower development of 1-TeV photons, 1-TeV protons, and 1-TeV iron nuclei in the atmosphere. It is clearly visible that transverse momenta of secondary particles fan out the hadron cascades.

The intensity of protons, electrons, and muons of all energies as a function of the altitude in the atmosphere is plotted in Fig. 7.9. The absorption of protons can be approximately described by an exponential function.

The electrons and positrons produced through π^0 decay with subsequent pair production reach a maximum intensity at an altitude of approximately 15 km and soon after are relatively quickly absorbed while, in contrast, the flux of muons is attenuated only relatively weakly.

Because of the steepness of the energy spectra the particle intensities are of course dominated by low-energy particles. These low-energy particles, however, are mostly of secondary origin. If only particles with energies in excess of 1 GeV are counted, a different picture emerges (Fig. 7.10).

Primary nucleons (protons and neutrons) with the initial high energies dominate over all other particle species down to altitudes of 9 km, where muons take over. Because of the low interaction probability of neutrinos these particles are practically not at all absorbed in the atmosphere. Their flux increases monotonically because additional neutrinos are permanently produced by particle decays.

Since the energy spectrum of primary particles is relatively steep, the energy distribution of secondaries also has to reflect this property.

"I hear they call them cosmetic rays. After all, they appear to have no dangerous side effects!"

7.2 Cosmic Rays at Sea Level

The joy of discovery is certainly the liveliest that the mind of man can ever feel.

Claude Bernard

A measurement of charged particles at sea level clearly shows that, apart from some protons, muons are the dominant component (Fig. 7.11).

Approximately 80% of the charged component of secondary cosmic rays at sea level are muons. Figure 7.12 shows the track of a cosmic-ray muon in a historical optical multiplate spark chamber.

The muon flux at sea level through a horizontal area amounts to roughly one particle per cm^2 and minute. These muons originate predominantly from pion decays, since pions as lightest mesons are produced in large numbers in hadron cascades. The muon spectrum at sea level is therefore a direct consequence of the pion source spectrum. There are, however, several modifications. Figure 7.13 shows the parent pion spectrum at the location of production in comparison to the observed sea-level

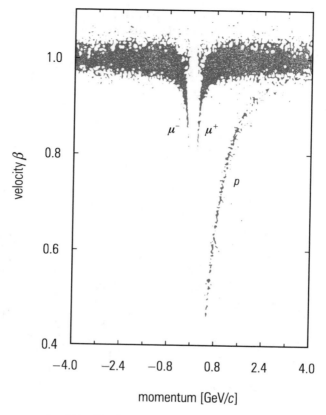

Fig. 7.11 Measurement and identification of charged particles at sea level [113]

muon spectrum. The shape of the muon spectrum agrees relatively well with the pion spectrum for momenta between 10 and 100 GeV/c. For energies below 10 GeV and above 100 GeV the muon intensity, however, is reduced compared to the pion source spectrum. For low energies the muon decay probability is increased. A muon of 1 GeV with a Lorentz factor of $\gamma = E/m_\mu c^2 = 9.4$ has a mean decay length of

$$s_\mu \approx \gamma \tau_\mu c = 6.2 \,\text{km}. \tag{7.2.1}$$

Since pions are typically produced at altitudes of 15 km and decay relatively fast (for $\gamma = 10$ the decay length is only $s_\pi \approx \gamma \tau_\pi c = 78$ m), the decay muons do not reach sea level but rather decay themselves or get absorbed in the atmosphere. At high energies the situation is changed. For pions of 100 GeV ($s_\pi = 5.6$ km, corresponding to a column density of 160 g/cm² measured from the production altitude) the interaction probability dominates ($s_\pi > \lambda$). Pions of these energies will therefore produce further, tertiary pions in subsequent interactions, which will also decay

Fig. 7.12 Cosmic-ray muon, recorded in a multiplate spark chamber, 1957 [114]

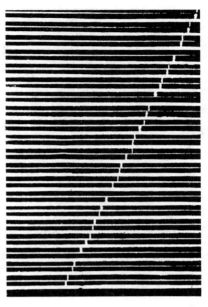

Fig. 7.13 Muon spectrum at sea level compared to the spectrum of parent pions at production

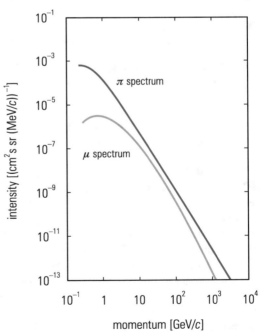

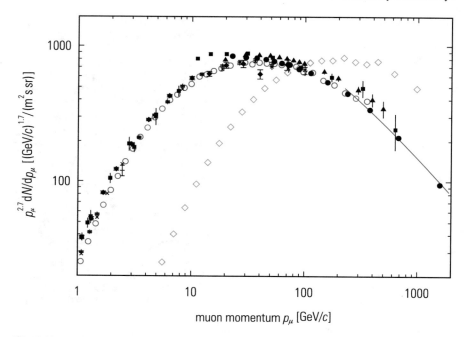

Fig. 7.14 Sea-level momentum spectrum of muons. The *full circles*, the *open circles*, and *crosses* refer to vertical muons; the *open diamonds* refer to muons from inclined directions (75°), and the *solid line at the right-hand side* is the theoretical description of the vertical muon spectrum for energies when the muon decay is negligible (above $100\,\mathrm{GeV}/(\cos\theta)$) and the curvature of the Earth can be neglected (θ is the zenith angle) [112]

eventually into muons, but providing muons of lower energy. Therefore, the muon spectrum at high energies is always slightly steeper compared to the parent pion spectrum.

If muons from inclined horizontal directions are considered, a further aspect has to be taken into account. For large zenith angles the parent particles of muons travel relatively long distances in rare parts of the atmosphere. Because of the low area density at large altitudes for inclined directions the decay probability is increased compared to the interaction probability. Therefore, for inclined directions pions will produce predominantly high-energy muons in their decay.

The result of these considerations is in agreement with observation (Fig. 7.14). For about $90\,\mathrm{GeV}/c$ the muon intensity at 75° zenith angle starts to outnumber that of the vertical muon spectrum. The intensity of muons from horizontal directions at low energies is naturally reduced because of muon decays and absorption effects in the thicker atmosphere at large zenith angles.

The sea-level muon spectrum for inclined directions has been measured with solid-iron momentum spectrometers up to momenta of approximately $20\,\mathrm{TeV}/c$. For higher energies the muon intensity decreases steeply. The spectrum of muons beyond

$20\,\text{TeV}/c$ can be measured indirectly by observing electromagnetic showers induced by these muons. Since the energy losses of muons at very high energies are dominated by bremsstrahlung and direct electron pair production, which are proportional to the muon energy, these muon-induced showers allow to infer the muon energy.

The total intensity of muons, however, is dominated by low-energy particles. Because of the increased decay probability and the stronger absorption of muons from inclined directions, the total muon intensity at sea level varies like

$$I_\mu(\theta) = I_\mu(\theta = 0)\,\cos^n\theta \qquad (7.2.2)$$

for not too large zenith angles θ. The exponent of the zenith-angle distribution is obtained to be $n = 2$. This exponent varies very little, even at shallow depths underground, if only muons exceeding a fixed energy are counted.

An interesting quantity is the *charge ratio of muons* at sea level. Since primary cosmic rays are positively charged, this positive charge excess is eventually also transferred to muons. If one assumes that primary protons interact with protons and neutrons of atomic nuclei in the atmosphere, where the multiplicity of produced pions can be quite large, the charge ratio of muons, $N(\mu^+)/N(\mu^-)$, can be estimated by considering the possible charge exchange reactions:

$$\begin{aligned} p + N &\to p' + N' + k\pi^+ + k\pi^- + r\pi^0\,, \\ p + N &\to n + N' + (k+1)\pi^+ + k\pi^- + r\pi^0\,. \end{aligned} \qquad (7.2.3)$$

In this equation k and r are the multiplicities of the produced particle species and N represents a target nucleon. If one assumes that for the reactions in (7.2.3) the cross sections are the same, the charge ratio of pions is obtained to be

$$R = \frac{N(\pi^+)}{N(\pi^-)} = \frac{2k+1}{2k} = 1 + \frac{1}{2k}\,. \qquad (7.2.4)$$

Under these simplifying assumptions one would get for low energies and $k = 2$ a value for the charge ratio of $R = 1.25$. However, the situation is not as simple as that: the cross sections for Eq. (7.2.3) are not the same. The inclusive cross section for the second reaction ($p + N \to n + N' + \cdots$) relative to the first one ($p + N \to p' + N' + \cdots$) is only 30%, i.e., in 70% of the cases the incident proton stays a proton, and only in 30% of the cases one gets a leading neutron. This would lead to a charge ratio smaller than $R = 1.25$. Still, the theoretically expected charge ratio is higher, close to $R = 1.25$, for the following reason: the charge ratio is, of course, linked to the primary spectrum. The flux of secondary particles (in this case muons) is related to the primary spectrum and to first approximation it is equal to the primary spectrum multiplied by the so-called spectrum-weighted moment $Z_{p\to\mu}$. This factor is the integral over the inclusive cross section multiplied with the weighting factor $x^{\gamma-1}$, where $x = E_\mu/E_p$ and $\gamma = 2.7$ for the primary spectrum. The energy distribution along with the steep spectrum now leads to a charge ratio of about $R = 1.25$, because the pairwise produced pions have a softer energy spectrum and

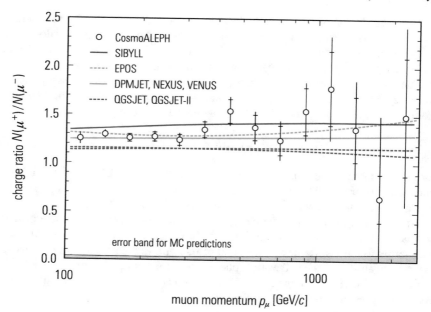

Fig. 7.15 Charge ratio of muons at sea level. The results from the CosmoALEPH experiment are compared to the predictions of various hadronization models [116]

enter with a lower weight into the estimate. This charge ratio is eventually transferred to the muons, which then get a similar ratio [115].

Experimentally one observes that the charge ratio of muons at sea level is constant over a wide momentum range and takes on a value of

$$N(\mu^+)/N(\mu^-) \approx 1.28 \,. \tag{7.2.5}$$

Figure 7.15 shows the charge ratio at sea level for muon momenta up to 2.5 TeV. Monte Carlo simulations of the current hadronization models describe the charge ratio rather well. However, the models SIBYLL and QGSJET show a small deviation from the experimental value of the charge ratio. One has to keep in mind that most of the models are based on accelerator data obtained from central production. When new data from accelerators also from forward production become available, the models are appropriately adjusted. Experiments at CERN at the LHC have obtained now also data for forward scattering, which are more appropriate to compare with cosmic ray results.

In addition to 'classical' production mechanisms of muons by pion and kaon decays, they can also be produced in semileptonic decays of charmed mesons (for example, $D^0 \to K^- \mu^+ \nu_\mu$ and $D^+ \to \bar{K}^0 \mu^+ \nu_\mu$, $D^- \to K^0 \mu^- \bar{\nu}_\mu$). Since these charmed mesons are very short-lived ($\tau_{D^0} \approx 0.4$ ps, $\tau_{D^\pm} \approx 1.1$ ps), they decay practically immediately after production without undergoing interactions themselves.

Therefore, they are a source of high-energy muons. Since the production cross section of charmed mesons in proton–nucleon interactions is rather small, D decays contribute significantly only at very high energies. Correspondingly, this is also true for the semileptonic decays of B mesons.

Figure 7.11 already showed that apart from muons also some nucleons can be observed at sea level. These nucleons are either remnants of primary cosmic rays, which, however, are reduced in their intensity and energy by multiple interactions, or they are produced in atmospheric hadron cascades.

About one third of the nucleons at sea level are neutrons. The proton/muon ratio varies with the momentum of the particles. At low momenta ($\approx 500\,\mathrm{MeV}/c$) a p/μ ratio $N(p)/N(\mu)$ of about 10% is observed decreasing to larger momenta ($N(p)/N(\mu) \approx 2\%$ at $1\,\mathrm{GeV}/c$, $N(p)/N(\mu) \approx 0.5\%$ at $10\,\mathrm{GeV}/c$). The pion flux is even suppressed with respect to the proton flux by a factor of 20–50, depending on the momentum.

In addition to muons and protons, one also finds electrons, positrons, and photons at sea level as a consequence of the electromagnetic cascades in the atmosphere. A certain fraction of electrons and positrons originates from muon decays. Electrons can also be liberated by secondary interactions of muons ('knock-on electrons', see also Fig. 7.16).

The few pions and kaons observed at sea level are predominantly produced in local interactions.

Apart from charged particles, electron and muon neutrinos are produced in pion, kaon, and muon decays. They constitute an annoying background, in particular, for neutrino astronomy. On the other hand, the propagation of atmospheric neutrinos has provided new insights for elementary particle physics, such as neutrino oscillations.

Since the parent particles of neutrinos are dominantly pions and kaons and the decay probability of pions and kaons is increased compared to the interaction probability at inclined directions, the horizontal neutrino spectra are also harder in comparison to the spectra from vertical directions. Altogether, muon neutrinos would appear to dominate, since the ($\pi \rightarrow e\nu$) and ($K \rightarrow e\nu$) decays are strongly suppressed due to *helicity conservation*. Therefore, pions and kaons almost exclusively produce muon neutrinos only. Only in muon decay equal numbers of electron and muon neutrinos are produced. At high energies also semileptonic decays of charmed mesons constitute a source for neutrinos.

Based on these 'classical' considerations the integral neutrino spectra yield a neutrino-flavour ratio of

$$\frac{N(\nu_\mu + \bar{\nu}_\mu)}{N(\nu_e + \bar{\nu}_e)} \approx 2\,. \tag{7.2.6}$$

This ratio, however, is modified by propagation effects like neutrino oscillations (see Sect. 6.3: 'Neutrino Astronomy').

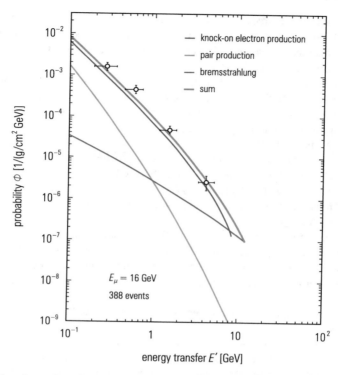

Fig. 7.16 Experimental results on muon interactions at average muon energies of $16\,\mathrm{GeV}/c$. The curves represent predictions of the yield of direct electron pair production (*green*), bremsstrahlung (*blue*), knock-on electron production (*red*), and their sum (*grey*). At these energies the knock-on production by muons is the dominant process. The vertical scale shows the interaction probability Φ per g/cm^2, per GeV, and per muon [117]

7.3   Cosmic Rays Underground

> *If your experiment needs statistics, then you ought to have done a better experiment.*
>
> Ernest Rutherford

Particle composition and energy spectra of secondary cosmic rays underground are of particular importance for neutrino astronomy. Experiments in neutrino astronomy are usually set up at large depths underground to provide a sufficient shielding against the other particles from cosmic rays. Because of the rarity of neutrino events even low fluxes of residual cosmic rays constitute an annoying background. In any case it is necessary to know precisely the identity and flux of secondary cosmic rays underground to be able to distinguish a possible signal from cosmic-ray sources from statistical fluctuations or systematical uncertainties of the atmospheric cosmic-ray background.

Long-range atmospheric muons, secondary particles locally produced by muons, and the interaction products created by atmospheric neutrinos represent the important background sources for neutrino astronomy.

Muons suffer energy losses by ionization, direct electron–positron pair production, bremsstrahlung, and nuclear interactions. These processes have been described in rather detail in Chap. 4. While the ionization energy loss at high energies is essentially constant, the cross sections for the other energy-loss processes increase linearly with the energy of the muon,

$$-\frac{dE}{dx} = a + b\,E. \tag{7.3.1}$$

The energy loss of muons as a function of their energy is shown in Fig. 7.17 for iron as absorber material. The energy loss of muons in rock in its dependence on the muon energy was already shown earlier (Fig. 4.6).

Equation (7.3.1) allows to work out the range R of muons by integration,

$$R = \int_E^0 \frac{dE}{-dE/dx} = \frac{1}{b}\ln(1 + \frac{b}{a}E), \tag{7.3.2}$$

if it is assumed that the parameters a and b are energy independent.

For not too large energies ($E < 100\,\text{GeV}$) the ionization energy loss dominates. In this case $bE \ll a$ and therefore

$$R = \frac{E}{a}. \tag{7.3.3}$$

The energy loss of a *minimum-ionizing muon* in the atmosphere is

$$\frac{dE}{dx} = 1.82\,\text{MeV}/(\text{g/cm})^2. \tag{7.3.4}$$

A muon of energy $100\,\text{GeV}$ has a range of about $40\,000\,\text{g/cm}^2$ in rock corresponding to $160\,\text{m}$ (or $400\,\text{m}$ water equivalent). An energy–range relation for standard rock is shown in Fig. 7.18. Because of the stochastic character of muon interaction processes with large energy transfers (e.g., bremsstrahlung) muons are subject to a considerable range straggling.

The knowledge of the sea-level muon spectrum and the energy-loss processes of muons allow one to determine the *depth–intensity relation* for muons. The integral sea-level muon spectrum can be approximated by a power law

$$N(>E) = A\,E^{-\gamma}. \tag{7.3.5}$$

Using the energy–range relation (7.3.2), the depth–intensity relation is obtained,

Fig. 7.17 Contributions to the energy loss of muons in iron

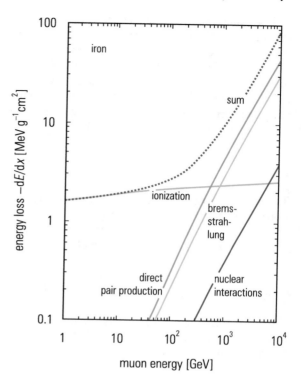

$$N(>E, R) = A \left[\frac{a}{b}(e^{bR} - 1) \right]^{-\gamma} . \tag{7.3.6}$$

For high energies ($E_\mu > 1\,\text{TeV}, bE \gg a$) the exponential dominates and one obtains

$$N(>E, R) = A \left(\frac{a}{b} \right)^{-\gamma} e^{-\gamma bR} . \tag{7.3.7}$$

For inclined directions the absorbing ground layer increases like $1/\cos\theta = \sec\theta$ (θ—zenith angle) for a flat overburden, so that for muons from inclined directions one obtains a depth–intensity relation of

$$N(>E, R, \theta) = A \left(\frac{a}{b} \right)^{-\gamma} e^{-\gamma bR \sec\theta} . \tag{7.3.8}$$

For shallower depths (7.3.6), or also (7.3.3), however, leads to a power law

$$N(>E, R) = A \, (aR)^{-\gamma} . \tag{7.3.9}$$

The measured depth–intensity relation for vertical directions is plotted in Fig. 7.19. From depths of 10 km water equivalent (≈ 4000 m rock) onwards muons induced

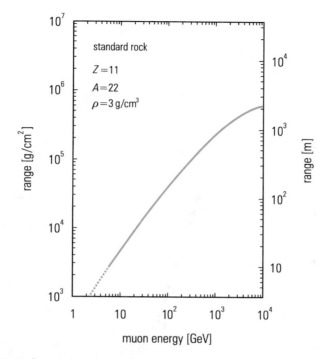

Fig. 7.18 Range of muons in rock

by atmospheric neutrinos dominate the muon rate. Because of the low interaction probability of neutrinos the neutrino-induced muon rate does not depend on the depth. At large depths (> 10 km w.e.) a neutrino telescope with a collection area of $100 \times 100\,\mathrm{m}^2$ and a solid angle of π would still measure a background rate of 10 events per day.

The zenith-angle distributions of atmospheric muons for depths of 1500 and 7000 meter water equivalent are shown in Fig. 7.20. For large zenith angles the flux decreases steeply, because the thickness of the overburden increases like $1/\cos\theta$. Therefore, at large depths and from inclined directions neutrino-induced muons dominate.

For not too large zenith angles and depths the zenith-angle dependence of the integral muon spectrum can still be represented by

$$I(\theta) = I(\theta = 0)\cos^n\theta \tag{7.3.10}$$

(Fig. 7.21). For large depths the exponent n in this distribution, however, gets very large, so that it is preferable to use (7.3.8) instead.

The average energy of muons at sea level is in the range of several GeV. Absorption processes in rock reduce predominantly the intensity at low energies. Therefore, the average muon energy of the muon spectrum increases with increasing depth.

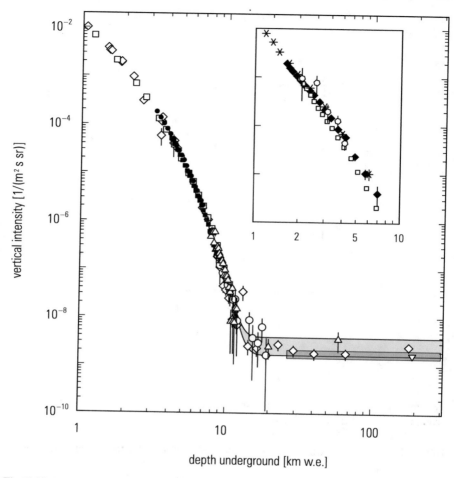

Fig. 7.19 Depth–intensity relation for muons from vertical directions. The *grey band* at large depths represents the flux of ν-induced muons with energies above 2 GeV. The *upper line* refers to horizontal ν-induced muons, the *lower one* for vertical upward ν-induced muons. The *stronger shadowed area* at very large depths shows the measurements of the Super-Kamiokande experiment. The *inset* shows the vertical depth–intensity relation for water and ice. The *left-hand vertical scale in the inset* can be read from the *corresponding vertical scale in the main diagram*. The *horizontal scale* is in km w.e. [118]

Muons of high energy can also produce other secondary particles in local interactions. Since low-energy muons can be identified by their ($\mu \to e\nu\nu$) decay with the characteristic decay time in the microsecond range, the measurement of stopping muons underground provides an information about local production processes.

A certain fraction of stopping muons is produced locally by low-energy pions, which decay relatively fast into muons. The flux of penetrating muons decreases

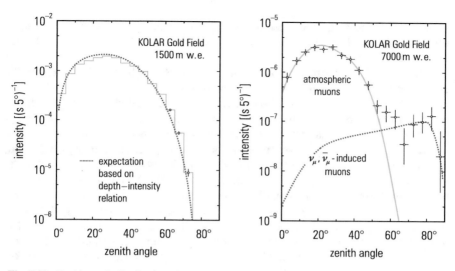

Fig. 7.20 Zenith-angle distribution of atmospheric muons at depths of 1500 and 7000 m w.e. [119]

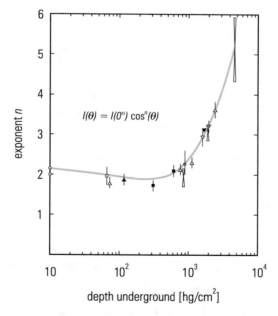

Fig. 7.21 Variation of the exponent n of the zenith-angle distribution of muons with depth [120]

rapidly with increasing depth, therefore the ratio of stopping to penetrating muons is strongly influenced by stopping muons induced by neutrino interactions for depths larger than 5000 m w.e.

The knowledge of the particle composition at large depths below ground represents an important information for neutrino astrophysics.

Also remnants of extensive air showers, which developed in the atmosphere, are measured underground. Electrons, positrons, photons, and hadrons are completely absorbed already in relatively shallow layers of rock. Therefore, only muons and neutrinos of extensive air showers penetrate to larger depths. The primary interaction vertex of particles that initiate the air showers is typically at an atmospheric altitude of 15 km. Since secondary particles in hadronic cascades have transverse momenta of typically 300 MeV/c or less, the high-energy muons essentially follow the shower axis. For primaries of energy around 10^{14} eV lateral displacements of energetic muons ($\approx$1 TeV) at shallow depths underground of less than a meter exclusively caused by transferred transverse momenta are obtained. Typical multiple-scattering angles for energetic muons ($\approx$100 GeV) in thick layers of rock (50–100 m) are on the order of a few mrad.

The multiplicity of produced secondary particles increases with energy of the initiating particle (for a 1-TeV proton the charged multiplicity of particles for proton–proton interactions is about 15). Since the secondaries produced in these interactions decay predominantly into muons, one observes bundles of nearly parallel muons underground in the cores of extensive air showers. Figure 7.22 shows such a shower with more than 50 parallel muons observed by the ALEPH experiment at a depth of 320 m w.e.

Apart from the numerous muons also a knock-on electron is visible in the central tracking chamber. The electron is knocked out of an atom by one of the muons, and—because of its low energy—is forced by the strong transverse magnetic field on a circular track in the time projection chamber. In contrast, the tracks of the energetic muons are hardly bent in the magnetic field of 1.5 tesla strength.

Figure 7.23 shows a relatively rare example of a muon pair production by a cosmic-ray muon in the CosmoALEPH experiment: a so-called muon-trident process ($\mu + N \rightarrow \mu + \mu^+ + \mu^- + N$). In precise Monte Carlo simulations of extended air showers also this rare reaction should be included, even though its cross section is quite small.

High-energy muons are produced by high-energy primaries and, in particular, muon showers correlate with even higher primary energies. Therefore, one is tempted to localize extraterrestrial sources of high-energy cosmic rays via the arrival directions of single or multiple muons. Since Cygnus X3 has been claimed to emit photons with energies up to 10^{16} eV, this astrophysical source also represents an excellent candidate for the acceleration of high-energy charged primary cosmic rays. Cygnus X3 at a distance of approximately 33000 light-years is an X-ray binary consisting of a superdense pulsar and a stellar companion. The material flowing from the companion into the direction of the pulsar forms an accretion disk around the pulsar. If apparently photons of very high energy can be produced, one would expect them to originate from the π^0 decay ($\pi^0 \rightarrow \gamma\gamma$). Neutral pions are usually produced in

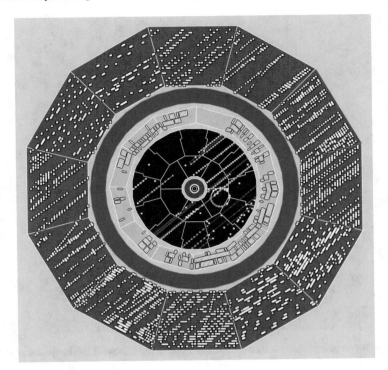

Fig. 7.22 Muon shower in the CosmoALEPH experiment (detector diameter about 10 m) [121]

proton interactions. Therefore, the source should also be able to produce charged pions and via their decay muons and muon neutrinos. Because of their short lifetime, muons would never survive the 33 000-light-year distance from Cygnus X3 to Earth, so that a possible muon signal must be caused by neutrino-induced muons. Unfortunately, muons and multi-muons observed in the Frejus experiment also from the directions of Cygnus X3 are predominantly of atmospheric origin and do not confirm that Cygnus X3 is a strong source of high-energy particles. The primary particles themselves accelerated in the source could in principle point back to the source when measured on Earth. However, the arrival direction of primary charged particles from Cygnus X3 could also have been completely randomized by the irregular galactic magnetic field. Muon production by neutrinos from Cygnus X3 would have been a rare event, which would have required an extremely massive detector to obtain a significant rate.

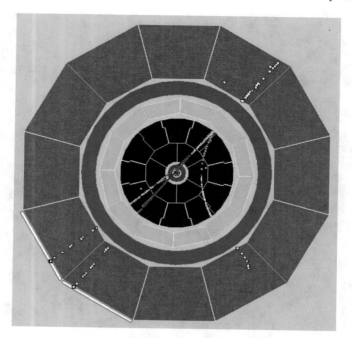

Fig. 7.23 Muon pair production by a cosmic-ray muon in the CosmoALEPH experiment [122]

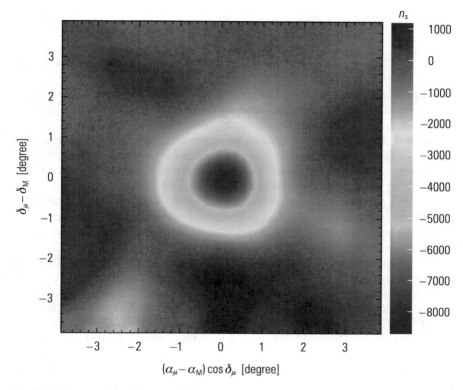

Fig. 7.24 Contour plot of the muon deficit as measured by ICECUBE in the region around the Moon's position. Since ICECUBE only measures muons, this plot shows the image of the Moon by the absence of muons. The significance of the deficit is more than 6σ. To obtain this map of the Moon shadow data from more than a year have been used [123]

Figure 7.24, on the other hand, presents an *anti source* of cosmic rays, namely, a shadow of the Moon in the light of TeV muons as measured in the ICECUBE experiment. High-energy muons are created in interactions of primary cosmic rays in the atmosphere. Because of their high energy they practically retain their original direction of incidence, which is more or less identical to the direction of the primary particles that have produced them. These muons represent an unwanted background for neutrino astronomy. The Moon, however, absorbs a certain amount of primary cosmic rays. Therefore, one expects a reduction of cosmic particles from this direction and thereby also a deficit of muons, which would have otherwise been created in the atmosphere. The width of the Moon shadow was in agreement with the expectations from Monte Carlo simulations.

The observation of the Moon shadow via TeV muons shows that the ICECUBE experiment is able to search for cosmic point sources. The measurement of the Moon shadow also allows to infer an angular resolution of ICECUBE resulting in an absolute pointing accuracy of about 0.2 degrees.

7.4 Extensive Air Showers

Science never solves a problem without creating ten more.

George Bernard Shaw

Extensive air showers are cascades initiated by energetic primary particles, which develop in the atmosphere. An extensive air shower (EAS) has an electromagnetic, a muonic, a hadronic, and a neutrino component (see Fig. 7.4). The air shower develops a shower nucleus consisting of energetic hadrons, which permanently inject energy into the electromagnetic and the other shower components via interactions and decays. Neutral pions, which are produced in nuclear interactions and whose decay photons produce electrons and positrons via pair production, supply the electron, positron, and photon component. Photons, electrons, and positrons initiate electromagnetic cascades through alternating processes of pair production and bremsstrahlung. The muon and neutrino components are formed by the decay of charged pions and kaons (see also Fig. 7.4).

The inelasticity in hadron interactions is on the order of 50%, i.e., 50% of the primary energy is transferred into the production of secondary particles. Since predominantly pions are produced ($N(\pi) : N(K) = 9 : 1$) and all charge states of pions (π^+, π^-, π^0) are produced in equal amounts, one third of the inelasticity is transferred into the formation of the electromagnetic component. Since most of the charged hadrons and the hadrons produced in hadron interactions also undergo multiple interactions, the largest fraction of the primary energy is eventually transferred into the electromagnetic cascade. Therefore, in terms of the number of particles, electrons and positrons constitute the main shower component. The particle number increases with shower depth t until absorptive processes like ionization for charged particles and Compton scattering and photoelectric effect for photons start to dominate and cause the shower to die out.

The development of electromagnetic cascades is shown in Fig. 7.25 for various primary energies. The particle intensity increases initially in a parabolic fashion and decays exponentially after the maximum of the shower has been reached. The longitudinal profile of the particle number can be parameterized by

$$N(t) \sim t^\alpha e^{-\beta t}, \tag{7.4.1}$$

where $t = x/X_0$ is the shower depth in units of the radiation length and α and β are free fit parameters. The position of the shower maximum varies only logarithmically with the primary energy, while the total number of shower particles increases linearly with the energy. The latter can therefore be used for the energy determination of the primary particle. One can imagine that the Earth's atmosphere represents a combined hadronic and electromagnetic calorimeter, in which the extensive air shower develops. The atmosphere constitutes approximately a target of 11 interaction lengths and 27 radiation lengths. The minimum energy for a primary particle to be reasonably well measured at sea level via the particles produced in the air shower is about $10^{14}\,\text{eV} = 100\,\text{TeV}$. As a rough estimate for the particle number N at sea level in its dependence on the primary energy E_0, one can use the relation

$$N = 10^{-10} \, E_0[\mathrm{eV}] \,. \tag{7.4.2}$$

Only about 10% of the charged particles in an extensive air shower are muons. The number of muons reaches a plateau already at an atmospheric depth of $200 \, \mathrm{g/cm^2}$ (see also Figs. 7.9 and 7.10). Its number is hardly reduced to sea level, since the probability for catastrophic energy-loss processes, like bremsstrahlung, is low compared to electrons because of the large muon mass. Muons also lose only a small fraction of their energy by ionization. Because of the relativistic time dilation the decay of energetic muons ($E_\mu > 3 \, \mathrm{GeV}$) in the atmosphere is strongly suppressed.

Cosmic Shower

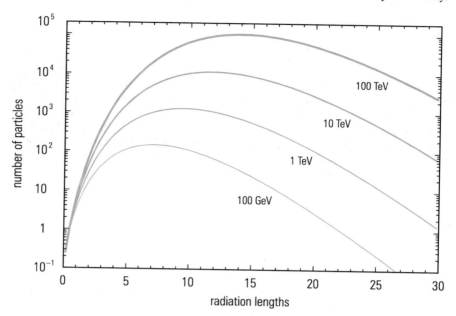

Fig. 7.25 Longitudinal development of electromagnetic cascades. The shower depth is given in units of the radiation length. The atmosphere comprises 27 radiation lengths [124]

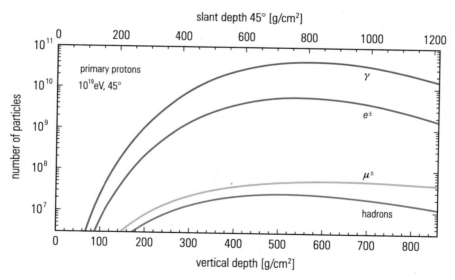

Fig. 7.26 Longitudinal development of different particle species in an air shower of 10 EeV energy for a zenith angle of 45° [125]

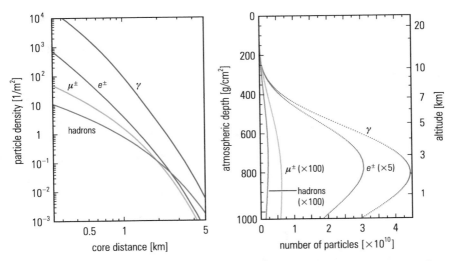

Fig. 7.27 Lateral and longitudinal shower profiles for vertical proton-induced showers of 10^{19} eV, simulated with the program CORSIKA-SIBYLL2.1. The lateral distribution of shower particles has been simulated for a shower depth of 870 g/cm^2, corresponding to the atmospheric depth of the Auger experiment for vertical incidence. The energy thresholds for secondary particles are 0.25 MeV for photons, electrons, and positrons and 0.1 GeV for muons and hadrons [126]

Figure 7.26 shows schematically the longitudinal development of the various components of an extensive air shower in the atmosphere for a primary energy of 10^{19} eV. The lateral spread of an extensive air shower is essentially caused by the transferred transverse momenta in hadronic interactions and by multiple scattering of low-energy shower particles. The muon component is somewhat flatter compared to the lateral distribution of electrons and hadrons. Figure 7.27 shows the lateral particle profile for the various shower components. Neutrinos essentially follow the shape of the muon component.

Even though an extensive air shower initiated by primary particles with energies below 100 TeV does not reach sea level, it can nevertheless be recorded via the Cherenkov light emitted by the shower particles (see Sect. 6.4 on gamma-ray astronomy). At higher energies one has the choice of various detection techniques.

The classical technique for the measurement of extensive air showers is the sampling of shower particles at sea level with typically 1 m^2 large scintillators or water Cherenkov counters. This technique is sketched in Fig. 7.28. In the Auger project in Argentina 1600 sampling detectors spread over an area of 3000 km^2 will be used for the measurement of the ground-level component of extensive air showers. However, the energy assignment for the primary particle using this technique is not very precise. The shower develops in the atmosphere, which acts as a calorimeter of 27 radiation lengths thickness. The information on this shower is sampled in only *one*, the last layer of this calorimeter and the coverage of this layer is on the order of much

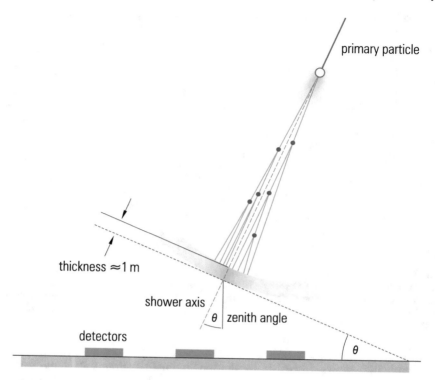

Fig. 7.28 Air-shower measurements with sampling detectors

less than 1%. The direction of incidence of the primary particle can be obtained from the arrival times of shower particles in the different sampling counters (see Fig. 7.28).

It would be much more advantageous to measure the total longitudinal development of the cascade in the atmosphere. This can be achieved using the technique of the Fly's Eye (Fig. 7.29). Apart from the directional Cherenkov radiation the shower particles also emit an isotropic scintillation light in the atmosphere. The Auger experiment in Argentina uses surface detectors and fluorescence telescopes, where the latter record the longitudinal development of the air shower. Figures 7.30 and 7.31 show two examples of high-energy showers of 2×10^{18} eV respectively 10^{19} eV energy, which develop their shower maxima at different atmospheric depths.

The energy domain covered by the Auger experiment is now also investigated by the large Telescope Array (TA) in Millard County, Utah. It also uses a ground array along with an air-fluorescence technique for the measurement of the highest cosmic-ray energies [128].

„We capture the cosmic particles and use them for energy supply!"

For particles with energies exceeding 10^{17} eV the fluorescence light of nitrogen is sufficiently intense to be recorded at sea level in the presence of the diffuse background of starlight. The actual detector consists of a system of mirrors and photomultipliers, which view the whole sky. An air shower passing through the atmosphere near such a Fly's Eye type detector activates only those photomultipliers whose field of view is hit. The fired photomultipliers allow to reconstruct the longitudinal profile of the air shower. The total recorded light intensity is used to determine the shower energy. Such a type of detector allows much more precise energy assignments, however, it has a big disadvantage compared to the classical air-shower technique that it can only be operated in clear moonless nights. In the Auger experiment the array of sampling detectors is complemented by a number of telescopes, which measure

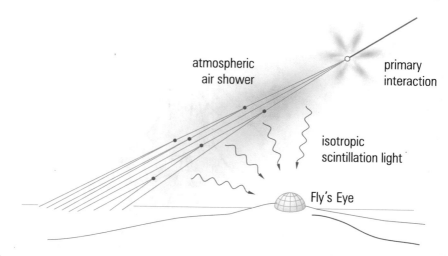

Fig. 7.29 Principle of the measurement of scintillation light of extensive air showers

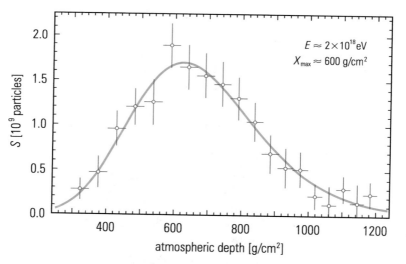

Fig. 7.30 Measured longitudinal development of an air shower with energy 2×10^{18} eV in the Auger experiment. The depth of the shower maximum is at about $600 \, \text{g/cm}^2$ [127]

the scintillation light produced in the atmosphere. Figure 7.32 shows one of the mirrors of the Auger telescopes along with its camera of photomultipliers mounted in its focal point. Much larger acceptances could be provided if such a Fly's Eye type detector would be installed in orbit ('Air Watch', Fig. 7.29).

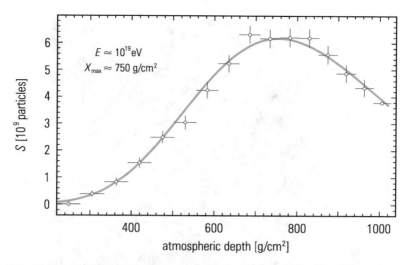

Fig. 7.31 Measured longitudinal development of an air shower of energy 10^{19} eV in the Auger experiment. The depth of the shower maximum is at around $750 \, \text{g/cm}^2$ [127]

There is in fact a proposal to observe air showers from space. Such an experiment would record the air showers from a detector at the ISS in the Earth's orbit (JEM-EUSO), see Fig. 7.33.

As a result of the different detection techniques, Fig. 7.34 shows the measured all-particle spectrum of primary cosmic rays [131].

Apart from these detection techniques it has also been tried to observe air showers via the electromagnetic radiation emitted in the radio band (see also Sect. 7.5 about more details on 'Radio Measurements of Air Showers'). It is generally believed that this radio signal is caused by shower electrons deflected in the Earth's magnetic field thereby creating synchrotron radiation. In spite of the strong background in practically all wavelength ranges these attempts have been quite successful, in particular, in the radio band from 40 to 80 MHz.

The possibility to detect large air showers via their muon content in underground experiments—possibly jointly with an air-shower detector on the surface in coincidence—has been followed up in recent experiments.

A more exotic technique would be the detection of the acoustic signal produced by high-energy particles in water or ice (see also Sect. 7.6: 'Acoustic Detection of Air Showers').

Apart from elementary-particle-physics aspects the purpose of the measurement of extensive air showers is the determination of the chemical composition of primary cosmic rays and the search for the sites of cosmic accelerators.

Both of these problems are hard to solve. The chemical composition up to the TeV range can be determined with balloons or satellites in direct measurements (see Fig. 6.3), but beyond that one has to resort to extensive air showers. The determination of the mass of a primary cosmic-ray particle with energy $> 10^{15}$ eV is rather indirect.

Fig. 7.32 Photo of a mirror and camera of a detector for the measurement of fluorescence radiation in the Auger experiment [129]

The position of the shower maximum or the muon content of an air shower provides some evidence of the nature of the primary particle that has initiated the shower. Figure 7.35 gives an idea of the problem: The mean logarithmic mass is plotted against the primary energy. There are indications that the masses of the primary cosmic rays get heavier beyond the knee ($\approx$ several PeV). At the highest energies—at least for the old Fly's Eye data—lighter particles seem to be dominant. It would not be a surprise if the chemical composition of extragalactic particles, which certainly come into play in this energy domain, is different from that of galactic cosmic rays. From the inspection of the figure one can appreciate that it is obviously not easy to determine the mass of a particle whose energy is more than a million times heavier than its rest mass. The scatter of the experimental results is substantial.

The arrival directions of the highest-energy particles ($>10^{19}$ eV), which for intensity reasons can only be recorded via air-shower techniques, practically show no correlation to the galactic plane. This clearly indicates that their origin must be

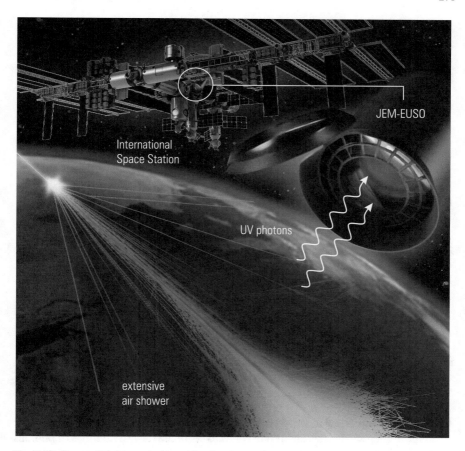

Fig. 7.33 Proposal for an experiment for the observation of air showers from an experiment in the Earth's orbit (JEM-EUSO; Extreme Universe Space Observatory in the Japanese module of the International Space Station) at the ISS [130]

extragalactic. If the highest-energy primary cosmic-ray particles are protons, then their energies must be below 10^{20} eV, if they originate from distances of more than 50 Mpc. Even if their original energy were much higher, they would lose energy by photoproduction of pions on photons of the blackbody radiation until they fall below the threshold of the Greisen–Zatsepin–Kuzmin cutoff ($\approx 6 \times 10^{19}$ eV). Protons of this energy would point back to the sources, because galactic and intergalactic magnetic fields only cause angular distortions on the order of one degree at these high energies. The irregularities of magnetic fields, however, could lead to significant time delays between neutrinos and photons on the one hand and protons, on the other hand, from such distant sources. This comes about because the proton trajectories are somewhat longer, even though their magnetic deflection is rather small. Depending on the distance from the source, time delays of months and even years can occur.

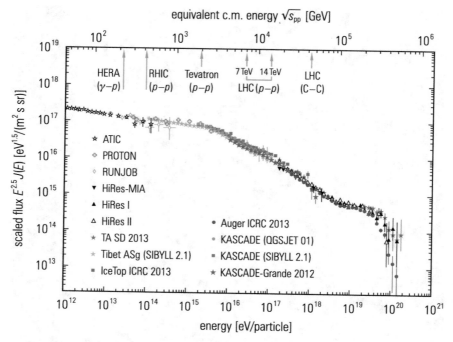

Fig. 7.34 All-particle spectrum of primary cosmic rays. The scatter at very high energies also shows the problems of an accurate energy assignment [131, 132]

This effect is of particular importance, if γ-ray bursters are also able to accelerate the highest-energy particles and if one wants to correlate the arrival times of photons from γ-ray bursts with those of extensive air showers initiated by charged primaries.

Out of the up to now 27 measured highest-energy showers (>57 EeV) in the Auger experiment one might see a tentative evidence of a clustering along the supergalactic plane (see Fig. 7.36). The supergalactic plane is a sheet-like structure containing the Local Supercluster, the Virgo Cluster, the Great Attractor, the Coma Supercluster, the Perseus–Pisces Supercluster, and the Shapley Concentration. The fact that the attenuation length for protons with energies $>6 \times 10^{19}$ eV in intergalactic space is ≈ 50 Mpc, could make an origin in the supercluster (maximum extension 30 Mpc) plausible. Two air-shower events arrive within the experimental angular resolution of Auger from the potential source Centaurus A. A correlation with active galactic nuclei observed in γ rays is not visible. Urgently better statistics are required. Astronomy with a handful of events is not really possible. The latest news from Auger do not improve the statistical evidence for showers coming from Centaurus A.

However, recently the Auger experiment has observed a significant ($5.2\,\sigma$) large-scale dipole anisotropy for showers of energy above 8×10^{18} eV. Figure 7.37 shows the fluxes of high-energy particles in galactic coordinates. The galactic center is at the origin. The cross indicates the measured dipole direction; the contours denote

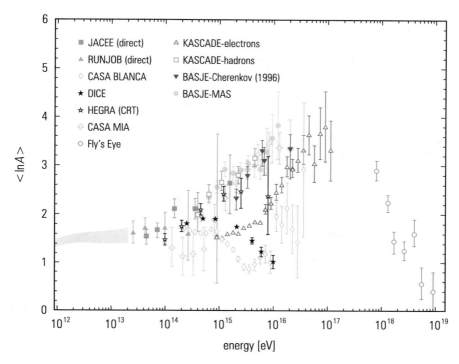

Fig. 7.35 Mean logarithmic mass ⟨ln A⟩ for primary cosmic rays for high energies [132]

the 68% and 95% confidence-level regions. The direction of the dipole anisotropy observed by the 2 Micron all-sky Redshift Survey (2MRS)) shown in the figure is assumed to be correlated with mass aggregations and centers of possible sites of acceleration. The position of the dipole anisotropy for the highest-energy primary particles indicates an extragalactic origin for these events [134].

Normal extensive air showers have lateral widths of at most 10 km, even at the highest energies. However, there are indications that *correlations between arrival times* of air showers over distances of more than 100 km exist. Such coincidences could be understood by assuming that energetic primary cosmic particles undergo interactions or fragmentations at large distances from Earth. The secondary particles produced in these interactions would initiate separate air showers in the atmosphere (Fig. 7.38).

Even moderate distances of only one parsec (3×10^{16} m) are sufficient to produce separations of air showers at Earth on the order of 100 km (primary energy 10^{20} eV, transverse momenta ≈ 0.3 GeV/c). Variations in arrival times of these showers could be explained by unequal energies of the fragments, which could cause different propagation times. Galactic or extragalactic magnetic fields could also affect the trajectories of the fragments in a different way thus also influencing the arrival times.

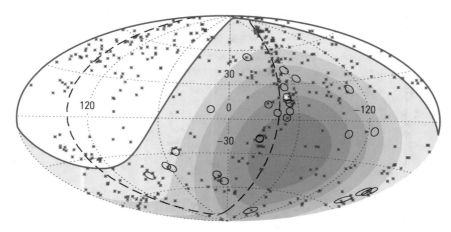

Fig. 7.36 Arrival directions of the 27 most energetic air showers measured from the Auger experiment in galactic coordinates. The energies of the air showers are larger than 57 EeV. They are shown as open circles. At the same time the positions of 471 active galactic nuclei (AGNs) within 75 Mpc are given as *red stars* ∗. The *blue region* defines—depending on the exposure time—the field of view of the Auger experiment. The *full line* shows the limit of the Auger acceptance. Centaurus A is marked as *white star* (∗). Two out of the 27 air-shower events arrive within the angular resolution of Auger from this direction. The *dashed line* indicates the position of the supergalactic plane [133]

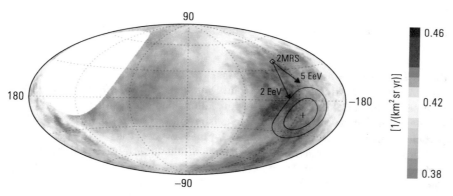

Fig. 7.37 Auger sky map in galactic coordinates showing the cosmic-ray flux for energies in excess of 8 EeV. The direction of the dipole anisotropy indicates an extragalactic origin for these ultra-high-energy particles. The direction of the dipole anisotropy observed with the near-infrared flux of photons (the 2 Micron all-sky Redshift Survey (2MRS)) is indicated. The arrows show the expected deflections of particles with $E/Z = 5$ EeV or 2 EeV from a possible origin in the 2MRS direction for a given typical galactic magnetic field [134]

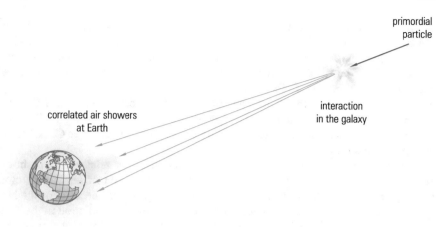

Fig. 7.38 Possible explanation for the origin of distant correlated air showers

7.5   Radio Measurement of Air Showers

> *Radio signals, which span nearly five decades of the electromagnetic spectrum, provide a unique diagnostic tool for probing the universe.*
>
> Kurt van der Heyden

High-energy cosmic rays produce a large number of charged particles when they induce an air shower in the atmosphere. A primary proton of 10^{18} eV creates about 10^8 secondary charged particles at ground level. At these primary energies mainly electrons and positrons are produced, in addition to a much lower number of hadrons and muons. The electrons and positrons of the electromagnetic shower component undergo various interactions in the atmosphere. Apart from ionization, bremsstrahlung, and the emission of Cherenkov radiation they also generate synchrotron radiation in the Earth's magnetic field. Due to their relatively low energy this geosynchrotron emission is in the radio range. This radio emission is practically not absorbed in the atmosphere, and it can be used as a fingerprint of the air shower, which can be recorded 24 h a day, in contrast to Cherenkov radiation and optical fluorescence emission, which can only be measured during moonless nights.

In addition to geosynchrotron emission there are essentially two more mechanisms for the production of radio emission in the atmosphere. As already mentioned the geosynchrotron emission is the dominant mechanism in the weak magnetic field of the Earth. This geosynchrotron emission is best studied in the high frequency range below the VHF window, that is in the range 40–80 MHz. At higher frequencies one would have to live with a considerable background from man-made radio waves. At lower frequencies the radio noise from the Milky Way, generated by synchrotron emission of spiraling electrons in the galactic disk, is a major background. It has also been tried to measure radio emission from air showers in the GHz regime.

Another mechanism, which is important also for dense media, is the Askaryan effect. In the course of the shower development a negative charge excess of 10–20% is generated. This is a consequence of the ionization of the air by the air-shower particles. The ionization electrons essentially follow the cascade, while the much heavier ions stay behind. In the course of shower development this negative charge excess increases up to the shower maximum and decreases later on. This time-dependent negative charge anisotropy creates the emission of radio waves, like from a time-dependent electric dipole. This radio emission is named after G. Askaryan, who postulated it in 1962.

A third mechanism is Cherenkov radiation produced by relativistic air-shower electrons and positrons. Even though the index of refraction of air is very close to unity, $n = 1.000\,292$, still Cherenkov radiation is generated because of the high velocities of the air-shower particles. This creates Cherenkov radiation in the radio regime and leads to Cherenkov rings with typical radii of about 150 m at ground level for vertical showers.

Other mechanisms, like bremsstrahlung of electrons and positrons, do not play any role for radio emission.

It has also been tried to measure air showers at frequencies different from the preferred 40-to-80-MHz range. Measurements in the GHz regime were quite successful, but did not achieve the quality and significance of the standard techniques in the 40-to-80-MHz range.

The big advantage of radio detection of air showers is that the radio signal is proportional to the energy of the primary particle. On top of that the complete longitudinal development of the cascade in the atmosphere is recorded. This also supplies the measurement of the position of the shower maximum, which is sensitive to the mass of the primary particle. A determination of the chemical composition of primary cosmic rays with classical surface detector arrays is particularly difficult.

The first measurements of radio emission of extensive air showers was already performed in the sixties and seventies of the last century, e.g., at Haverah Park in England. A most comprehensive early review article about radio emission was published by Harold Allan in 1971 [135]. In these measurements the radio antennas were rather simple providing only analogue information. The availability of modern fast digital electronics at affordable prices around the year 2000 revived the radio technique and has led to a renaissance of radio detection of air showers [136].

A typical detector for radio measurements of extensive air showers consists of a (large) number of antennas, which pick up individual radio signals. If possible, the antennas should also measure all three polarization components of the radio field. The lateral width of the radio signal is quite limited because the radio photons are emitted under relatively small angles with respect to the shower axis. Therefore separations of the radio antennas of about 100 meter are favoured. The antennas can be comparatively simple, they must, however, provide a very accurate timing to enable a good reconstruction of the shower axis (see Fig. 7.39). The reconstruction of the shower axis is based on a correlation technique using the individual signals of all antennas. This beam-forming method also permits to suppress effectively the noise generated by background sources. A big advantage is obviously to operate the radio array in radio-quiet surroundings.

It is particularly important to calibrate the radio antennas precisely. Commercial radio transmitters covering the complete frequency range are adequate for this purpose. With such a setup the whole frequency-dependent properties of the complete analysis chain are recorded. A different calibration possibility consists in measuring the galactic noise, which is precisely known from radio astronomy.

Usually such radio arrays are operated jointly with classical air-shower detectors, which can also provide a trigger for the readout of the radio antennas. A self-triggering radio experiment has to meet the challenge of overcoming the generally high background noise level. In radio-quiet surroundings such a self-triggering mode should be possible.

Figure 7.40 shows the radio signal of an air shower as it is obtained from the optimized correlation (beam forming) of 10 antennas of the LOPES experiment as part of the KASCADE-Grande air-shower array [138].

Figure 7.41 shows a radio map of an air shower event. The bright central blob has been reconstructed from the signals of the LOPES experiment operated jointly with KASCADE-Grande. Weak signals in the upper part of the figure are artefacts of the reconstruction procedure.

Figure 7.42 shows the variation of the radio signal with the primary energy. The obtained linear relation allows a clear determination of the primary energy, which initiates the radio shower.

A very useful advantage of radio measurements of air showers is that they can be easily modeled. The different production and propagation processes of radio waves are well understood, and can be reliably described at microscopic level. The possibility of modeling also simplifies the planning and optimization of new radio experiments. There are a number of radio arrays taking data, and extensions of existing air-shower experiments in the planning stage, like Auger in Argentina and

Fig. 7.39 Inverted dipole of the LOPES experiment in the KASCADE-Grande air-shower detector. In the background several measurement stations are visible, which were used to trigger the readout of the radio signals [137]

at ICECUBE at the South Pole. A recent review article on radio measurements has been published by Schröder [139].

The advantage of radio measurements of air showers is the excellent understanding of simulations, which guarantees reasonable comparability of different experiments. In this way also the uncertainties of antenna calibration can be largely excluded. Another big point is the full-time availability of radio detection of air showers—in contrast to fluorescence and Cherenkov measurements. The good energy resolution, and the possibility to identify the nature of the particles initiating the showers via the determination of the position of the shower maximum is also an advantage compared to the measurements with ground-based scintillation or water Cherenkov counters. A disadvantage is the difficulty to operate self-triggering arrays. This is really a challenge, which might bear fruit in radio-quiet surroundings. In most cases one still needs a trigger from classical air-shower detectors to enable a reliable correlation of background-free low-noise radio signals.

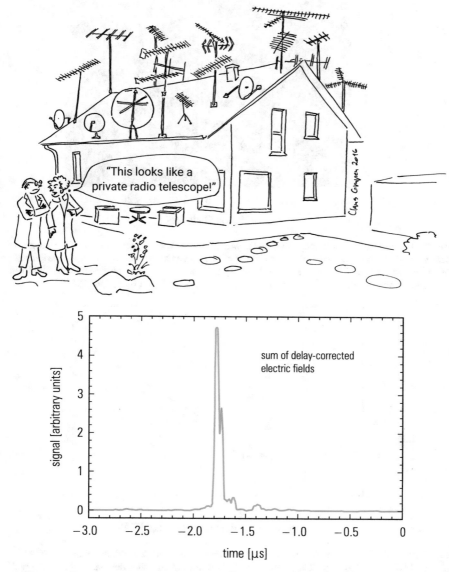

Fig. 7.40 Sum of the ten LOPES antennas by synchronizing and correlating the radio signals in the KASCADE-Grande experiment [137, 138]

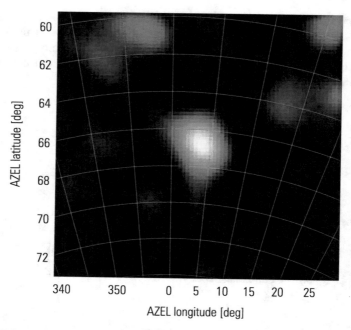

Fig. 7.41 False-colour radio map of an air-shower event. The reconstructed image of the shower is seen as bright blob at the center of the figure. Other weak signals surrounding the central brightness maximum result from interferometer side lobes by the sparse radio array of ten antennas and from background noise in the radio signals. AZEL stands for Azimuth and Elevation [136–138]

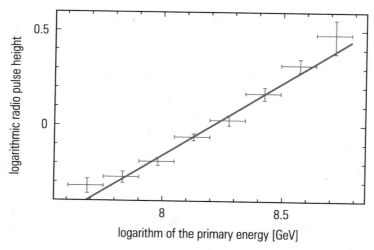

Fig. 7.42 Variation of the radio signal with the primary energy, as determined by the KASCADE-Grande/LOPES experiments. The shower energy was measured using the classical method with surface scintillation counters. The radio signal from the LOPES experiment was corrected for the angle of the shower axis with respect to the geomagnetic field and to the distance from the shower axis [137, 139]

7.6   Acoustic Detection of Air Showers

Cosmic sounds bring the message of big events in the sky.
Anonymous

An inexpensive alternative to large ground-based or fluorescence air-shower arrays would be the acoustic detection of air showers. Such a detection technique was already planned for the DUMAND (Deep Underwater Muon And Neutrino Detector) ocean water detector near Hawaii in the seventies. Presently there is a large number of prototype experiments studying the feasibility of this detection technique for the measurement of high-energy air showers or neutrinos, also as extension for existing air-shower arrays [140].

It has been reported that the US Navy was operating sophisticated hydrophones in the ocean for military purposes. These stations have probably detected acoustic signals from extensive air showers. It is assumed that large showers with energies in excess of, say, 10^{18} eV will produce a thermoacoustic shock wave in water or ice that could be detected by appropriate hydrophones. Accelerator experiments have

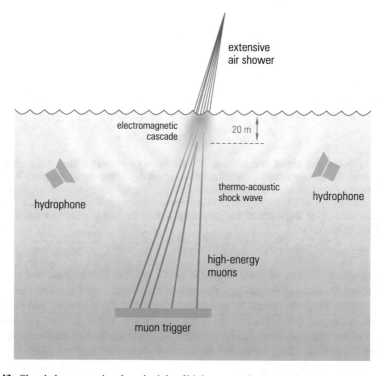

Fig. 7.43   Sketch demonstrating the principle of high-energy air-shower detection via thermoacoustic shock waves. The electromagnetic component of the shower is absorbed in relatively shallow depths. A possible trigger using energetic muons is indicated [141, 142]

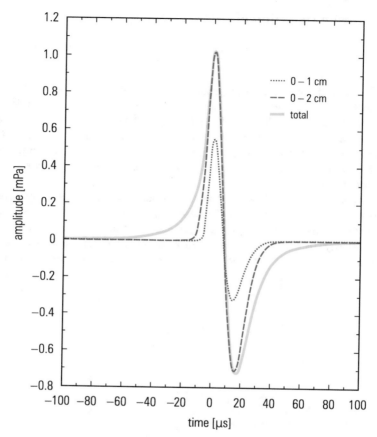

Fig. 7.44 The acoustic signal at a distance of 1 km from the shower axis in the median plane computed from the average of 100 CORSIKA showers each depositing a total energy of 10^9 GeV in water. The *dotted*, *dashed*, and *solid* curves show the signals computed from the deposited energies within cores of radius 1.025, 2.05 g/cm^2, and the whole shower (*solid curve*), respectively. It can be seen that most of the amplitude of the signal comes from the energy within a rather small core of radius 2.05 g/cm^2 [144]

confirmed the existence of such acoustic shock waves. The signals are supposed to be created by the sudden energy deposit of relativistic particles. In the case of air showers such signals would be generated by relativistic electrons and positrons, which are absorbed over a relatively short distance in water or ice. The acoustic detectors would therefore have to be installed near the surface of such arrays.

Figure 7.43 shows the principle of the acoustic detection method [141, 142], and Fig. 7.44 gives an idea, what an acoustic signal induced by a high-energy neutrino in water or ice might look like [143, 144].

The advantage of acoustic detectors is the very long attenuation length of acoustic signals in water or ice, which would allow large effective volumes to be instrumented.

The pressure amplitude of such air showers or neutrino events is expected to be proportional to the shower energy and inversely proportional to the distance of the hydrophones from the shower core. The shower electrons will deposit their energy over a short distance, which is about 50 radiation lengths, corresponding roughly to 20 m in water or ice. The energy deposition extends over a period of $\approx$50 ns, which is instantaneous compared to the signal propagation time of sound waves ($\approx$1500 m/s speed of sound in water at 20 degrees centigrade, and $\approx$3850 m/s in ice at -5 degrees centigrade). One of the problems of acoustic detection is the noise level. For this reason the acoustic detection requires high-energy showers beyond 10^{18} eV or even higher. To pick up the acoustic signal in the presence of background a trigger based of other remnant shower particles like penetrating muons would be very helpful for noise suppression. A survey of existing installations and extension of air-shower and neutrino detectors for acoustic detection is given in [144].

7.7 Some Thoughts on the Highest Energies

No object is mysterious. The mystery is your eye.
Elizabeth Bowen

As already explained in Sect. 6.2, the highest-energy particles of cosmic rays appear to be of extragalactic origin. The problem of the sources of these particles is closely related to the identity of these particles. Up to the present time one had always assumed that the chemical composition of primary cosmic rays might change with energy. However, one always anticipated that the highest-energy particles were either protons, light, or possibly medium heavy nuclei (up to iron). For particles with energies exceeding 10^{20} eV this problem is completely open. In the following the candidates, which might be responsible for cosmic-ray events with energies $> 10^{20}$ eV, will be critically reviewed.

Up to now just over a dozen events with energies exceeding 10^{20} eV have been observed (Auger alone has recorded 14 events with energies beyond 10^{20} eV until the beginning of 2018) [145]. Due to the measurement technique via extensive-air-shower experiments the energy assignments are connected with an experimental error of typically $\pm 30\%$. For the accelerated parent particles of these high-energy particles the gyroradii must be smaller than the size of the source. This leads under rather generous assumptions on the galactic magnetic field and size of our galaxy to a maximum energy of

$$E_{\max} = 10^7 \, \text{TeV} = 10^{19} \, \text{eV} \,. \tag{7.7.1}$$

This equation implies that our Milky Way can hardly accelerate or store particles of these energies, so that for particles with energies exceeding 10^{19} eV one has to assume that they are of extragalactic origin, and for very high energies it also seems to be confirmed by events measured by ICECUBE.

For protons the Greisen–Zatsepin–Kuzmin cutoff (GZK) of photoproduction of pions off blackbody photons through the Δ resonance takes an important influence on the propagation. The energy threshold for this process is at 6×10^{19} eV. Protons exceeding this energy lose rapidly their energy by such photoproduction processes. Estimates on the mean free path of protons lead to a distance, from which one might get protons with energies exceeding the GZK cutoff of about 50 Mpc. The Markarian galaxies Mrk 421 and Mrk 501, which have been shown to be sources of photons of the highest energies, would be candidates for the production of high-energy protons. Since they are residing at distances of approximately 100 Mpc, protons from these distances with energies exceeding 10^{20} eV might arrive at Earth. Normally protons can initiate such high-energy air-shower events only if they come from relatively nearby sources. The giant elliptical galaxy M87 lying in the heart of the Virgo cluster (distance $\approx$20 Mpc) is one of the most remarkable objects in the sky. It meets all of the conditions for being an excellent candidate for a high-energy cosmic-ray source.

Another possible candidate source is the local supergalaxy. It is a kind of 'Milky Way' of galaxies whose center lies in the direction of the Virgo cluster. The local group of galaxies, of which our Milky Way is a member, has a distance of about 20 Mpc from the center of this local supergalaxy and the members of this supergalaxy scatter around the supergalactic center only by about 20 Mpc.

There are actually some hints that the sources for some high-energy events really might lie in the supergalactic plane. The Auger experiment in Argentina has some tentative indications for such a correlation of events with the supergalactic plane at the highest energies. Certainly more events are required to confirm in detail that such a correlation really exists.

It is, however, conceivable to shift the effect of the Greisen–Zatsepin–Kuzmin cutoff to higher energies by assuming that primary particles are nuclei. Since the threshold energy must be available per nucleon, the corresponding threshold energy, for example, for carbon nuclei ($Z = 6$, $A = 12$) would be correspondingly higher,

$$E = E_{\text{cutoff}}^{p} A = 7.2 \times 10^{20} \text{ eV} , \tag{7.7.2}$$

or correspondingly for iron primaries

$$E = E_{\text{cutoff}}^{\text{Fe}} A = 3.4 \times 10^{21} \text{ eV} , \tag{7.7.3}$$

so that the observed events would not be in conflict with the Greisen–Zatsepin–Kuzmin cutoff. It is difficult to understand, how atomic nuclei can be accelerated to such high energies, without being disintegrated by photon interactions or by fragmentation or spallation processes. However, one might argue that the iron fraction in the cosmic-ray beam at the highest energies actually increases, at least from the evidence of some experiments. On the other hand, the Auger experiment finds no significant fraction of iron nuclei at the very highest energies [146].

Already in ancient times jets have been worshipped!

"Is our planet possibly the preferred target for energetic cosmic rays?"

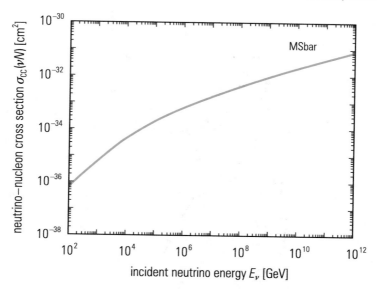

Fig. 7.45 The neutrino–nucleon cross section for the charged-current process as a function of the incident neutrino energy. The cross section is evaluated using the massless MSbar scheme [147]

Another way out would be the rather drastic assumption to assume Lorentz invariance violation. If Lorentz transformations would not only depend on the relative velocity difference of inertial frames, but also on the absolute velocities, the threshold energy for γp collisions for interactions of blackbody photons with high-energy protons would be washed out.

Photons as possible candidates for the observed high-energy cascades are even more problematic. Photon interactions with all kinds of photons from blackbody photons to photons from the radio band attenuate energetic photons considerably. So, just as for protons, photon sources must be rather near for their photons to survive at Earth.

Neutrinos are sometimes discussed as possible candidates for high-energy events. But neutrinos also encounter severe problems in explaining such events. The interaction cross section for neutrino–air interactions at 10^{20} eV is quite low. Enormous neutrino fluxes are required to explain the events with energies $> 10^{20}$ eV. It has been argued that the measurements of the structure function of the proton at HERA[1] have shown that protons have a rich structure of partons at low x ($x = E_{\mathrm{parton}}/E_{\mathrm{proton}}$). Even in view of these results showing evidence for a large number of gluons in the proton, one believes that the neutrino interaction cross section with nuclei of air cannot exceed $0.3\,\mu$b (see Fig. 7.45). This makes interactions of extragalactic neutrinos in the atmosphere very unlikely.

---

[1] HERA—Hadron Elektron Ring Anlage at the Deutsches Elektronensynchrotron (DESY) in Hamburg.

To obtain a reasonable interaction rate only neutrino interactions for inclined directions of incidence or in the Earth can be considered. The resulting expected distribution of primary vertices due to neutrino interactions is in contrast to observation. Therefore, neutrinos as well can very likely be excluded as candidates for the highest-energy cosmic air-shower events.

It has been demonstrated that a large fraction of matter is in the form of dark matter. A possible way out concerning the question of high-energy particles in cosmic rays would be to assume that weakly interacting massive particles (WIMPs) could also be responsible for the observed showers with energies $> 10^{20}$ eV. It has to be considered that all these particles have only weak or even superweak interactions so that their interaction rate can only be on the order of magnitude of neutrino interactions.

The events with energies exceeding 10^{20} eV therefore represent a particle physics dilemma. One tends to assume that protons are the favoured candidates. They must come from relative nearby distances (<50 Mpc), because otherwise they would lose energy by photoproduction processes and fall below the energy of 6×10^{19} eV. It is, however, true that up to these distances there are quite a number of galaxies (e.g., M87). The fact that the observed events do not clearly point back to a nearby source can be explained by the assumption that the extragalactic magnetic fields are so strong that the directional information can be lost, even if the protons are coming from comparably close distances. Actually, there are hints showing that these fields are more in the μgauss rather than in the ngauss region [148].

As already shown, our Milky Way is too small to accelerate and store particles with energies exceeding 10^{19} eV. Furthermore, the arrival directions of the high-energy particles show practically no correlation to the galactic plane. Therefore, one has to assume that they are of extragalactic origin. This is actually indicated by recent data from the Auger experiment, even though this analysis concerns particles with energies above 8 EeV, where the GZK cutoff does not apply, because—due to the steepness of the primary spectrum—most particles will be below the GZK cutoff.

Active galactic nuclei (AGNs) are frequently discussed as possible sources for the highest cosmic-ray energies. In this group of galaxies *blazars* play an outstanding role. Blazar is a short for sources belonging to the class of BL-Lacertae objects and quasars. BL-Lacertae objects, equally as quasars, are Milky Way-like sources, whose nuclei outshine the whole galaxy making them to appear like stars. While the optical spectra of quasars exhibit emission and absorption lines, the spectra of BL-Lacertae objects show no structures at all. This is interpreted in such a way that the galactic nuclei of quasars are surrounded by dense gas, while BL-Lacertae objects reside in low-gas-density elliptical galaxies.

A characteristic feature of blazars is their high variability. Considerable brightness excursions have been observed on time scales as short as a few days. Therefore, these objects must be extremely compact, because the size of the sources can hardly be larger than the time required for light to travel across the diameter of the source. It is generally assumed that blazars are powered by black holes at their center. The matter falling into a black hole liberates enormous amounts of energy. While in nuclear fission only 0.1% and in nuclear fusion still only 0.7% of the mass is transformed into energy, it is assumed that an object of mass m can liberate a substantial part of its rest energy mc^2 if it is swallowed by a black hole. It is known from the black-hole mergers observed by the LIGO Collaboration that a considerable amount of mass can be transformed into the emission of gravitational radiation in such events. Certainly also a large fraction of mass can be converted in polar jets into the acceleration of particles. Such giant sources might be capable to accelerate particles to ZeV energies (10^{21} eV) in galactic jets.

Many high-energy γ-ray sources, which were found by the CGRO (Compton Gamma Ray Observatory) satellite, could be correlated with blazars. This led to the conjecture that these blazars could also be responsible for the acceleration of the highest-energy particles. The *particle jets* produced by blazars exhibit magnetic fields of more than 10 gauss and extend over 10^{-2} pc and more. Therefore, particles could be accelerated to energies exceeding 10^{20} eV. If protons are accelerated in such sources, they could easily escape from these galaxies, because their interaction strength is smaller than that of the electrons, which must certainly be accelerated as well. If these arguments are correct, blazars should also be a rich source of high-energy neutrinos. This prediction can be tested with the large water (or ice) Cherenkov counters.

Recently, a blazar at a distance of about 13 billion years was discovered by a team of international astronomers working at the Keck Observatory in Hawaii [149]. This is so far the oldest galaxy known. The estimated mass of the black hole at the heart of this active galactic nucleus is about 800 million solar masses. Such a giant blazar might well be able to accelerate particles up to the highest energies.

Finally, ideas have also been put forward that the extreme-energy cosmic rays are not the result of the acceleration of protons or nuclei but rather decay products of unstable primordial objects. Candidates discussed as possible sources are decays of massive GUT particles spread through the galactic halo, topological defects produced in the early stages of the universe like domain walls, 'necklaces' of magnetic monopoles connected by cosmic strings, closed cosmic loops containing a superconducting circulating current, or cryptons—relic massive metastable particles born during cosmic inflation. Roger Penrose has also speculated that a lot of energy can be extracted from rotating black holes [150]. The rotational energy is not located inside the event horizon of the black hole but rather outside of it in the ergosphere, a region of the Kerr space-time.

7.8 Summary

Secondary cosmic rays result from the interaction of primary cosmic rays in the atmosphere. For low primary energies ($\approx$GeV region) penetrating muons dominate at sea level. In the high-energy regime ($\gg 1$ TeV) primary particles and also γ rays induce particle cascades in the atmosphere. Depending on the particle type, the secondary particles are differently attenuated. For these high energies secondary electrons, positrons, and γ rays are the dominant particle species at ground level. Muons—depending on their energy—reach also to large depths underground. At the highest energies ($\gg 10^{15}$ eV) the primary particles initiate large air showers producing millions of secondary particles. These secondaries can be recorded with different detector techniques, like surface detectors (scintillation counters, Cherenkov counters, fluorescence telescopes, Cherenkov telescopes, or via radio emission or acoustic detection, ...). A pending problem is still the determination of the chemical composition of primary cosmic rays. Direct measurements on the chemical composition are only possible outside the atmosphere. Due to the steeply falling primary energy spectrum this is only possible for energies up to several TeV. At higher energies one has to rely on indirect techniques by using the method of extensive air showers. This, however, allows only a very crude distinction between different primary particles. One tries to distinguish heavy from light primary particles, i.e., the periodic table of elements at high energies has essentially only protons and iron nuclei, and even this distinction is sometimes only marginal. Because of irregular galactic and intergalactic magnetic fields it is also very difficult to determine the origin of cosmic rays. This might work at very high energies ($\gg 10^{18}$ eV), or with neutrinos, but there the numbers of detected particles is quite low, and it is difficult to do astronomy with a handful of particles.

7.9 Problems

1. The pressure at sea level is 1013 hPa. Convert this pressure into a column density in kg/cm^2!

2. The barometric pressure varies with altitude h in the atmosphere (assumed to be isothermal) like

$$p = p_0\, e^{-h/7.99\,km}\ .$$

What is the residual pressure at 20 km altitude and what column density of residual gas does this correspond to?

3. For not too large zenith angles the angular distribution of cosmic-ray muons at sea level can be parameterized as $I(\theta) = I(0)\cos^2\theta$. Motivate the $\cos^2\theta$ dependence!

4. The integral spectrum of atmospheric muons underground for shallow depths can be parametrized by Eq. (7.3.9):

$$N(>E,\,R) = A\,(aR)^{-\gamma}\ .$$

Work out the rate of stopping atmospheric muons as a function of depth underground for shallow depths!

5. Figure 7.22 shows a muon shower in the ALEPH experiment. Typical energies of muons in this shower are 100 GeV. What is the r.m.s. scattering angle of muons in rock for such muons (overburden 320 m w.e., radiation length in rock $X_0 = 25\,g/cm^2 \cong 10\,cm$)?

6. Narrow muon bundles with muons of typically 100 GeV originate in interactions of primary cosmic rays in the atmosphere. Estimate the typical lateral separation of cosmic-ray muons in a bundle at a depth of 320 m w.e. underground.

7. Due to the dipole character of the Earth's magnetic field the geomagnetic cut-off varies with geomagnetic latitude. The minimum energy for cosmic rays to penetrate the Earth's magnetic field and to reach sea level can be worked out to be

$$E_{min} = \frac{ZeM}{4R^2} \cos^4 \lambda \,,$$

where Z is the charge number of the incident particle, $M = 6.7 \times 10^{22}$ A m^2 is the moment of the Earth's magnetic dipole, R is the Earth radius, λ is the geomagnetic latitude ($0°$ at the Equator). For protons one gets $E_{min} = 15$ GeV $\cos^4 \lambda$.

The Earth's magnetic field has reversed several times over the history of our planet. In those periods when the dipole changed polarity, the magnetic field went through zero. In these times when the magnetic shield decayed, more cosmic-ray particles could reach the surface of the Earth causing a higher level of radiation for life developing on our planet. Whether this had a positive effect on the biological evolution or not is the object of much debate. The estimation of the increased radiation level for periods of zero field can proceed along the following lines:

- the differential energy spectrum of primary cosmic rays can be represented by a power law $N(E) \sim E^{-\gamma}$ with $\gamma = 2.7$.
- in addition to the geomagnetic cutoff there is also an atmospheric cutoff due to the energy loss of charged particles in the atmosphere of ≈ 2 GeV.

Work out the increase in the radiation level using the above limits!

If you want to work out a number for the cutoff energy, you will realize that—to get the dimension right—you will need a factor of $(4\pi \varepsilon_0 c)$ in the denominator, where ε_0 is the permittivity of free space and c the velocity of light. These numerical constants frequently occur in such formulae. Then you have to convert joule to GeV.

8. How many Cherenkov photons in the radio domain (40–80 MHz) are emitted by a relativistic electron in the atmosphere per meter?

9. Radio astronomers have occasionally been fooled by noise from a microwave oven in the astronomy building. It would be interesting to know what happens if astronauts on the Moon would cook their meal with such a microwave oven. Suppose if one considers a microwave oven with a power of 1 kW on the Moon and that its power is emitted isotropically across a bandwidth of 200 MHz centered at 2.7 GHz. What is the microwave flux density on Earth of this microwave oven? Would this oven be detectable with a typical radio telescope driven with a bandwidth of 100 MHz?

10. Work out the number of synchrotron photons in the radio range (around 60 MHz), which are produced by a 100-MeV air-shower electron over a distance of 100 m in the Earth's magnetic field. Hint(s): See [151].

11. The speed of sound in sea water depends on the temperature (T in degree Celsius), salinity (S in parts per thousand), and depth (z in m). It can very approximately be parametrized by

$$c(T, S, z) \approx a_1 + a_2 \cdot T + a_3 \cdot T^2 + a_5 \cdot (S - 35) + a_6 \cdot z. \qquad (7.9.1)$$

The sound velocity for an air-shower event slightly below the surface or at larger depths (e.g., $z = 1000$ m) is different. This could lead to an error in the determination of the angle of incidence, if this is not taken into account. What kind of systematic error could be introduced if the variation of the sound velocity with depth is neglected for the propagation of the thermoacoustic signal of the air shower? Hint: see [152].

12. In the attenuation of a thermoacoustic signal, the energy is essentially converted into heat. For low-frequency sound the attenuation is

$$\alpha(\nu) = 1.2 \times 10^{-7} \cdot \nu^2 \; \mathrm{dB/km/Hz^2}. \qquad (7.9.2)$$

The attenuation factor α is defined by

$$\alpha = 20 \cdot \lg \left(\frac{V_2}{V_1} \right) \; \mathrm{dB}, \qquad (7.9.3)$$

where V_1 is the reference voltage.

What is a reasonable distance of hydrophones for a frequency of a 100-Hz signal?

13. Acoustic detection of air showers in sea water is troubled with ambient noise of the ocean. Upon the impact of an energetic air shower instantaneously a pressure wave is generated following the sudden deposition of energy. The pressure-wave generation is essentially a thermoacoustic process. In the ocean the resulting pressure amplitude p can be estimated from

$$p[\mathrm{Pa}] \approx 6 \times 10^{-21} \; E[\mathrm{eV}]. \qquad (7.9.4)$$

The ambient sea noise for frequencies of interest for acoustic detection is approximately 50 dB. This noise level sets an energy threshold for the detection of high-energy particles. What is the minimum energy to create a pressure wave that exceeds the ambient noise level? Hint: look at [153].

Chapter 8
Cosmology

As far as the laws of mathematics refer to reality, they are not certain; as far as they are certain, they do not refer to reality.
Albert Einstein 1921

In the following chapters the application of our knowledge of particle physics to the very early universe in the context of the Hot Big Bang model of cosmology will be explored. The basic picture is that the universe emerged from an extremely hot, dense phase about 14 billion years ago. The earliest time, about which one can meaningfully speculate, is about 10^{-43} s after the Big Bang (the Planck time). To go earlier requires a quantum-mechanical theory of gravity, and this is not yet available.

At early times the particle densities and typical energies were extremely high, and particles of all types were continually being created and destroyed. For the first 10^{-38} s or so, it appears that all of the particle interactions could have been 'unified' in a theory containing only a single coupling strength. It was not until after this, when typical particle energies dropped below around 10^{16} GeV, that the strong and electroweak interactions became distinct. At this time, from perhaps 10^{-38} to 10^{-36} s after the Big Bang, the universe may have undergone a period of *inflation*, a tremendous expansion, where the distances between any two elements of the primordial plasma increased by a factor of perhaps e^{100}. When the temperature of the universe dropped below 100 GeV, the electroweak unification broke apart into separate electromagnetic and weak interactions.

Until around 1 microsecond after the Big Bang, quarks and gluons could exist as essentially free particles. After this point, energies dropped below around 1 GeV and the partons became bound into hadrons, namely, protons, neutrons, and their antiparticles. Had the universe contained at this point equal amounts of matter and antimatter, almost all of it would have annihilated, leaving us with photons, neutrinos, and little else. For whatever reason, nature apparently made one a bit more abundant than the other, so there was some matter left over after the annihilation phase to make the universe as it is now. Essentially all of the positrons had annihilated with electrons within the first couple of seconds.

© Springer Nature Switzerland AG 2020
C. Grupen, *Astroparticle Physics*, Undergraduate Texts in Physics,
https://doi.org/10.1007/978-3-030-27339-2_8

Around three minutes after the Big Bang, the temperature had dropped to the point, where protons and neutrons could fuse to form deuterons. In the course of the next few minutes these combined to form helium, which makes up a quarter of the universe's nuclear matter by mass, and smaller quantities of a few light elements such as deuterium, lithium, and beryllium. The model of Big Bang Nucleosynthesis (BBN) is able to correctly predict the relative abundances of these light nuclei and this is one of the cornerstones of the Hot Big Bang model.

As the universe continued to expand over the next several hundred thousand years, the temperature finally dropped to the point, where electrons and protons could join to form neutral atoms. After this the universe became essentially transparent to photons, and those, which existed at that time, have been drifting along unimpeded ever since. They can be detected today as the cosmic microwave background radiation. Only small variations in the temperature of the radiation, depending on the direction, at a level of one part in 10^5 are observed. These are thought to be related to small density variations in the universe left from a much earlier period, perhaps as early as the inflationary epoch only 10^{-36} s after the Big Bang. The details of these temperature variations have been determined with great accuracy by the satellites COBE, WMAP, and Planck. At present one can still watch what happened at the Big Bang by tuning an old television set to a frequency not occupied by a TV station: one can see a kind of snowy picture, in which about one percent of the snow dots are caused by cosmological blackbody photons like an echo of the Big Bang.

Studies of the cosmic microwave background radiation (CMB) also lead to a determination of the total density of the universe, and one finds a value very close to the so-called critical density, above which the universe should recollapse in a 'Big Crunch'. The same CMB data and also observations of distant supernovae, however, show that about 70% of this is not what one would call matter at all, but rather a sort of energy density associated with empty space—a vacuum energy density.

The remaining 30% appears to be gravitating matter, but of what sort? One of the indirect consequences of the Big Bang Nucleosynthesis is that only a small fraction of the matter in the universe appears to be composed of known particles. The remainder of the *dark matter* may consist of *neutralinos*, particles predicted by a theory called supersymmetry.

The framework, in which the early universe will be studied, is based on the 'Standard Cosmological Model' or 'Hot Big Bang'. The basic ingredients are Einstein's theory of general relativity and the hypothesis that the universe is isotropic and homogeneous when viewed over sufficiently large distances. It is in the context of this model that the laws of particle physics will be applied in an attempt to trace the evolution of the universe at very early times. In this chapter the important aspects of cosmology that one needs will be reviewed.

8.1 The Hubble Expansion

The history of astronomy is a history of receding horizons.

Edwin Powell Hubble

Inflation

Big Crunch

The first important observation that leads to the Standard Cosmological Model is Hubble's discovery that all but the nearest galaxies are receding away from us (i.e., from the Milky Way) with a speed proportional to their distance. The speeds are determined from the Doppler shift of spectral lines. Suppose a galaxy receding from us (i.e., from the Milky Way) with a speed $v = \beta c$ emits a photon of wavelength λ_{em}. When the photon is observed, its wavelength will be shifted to λ_{obs}. To quantify this, the *redshift* z is defined as

$$z = \frac{\lambda_{obs} - \lambda_{em}}{\lambda_{em}} . \tag{8.1.1}$$

From relativity one obtains the relation between the redshift and the speed,

$$z = \sqrt{\frac{1 + \beta}{1 - \beta}} - 1 , \tag{8.1.2}$$

which can be approximated by

$$z \approx \beta \tag{8.1.3}$$

for $\beta \ll 1$.

To measure the distance of a galaxy one needs in it a light source of a calibrated brightness (a 'standard candle'). The light flux from the source falls off inversely as the square of the distance r, so if the absolute luminosity L is known and the source radiates isotropically, then the measured light flux is $F = L/4\pi r^2$. The *luminosity distance* can therefore be determined from

$$r = \sqrt{\frac{L}{4\pi F}} . \tag{8.1.4}$$

Figure 8.1 shows the problems to determine a precise relation between the velocity and distance of galaxies, i.e., to find an exact value for the Hubble constant, which describes the expansion of the universe. The reason are random galaxy motions and observational difficulties in distance measurements related to problems to know the absolute luminosity of stars and galaxies. The large uncertainties and systematic problems in the history of the determination of the Hubble constant are clearly demonstrated in Fig. 8.2. Therefore, *standard candles* of known luminosity must be found, such as Cepheid variable stars, used by Hubble or type Ia supernovae. A plot of speed versus distance determined from type Ia supernovae is shown in Fig. 8.3 [154–156]. The spectra of SN Ia are hydrogen poor. The absence of planetary nebulae allows to reconstruct the genesis of these events. It is generally believed that the progenitor of a SN Ia is a binary consisting of a white dwarf and a red giant companion. Both members are gravitationally bound. In white dwarves the electron degeneracy pressure compensates the inward-bound gravitational pressure. The strong gravitational potential of the white dwarf overcomes the weaker gravity of the red giant. At the periphery of the red giant the gravitational force of the white

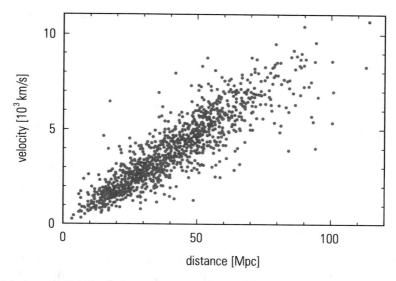

Fig. 8.1 An early Hubble relation between velocity and distance for a sample of galaxies [159]. The data have been obtained for 1355 galaxies. The scatter is due to observational uncertainties and random galaxy motions. For the distance scale a Hubble constant of about 70 km/s per Mpc was used

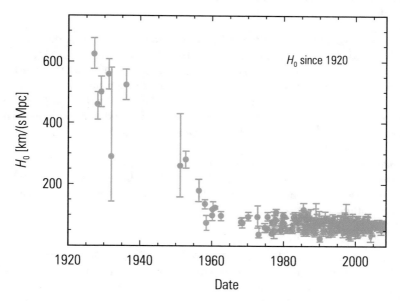

Fig. 8.2 The evolution of the Hubble constant as a function of time [160]

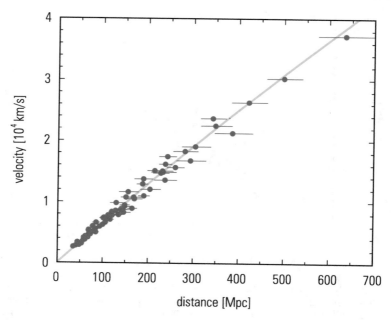

Fig. 8.3 The Hubble relation between velocity and distance for a sample of type Ia supernovae [154–156]

dwarf is stronger than that of the red giant causing mass from its outer envelope to be accreted onto the white dwarf. Since for white dwarves the product of mass times volume is constant, it decreases in size during accretion. When the white dwarf reaches the Chandrasekhar limit (1.44 $M_\odot$), the electron degeneracy pressure can no longer withstand the gravitational pressure. It will collapse under its own weight. This goes along with an increase in temperature causing hydrogen to fuse to helium and heavier elements. This sudden burst of energy leads to a thermonuclear explosion that destroys the star.

Since the Chandrasekhar limit is a universal quantity, all SN Ia explode in the same way. Therefore, they can be considered as standard candles.

For the distances covered in the plot, the data are clearly in good agreement with a linear relation,

$$v = H_0\, r \,, \tag{8.1.5}$$

which is Hubble's Law. The parameter H_0 is Hubble's constant, which from the data in Fig. 8.3 is determined to be 64 (km/s)/Mpc. Here the subscript 0 is used to indicate the value of the parameter today. The relation between speed and distance is, however, not constant in time.

Determinations of the Hubble constant based on different observations have yielded somewhat inconsistent results, although less so now than a few decades ago. The most recent value (2018) based on data from the Planck satellite gives a value of

$$H_0 = 67.5 \pm 0.6 \,(\mathrm{km/s})/\mathrm{Mpc}\,, \qquad (8.1.6)$$

with further systematic uncertainties at the percent level [157, 158]. In addition, one usually defines the parameter h by

$$H_0 = h \times 100 \,(\mathrm{km/s})/\mathrm{Mpc}\,. \qquad (8.1.7)$$

Quantities that depend on H_0 are then written with the corresponding dependence on h. To obtain a numerical value one substitutes for h the most accurate estimate available at the time ($h = 0.675 \pm 0.012$ in 2018). For purposes of this course one can use $h \approx 0.7 \pm 0.1$.

8.2 The Isotropic and Homogeneous Universe

> *The center of the universe is everywhere, and the circumference is nowhere.*
>
> Blaise Pascal

The assumption of an isotropic and homogeneous universe, sometimes called the *cosmological principle*, was initially made by Einstein and others because it simplified the mathematics of general relativity. Today there is ample observational evidence in favour of this hypothesis. The cosmic microwave background radiation, for example, is found to be isotropic to a level of around one part in 10^5. Still, one has to keep in mind that the assumption of isotropy and homogeneity and the corresponding simplifying symmetry is a very important input to the derivation of the Friedmann equation, and it should be tested experimentally. The clumpy universe around us may cast some doubt on the idea of isotropy and homogeneity. On the other hand, the idea of isotropy and homogeneity appears today to hold at sufficiently large distances, say, greater than around 100 Mpc. At smaller distances, galaxies are indeed found to clump together forming clusters and voids. A typical intergalactic distance is on the order of 1 Mpc, so a cube with sides of 100 Mpc could have a million galaxies. Therefore, one should think of the galaxies as the 'molecules' of a gas, which is isotropic and homogeneous when a volume large enough to contain large numbers of them is considered.

If the universe is assumed to be isotropic and homogeneous, then the only possible motion is an overall expansion or contraction. Consider, for example, two randomly chosen galaxies at distances $r(t)$ and $R(t)$ from ours, as shown in Fig. 8.4.

An isotropic and homogeneous expansion (or contraction) means that the ratio

$$\chi = r(t)/R(t) \qquad (8.2.1)$$

is constant in time. Therefore, $r(t) = \chi R(t)$ and

$$\dot{r} = \chi \dot{R} = \frac{\dot{R}}{R} r \equiv H(t) r\,, \qquad (8.2.2)$$

Fig. 8.4 Two arbitrary
galaxies at distances $r(t)$ and
$R(t)$ from our Milky Way

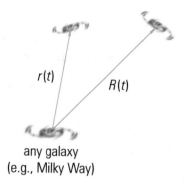

any galaxy
(e.g., Milky Way)

where dots indicate derivatives with respect to time. The ratio

$$H(t) = \dot{R}/R \tag{8.2.3}$$

is called the *Hubble parameter*. It is the fractional change in the distance between any pair of galaxies per unit time. H is often called the *expansion rate* of the universe.

Equation (8.2.2) is exactly Hubble's law, where $H(t)$ at the present time is identified as the Hubble constant H_0. So the hypothesis of an isotropic and homogeneous expansion explains why the speed, with which a galaxy moves away from us, is proportional to its distance (s. Fig. 8.4).

8.3 The Friedmann Equation from Newtonian Gravity

No theory is sacred.

Edwin Powell Hubble

The evolution of an isotropic and homogeneous expansion is completely determined by giving the time dependence of the distance between any representative pair of galaxies. One can denote this distance by $R(t)$, which is called the *scale factor*. An actual numerical value for R is not important. For example, one can define $R = 1$ at a particular time (e.g., now). It is the time dependence of R that gives information about how the universe as a whole evolves.

The rigorous approach would now be to assume an isotropic and homogeneous matter distribution and to apply the laws of general relativity to determine $R(t)$. By fortunate coincidence, in this particular problem, Newton's theory of gravity leads to the same answer, namely, to the Friedmann equation for the scale factor R. This approach will now be briefly reviewed.

Consider a spherical volume of the universe with a radius R sufficiently large to be considered homogeneous, as shown in Fig. 8.5. In today's universe this would mean taking R at least 100 Mpc. If one assumes that the universe is electrically neutral, then the only force that is significant over these distances is gravity. As a test mass,

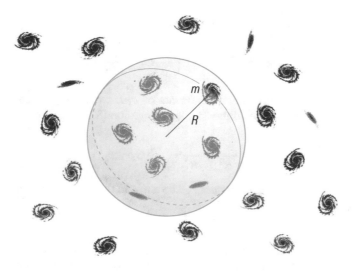

Fig. 8.5 A sphere of radius R, containing many galaxies, along with a test galaxy of mass m at its edge

consider a galaxy of mass m at the edge of the volume. It feels the gravitational attraction from all of the other galaxies inside. As a consequence of the inverse-square nature of gravity, this force is the same as what one would obtain if all of the mass inside the sphere were placed at the center.

Another non-trivial consequence of the $1/r^2$ force is that the galaxies outside the sphere do not matter. Their total gravitational force on the test galaxy is zero. In Newtonian gravity these properties of isotropically distributed matter inside and outside a sphere follow from Gauss' law for a $1/r^2$ force. The corresponding law holds in *general relativity* as well, where it is known as Birkhoff's theorem.

If one assumes that the mass of the galaxies is distributed in space with an average density ϱ, then the mass inside the sphere is

$$M = \frac{4}{3}\pi R^3 \varrho .\tag{8.3.1}$$

The gravitational potential energy V of the test galaxy is therefore

$$V = -\frac{GmM}{R} = -\frac{4\pi}{3}GmR^2\varrho .\tag{8.3.2}$$

The sum of the kinetic energy T and potential energy V of the test galaxy gives its total energy E,

$$E = \frac{1}{2}m\dot{R}^2 - \frac{4\pi}{3}GmR^2\varrho = \frac{1}{2}mR^2\left(\frac{\dot{R}^2}{R^2} - \frac{8\pi}{3}G\varrho\right) .\tag{8.3.3}$$

The *curvature parameter* k is now defined by

$$k = \frac{-2E}{m} = R^2 \left(\frac{8\pi}{3} G\varrho - \frac{\dot{R}^2}{R^2} \right) . \tag{8.3.4}$$

If one were still dragging along the factors of c, k would have been defined as $-2E/mc^2$; in either case k is dimensionless. Equation (8.3.4) can be written as

$$\frac{\dot{R}^2}{R^2} + \frac{k}{R^2} = \frac{8\pi}{3} G\varrho , \tag{8.3.5}$$

which is called the *Friedmann equation*. The terms in this equation can be identified as representing

$$T - E = -V , \tag{8.3.6}$$

i.e., the Friedmann equation is simply an expression of conservation of energy applied to our test galaxy. Since the sphere and test galaxy could be anywhere in the universe, the equation for R applies to any pair of galaxies sufficiently far apart that one can regard the intervening matter as being homogeneously distributed.

The Friedmann equation can also be applied to the early universe, before the formation of galaxies. It will hold even for an ionized plasma as long as the universe is electrically neutral overall and one averages over large enough distances. The scale factor R in that case represents the distance between any two elements of matter sufficiently far apart such that gravity is the only force that does not cancel out.

The general theory of relativity essentially leads to the same result, even though with some sophistication. In the general formulation of the Friedmann equation ϱ is now no longer the matter density, but, because of the equivalence of mass and energy, the energy density. ϱ includes now all forms of energy, e.g., also photons. The curvature parameter k describes the curvature of space, as its name already indicates. In the Newtonian treatment k was only a measure for the total energy. A very important modification is the addition of the cosmological constant Λ, which has been introduced by Einstein by hand to the field equations of general relativity. At the time when Einstein came up with the general relativity the universe was assumed to be static and constant in time. To prevent a gravitational collapse, a repulsive force had to be introduced that would allow a static universe. When Hubble discovered the expansion of the universe, Einstein considered his idea of introducing a cosmological constant as his biggest blunder. In the present form of cosmology, however, there is a new justification of having a repulsive force, and the cosmological constant has reappeared in the theory.

With these modifications and extensions the Friedmann equation, describing the evolution of the scale factor, now reads

$$\frac{\dot{R}^2}{R^2} + \frac{k}{R^2} = \frac{8\pi}{3} G\varrho + \frac{\Lambda}{3} . \tag{8.3.7}$$

Fig. 8.6 Pure vacuum energy?

The cosmological constant Λ can also be expressed by the *vacuum energy density* ϱ_v, like:

$$\varrho_v = \frac{\Lambda}{8\pi G}. \tag{8.3.8}$$

The total energy density now also includes ϱ_v. ϱ_v is called vacuum energy because such a quantity is also predicted in quantum mechanics. In quantum mechanics the vacuum energy originates from virtual particle pairs, which are permanently created and destroyed by decay (Fig. 8.6).

In a quantum-mechanical treatment of the harmonic oscillator the energy eigenvalues are obtained to $E = (n + 1/2) \cdot \hbar \cdot \omega$. Even the lowest-energy state, i.e., $n = 0$, the ground state, has an energy of $E = 1/2 \cdot \hbar \cdot \omega$: this is the zero-point energy. Also in quantum field theories the vacuum is not empty. According to Heisenberg's uncertainty relation the vacuum contains infinitely many particle–antiparticle pairs. A clear evidence for this statement is given by the Casimir effect (see also Fig. 8.7). Consider two parallel metal plates at a very small separation in vacuum. Because the distance is so small, not every possible wavelength can exist in the space between the two plates, quite in contrast to the surrounding vacuum. The effect of this limited

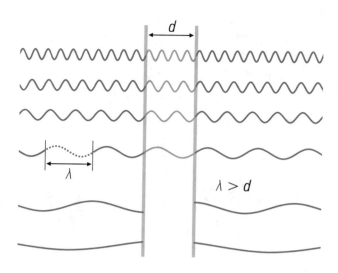

Fig. 8.7 Illustration of the Casimir effect: Only certain wavelengths fit into the space between the plates. The outside of the plates does not limit the number of possible frequencies, resp. wavelengths [161]

choice of quanta between the plates leads to a small attractive force of the plates. This pressure of the surrounding vacuum was experimentally confirmed. This force can be imagined in such a way that the reduced number of field quanta in the space between the plates cannot resist the pressure of the unlimited number of field quanta in the surrounding vacuum. Simple estimations about the order of magnitude of this vacuum energy Λ from quantum field theories suggest a value, which is 120 orders of magnitude larger compared to the vacuum energy in cosmology.

There is clearly much that is not understood about the cosmological constant, and Einstein famously regretted proposing it. The more modern view of physical laws, however, leads one to believe that when a term is absent from an equation, it is usually because its presence would violate some symmetry principle. The cosmological constant is consistent with the symmetries, on which general relativity is based. And in the years since 1998 clear evidence from the redshifts of distant supernovae has been presented that the expansion of the universe is accelerating. This accelerated expansion can be understood as the effect of a repulsive gravity, which can be described by a cosmological constant. This means that the cosmological constant is indeed non-zero, and that the vacuum energy makes a large contribution to the total energy density of the universe. In fact, it appears that the energy density of the vacuum constitutes about 70% of the total energy density of the universe. In Chaps. 12 and 13 this will be discussed in more detail.

8.4 The Fluid Equation

The universe is like a safe to which there is a combination – but the combination is locked in the safe.

Peter de Vries

The Friedmann equation cannot be solved yet because one does not know how the energy density ϱ varies with time. Instead, in the following a relation between ϱ, its time derivative $\dot{\varrho}$, and the pressure P will be derived. This relation, called *fluid equation*, follows from the *first law of thermodynamics* for a system with energy U, temperature T, entropy S, and volume V,

$$dU = T\, dS - P\, dV \, . \tag{8.4.1}$$

The first law of thermodynamics will now be applied to a volume R^3 in our expanding universe. Since by symmetry there is no net heat flow across the boundary of the volume, one has $dQ = T\, dS = 0$, i.e., the expansion is adiabatic. Dividing (8.4.1) by the time interval dt then gives

$$\frac{dU}{dt} + P\frac{dV}{dt} = 0 \, . \tag{8.4.2}$$

The total energy U is

$$U = R^3\varrho \, . \tag{8.4.3}$$

The derivative dU/dt is therefore

$$\frac{dU}{dt} = \frac{\partial U}{\partial R}\dot{R} + \frac{\partial U}{\partial \varrho}\dot{\varrho} = 3R^2\varrho\dot{R} + R^3\dot{\varrho} \, . \tag{8.4.4}$$

For the second term in (8.4.2) one gets

$$\frac{dV}{dt} = \frac{d}{dt}R^3 = 3R^2\dot{R} \, . \tag{8.4.5}$$

Putting (8.4.4) and (8.4.5) into (8.4.2) and rearranging terms gives

$$\dot{\varrho} + \frac{3\dot{R}}{R}(\varrho + P) = 0 \, , \tag{8.4.6}$$

which is the *fluid equation*. Unfortunately, this is still not enough to solve the problem, since an *equation of state* relating ϱ and P is needed. This can be obtained from the laws of statistical mechanics as will be shown in Sect. 9.3. With these ingredients the Friedmann equation can then be used to find $R(t)$ (see also Sect. 9.4).

8.5 The Acceleration Equation

It is Nature you always have to ask for advice.

Hector Guimard

In a number of cases it can be useful to combine the Friedmann and fluid equations to obtain a third equation involving the second derivative $\ddot{R}$. Here only the relevant steps will be outlined, and the results will be given; the derivation is straightforward. The Friedmann equation without cosmological constant was

$$\frac{\dot{R}^2}{R^2} + \frac{k}{R^2} = \frac{8\pi}{3}G\varrho \tag{8.5.1}$$

or rephrased

$$\dot{R}^2 + k = \frac{8\pi}{3}G\varrho R^2 . \tag{8.5.2}$$

If this expression is derived with respect to time, one gets

$$2\dot{R}\ddot{R} = \frac{8\pi}{3}G(\dot{\varrho}R^2 + 2R\dot{R}\varrho) ; \tag{8.5.3}$$

if now $\dot{\varrho}$ from the fluid equation is used, one obtains

$$2\dot{R}\ddot{R} = \frac{8\pi}{3}G\left(2R\dot{R}\varrho - \frac{3\dot{R}}{R}(\varrho + P)R^2\right) . \tag{8.5.4}$$

This equation is equivalent to

$$\ddot{R} = \frac{4\pi}{3}G(2R\varrho - 3R\varrho - 3RP) \quad \text{or} \quad \frac{\ddot{R}}{R} = -\frac{4\pi}{3}G(\varrho + 3P) , \tag{8.5.5}$$

and this is called the *acceleration equation*. We did not get any additional insight compared to the Friedmann equation or fluid equation, but there is a number of applications and problems, where the acceleration equation turns out to be very useful.

8.6 Solution of the Friedmann Equation Without Vacuum Energy

My view is that if your philosophy is not unsettled daily then you are blind to all the universe has to offer.

Neil deGrasse Tyson

Without explicitly solving the Friedmann equation one can already make some general statements about the nature of possible solutions. The Friedmann equation (8.3.5) can be written as

$$H^2 = \frac{8\pi G}{3}\varrho - \frac{k}{R^2}, \tag{8.6.1}$$

where, as always, $H = \dot{R}/R$. From the observed redshifts of galaxies, it is known that the current expansion rate H is positive. One expects, however, that the galaxies should be slowed down by their gravitational attraction. One can therefore ask whether this attraction will be sufficient to slow the expansion to a halt, i.e., whether H will ever decrease to zero.

If the curvature parameter k is negative, then this cannot happen, since everything on the right-hand side of (8.6.1) is positive. Recall that $k = -2E/m$ basically gives the total energy of the test galaxy on the edge of the sphere of galaxies. Having $k < 0$ means that the total energy of the test galaxy is positive, i.e., it is not gravitationally bound. In this case the universe is said to be *open*; it will continue to expand forever.

If, for example, the energy density of the universe is dominated by non-relativistic matter, then one will find that ϱ decreases as $1/R^3$. So, eventually, the term with the curvature parameter will dominate on the right-hand side of (8.6.1) leading to

$$\frac{\dot{R}^2}{R^2} = -\frac{k}{R^2}, \tag{8.6.2}$$

which means $\dot{R}$ is constant or $R \sim t$.

If, on the other hand, one has $k > 0$, then as ϱ decreases, the first and second terms on the right-hand side of (8.6.1) will eventually cancel with the results of $H = 0$. That is, the expansion will stop. At this point, all of the kinetic energy of the galaxies is converted to gravitational potential energy, just as when an object thrown vertically into the air reaches its highest point. And just as with the thrown object, the motion then reverses and the universe begins to contract. In this case the universe is said to be *closed*.

One can also ask what happens if the curvature parameter k and hence also the energy E of the test galaxy are zero, i.e., if the universe is just on the borderline between being open and closed. In this case the expansion will be decelerated but H will approach zero asymptotically. The universe is then said to be *flat*. This is analogous to throwing a projectile upwards with a speed exactly equal to the escape velocity.

Which of the three scenarios one obtains—open, closed, or flat—depends on what is in the universe to slow or otherwise affect the expansion. To see how this corresponds to the energy density, the Friedmann equation (8.3.5) can be solved for the curvature parameter k, which gives

$$k = R^2 \left(\frac{8\pi}{3} G\varrho - H^2 \right). \tag{8.6.3}$$

Fig. 8.8 The scale factor R as function of time for $\Omega < 1$, $\Omega > 1$, and $\Omega = 1$

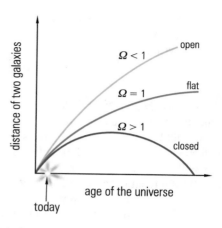

One can then define the *critical density* ϱ_c by

$$\varrho_c = \frac{3H^2}{8\pi G} \tag{8.6.4}$$

and express the energy density ϱ by giving the ratio Ω,

$$\Omega = \frac{\varrho}{\varrho_c} . \tag{8.6.5}$$

Using this with $H = \dot{R}/R$, (8.6.3) becomes

$$k = R^2(\Omega - 1)H^2 . \tag{8.6.6}$$

So, if $\Omega < 1$, i.e., if the density is less than the critical density, then $k < 0$ and the universe is open; the expansion continues forever. Similarly, if $\Omega > 1$, the universe is closed; H will decrease to zero and then become negative. If the density ϱ is exactly equal to the critical density ϱ_c, then one has $k = 0$, corresponding to $\Omega = 1$, and the universe is flat.

Figure 8.8 illustrates schematically how the scale factor R depends on time for the three scenarios. These curves only apply if the cosmological constant is zero. The names closed, open, and flat refer to the geometrical properties of space-time that one finds in the corresponding solutions using general relativity.

A completely different type of solution will be found if the total energy density is dominated by vacuum energy. In that case the expansion increases exponentially. This scenario will be dealt with in Chap. 12 when inflation will be discussed (see also Fig. 8.9).

Fig. 8.9 Difficult question

8.7 Experimental Evidence for the Vacuum Energy

> *It happened five billion years ago. That was when the Universe stopped slowing down and began to accelerate, experiencing a cosmic jerk.*
>
> Adam Riess

For a given set of contributions to the energy density of the universe, the Friedmann equation predicts the scale factor, $R(t)$, as a function of time. From an observational standpoint, one would like to turn this around: from measurements of $R(t)$ one can make inferences about the contents of the universe. Naively, one would expect the attractive force of gravity to slow the Hubble expansion, leading to a deceleration, i.e., $\ddot{R} < 0$. One of the most surprising developments of the recent past has been the discovery that the expansion is, in fact, accelerating, and apparently has been for several billion years. This can be predicted by the Friedmann equation if one assumes a contribution to the energy density with *negative pressure*, such as the vacuum energy previously mentioned. Such a contribution to ϱ is called *dark energy*.

A negative pressure and its cosmological effect can be illustrated as follows: Einstein has shown that all forms of energy are equivalent and all act gravitationally. But not only matter, but also forces create gravitation. Even forces that resist gravitation produce a certain gravitational effect. A celestial body resists a gravitative collapse by pressure forces. These pressure forces contribute energy, which in turn are equivalent to a gravitational effect. This means that a positive pressure leads to a vicious circle: the pressure forces defeat themselves. The more they grow, the more their effect increases the gravitation. In contrast, a negative pressure is like a tension, which is a force that pulls things together. In the same way as a positive pressure leads to an increased attraction, the negative pressure starts a vicious circle in the opposite direction. A positive pressure, which gets out of hand, leads eventually to a compact object, like a neutron star or even a black hole, while a negative pressure results in an increasing expansion or even an inflationary scenario.

This can also be seen formally by a simple dimensional argument. The physical dimension of the energy density is energy per volume

$$[E/V] = \frac{\text{kg m}^2}{\text{s}^2} \cdot \frac{1}{\text{m}^3} = \frac{\text{kg m}}{\text{s}^2} \cdot \frac{1}{\text{m}^2}, \tag{8.7.1}$$

where the last expression is newton per area, i.e., this is a pressure. A negative pressure therefore corresponds to a negative energy density.

Because of the finite velocity of light a view into the sky also provides a view into the past. A comparison of the velocity of a distant galaxy with the one in our neighbourhood allows to say something whether the expansion of the universe is slowed down or is accelerated. To judge on this we need the escape velocity of the galaxies and their distance. Supernovae of type Ia are considered as standard candles with known absolute luminosity. Therefore, their observed apparent brightness on Earth allows to determine their distance. Their escape velocity tells us their redshift.

A word of caution, however, is in order here. Type Ia supernovae are considered as standard candles because they seem to explode always in the same way. But detailed studies have revealed that they are only on average good as standard candles, still there is some scatter in detail, which might be important for the distance determination [162].

The relation between the apparent brightness of a supernova and its redshift can be derived from the time dependence of the scale factor in the Friedmann equation, if all contributions to the energy density of the universe are known. We have to consider that non-relativistic matter and vacuum energy contribute to the energy density, and that the vacuum energy is described by the cosmological constant. Both contributions to the energy density can be referred to the critical density ϱ_c, and we obtain $\Omega_{m,0}$ as the contribution of matter, and $\Omega_{\Lambda,0}$ as the effect of the vacuum energy. The index 0 refers to the present-day state of the universe. However, one also has to take into account the possible contribution of photons and neutrinos to $\Omega_{r,0}$. According to the results from the Planck satellite this contribution seems to be very small.

The relative brightness of stars and galaxies is described by the comparatively complicated quantity called *magnitude* [163, 164]. The brightness difference of two stars with intensities I_1 and I_2 is given by the difference in magnitudes $m_1 - m_2 = -2.5 \lg(I_1/I_2)$. The zero-point of the magnitude is fixed by the magnitude of the polestar with $2^{\text{m}}_{.}12$. An apparent brightness difference of $\Delta m = 2.5$ corresponds to an

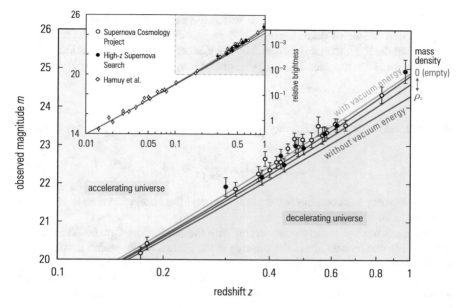

Fig. 8.10 Magnitudes m of type Ia supernovae in their dependence on the redshift z of their host galaxies compared to the expectations of various models, image credit: Saul Perlmutter [166]

intensity ratio of $10 : 1$. Using this scale, Venus gets the magnitude $m = -4.4$ and the Sun even $m = -26$. With the naked eye one can see stars up to a magnitude of $m = 6$. The Hubble telescope can observe stars up to $m = 31$. The Hubble successor, the James Webb Space Telescope, is expected to be able to see objects with a magnitude of $m = 34$ [165] in the infrared. The larger the magnitude, the fainter the celestial object. The relation of the apparent brightness m of distant supernovae in their dependence on the redshift z shows Fig. 8.10.

Qualitatively the behaviour shown in Fig. 8.10 can be understood in the following way. The data points at high redshift lie above the curve for zero vacuum energy, i.e., at higher magnitudes, which means that they are dimmer than expected. Thus, the supernovae one sees at a given z are farther away than one would expect, therefore, the expansion must be speeding up. At high z the magnitude–redshift relation, however, depends on details of the energy density. The lower curve without vacuum energy, corresponding to $\Omega_{m,0} = 1$, i.e., $\Omega_{\Lambda,0} = 0$, clearly is in conflict with the data. On the other hand, the measurements are best described by $\Omega_{m,0} \approx 0.25$ and $\Omega_{\Lambda,0} \approx 0.75$. It seems obvious that the data can only be described by an accelerated expansion of the universe. More detailed comparisons will be given in subsequent chapters.

The exact physical origin of the vacuum energy remains a mystery. On the one hand, vacuum energy is expected in a quantum field theory such as the Standard Model of elementary particles. Naively, one would expect its value to be of the order

$$\varrho_v \approx E_{max}^4 , \tag{8.7.2}$$

where E_{max} is the maximum energy, at which the field theory is valid. One expects the Standard Model of particle physics to be incomplete, for example, at energies higher than the Planck energy, $E_{Pl} \approx 10^{19}$ GeV, where quantum-gravitational effects come into play. So the vacuum energy might be roughly

$$\varrho_v \approx E_{Pl}^4 \approx 10^{76} \, \text{GeV}^4 \,. \tag{8.7.3}$$

But from the observed present-day value $\Omega_{\Lambda,0} \approx 0.7$, the vacuum energy density is

$$\varrho_{\Lambda,0} = \Omega_{\Lambda,0} \varrho_{c,0} = \Omega_{\Lambda,0} \frac{3H_0^2}{8\pi G} \approx 10^{-46} \, \text{GeV}^4 \,. \tag{8.7.4}$$

The discrepancy between the naive prediction and the observed value is 122 orders of magnitude.

So something has clearly gone wrong with the prediction. One could argue that not much is understood about the physics at the Planck scale, which is surely true, and so a lower energy cutoff should be tried. Suppose one takes the maximum energy at the electroweak scale, $E_{EW} \approx 100$ GeV, roughly equal to the masses of the W and Z bosons. At these energies the Standard Model has been tested to high accuracy. The prediction for the vacuum energy density then becomes

$$\varrho_v \approx E_{EW}^4 \approx 10^8 \, \text{GeV}^4 \,. \tag{8.7.5}$$

Now the discrepancy with the observational limit is 'only' 54 orders of magnitude—perhaps an improvement but clearly not enough. On the other hand, the existence of the vacuum energy in field theories cannot be denied, as is clearly demonstrated by the Casimir effect. There are obviously more fundamental problems combining quantum field theories and results from cosmology. The data suggest that the discrepancy probably rests at the inability to formulate a quantum theory of relativity.

After the discovery of accelerated expansion of the universe and looking at Fig. 8.8, we now have to reconsider also the evolution of the scale factor. Figure 8.11 shows the size of the universe for various assumptions on the energy content. Presently one favours the time evolution of the scale factor using the values $\Omega_{m,0} \approx 0.3$ and $\Omega_{\Lambda,0} \approx 0.7$. The universe without vacuum energy is at conflict with the supernova measurements at high redshifts. The conclusion of an accelerated expansion, already over a period of several billion years, seems unavoidable.

According to the result of the Planck satellite, in accordance with WMAP and COBE data, the universe is now 13.8 billion years old. This does not mean that the size of the universe corresponds to 13.8 billion years. This age only tells us that we are only able to see objects, which have emitted light 13.8 billions years ago. Because, however, the universe is expanding since the Big Bang, the stars and galaxies, which have emitted light 13.8 billions years ago, are now at a much larger distance from us. Since the universe seems to be flat, one can estimate its size based on the curvature parameter, including its uncertainty of about 2%, to be about 46 billion light-years [168].

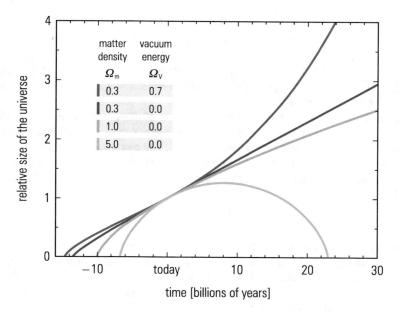

Fig. 8.11 Relative size of the universe in its dependence on time for various assumptions on its energy density [167]

8.8 Summary

Cosmology is based on the field equations of general relativity and tries to describe the evolution of our universe. We have derived the history of the universe starting from a semi-classical treatment of the Friedmann equation. The effect of the cosmological constant, already introduced by Einstein in 1915—albeit for other reasons—receives new support from the observed accelerated expansion of the universe. In the light of this measurement we have to reconsider the evolution of the universe. Obviously we are living in a flat universe, which is described by a matter density dominated by dark energy and dark matter, and there is little room for the known and accustomed baryonic form of matter (less than 5%).

8.9 Problems

1. Derive the relativistic relation (8.1.2) between redshift and velocity of a receding galaxy.
2. A gas cloud gets unstable if the gravitational energy exceeds the thermal energy of the molecules constituting the cloud, i.e.,

$$\frac{GM^2}{R} > \frac{3}{2}kT\frac{M}{\mu},$$

where $\frac{3}{2}kT$ is the thermal energy of a molecule, $\frac{M}{\mu}$ is the number of molecules, and μ is the molecule mass. Derive from this condition the stability limit of a gas cloud (Jeans criterion).

3. Let us assume that a large astrophysical object of constant density not stabilized by internal pressure is about to contract due to gravitation. Estimate the minimum rotational velocity so that this object is stabilized against gravitational collapse! How does the rotational velocity depend on the distance from the galactic center?

4. Show that the gravitational redshift of light emitted from a massive star (mass M) of radius R is

$$\frac{\Delta v}{v} = \frac{GM}{c^2 R}.$$

5. Estimate the classical value for the deflection of starlight passing near the Sun.

6. Clocks in a gravitational potential run slow relative to clocks in empty space. Estimate the slowing-down rate for a clock on a pulsar ($R = 10\,\text{km}$, $M = 10^{30}\,\text{kg}$)!

7. Estimate the gravitational pressure at the center of the Sun (average density $\varrho = 1.4\,\text{g/cm}^3$) and the Earth ($\varrho = 5.5\,\text{g/cm}^3$).

8. Estimate the average density of

   (a) a large black hole residing at the center of a galaxy of 10^{11} solar masses,
   (b) a solar-mass black hole,
   (c) a mini black hole ($m = 10^{15}\,\text{kg}$).

9. The orbital velocity v of stars in our galaxy varies up to distances of $20\,000$ light-years as if the density were homogeneous and constant ($\varrho = 6 \times 10^{-21}\,\text{kg/m}^3$). For larger distances the velocities of stars follow the expectation from Keplerian motion.

   (a) Work out the dependence of $v(R)$ for $R < 20\,000$ light-years.
   (b) Estimate the mass of the Milky Way.
   (c) The energy density of photons is on the order of $0.3\,\text{eV/cm}^3$. Compare the critical density to this number!

Chapter 9
The Early Universe

Who cares about half a second after the Big Bang; what about the half second before?

Fay Weldon quoted after Paul Davies

In this chapter the history of the universe through the first ten microseconds of its existence will be described. First, in Sect. 9.1 the *Planck scale*, where quantum-mechanical and gravitational effects both become important, will be defined. This sets the starting point for the theory to be described. In Sects. 9.2 and 9.3 some formulae from statistical and thermal physics will be assembled, which are needed to describe the hot dense phase, out of which the universe then evolved. Then, in Sects. 9.4 and 9.5 these formulae will be used to solve the Friedmann equation and to investigate the properties of the universe at very early times. Finally, in Sect. 9.6 one of the outstanding puzzles of the Hot Big Bang model will be presented, namely, why the universe appears to consist almost entirely of matter rather than a mixture of matter and antimatter.

9.1 The Planck Scale

A new scientific truth does not triumph by convincing its opponents and making them see the light, but rather because its opponents eventually die, and a new generation grows up that is familiar with it.

Max Planck

The earliest time, about which one can meaningfully speculate with our current theories, is called the *Planck era*, around 10^{-43} s after the Big Bang. Before then, the quantum-mechanical aspects of gravity are expected to be important, so one would need a quantum theory of gravity to describe this period. Although superstrings could perhaps provide such a theory, it is not yet in such a shape that one can use it to make specific predictions.

© Springer Nature Switzerland AG 2020
C. Grupen, *Astroparticle Physics*, Undergraduate Texts in Physics,
https://doi.org/10.1007/978-3-030-27339-2_9

To see how the Planck scale arises, consider the Schwarzschild radius. The Schwarzschild radius describes the radius of a black hole, from which nothing, not even light can escape. The escape velocity from a massive object can be worked out from

$$\frac{1}{2}mv^2 = G\frac{m\,M}{R} ,$$

where m is the mass of the escaping particle and M and R are the mass, respectively the radius of the object, from which the particle wants to escape. Let us now assume that this relation is also valid for light—which is outrageously wrong—and replace v with c, and we get

$$c^2 = \frac{2\,G\,M}{R} \quad \text{or} \quad R = \frac{2\,G\,M}{c^2} .$$

This result derived under unacceptable assumptions luckily agrees with the result of the correct derivation based on general relativity.

Let us start now from this result

$$R_S = \frac{2MG}{c^2} , \tag{9.1.1}$$

where for the moment factors of c and $\hbar$ will be explicitly inserted. The distance R_S gives the event horizon of a black hole. It represents the distance, at which the effects of space-time curvature become significant.

Now consider the Compton wavelength of a particle of mass m,

$$\lambda_C = \frac{h}{mc} . \tag{9.1.2}$$

This represents the distance, at which quantum effects become important. The Planck scale is thus defined by the condition $\lambda_C/2\pi = R_S/2$, where we now consider an object of mass m, i.e.,

$$\frac{\hbar}{mc} = \frac{mG}{c^2} . \tag{9.1.3}$$

Solving for the Planck mass gives

$$m_{Pl} = \sqrt{\frac{\hbar c}{G}} \approx 2.2 \times 10^{-5}\,\text{g} , \tag{9.1.4}$$

which corresponds to the mass of a water droplet about $1/3$ mm in diameter. The rest-mass energy of m_{Pl} is the Planck energy,

$$E_{Pl} = \sqrt{\frac{\hbar c^5}{G}} \approx 1.22 \times 10^{19}\,\text{GeV} , \tag{9.1.5}$$

which is about 2 GJ or 650 kg TNT equivalent. Using the Planck mass in the reduced Compton wavelength, $\hbar/mc$, gives the Planck length,

$$l_{Pl} = \sqrt{\frac{\hbar G}{c^3}} \approx 1.6 \times 10^{-35} \text{ m}. \tag{9.1.6}$$

The time that it takes light to travel l_{Pl} is the Planck time,

$$t_{Pl} = \frac{l_{Pl}}{c} = \sqrt{\frac{\hbar G}{c^5}} \approx 5.4 \times 10^{-44} \text{ s}. \tag{9.1.7}$$

The Planck mass, length, time, etc. are the unique quantities with the appropriate dimension that can be constructed from the fundamental constants linking quantum mechanics and relativity: $\hbar$, c, and G. Since henceforth $\hbar$ and c will be set equal to one, and one has

$$m_{Pl} = E_{Pl} = 1/\sqrt{G}, \tag{9.1.8}$$
$$t_{Pl} = l_{Pl} = \sqrt{G}. \tag{9.1.9}$$

So the Planck scale basically characterizes the strength of gravity. As the Planck mass (or energy), 1.2×10^{19} GeV, is a number people tend to memorize, often $1/m_{Pl}^2$ will be used as a convenient replacement for G.

9.2 Thermodynamics of the Early Universe

We are startled to find a universe we did not expect.
Walter Bagehot

In this section some results from statistical and thermal physics will be collected that will be needed to describe the early universe. Some of the relations presented may differ from those covered in a typical course in statistical mechanics. This is for two main reasons. First, the particles in the very hot early universe typically have speeds comparable to the speed of light, so the relativistic equation $E^2 = p^2 + m^2$ must be used to relate energy and momentum. Second, the temperatures will be so high that particles are continually being created and destroyed, e.g., through reactions such as $\gamma\gamma \leftrightarrow e^+e^-$. This is in contrast to the physics of low-temperature systems, where the number of particles in a system is usually constrained to be constant. The familiar exception is blackbody radiation, since massless photons can be created and destroyed at any non-zero temperature. For a gas of relativistic particles expressions for ϱ, n, and P will be found that are similar to those for blackbody radiation. Here ϱ, n, and P are the density or—more generally—the energy density, the number density, and the pressure.

The formulae in this section are derived in standard texts on statistical mechanics. Here merely the results in a form appropriate for the early universe will be quoted. In a more rigorous treatment one would need to consider conservation of various quantum numbers such as charge, baryon number, and lepton number. For each conserved quantity one has a chemical potential μ, which enters into the expressions for the energy and number densities. For most of the treatment of the very early universe one can neglect the *chemical potentials*, and thus they will not appear in the formulae that are given here.

In the limit, where the particles are relativistic, we have $T \gg m$.[1]

Using the Stefan–Boltzmann law one gets for the energy density for relativistic particles, just as for photons

$$\varrho \sim T^4 . \tag{9.2.1}$$

The constant of proportionality in this relation depends among others on the spin of the particles and on the colour states of quarks and gluons. Therefore, it is different for bosons and fermions. Using dimensional arguments and considering that an inverse length corresponds to an energy (like in the Compton wavelength, $\lambda = h/p$), the number density can be expressed as

$$n \sim T^3 \tag{9.2.2}$$

because of $n \sim 1/V$ and $V \sim R^3$.

From the energy and number density one obtains the average energy per particle $\langle E \rangle = \varrho/n$. For $T \gg m$ one immediately finds

$$\langle E \rangle \sim T . \tag{9.2.3}$$

In the non-relativistic case the energy density is just

$$\varrho = mn , \tag{9.2.4}$$

where n is the number density. This is given in the non-relativistic limit (compare the Maxwell–Boltzmann distribution and Eq. (9.2.2)), and one gets

$$n = g \left(\frac{mT}{2\pi} \right)^{3/2} e^{-m/T} . \tag{9.2.5}$$

The factor g considers spin and colour states of the particles. This equation is also obtained in the limit of low energies from the Fermi–Dirac and also from the Bose–Einstein distribution. The number density of non-relativistic particles varies then like $e^{-m/T}$. Particles of high mass are correspondingly suppressed. In the classical limit the average energy is therefore the sum of the mass and the kinetic energy ($\frac{3}{2}kT$):

---

[1]Note that Boltzmann's constant k has also been set equal to one.

Table 9.1 Particles of the Standard Model along with their properties (Particle Data Book). The mass limits given for neutrinos have been taken from direct measurements. Results from the Planck satellite suggest a limit for the sum of neutrino masses like $m_{\nu_e} + m_{\nu_\mu} + m_{\nu_\tau} < 1$ eV. With this information we have for each neutrino flavour $m_\nu < 1$ eV

Particle	Mass	Spin states	Colour states
Bosons			
Photon (γ)	0	2	1
W^+, W^-	80.4 GeV	3	1
Z	91.2 GeV	3	1
Gluon (g)	0	2	8
Higgs	125.2 GeV	1	1
Fermions			
$u, \bar{u}$	2.2 MeV	2	3
$d, \bar{d}$	4.7 MeV	2	3
$s, \bar{s}$	95 MeV	2	3
$c, \bar{c}$	1.3 GeV	2	3
$b, \bar{b}$	4.2 GeV	2	3
$t, \bar{t}$	173 GeV	2	3
e^+, e^-	0.511 MeV	2	1
μ^+, μ^-	105.7 MeV	2	1
τ^+, τ^-	1.777 GeV	2	1
$\nu_e, \bar{\nu}_e$	<2 eV	1	1
$\nu_\mu, \bar{\nu}_\mu$	<0.19 MeV	1	1
$\nu_\tau, \bar{\nu}_\tau$	<18.2 MeV	1	1

$$\langle E \rangle = m + \frac{3}{2}T \approx m \,. \tag{9.2.6}$$

This approximation only holds for $T \ll m$.

To solve the Friedmann equation one naturally needs the total energy density of all particles in the early universe. The constant of proportionality for the energy density will include all bosons and fermions with their corresponding spin and colour states, including also all particles even if they have not yet been discovered. Not all particles in the early universe were necessarily relativistic. It could have been possible that the hypothetical supersymmetric partners of 'normal' particles were very heavy (>1 TeV). The masses and numbers of spin and colour states of the particles of the Standard Model are listed in Table 9.1.

Here the neutrinos will be treated as only being left-handed and antineutrinos as only right-handed, i.e., they only have one spin state each, which again is related to them being considered massless.

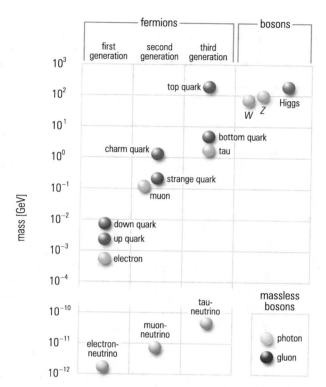

Fig. 9.1 Hierarchy of particle masses. The masses of neutrinos are not yet known. From direct measurements together with cosmology one gets a limit for each neutrino flavour of below 10^{-9} GeV. Neutrino oscillations are only sensitive to the difference of the squares of the masses. Assuming a mass hierarchy similar to the sector of charged leptons one can have a guess at the neutrino masses [169]

In fact, recent evidence (see Sect. 6.3 on neutrino astronomy) indicates that neutrinos do have non-zero mass, but the coupling to the additional spin states is so small that their effect on g (number of internal degrees of freedom for the particle) can be ignored.[2] Neutrino oscillations so far have not been able

---

[2]The neutrino within the Standard Model is assumed to behave as Dirac fermion. But the nature of the neutrino is not yet settled; it may be either a Dirac or Majorana particle. A Majorana particle is a fermion that is its own antiparticle. Such type of particles were hypothesized by Ettore Majorana in 1937. If neutrinos were of Majorana type a neutrinoless double beta decay should be possible. Such a decay can be considered as two subsequent beta decays, where the produced antineutrino in the first decay will immediately transform into its (own) antiparticle, which is absorbed in the second decay. Even though many searches for Majorana-type neutrinos have been performed, such a decay mode has never been observed, and the limit for the lifetime of such a decay, e.g., of ^{76}Ge is $T_{1/2} = 5.3 \times 10^{25}$ years, which is fairly long. Also sterile neutrinos, which are assumed to be hypothetical particles that interact only via gravity and not via the known interactions of the Standard Model, are not considered here. Sterile neutrinos are considered to be right-handed, and sterile antineutrinos should be left-handed. Recent results from the ICECUBE Neutrino Observatory did not find any evidence for sterile neutrinos.

to measure directly neutrino masses but seem to indicate that they are below 50 meV (Fig. 9.1).

9.3 Equation of State

Next the required equations of state will be recalled, that is, relations between energy density ϱ and pressure P. They are needed in conjunction with the acceleration and fluid equations in order to solve the Friedmann equation for $R(t)$. For the pressure for a gas of relativistic particles one finds

$$P = \frac{\varrho}{3}. \qquad (9.3.1)$$

This is the well-known result from blackbody radiation, but in fact it applies for any particle type in the relativistic limit $T \gg m$. In the non-relativistic limit, the pressure is given by the ideal-gas law, $P = nT$. In this case, however, one has an energy density $\varrho = mn$, so for $T \ll m$ one has $P \ll \varrho$ and in the acceleration and fluid equations one can approximate

$$P \approx 0. \qquad (9.3.2)$$

In addition, one can show that for the case of the vacuum energy density from a cosmological constant,

$$P = -\varrho_{\rm v}. \qquad (9.3.3)$$

That is, a vacuum energy density leads to a negative pressure. In general, the equation of state can be expressed as

$$P = w\varrho, \qquad (9.3.4)$$

where the parameter w is $1/3$ for relativistic particles, 0 for (non-relativistic) matter, and -1 for vacuum energy.

Finally, in this section a general relation between the temperature T and the scale factor R will be noted. All lengths, when considered over distance scales of at least 100 Mpc, increase with R. Since the de Broglie wavelength of a particle, $\lambda = h/p$, is inversely proportional to the momentum, one sees that particle momenta decrease as $1/R$. For photons, one has $E = p$, and so their energy decreases also as $1/R$. Furthermore, the temperature of photons in thermal equilibrium is simply a measure of the photons' average energy, so one gets the important relation

$$T \sim R^{-1}. \qquad (9.3.5)$$

This relation holds as long as T is interpreted as the photon temperature and as long as the Hubble expansion is what provides the change in T. In fact this is not exact, because there are other processes that affect the temperature as well. For example, as electrons and positrons become non-relativistic and annihilate into photons, the photon temperature receives an extra contribution. These effects can be taken into account by thermodynamic arguments using conservation of entropy. The details of this are not critical for the present treatment, and one will usually be able to assume (9.3.5) to hold.

Cosmology is like modern art. Nobody understands it!

Gun 2010

Difficult question about cosmology

9.4   Refining the Solution of the Friedmann Equation

No one will be able to read the great book of the universe if he does not understand its language which is that of mathematics.

Galileo Galilei

Now enough information is available to solve the Friedmann equation. This will allow to derive the time dependence of the scale factor R, temperature T, and energy density ϱ. If, for the start, very early times will be considered, one can simplify the problem by seeing that the term in the Friedmann equation (8.3.5) with the curvature parameter, k/R^2, can be neglected. To show this, recall from Sect. 8.4 the fluid equation (8.4.6),

$$\dot{\varrho} + \frac{3\dot{R}}{R}(\varrho + P) = 0 \,, \tag{9.4.1}$$

which relates the time derivative of the energy density ϱ and the pressure P. One can suppose that ϱ is dominated by radiation, so that the equation of state (9.3.1) can be used,

$$P = \frac{\varrho}{3} \,. \tag{9.4.2}$$

Substituting this into the fluid equation (9.4.1) gives

$$\dot{\varrho} + \frac{4\varrho\dot{R}}{R} = 0 \,. \tag{9.4.3}$$

The left-hand side is proportional to a total derivative, so one can write

$$\frac{1}{R^4}\frac{d}{dt}\left(\varrho R^4\right) = 0 \,. \tag{9.4.4}$$

This implies that ϱR^4 is constant in time, and therefore

$$\varrho \sim \frac{1}{R^4} \,. \tag{9.4.5}$$

If, instead, one would have assumed that ϱ was dominated by non-relativistic matter, one would have used $P = 0$ in the equation of state, and in a similar way one would have found

$$\varrho \sim \frac{1}{R^3} \,. \tag{9.4.6}$$

In either case the dependence of ϱ on R is such that for very early times, that is, for sufficiently small R, the term $8\pi G\varrho/3$ on the right-hand side of the Friedmann equation will be much larger than k/R^2. One can then ignore the curvature parameter and effectively set $k = 0$; this is definitely valid at the very early times that will be considered in this chapter and it is still a good approximation today, 14 billion years later. The Friedmann equation then becomes

$$\frac{\dot{R}^2}{R^2} = \frac{8\pi}{3}G\varrho \,. \tag{9.4.7}$$

   In the following (9.4.7) will be solved for the case, where ϱ is radiation dominated. One can write (9.4.5) as

$$\varrho = \varrho_0 \left(\frac{R_0}{R}\right)^4 \,, \tag{9.4.8}$$

where here ϱ_0 and R_0 represent the values of ϱ and R at some particular (early) time.

One can guess a solution of the form $R = At^p$ and substitute this along with (9.4.8) for ϱ into the Friedmann equation (9.4.7). With this ansatz the Friedmann equation can only be satisfied if $p = 1/2$, i.e.,

$$R \sim t^{1/2} . \tag{9.4.9}$$

The expansion rate H is therefore

$$H = \frac{\dot{R}}{R} = \frac{1}{2t} . \tag{9.4.10}$$

If, on the other hand, ϱ were dominated by non-relativistic matter, i.e., $P = 0$, one would find, using the ansatz

$$R = A t^p , \quad \dot{R} = p A t^{p-1} \tag{9.4.11}$$

the relation

$$\frac{\dot{R}^2}{R^2} = \frac{p^2 A^2 t^{2p} t^{-2}}{A^2 t^{2p}} = \frac{p^2}{t^2} = \frac{8\pi}{3} G \varrho_0 \left(\frac{R_0}{R}\right)^3 = \frac{8\pi}{3} G \varrho_0 R_0^3 A^{-3} t^{-3p} . \tag{9.4.12}$$

Comparing the t dependence on both sides of this equation, one gets

$$p = \frac{2}{3} \tag{9.4.13}$$

and thereby

$$R \sim t^{2/3} \tag{9.4.14}$$

and

$$H = \frac{2}{3t} . \tag{9.4.15}$$

Now, one can combine the Friedmann equation (9.4.7) with the energy density (9.2.1) and use $G = 1/m_{\text{Pl}}^2$ to replace the gravitational constant. Taking the square root then gives the expansion rate H as a function of the temperature,

$$H \sim \frac{T^2}{m_{\text{Pl}}} . \tag{9.4.16}$$

Combining this result with Eq. (9.4.10) one finds a relation between temperature and time according to

$$t \sim \frac{m_{\text{Pl}}}{T^2} . \tag{9.4.17}$$

If one combines the Friedmann equation with the solution $H = 1/2t$, one finds after a relatively short and simple calculation the time-dependent energy density

$$\varrho = \frac{3m_{\text{Pl}}^2}{32\pi} \frac{1}{t^2}.$$ (9.4.18)

9.5 Thermal History of the First Ten Microseconds

> *An elementary particle that does not exist in particle theory should also not exist in cosmology.*
>
> Anonymous

The relations from the previous sections can now be used to work out the energy density and temperature of the universe as a function of time. As a start, one can use (9.4.18) to give the energy density at the Planck time, although one needs to keep in mind from the previous section that the assumption of thermal equilibrium may not be valid. In any case the formula gives

$$\varrho(t_{\text{Pl}}) = \frac{3m_{\text{Pl}}^2}{32\pi} \frac{1}{t_{\text{Pl}}^2} = \frac{3}{32\pi} m_{\text{Pl}}^4 \approx 6 \times 10^{74} \, \text{GeV}^4,$$ (9.5.1)

where $m_{\text{Pl}} = 1/t_{\text{Pl}} \approx 1.2 \times 10^{19} \, \text{GeV}$ has been used. One can convert this to normal units by dividing by $(\hbar c)^3$,

$$\varrho(t_{\text{Pl}}) \approx 6 \times 10^{74} \, \text{GeV}^4 \times \frac{1}{(0.2 \, \text{GeV fm})^3}$$
$$\approx 8 \times 10^{76} \, \text{GeV/fm}^3.$$ (9.5.2)

This density corresponds to about 10^{77} proton masses in the volume of a single proton!

Proceeding now more systematically, one can find the times and energy densities, at which different temperatures were reached. By combining this with the knowledge of particle physics, one will see what types of particle interactions were taking place at what time. To do this numerically, we need all particles, which have contributed to the energy density, including all their spin and colour states. In good approximation we can treat all particles of the Standard Model as being relativistic (T larger than a few hundred GeV). Possibly existing heavier particles from supersymmetric theories will change the results only insignificantly.

Table 9.2 shows values for the temperature and energy density at several points within the first 10 ms after the Big Bang, where most of the values have been rounded to the nearest order of magnitude.

At the Planck scale, i.e., with energies on the order of 10^{19} GeV, and earlier times than 10^{-39} s, the limit of our ability to speculate about cosmology and cosmoparticle

Table 9.2 Thermal history of the first 10 ms

Scale	T (GeV)	ϱ (GeV^4)	t (s)
Planck	10^{19}	10^{77}	10^{-43}
GUT	10^{16}	10^{66}	10^{-39}
Electroweak	10^2	10^{10}	10^{-11}
QCD	0.2	0.01	10^{-5}

physics has been reached. At temperatures or corresponding energies of 10^{16} GeV the different interactions separate into strong and electroweak interactions with their own couplings strengths. This is the scale of Grand Unified Theories (GUT scale). One expects that this phase transition is related to the Higgs field. The vacuum expectation value of the Higgs field above the GUT scale should be zero. Up to this moment the masses of elementary particles were zero. During the transition the vacuum expectation value of the Higgs field will adopt non-zero values. This phenomenon is called spontaneous symmetry breaking (SSB). As a result, the hypothetical X and Y bosons from earlier times go from being massless to having very high masses on the order of the GUT scale. So, at lower temperatures, baryon-number-violating processes mediated by exchange of X and Y bosons are highly suppressed.

After around 10^{-11} s, the temperature is on the order of 100 GeV; this is called the 'electroweak scale'. Here another SSB phase transition is expected to occur, whereby the electroweak Higgs field acquires a non-zero vacuum expectation value. As a result, W and Z bosons as well as the quarks and leptons acquire their masses. At temperatures significantly lower than the electroweak scale, the masses $M_W \approx 80$ GeV and $M_Z \approx 91$ GeV are large compared to the kinetic energies of other colliding particles, and the W and Z propagators effectively suppress the strength and the range of the weak interaction.

At temperatures around 0.2 GeV (the 'QCD scale'), the effective coupling strength of the strong interaction, α_s, becomes very large. At this point quarks and gluons become confined into colour-neutral hadrons: protons, neutrons, and their antiparticles. This process, called *hadronization*, occurs around $t \approx 10^{-5}$ s. Here one has to keep in mind that the assumption of highly relativistic particles becomes a little questionable.

It is interesting to convert the energy density at the QCD scale to normal units, which gives about 1 GeV/fm^3. This is about seven times the density of ordinary nuclear matter. Experiments at the Relativistic Heavy Ion Collider (RHIC) at the Brookhaven National Laboratory near New York are currently underway to recreate these conditions by colliding together heavy ions at very high energies [170]. This will allow more detailed studies of the 'quark–gluon plasma' and its transition to colour-neutral hadrons.

This short sketch of the early universe has ignored at least two major issues. First, it has not yet been explained why the universe appears to be composed of matter, rather than a mixture of matter and antimatter. There is no definitive answer to this question but there are plausible scenarios, whereby the so-called baryon asymmetry of the universe could have arisen at very early times, perhaps at the GUT scale or later at the electroweak scale. This question will be looked at in greater detail in the next section.

One will also see that the model that has been developed thus far fails to explain several observational facts, the most important of which are related to the cosmic microwave background radiation. A possible remedy to these problems will be to suppose that the energy density at some very early time was dominated by vacuum energy. In this case one will not find $R \sim t^{1/2}$ but rather an exponential increase, known as *inflation*.

9.6 The Baryon Asymmetry of the Universe

> *Astronomy, the oldest and one of the most juvenile of the sciences, may still have some surprises in store. May anti-matter be commended to its case.*
>
> Arthur Schuster

For several decades after the discovery of the positron, it appeared that the laws of nature were completely symmetric between matter and antimatter. The universe known to us, however, seems to consist of matter only. The relative abundance of baryons will now be looked at in detail and therefore this topic is called the *baryon asymmetry of the universe*. One has to examine how this asymmetry could evolve from a state that initially contained equal amounts of matter and antimatter, a process called *baryogenesis*. For every proton there is an electron, so the universe also seems to have a non-zero lepton number. This is a bit more difficult to pin down, however, since the lepton number could in principle be compensated by unseen antineutrinos. In any case, models of baryogenesis generally incorporate in some way lepton production (*leptogenesis*) as well.

Baryogenesis provides a nice example of the interplay between particle physics and cosmology. In the final analysis it will be seen that the Standard Model as it stands cannot explain the observed baryon asymmetry of the universe. This is a compelling indication that the Standard Model is incomplete.

9.6.1 Experimental Evidence for the Baryon Asymmetry

If antiparticles were to exist in significant numbers locally, one would see evidence of this from proton–antiproton or electron–positron annihilation. $p\bar{p}$ annihilation produces typically several mesons including neutral pions, which decay into two photons. So, one would see γ rays in an energy range up to around $100\,\text{MeV}$. No such gamma rays resulting from asteroid impacts on other planets are seen, man-made space probes landing on Mars survived, and Neil Armstrong landing on the Moon did not annihilate, so one can conclude that the entire solar system is made of matter.

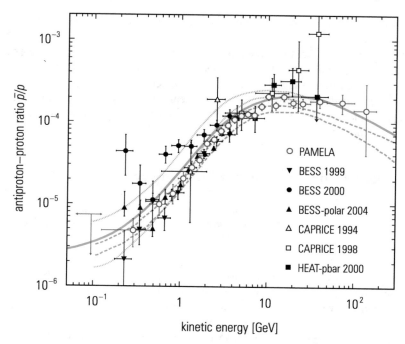

Fig. 9.2 Antiproton–proton ratio in primary cosmic rays according to the PAMELA experiment. The different calculations (*curves*) clearly show that antiproton production can be understood in terms of secondary production [171]

One actually finds some antiprotons bombarding the Earth as cosmic rays at a level of around 10^{-4} compared to cosmic-ray protons (see Sect. 6.1 and Fig. 9.2), so one may want to leave open the possibility that more distant regions are made of antimatter. But the observed antiproton rate is compatible with production in collisions of ordinary high-energy protons with interstellar gas or dust through reactions of the type

$$p + p \rightarrow 3p + \bar{p}.\tag{9.6.1}$$

There is currently no evidence of antinuclei in cosmic rays. The PAMELA satellite and the Alpha Magnetic Spectrometer on board the International Space Station failed to find antinuclei, like antihelium or even heavier antinuclei, even though heavy antinuclei, like antihelium, have been created at CERN and RHIC (Relativistic Heavy Ion Collider in Brookhaven) in heavy-ion collisions. The STAR experiment was able to produce 18 antihelium-4 events at the relativistic heavy ion collider RHIC [172]. This indicates that also heavier antinuclei might be produced in cosmic rays.

If there would exist appreciable numbers of positrons in the universe and in the Milky Way, one would also expect to find evidence for the 511-keV annihilation line according to the reaction $e^+e^- \rightarrow \gamma\gamma$. Actually one finds such a line in the gamma-ray spectrum from our galaxy (s. Fig. 9.3).

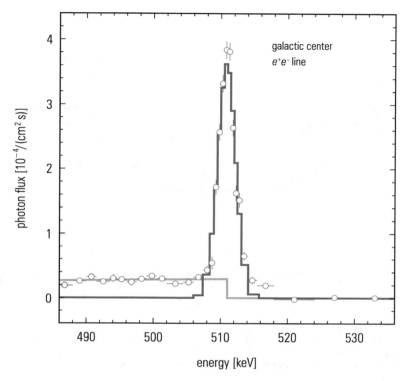

Fig. 9.3 The 511-keV positron annihilation line (*red curve*), observed from the galactic center. The *blue continuum line* originates from the positron annihilation into three photons [173, 174]

However, the intensity of this line can also be understood by the annihilation of locally produced positrons. On the other hand, one finds in the high-energy regime a positron flux, which cannot easily be understood by local positron production (s. Figs. 9.4, 9.5). But then it is conceivable that a relatively close supernova has injected cosmic particles, and thereby also positrons, into the Milky Way a long time ago. On the other hand, the positron excess has also been attributed to the decay of candidates of heavy dark-matter particles.

It is also conceivable that the sources of positrons are in very distant regions even beyond the local group of galaxies. In that case one would expect deviations in the observed gamma-ray spectrum from expectation. Model calculations for a matter–antimatter symmetric universe are, however, incompatible with data from the Compton Gamma Ray Observatory. The calculations refer to very distant domains of 20 Mpc or 1 Gpc in size, respectively (s. Fig. 9.6). If there would exist in the universe, or at least in part of it, regions dominated by antimatter they should be at very large distances (> some Gpc).

So one can conclude that, if antimatter regions of the universe exist, they must be separated by distances on the order of a gigaparsec or larger, which is already

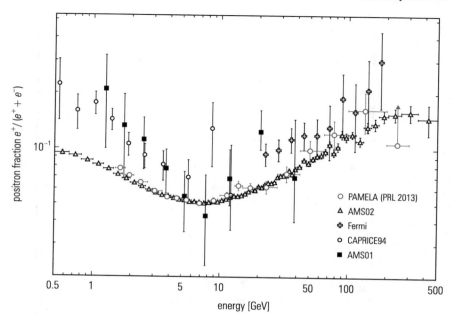

Fig. 9.4 The positron fraction in primary cosmic rays according to the PAMELA experiment [175]. At low energies the positron fraction follows the expectation on secondary local production. In contrast, at energies beyond 10 GeV there are significant deviations, which are not yet understood. The AMS experiment has confirmed the PAMELA result and extended the measurements to higher energies [176]

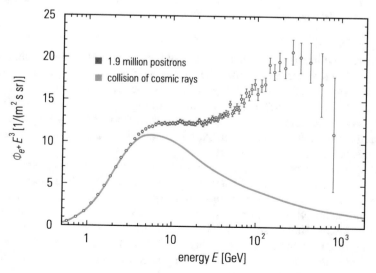

Fig. 9.5 The most recent data from the AMS Collaboration on the positron flux up to energies of 1 TeV. The flux exhibits a sharp drop at energies around 1 TeV [60]

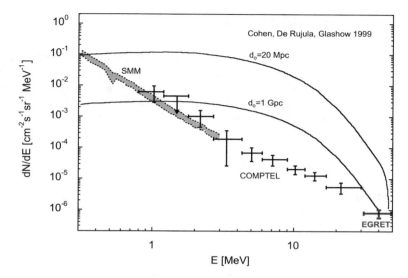

Fig. 9.6 The gamma-ray spectrum of the Compton telescope (COMPTEL) and EGRET calorimeters on board the Compton Gamma Ray Observatory, and the gamma-ray spectrometer for the Solar Maximum Mission (SMM) in comparison with model calculations for a matter–antimatter-symmetric universe at domains of size 20 Mpc resp. 1 Gpc separated from us [177]

a significant fraction of the observable universe. Given that there is no plausible mechanism for separating matter from antimatter over such large distances, it is far more natural to assume that the universe is made of matter, i.e., that it has a net non-zero baryon number. Also the absence of a significant flux of 511-keV γ rays from electron–positron annihilation adds to this conclusion.

If one then takes as working hypothesis that the universe contains much more matter than antimatter, one needs to ask how this could have come about. One possibility is that the non-zero baryon number existed as an initial condition, and that this was preserved up to the present day. This is not an attractive idea for several reasons. First, although the asymmetry between baryons and antibaryons today appears to be large, i.e., lots of the former and almost none of the latter, at times closer to the Big Bang, there were large amounts of both and the relative imbalance was very small. This will be quantified in Sect. 9.6.2. Going back towards the Big Bang one would like to think that nature's laws become in some sense more fundamental, and one would prefer to avoid the need to impose any sort of small asymmetry by hand.

Furthermore, it now appears that the laws of nature allow, or even require, that a baryon asymmetry would arise from a state that began with a net baryon number of zero. The conditions needed for this will be discussed in Sect. 9.6.3.

9.6.2 Size of the Baryon Asymmetry

Although the universe today seems completely dominated by baryons and not antibaryons, the relative asymmetry was very much smaller at earlier times. This can be seen roughly by considering a time when quarks and antiquarks were all highly relativistic, at a temperature of, say, $T \approx 1\,\text{TeV}$, and suppose that since that time there have been no baryon-number-violating processes. The net baryon number in a comoving volume R^3 is then constant, so one has

$$(n_b - n_{\bar{b}})R^3 = (n_{b,0} - n_{\bar{b},0})R_0^3 , \qquad (9.6.2)$$

where the subscript 0 on the right-hand side denotes present values. Today, however, there are essentially no antibaryons, so one can approximate $n_{\bar{b},0} \approx 0$. The *baryon–antibaryon asymmetry* A is therefore

$$A \equiv \frac{n_b - n_{\bar{b}}}{n_b} = \frac{n_{b,0}}{n_b} \frac{R_0^3}{R^3} . \qquad (9.6.3)$$

One can now relate the ratio of scale factors to the ratio of temperatures, using the relation $R \sim 1/T$. Therefore, one gets

$$A \approx \frac{n_{b,0}}{n_b} \frac{T^3}{T_0^3} . \qquad (9.6.4)$$

Now one can use the fact that the number densities are related to the temperature. Equation (9.2.2) had shown

$$n_b \approx T^3 , \qquad (9.6.5)$$
$$n_{\gamma,0} \approx T_0^3 , \qquad (9.6.6)$$

where these are rough approximations with the missing factors of order unity. Using these ingredients one can express the asymmetry as

$$A \approx \frac{n_{b,0}}{n_{\gamma,0}} . \qquad (9.6.7)$$

Further, the baryon-number-to-photon ratio can be defined:

$$\eta = \frac{n_b - n_{\bar{b}}}{n_\gamma} . \qquad (9.6.8)$$

One expects this ratio to remain constant as long as there are no further baryon-number-violating processes and there are no extra influences on the photon temperature beyond the Hubble expansion. So one can also assume that η refers to the current value, $(n_{b,0} - n_{\bar{b},0})/n_{\gamma,0} \approx n_{b,0}/n_{\gamma,0}$, although—strictly speaking—one should call this η_0. So one finally obtains that the baryon–antibaryon asymmetry A is roughly equal to the current baryon-to-photon ratio η. A more careful analysis, which keeps track of all the missing factors, gives $A \approx 6\eta$.

The current photon density $n_{\gamma,0}$ is well determined from the CMB temperature to be 410.4 cm^{-3}. In principle, one could determine $n_{b,0}$ by adding up all of the baryons that one finds in the universe. This in fact is expected to be an underestimate, since some matter such as gas and dust will not be visible and these will also obscure stars further away. A more accurate determination of η comes from the model of Big Bang Nucleosynthesis combined with measurements of the ratio of abundances of deuterium to hydrogen. From this one finds $\eta \approx 5 \times 10^{-10}$. So, finally, the baryon asymmetry can be expressed as

$$A \approx 6\eta \approx 4 \times 10^{-9}. \tag{9.6.9}$$

This means that at early times, for every billion antiquarks there were a billion and four quarks. The matter in the universe one sees today is just the tiny amount left over after essentially all of the antibaryons annihilated and this led to the large photon density (s. also Fig. 9.7).

9.6.3 The Sakharov Conditions

In 1967 Andrei Sakharov pointed out that three conditions must exist in order for a universe with non-zero baryon number to evolve from an initially baryon-symmetric state [178]. Nature must provide:

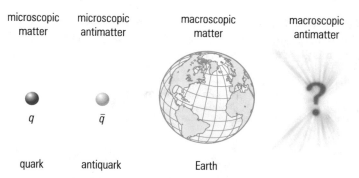

| microscopic matter | microscopic antimatter | macroscopic matter | macroscopic antimatter |

q $\bar{q}$

quark   antiquark   Earth

Fig. 9.7 The matter–antimatter symmetry on microscopic scales is obviously broken at macroscopic level

1. baryon-number-violating processes;
2. violation of C and CP symmetry;
3. departure from thermal equilibrium.

The first condition must clearly hold, or else a universe with $B = 0$ will forever have $B = 0$. In the second condition, C refers to charge conjugation and P to parity. C and CP symmetry roughly means that a system of particles behaves the same as the corresponding system made of antiparticles. If all matter and antimatter reactions proceed at the same rate, then no net baryon number develops; thus, violation of C and CP symmetry is needed. The third condition on departure from equilibrium is necessary in order to obtain unequal occupation of particle and antiparticle states, which necessarily have the same energy levels. A given theory of the early universe that satisfies at some level the Sakharov conditions will in principle predict a baryon density or, equivalently, a baryon-to-photon ratio η. One wants the net baryon number to be consistent with the measured baryon-to-photon ratio $\eta = n_b/n_\gamma \approx 5 \times 10^{-10}$. It is not entirely clear how this can be satisfied, and a detailed discussion goes beyond the scope of this book. Here only some of the currently favoured ideas will be mentioned.

Baryon-number violation is predicted by Grand Unified Theories, but it is difficult there to understand how the resulting baryon density could be preserved when this is combined with other ingredients such as inflation. Surprisingly, a non-zero baryon number is also predicted by *quantum anomalies*[3] in the usual Standard Model, and this is currently a leading candidate for baryogenesis.

---

[3] Quantum anomalies can arise if a classical symmetry is broken in the process of quantization and renormalization. The perturbative treatment of quantum field theories requires a renormalization, and this adds non-invariant counter terms to the invariant Lagrange density that one gets at the classical level.

CP violation is observed in decays of *K* and *B* mesons and it is predicted by the Standard Model of particle physics but at a level far too small to be responsible for baryogenesis. If nature includes further *CP*-violating mechanisms from additional Higgs fields, as would be present in supersymmetric models, then the effect could be large enough to account for the observed baryon density. This is one of the clearest indications from cosmology that the Standard Model is incomplete and that other particles and interactions must exist. It has been an important motivating factor in the experimental investigation of *CP*-violating decays of *K* and *B* mesons. These experiments have not, however, revealed any effect incompatible with Standard Model predictions.

The departure from thermal equilibrium could be achieved simply through the expansion of the universe, i.e., when the reaction rate needed to maintain equilibrium falls below the expansion rate: $\Gamma \ll H$. Alternatively, it could result from a phase transition such as those associated with spontaneous symmetry breaking.

So, the present situation with the baryon asymmetry of the universe is a collection of incomplete and inconclusive experimental observations and partial theories, which point towards the creation of a non-zero baryon density at some point in the early universe. The details of baryogenesis are still murky and remain an active topic of research. It provides one of the closest interfaces between particle physics and cosmology. Until the details are worked out one needs to take the baryon density of the universe or, equivalently, the baryon-to-photon ratio, as a free parameter that must be obtained from observation, see also Fig. 9.7.

Recently a proposal has been made that at the Big Bang not only our universe but also in addition a universe made of antimatter has been created. This could certainly explain the matter–antimatter asymmetry in our part of the universe. The details of such a simultaneous creation of a matter-dominated universe ('ours') and an antimatter universe and its influence on problems on, e.g., inflation still have to be worked out. It is also conceivable that our own universe might have also had a mirror image as an anti-universe that could have existed before the Big Bang [179].

9.7   Summary

The Friedmann equation is able to describe the thermal behaviour of the early universe reasonably well. The universe proceeds through various phases during its expansion, in which the different interactions—depending on the temperature—break apart into diverging types of interaction with the result that we now have four different interactions: strong, electroweak, weak and gravitation. The nearly complete disappearance of antimatter is still a puzzle, if one assumes that there was a matter-antimatter symmetry to start with at the time of the Big Bang. A violation of charge and parity symmetry known from particle physics is insufficient to explain the matter dominance in the universe.

9.8   Problems

1. Derive the relation between the scale factor R and the energy density ϱ for a universe dominated by non-relativistic matter!
2. Derive the relation between the scale factor R and time for an early universe dominated by non-relativistic matter!
3. The total energy density of the early universe varies with the temperature like

$$\varrho = \frac{\pi^2}{30} g T^4 . \tag{9.8.1}$$

   g is the number of internal degrees of freedom for the particle. Work out the expansion rate H as a function of the temperature and Planck mass using the Friedmann equation!
4. In Sect. 9.1 the Planck length has been derived using the argument that quantum and gravitational effects become equally important. One can also try to combine the relevant constants of nature $(G, \hbar, c)$ in such a way that a characteristic length results. What would be the answer?
5. Estimate a value for the Schwarzschild radius assuming weak gravitational fields (which do not really apply to the problem). What is the Schwarzschild radius of the Sun and the Earth?
   Hint: Consider the non-relativistic expression for the escape velocity and replace formally v by c.

6. A gas cloud becomes unstable and collapses under its own gravity if the gravitational energy exceeds the thermal energy of its molecules. Estimate the critical density of a hydrogen cloud at $T = 1000$ K, which would have collapsed into our Sun! Refer to Problem 8.2 for the stability condition (Jeans criterion).

7. Occasionally it has appeared that stars move at superluminal velocities. In most cases this is due to a motion of the star into the direction of Earth during the observation time. Figure out an example, which would give rise to a 'superluminal' speed!

Chapter 10
Big Bang Nucleosynthesis

*Evolution of nature is not inconsistent with the notion of
creation because evolution presupposes the creation of beings
which evolve. The Big Bang, which is today posited as the origin
of the world, does not contradict the divine act of creation;
rather, it requires it.*

Pope Francis

At times from around 10^{-2} s through the first several minutes after the Big Bang, the temperature passed through the range from around 10 to below 10^{-1} MeV. During this period protons and neutrons combined to produce a significant amount of ^{4}He—one quarter of the universe's nuclei by mass—plus smaller amounts of deuterium (D, i.e., ^{2}H), tritium (^{3}H), ^{3}He, ^{6}Li, ^{7}Li, and ^{7}Be. Further synthesis of nuclei in stars accounts for all of the heavier elements plus only a relatively small (1 to 2%) additional amount of helium [180]. The predictions of Big Bang Nucleosynthesis (BBN) are found to agree remarkably well with observations and provide one of the most important pillars of the Big Bang model.

The two main ingredients of BBN are the equations of cosmology and thermal physics that have already been described, plus the rates of nuclear reactions. Although the nuclear cross sections are difficult to calculate theoretically, they have for the most part been well measured in laboratory experiments. Of crucial importance is the rate of the reaction $\nu_e n \leftrightarrow e^- p$, which allows transformations between neutrons and protons. The proton is lighter than the neutron by $\Delta m = m_n - m_p \approx 1.3$ MeV, and as long as this reaction proceeds sufficiently quickly, one finds that the neutron-to-proton ratio is suppressed by the Boltzmann factor $e^{-\Delta m/T}$.[1] At a temperature around 0.7 MeV the reaction is no longer fast enough to keep up and the neutron-to-proton ratio 'freezes out' at a value of around 1/6. To first approximation one can estimate the helium abundance simply by assuming that all of the available neutrons end up in ^{4}He.

---

[1] In the following the natural constants c, $\hbar$, and k are set to unity.

© Springer Nature Switzerland AG 2020
C. Grupen, *Astroparticle Physics*, Undergraduate Texts in Physics,
https://doi.org/10.1007/978-3-030-27339-2_10

The one free parameter of BBN is the baryon density Ω_b or, equivalently, the baryon-to-photon ratio η. By comparing the observed abundances of the light elements with those predicted by BBN, the value of η can be estimated. The result will turn out to be of fundamental importance for the dark-matter problem, which will be dealt with in Chap. 13.

10.1   Some Ingredients for the Big Bang Nucleosynthesis

The point of view of a sinner is that the church promises him hell in the future, but cosmology proves that the glowing hell was in the past.

Ya. B. Zel'dovich

To model the synthesis of light nuclei one needs the equations of cosmology and thermal physics relevant for temperatures in the MeV range. At this point the total energy density of the universe is still dominated by radiation (i.e., relativistic particles), so the pressure and energy density are related by $P = \varrho/3$. In Chap. 9 it was shown that this led to relations for the expansion rate and time as a function of temperature (see Eqs. (9.4.16) and (9.4.17)),

$$H \sim \frac{T^2}{m_{\rm Pl}}, \tag{10.1.1}$$

$$t \sim \frac{m_{\rm Pl}}{T^2}. \tag{10.1.2}$$

To use these equations numerically one needs to know the effective number of degrees of freedom. Recall that quarks and gluons have already become bound into protons and neutrons at around $T = 200\,\mathrm{MeV}$, and since the nucleon mass is around $m_{\rm N} \approx 0.94\,\mathrm{GeV}$, these are no longer relativistic. Nucleons and antinucleons can remain in thermal equilibrium down to temperatures of around 50 MeV, and during this period their number densities are exponentially suppressed by the factor $e^{-m_N/T}$. At temperatures below several tens of MeV the antimatter has essentially disappeared and the resulting nucleon density therefore does not make a significant contribution to the energy density.

At temperatures in the MeV range, the relativistic particles are photons, e^-, ν_e, ν_μ, ν_τ, and their antiparticles. Using appropriate numbers for the degrees of freedom and using $N_\nu = 3$, (10.1.2) can be written in a form convenient for description of the BBN era, see also [181],

$$t T^2 \approx 0.74 \, s \, MeV^2 . \tag{10.1.3}$$

10.2   Start of the BBQ Era

> *By the word of the Lord were the heavens made. For he spoke, and it came to be; he commanded, and it stood firm.*
> The Bible: Psalm 33:6,9

From (10.1.3) a temperature of $T = 10$ MeV is reached at a time $t \approx 0.007$ s. At this temperature, all of the relativistic particles—γ, e^-, ν_e, ν_τ, and their antiparticles—are in thermal equilibrium through reactions of the type $e^+e^- \leftrightarrow \nu\bar{\nu}$, $e^+e^- \leftrightarrow \gamma\gamma$, etc. The number density of the neutrinos, for example, is given by the equilibrium formula appropriate for relativistic fermions, see (9.2.2),

$$n_\nu \sim T^3 . \tag{10.2.1}$$

Such a proportionality also holds for the electron density.

Already at temperatures around 20 MeV, essentially all of the antiprotons and antineutrons annihilated. The baryon-to-photon ratio is a number that one could, in principle, predict, if a complete theory of baryogenesis were available. Since this is not the case, however, the baryon density has to be treated as a free parameter. Since one does not expect any more baryon-number-violating processes at temperatures near the BBN era, the total number of protons and neutrons in a comoving volume remains constant. That is, even though protons and neutrons are no longer relativistic, baryon-number conservation requires that the sum of their number densities follows

$$n_n + n_p \sim \frac{1}{R^3} \sim T^3 . \tag{10.2.2}$$

At temperatures much greater than the neutron–proton mass difference, $\Delta m = m_n - m_p \approx 1.3$ MeV, one has $n_n \approx n_p$.

10.3   The Neutron-to-Proton Ratio

> *The most serious uncertainty affecting the ultimate fate of the universe is the question whether the proton is absolutely stable against decay into lighter particles. If the proton is unstable, all matter is transitory and must dissolve into radiation.*
>
> Freeman J. Dyson

Although the total baryon number is conserved, protons and neutrons can be transformed through reactions like $n\nu_e \leftrightarrow pe^-$ and $ne^+ \leftrightarrow p\bar{\nu}_e$. A typical Feynman diagram is shown in Fig. 10.1.

The crucial question is whether these reactions proceed faster than the expansion rate so that thermal equilibrium is maintained. If this is the case, then the ratio of neutron-to-proton number densities is given by

$$\frac{n_n}{n_p} = \left(\frac{m_n}{m_p}\right)^{3/2} e^{-(m_n - m_p)/T} \approx e^{-\Delta m/T}, \tag{10.3.1}$$

where $m_p = 938.272\,\text{MeV}, m_n = 939.565\,\text{MeV}$, and $\Delta m = m_n - m_p = 1.293\,\text{MeV}$. To find out whether equilibrium is maintained, one needs to compare the expansion rate H from (10.1.1) to the reaction rate Γ. This rate depends on the number density, the cross section, and the velocity, and is given by

$$\Gamma = n\langle\sigma v\rangle. \tag{10.3.2}$$

Γ is the reaction rate per neutron for $n\nu_e \leftrightarrow pe^-$, where the brackets denote an average of σv over a thermal distribution of velocities. The number density n in (10.3.2) refers to the target particles, i.e., neutrinos, which is therefore given by (10.2.1).

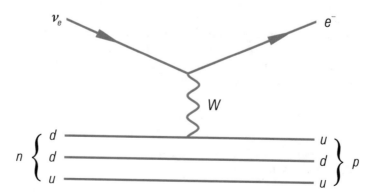

Fig. 10.1 Feynman diagram for the reaction $\nu_e n \leftrightarrow e^- p$

The cross section for the reaction $\nu_e n \leftrightarrow e^- p$ can be predicted using the Standard Model of electroweak interactions. An exact calculation is difficult but to a good approximation one finds for the thermally averaged speed times cross section

$$\langle \sigma v \rangle \approx G_{\mathrm{F}}^2 T^2 \,. \tag{10.3.3}$$

Here $G_{\mathrm{F}} = 1.166 \times 10^{-5}\,\mathrm{GeV}^{-2}$ is the Fermi constant, which characterizes the strength of weak interactions.

To find the reaction rate, one has to multiply (10.3.3) by the number density n from (10.2.1). But since the main interest here is to get a rough approximation, factors of order unity can be ignored and one can take $n \approx T^3$ to obtain

$$\Gamma(\nu_e n \to e^- p) \approx G_{\mathrm{F}}^2 T^5 \,. \tag{10.3.4}$$

A similar expression is found for the inverse reaction $e^- p \to \nu_e n$.

Now the question of whether this reaction proceeds quickly enough to maintain thermal equilibrium can be addressed. The point, where the reaction rate Γ equals the expansion rate H, $\Gamma = H$, determines the decoupling or *freeze-out temperature* T_{f}.

Equating the expressions for Γ and H,

$$G_{\mathrm{F}}^2 T^5 \sim \frac{T^2}{m_{\mathrm{Pl}}} \,, \tag{10.3.5}$$

and solving for T gives

$$T_{\mathrm{f}} \sim \left(\frac{1}{G_{\mathrm{F}}^2 m_{\mathrm{Pl}}} \right)^{1/3} \,. \tag{10.3.6}$$

A careful analysis of the numerical factors including all relevant degrees of freedom gives

$$T_{\mathrm{f}} \approx 0.7\,\mathrm{MeV} \,. \tag{10.3.7}$$

At temperatures below T_{f} the reaction $\nu_e n \leftrightarrow e^- p$ can no longer proceed quickly enough to maintain the equilibrium number densities. The neutron density is said to *freeze out*, i.e., the path, by which neutrons could be converted to protons, is effectively closed. If neutrons were stable, this would mean that the number of them in a comoving volume would be constant, i.e., their number density would follow $n_n \sim 1/R^3$. Actually, this is not quite true because free neutrons can still decay. But the neutron has a mean lifetime $\tau_n \approx 886\,\mathrm{s}$, which is relatively long, but not entirely negligible, compared to the time scale of nucleosynthesis.

Ignoring for the moment the effect of neutron decay, the neutron-to-proton ratio at the freeze-out temperature is

$$\frac{n_n}{n_p} = \mathrm{e}^{-(m_n - m_p)/T_{\mathrm{f}}} \approx \mathrm{e}^{-1.3/0.7} \approx 0.16 \,. \tag{10.3.8}$$

According to (10.1.3), this temperature is reached at a time $t \approx 1.5\,\text{s}$. By the end of the next five minutes, essentially all of the neutrons become bound into ^{4}He, and thus the neutron-to-proton ratio at the freeze-out temperature is the dominant factor in determining the amount of helium produced. To obtain a precise value for the n/p ratio, among others, also the neutron decay must be taken into account.

In addition, it is important to note the fact that in thermal equilibrium the reaction $e^+e^- \to \gamma\gamma$ provides energetic photons. Therefore, the photon temperature decreases more slowly than one would expect from the normal expansion. This effect is irrelevant for neutrinos, since the comparable reaction $e^+e^- \leftrightarrow \nu\bar{\nu}$ proceeds through weak interactions and is rather rare in comparison to the electromagnetic annihilation of electrons and positrons. A consequence of this is that the temperature of the present primordial neutrinos is slightly lower ($1.9\,\text{K}$, $336/\text{cm}^3$) than that of the blackbody photons ($2.725\,\text{K}$). This effect slightly changes the product of time and temperature, Eq. (10.1.3), to obtain

$$tT^2 = 1.32\,\text{s}\,\text{MeV}^2 \,. \tag{10.3.9}$$

If further neutron decay after decoupling is considered, according to

$$\frac{n_n}{n_p} = e^{-(m_n - m_p)/T_f}\, e^{-t/\tau_n} \,, \tag{10.3.10}$$

one finally obtains for the neutron-to-proton ratio a value of

$$\frac{n_n}{n_p} \approx 0.13 \,, \tag{10.3.11}$$

if one assumes that the nucleosynthesis is completed after three minutes ($t = 180\,\text{s}$).

10.4   Synthesis of Light Elements

Give me matter and I will construct a world out of it.

Immanuel Kant

The synthesis of ^{4}He proceeds through a chain of reactions, which includes, for example,

$$p\,n \to d\,\gamma \,, \tag{10.4.1}$$
$$d\,p \to {}^3\text{He}\,\gamma \,, \tag{10.4.2}$$
$$d\,{}^3\text{He} \to {}^4\text{He}\,p \,. \tag{10.4.3}$$

The binding energy of deuterium is $E_{\text{bind}} = 2.2\,\text{MeV}$, so if the temperature is so high that there are many photons with energies higher than this, then the deuterium will be broken apart as soon as it is produced. One might naively expect that the reaction (10.4.1) would begin to be effective as soon as the temperature drops to around 2.2 MeV. In fact this does not happen until a considerably lower temperature. This is because there are so many more photons than baryons, and the photon energy distribution, i.e., the Planck distribution, has a long tail towards high energies.

The nucleon-to-photon ratio is at this point essentially the same as the baryon-to-photon ratio $\eta = n_b/n_\gamma$, which is around 10^{-9}. One can estimate roughly when deuterium production can begin to proceed by finding the temperature, where the number of photons with energies greater than 2.2 MeV is equal to the number of nucleons. For a nucleon-to-photon ratio of 10^{-9}, this occurs at $T = 0.086$ MeV, which is reached at a time of $t \approx 3$ min. Over the next several minutes, essentially all of the neutrons, except those that decay, are processed into ^{4}He.

The abundance of ^{4}He is usually quoted by giving its mass fraction,

$$Y_P = \frac{\text{mass of } {}^4\text{He}}{\text{mass of all nuclei}} = \frac{m_{He}n_{He}}{m_N(n_n + n_p)}, \qquad (10.4.4)$$

where the neutron and proton masses have both been approximated by the nucleon mass, $m_N \approx m_n \approx m_p \approx 0.94$ GeV. There are four nucleons in ^{4}He, so, neglecting the binding energy, one has $m_{He} \approx 4m_N$. Furthermore, there are two neutrons per ^{4}He nucleus, so if one assumes that all of the neutrons end up in ^{4}He, one has $n_{He} = n_n/2$. This will turn out to be a good approximation, as the next most common nucleus, deuterium, ends up with an abundance four to five orders of magnitude smaller than that of hydrogen. The ^{4}He mass fraction is therefore

$$\begin{aligned} Y_P &= \frac{4m_N(n_n/2)}{m_N(n_n + n_p)} = \frac{2(n_n/n_p)}{1 + n_n/n_p} \\ &\approx \frac{2 \times 0.13}{1 + 0.13} \approx 0.23. \end{aligned} \qquad (10.4.5)$$

This rough estimate turns out to agree quite well with more detailed calculations. 23% ^{4}He mass fraction means that the number fraction of ^{4}He is about 6% with respect to hydrogen.

The fact that the universe finally ends up containing around one quarter ^{4}He by mass can be seen as the result of a number of rather remarkable coincidences. For example, mean lifetimes for weak decays can vary over many orders of magnitude. The fact that the mean neutron lifetime turns out to be 885.7 s is a complicated consequence of the rather close neutron and proton masses combined with strong and weak interaction physics. If it had turned out that τ_n were, say, only a few seconds or less, then essentially all of the neutrons would decay before they could be bound up into deuterium, and the chain of nuclear reactions could not have started. This would have lead to a completely different universe, and surely life could not have been created. The exact value of the decoupling temperature is also a complicated mixture of effects, being sensitive, for example, to the degrees of freedom, which depends on the number of relativistic particle species in thermal equilibrium. If, for example, the decoupling temperature T_f had been not 0.7 MeV but, say, 0.1 MeV, then the neutron-to-proton ratio would have been $e^{-1.3/0.1} \approx 2 \times 10^{-6}$, and essentially no helium would have formed. On the other hand, if it had been at a temperature much higher than $m_n - m_p$, then there would have been equal amounts of protons and neutrons. Then the entire universe would have been made of helium. Usual hydrogen-burning stars would be impossible, and the universe would certainly be a very different place (see also Chap. 14: 'Astrobiology)'.

For Germans only:

U(h)rknall = Big Bang

10.5   Detailed Big Bang Nucleosynthesis

> *The effort to understand the universe is one of the very few things that lifts human life above the level of farce, and gives it some of the grace of tragedy.*
>
> Steven Weinberg

A detailed modeling of nucleosynthesis uses a system of differential equations involving all of the abundances and reaction rates. The rates of nuclear reactions are parameterizations of experimental data.

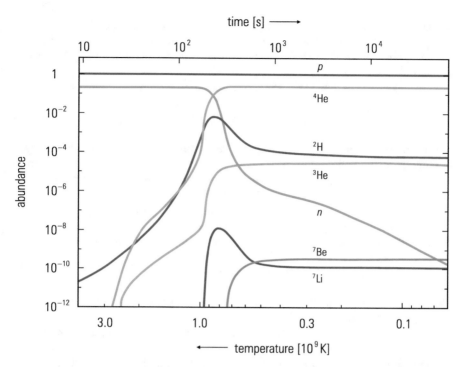

Fig. 10.2 Evolution of the elemental abundance in the primordial nucleosynthesis. ^{4}He is usually given as mass fraction, while for all the other elements the values are given as number fractions, i.e., a mass fraction of 25% ^{4}He corresponds to a number fraction of about 6% (four nucleons form a helium nucleus). The mass fraction of hydrogen is around 75% and its number fraction about 94%. This also implies that all other isotopes are extremely rare [183]

Several computer programs that numerically solve the system of rate equations are publicly available [182]. An example of predicted mass fractions versus temperature and time is shown in Fig. 10.2.

Around $t \approx 1000$ s, the temperature has dropped to $T \approx 0.03$ MeV. At this point the kinetic energies of nuclei are too low to overcome the Coulomb barriers and the fusion processes stop. So, in the early universe heavy elements cannot be produced because after about several minutes the temperatures are insufficient to fuse heavier nuclei.

In order to compare the predictions of Big Bang Nucleosynthesis with observations, one needs to measure the *primordial* abundances of the light elements, i.e., as they were just after the BBN era. This is complicated by the fact that the abundances change as a result of *stellar* nucleosynthesis. For example, helium is produced in stars and deuterium is broken apart.

To obtain the most accurate measurement of the ^{4}He mass fraction, for example, one tries to find regions of hot ionized gas from 'metal-poor' galaxies, i.e., those, where relatively small amounts of heavier elements have been produced through stellar burning of hydrogen. For standard, normal stars like our Sun the helium mass fraction from stellar hydrogen burning is on the order of 1 to 2% [180]. In the determination of the primordial helium abundance this is naturally taken into account.

A recent survey of data [157, 158] concludes for the primordial ^{4}He mass fraction

$$Y_P = 0.238 \pm 0.002 \pm 0.005 \,, \tag{10.5.1}$$

where the first error is statistical and the second reflects systematic uncertainties. In contrast to the helium-4 content of the universe, which is traditionally given as a mass fraction, the abundances of the other primordial elements are presented as number fractions, e.g., $n_{^7\mathrm{Li}}/n_p \equiv n_{^7\mathrm{Li}}/n_\mathrm{H}$ for ^{7}Li.

The best determinations of the ^{7}Li abundance come from hot metal-poor stars from the galactic halo. As with ^{4}He, one extrapolates to zero metallicity to find the primordial value. Recent data [157, 158] give a lithium-to-hydrogen ratio of

$$n_{^7\mathrm{Li}}/n_\mathrm{H} = 1.23 \times 10^{-10} \,. \tag{10.5.2}$$

The systematic uncertainty on this value is quite large, however, corresponding to the range from about 1 to 2×10^{-10}.

Although deuterium is produced by the first reaction in hydrogen-burning stars through $pp \to de^+\nu_e$, it is quickly processed further into heavier nuclei. Essentially, no net deuterium production takes place in stars and any present would be quickly fused into helium. So, to measure the primordial deuterium abundance, one needs to find gas clouds at high redshift, hence far away and far back in time, that have never been part of stars. These produce absorption spectra in light from even more distant quasars. The hydrogen Lyman-α line at $\lambda = 121.6$ nm appears at very high redshift ($z \geq 3$) in the visible part of the spectrum. The corresponding line from deuterium has a small isotopic shift to shorter wavelengths. Comparison of the two components gives an estimate of the deuterium-to-hydrogen ratio $\mathrm{D/H} \hat{=} n_d/n_p$. A recent measurement finds [184, 185]

$$n_d/n_p = (3.40 \pm 0.25) \times 10^{-5} \,. \tag{10.5.3}$$

A measurement of the primordial ^{3}He abundance is even more complicated. Estimations suggest a range of [186]

$$n_{^3\mathrm{He}}/n_\mathrm{H} = 2 \times 10^{-5}\text{--}3 \times 10^{-4} \,. \tag{10.5.4}$$

Other sources favour a slightly lower value of 1×10^{-5} [187]. However, presently the measured ^{3}He abundance does not allow to effectively test the Big Bang nucleosynthesis.

Finally, the predicted abundances of light nuclei are confronted with measurements. The predictions depend, however, on the baryon density n_b or, equivalently, on $\eta = n_b/n_\gamma$. The predicted mass fraction of ^{4}He as well as the numbers relative to hydrogen for D, ^{3}He, and ^{7}Li are shown as a function of η in Fig. 10.3 [158, 188].

The deuterium fraction D/H decreases for increasing η because a higher baryon density means that deuterium is processed more completely into helium. Since the resulting prediction for D/H depends quite sensitively on the baryon density, the measured D/H provides the most accurate determination of η. The measured value from (10.5.3), namely, $n_d/n_p = (3.40 \pm 0.25) \times 10^{-5}$, gives the range of allowed η values, which are shown by the vertical band in Fig. 10.3. This corresponds to

$$\eta = (5.1 \pm 0.5) \times 10^{-10}. \tag{10.5.5}$$

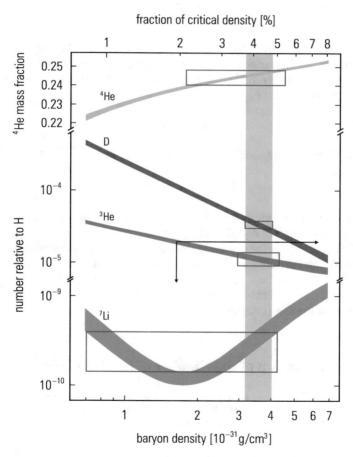

Fig. 10.3 Predictions for the abundance of ^{4}He, D, ^{3}He, and ^{7}Li in their dependence on the baryon density. The ^{4}He fraction is traditionally given as mass fraction. For the other primordial elements their number ratio relative to hydrogen is shown (the *vertical scale* is *broken* at $Y_P = 0.22$ and at $\approx 10^{-6}$). The *rectangles* for the number fractions of the primordial isotopes represent the added statistical and systematic errors. The trend for possible uncertainties of the ^{3}He fraction is indicated [188]

The boxes in Fig. 10.3 indicate the measured abundances of ^{4}He and ^{7}Li (from slightly different analyses than those mentioned above). The size of the boxes shows the measurement uncertainty. These measurements agree remarkably well with the predictions, especially when one considers that the values span almost 10 orders of magnitude.

The value of $\eta = n_b/n_\gamma$ determines the baryon density, since the photon density is well-known from the measured CMB temperature, $T = 2.725$ K, and this is related to the photon density according to $n_\gamma \sim T^3$. The constant of proportionality in this equation can be determined from the spin factors. The value of η can be converted into a prediction for the energy density of baryons divided by the critical density,

$$\Omega_b = \frac{\varrho_b}{\varrho_c}. \tag{10.5.6}$$

The critical density is given by (8.6.4) as $\varrho_c = 3H_0^2/8\pi G$. The baryons today are non-relativistic, so their energy density is simply the number of nucleons per unit volume times the mass of a nucleon, i.e., $\varrho_b = n_b m_N$, where $m_N \approx 0.94$ GeV. Putting these ingredients together gives

$$\Omega_b = 3.67 \times 10^7 \times \eta h^{-2}, \tag{10.5.7}$$

where h is defined by $H_0 = 100\,h$ km s^{-1} Mpc^{-1}. Using $h = 0.71^{+0.04}_{-0.03}$ and η from (10.5.5) gives

$$\Omega_b = 0.036 \pm 0.005, \tag{10.5.8}$$

where the uncertainty originates both from that of the Hubble constant and also from that of η. Recently η and Ω_b have been measured to higher accuracy using the temperature variations in the cosmic microwave background radiation as determined by the WMAP and mainly Planck satellites, leading to $\Omega_b = 0.0483 \pm 0.0004$; this will be followed up in Chap. 11 (see also Table 11.1). The values from the BBN and CMB studies are consistent with each other and, taken together, provide a convincing confirmation of the Big Bang model.

10.6 Determination of the Number of Neutrino Families

> *If there were many neutrino generations, we would not be here to count them, because the whole universe would be made of helium and no life could develop.*
>
> Anonymous

In this section it will be shown how the comparison of the measured and predicted ^{4}He mass fractions can result in constraints on the particle content of the universe at

BBN temperatures. For example, the Standard Model has $N_\nu = 3$, but one can ask whether additional families exist. It will be seen that BBN was able to constrain N_ν to be quite close to three—a number of years earlier than accelerator experiments were able to determine the same quantity to high precision using electron–positron collisions at energies near the Z resonance.

Once the parameter η has been determined, the predicted ^{4}He mass fraction is fixed to a narrow range of values close to $Y_P = 0.24$. As was noted earlier, this prediction is in good agreement with the measured abundance. The prediction depended, however, on the effective number of degrees of freedom.

The number of effective degrees of freedom g describes the expansion rate according to

$$H \sim \frac{T^2}{m_{Pl}}, \tag{10.6.1}$$

where the constant of proportionality for (10.6.1) contains this factor g. The effective number of degrees of freedom therefore also has an impact on the freeze-out temperature,

"Did you see it?"
"No, nothing."
"Then it was a neutrino!"

The freeze-out temperature is larger for higher values of g, and also with N_ν, since g increases with N_ν. At T_f the neutron-to-proton ratio freezes out to $n_n/n_p = e^{-(m_n - m_p)/T_f}$. If this occurs at a higher temperature, then the ratio is higher, i.e., there are more neutrons available to make helium, and the helium abundance will come out higher. A higher freeze-out temperature also corresponds to earlier times after the Big Bang, up to which protons and neutrons are in equilibrium, thus they are still quite numerous.

Figure 10.4 shows the predicted helium abundance as a function of the baryon density. The three diagonal bands show the predicted Y_P for different values of an equivalent number of neutrino families $N_\nu = 3.0$, 3.2, and 3.4. Of course, this no longer represents the (integer) number of neutrino flavours but rather an effective parameter. The data are consistent with $N_\nu = 3$ and are clearly incompatible with values much higher than this [184, 185, 189].

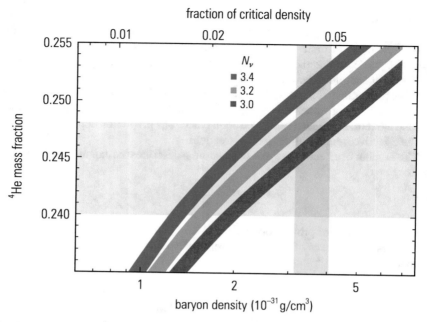

Fig. 10.4 The predicted ^{4}He mass fraction as a function of η for different values of N_{eff} [184, 189, 190]

Figure 10.5 shows the expected number of neutrino generations as a function of the baryon density. To determine the effective number of neutrino generations from cosmological considerations one uses the relation between the baryon density of the universe and helium isotope ratio, the deuterium-to-hydrogen ratio and the results from the Big Bang blackbody radiation as well as the primordial nucleosynthesis. These results clearly favour a value of $N_{\text{eff}} = 3$. With these cosmological data $N_{\text{eff}} = 4$ is strongly disfavoured.

A recent analysis of the most metal-poor damped Lyman-α system, which also shows the Lyman-series absorption lines of neutral deuterium, can also be used to determine the effective number of neutrino generations from the deuterium-to-hydrogen ratio and the CMB data. The Lyman-α data were based on observations made with the Keck HIRES spectrograph [192]. The results of this analysis are shown in Fig. 10.6.

These data clearly favour $N_{\text{eff}} = 3$ and exclude $N_{\text{eff}} = 4$, but they also show the scatter of the results, which include statistical and systematical errors.

Even as early as 1990 a number on the neutrino generations had been derived from the helium abundance, indicating that very likely only three or at most four neutrino generations would exist. The modern cosmological results confirm these conjectures and state more precisely $N_{\text{eff}} = 3.046$ [194].

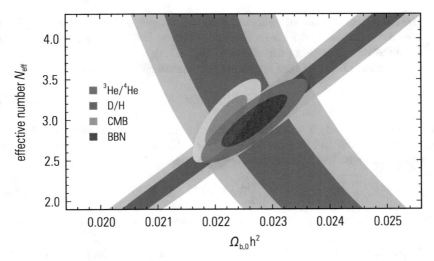

Fig. 10.5 Dependence of the effective number of neutrino generations N_{eff} on the baryon density, the helium isotope ratio, the deuterium-to-hydrogen ratio, and the results from measurements of the cosmic blackbody radiation and the Big Bang nucleosynthesis. The *shaded and dark contours* for the D/H and ^{3}He/^{4}He ratio represent the 68% and 95% confidence levels, respectively. The measurements of the blackbody radiation (CMB) are from the Planck satellite. The *red contours* show the confidence limits for the combined D/H and ^{3}He/^{4}He ratio (BBN). $\Omega_{b,0}$ is the cosmic baryon density and h is the Hubble constant in units of $100\,km/(s\,Mpc)$ [191]

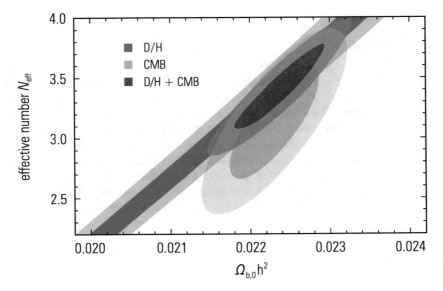

Fig. 10.6 Comparison of the effective number of neutrino generations N_{eff} from the expansion rate with the cosmic baryon density $\Omega_{b,0}h^2$ from Big Bang nucleosynthesis (*blue contours*) and the results from the Cosmological Microwave Background (CMB) (*grey contours*). The *dark and light shades* illustrate the 68% and 95% confidence contours, respectively, for the deuterium/hydrogen (D/H) ratio and the Big Bang nucleosynthesis (CMB) [193]

The equivalent number of light (i.e., with masses $\leq m_Z/2$) neutrino families has also been determined at the Large Electron–Positron (LEP) collider from the total width of the Z resonance, as was shown in Fig. 2.1.

From a combination of data from the LEP experiments one finds [195]

$$N_\nu = 2.9835 \pm 0.0083 . \tag{10.6.2}$$

Although this is nearly 2 times the quoted error bar below 3, it is clear that $N_\nu = 3$ fits reasonably well and that any other integer value is excluded. This number is to be compared to the result from cosmology, given as

$$N_\nu = 3.046 \tag{10.6.3}$$

with a small error related to the uncertainties due to the neutrino mixing parameters [194].

The interplay between these two different determinations of N_ν played an important role in alerting particle physicists to the relevance of cosmology. Although the example that has just been discussed is for the number of neutrino families, the same arguments apply to any particles that would contribute to g such as to affect the neutron freeze-out temperature. Thus the abundances of light elements provide important constraints for any theory involving new particles that would contribute significantly to the energy density during the BBN era.

10.7 Summary

Essential aspects of primordial nucleosynthesis have already been derived in the fifties of the last century by George Gamow. The light elements have been created in the first three minutes. These elements are mainly hydrogen and helium along with their isotopes and some traces of lithium, beryllium, and possibly boron. Heavy elements are only produced much later in supernova explosions. It is important to see that the observed helium abundance and other results from cosmology allow to draw conclusions on, e.g., the number of neutrino generations, which are confirmed by storage-ring experiments from the Large Electron–Positron collider LEP at CERN, indicating the fruitful interplay between cosmology and accelerators.

10.8 Problems

1. Equation (10.1.3) shows that

$$tT^2 \approx 0.74 \, \text{s} \, \text{MeV}^2 .$$

How can this numerical result be derived from the previous relations?

2. In the determination of the neutron-to-proton ratio the cross section for the $(n \leftrightarrow p)$ conversion reaction

$$\nu_e + n \rightarrow e^- + p$$

was estimated from

$$\langle \sigma \, v \rangle \approx G_F^2 T^2 \,.$$

How can the T^2 dependence of the cross section be motivated?

3. The measurement of the total width Γ_Z of the Z resonance at LEP led to a precise determination of the number of light neutrinos N_ν. How can N_ν be obtained from the Z width?

The solution of this problem requires some intricate details of elementary particle physics.

Chapter 11
The Cosmic Microwave Background

God created two acts of folly. First, He created the universe in a Big Bang. Second, he was negligent enough to leave behind evidence for this act, in the form of the microwave radiation.

Paul Erdös

In this chapter the description of the early universe to cover the first several hundred thousand years of its existence will be presented. This leads to one of the most important pillars of the Big Bang model: the cosmic microwave background radiation or CMB. It will be seen how and when the CMB was formed and what its properties are. Most important among these are its blackbody energy spectrum characterized by an average temperature of $T = 2.725$ K, and the fact that one sees very nearly the same temperature independent of direction. Recent measurements of the slight dependence of the temperature on direction have been used to determine a number of cosmological parameters to a precision of several percent.

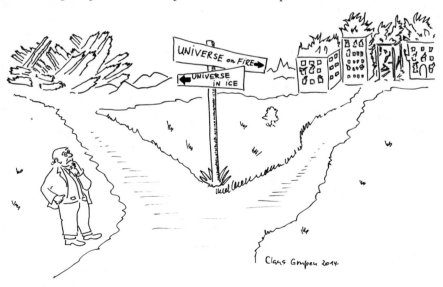

© Springer Nature Switzerland AG 2020

C. Grupen, *Astroparticle Physics*, Undergraduate Texts in Physics,
https://doi.org/10.1007/978-3-030-27339-2_11

11.1 Prelude: Transition to a Matter-Dominated Universe

We know too much about matter today to be materialists any longer.

Harides Chaudhuri

Picking up the timeline from the last chapter, Big Bang nucleosynthesis was completed by around $t \approx 10^3$ s, i.e., when the temperature had fallen to several hundredths of an MeV. Any neutrons that by this time had not become bound into heavier nuclei soon decayed.

The break-neck pace of the early universe now shifts gears noticeably. The next interesting event takes place when the energy density of radiation, i.e., relativistic particles (photons and neutrinos), drops below that of the matter (non-relativistic nuclei and electrons), called the time of 'matter–radiation equality'. In order to trace the time evolution of the universe, one needs to determine when this occurred, since the composition of the energy density influences the time dependence of the scale factor $R(t)$.

Estimates of the time of matter–radiation equality depend on what is assumed for the contents of the universe. For matter it will be seen that estimates based on the CMB properties as well as on the motion of galaxies in clusters give $\Omega_{m,0} = \varrho_{m,0}/\varrho_{c,0} \approx 0.3$, where as usual the subscript 0 indicates a present-day value. For photons it will be found from the CMB temperature: $\Omega_{\gamma,0} = 5.0 \times 10^{-5}$. Taking into account neutrinos brings the total for radiation to $\Omega_{r,0} = 8.4 \times 10^{-5}$, so currently, matter contributes some 3600 times more to the total energy density than does radiation.

In Chap. 8 it has been derived by solving the Friedmann equation how to predict the time dependence of the different components of the energy density. For radiation $\varrho_r \sim 1/R^4$ was obtained, whereas for matter $\varrho_m \sim 1/R^3$ was found. Therefore, the ratio follows $\varrho_m/\varrho_r \sim R$, and it was thus equal to unity when the scale factor R was 3600 times smaller than its current value.

In order to pin down when this occurred, one needs to know how the scale factor varies in time. If it is assumed that the universe has been matter dominated from the time of matter–radiation equality up to now, then one has $R \sim t^{2/3}$, and this leads to a time of matter–radiation equality, t_{mr}, of around 66 000 years. In fact, there is now overwhelming evidence that vacuum energy makes up a significant portion of the universe, with $\Omega_\Lambda \approx 0.7$. Taking this into account leads to a somewhat earlier time of matter–radiation equality of around $t_{mr} \approx 50\,000$ years.

When the dominant component of the energy density changes from radiation to matter, this alters the relation between the temperature and time. For non-relativistic particle types with mass m_i and number density n_i, one now gets for the energy density

$$\varrho \approx \sum_i m_i n_i \,, \tag{11.1.1}$$

where the sum includes at least baryons and electrons, and perhaps also 'dark-matter' particles. The Friedmann equation (neglecting the curvature term) then reads

$$H^2 = \frac{8\pi G}{3} \sum_i m_i n_i \,. \tag{11.1.2}$$

Assuming the particles that contribute to the energy density are stable, one obtains

$$n_i \sim 1/R^3 \sim T^3 \,. \tag{11.1.3}$$

Furthermore, since $R \sim t^{2/3}$, the expansion rate is now given by

$$H = \frac{\dot{R}}{R} = \frac{2}{3t} \,. \tag{11.1.4}$$

Combining (11.1.3) with (11.1.4) or $R \sim t^{2/3}$ then leads to

$$T^3 \sim \frac{1}{t^2} \,. \tag{11.1.5}$$

This is to be contrasted with the relation $T^2 \sim t^{-1}$ valid for the era when the energy density was dominated by relativistic particles.

11.2 Discovery and Basic Properties of the CMB

> *What we have found is evidence for the birth of the universe. It's like looking at God.*
>
> George Smoot

The existence of the CMB was predicted by Gamow [196] in connection with Big Bang nucleosynthesis. It was shown in Chap. 10 that BBN requires temperatures around $T \approx 0.08\,\text{MeV}$, which are reached at a time $t \approx 200\,\text{s}$. By knowing the cross section for the first reaction, $p + n \rightarrow d + \gamma$, and the number density n of neutrons and protons, one can predict the reaction rate $\Gamma = n \langle \sigma v \rangle$.

In order for BBN to produce the observed amount of helium, one needs a sufficiently high rate for the deuterium fusion reaction over the relevant time scale. This corresponds to requiring Γt to be at least on the order of unity at $t \approx 200\,\text{s}$, when the temperature passes through the relevant range. This assumption determines the nucleon density during the BBN phase.

Since the BBN era, the nucleon and photon densities have both followed $n \sim 1/R^3 \sim T^3$. So, by comparing the nucleon density in the BBN era to what one finds today, the current temperature of the photons can be predicted. Reasoning along these lines, Alpher and Herman [197] estimated a CMB temperature of around 5 K, which turned out to be not far off.

Even without invoking BBN one can argue that the contribution of photons to the current energy density cannot exceed by much the critical density. If one assumes, say, $\Omega_\gamma \leq 1$, then this implies $T \leq 32$ K. Gamow's prediction of the CMB was not pursued for a number of years. In the 1960s, a team at Princeton (Dicke, Peebles, Roll, and Wilkinson) did take the prediction seriously and set about building an experiment to look for the CMB. Unknown to them, a pair of radio astronomers, A. Penzias and R. Wilson at Bell Labs in New Jersey, were calibrating a radio antenna in preparation for studies unrelated to the CMB. They reported finding an "effective zenith noise temperature ... about 3.5 °K higher than expected. This excess temperature is, within the limits of our observations, isotropic, unpolarized, and free from seasonal variations ..." [198]. The Princeton team soon found out about Penzias' and Wilson's observation and immediately supplied the accepted interpretation [199].

Although the initial observations of the CMB were consistent with a blackbody spectrum, the earth-based observations were only able to measure accurately the radiation at wavelengths of several cm; shorter wavelengths are strongly absorbed by the water in the atmosphere. The peak of the blackbody spectrum for a temperature of 3 K, however, is around 2 mm. It was not until 1992 that the COBE satellite made accurate measurements of the CMB from space. This showed that the form of the energy distribution is extremely close to that of blackbody radiation, i.e., to a Planck distribution, as shown in Fig. 11.1.

The value for the blackbody temperature was confirmed by the results from the WMAP and Planck satellites [201].

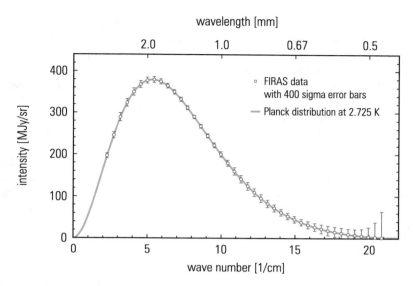

Fig. 11.1 Spectrum of the CMB measured with the COBE satellite in comparison to the blackbody curve for the temperature of 2.725 K; the unit 'Jy' used on the ordinate is 1 jansky $= 1\,\mathrm{Jy} = 10^{-26}\,\mathrm{W}$ $\mathrm{Hz}^{-1}\,\mathrm{m}^{-2}$. The *error bars* have been enlarged *for better visibility* by a factor of 400. The deviations from the Planck distribution are less than 0.005% [200]

11.3   Formation of the CMB

> *The old dream of wireless communication through space has now been realized in an entirely different manner than many had expected. The cosmos' short waves bring us neither the stock market nor jazz from different worlds. With soft noises they rather tell the physicists of the endless love play between electrons and protons.*
>
> Albrecht Unsöld

At very early times, any protons and electrons that managed to bind together into neutral hydrogen would be dissociated very quickly by collision with a high-energy photon. As the temperature decreased, the formation of hydrogen eventually became possible and the universe transformed from an ionized plasma to a gas of neutral atoms; this process is called *recombination*. The reduction of the free electron density to almost zero meant that the mean free path of a photon soon became so long that most photons have not scattered since. This is called the *decoupling* of photons from matter. By means of some simple calculations one can estimate when recombination and decoupling took place.

Neutral hydrogen has an electron binding energy of 13.6 eV and is formed through the reaction

$$p + e^- \rightarrow \mathrm{H} + \gamma \,. \tag{11.3.1}$$

Naively, one would expect the fraction of neutral hydrogen to become significant when the temperature drops below 13.6 eV. But because the baryon-to-photon ratio $\eta \approx 5 \times 10^{-10}$ is very small, the temperature must be significantly lower than this before the number of photons with $E > 13.6$ eV is comparable to the number of baryons. (This is the same basic argument for why deuterium production began not around $T = 2.2$ MeV, the binding energy of deuterium, but rather much lower.) One finds that the numbers of neutral and ionized atoms become equal at a *recombination temperature* of $T_{\text{rec}} \approx 0.3$ eV (3500 K). At this point the universe transforms from an ionized plasma to an essentially neutral gas of hydrogen and helium.

It can be estimated when recombination took place by comparing the temperature of the CMB one observes today, $T_0 \approx 2.73$ K, to the value of the recombination temperature $T_{\text{rec}} \approx 0.3$ eV. Recall from Sect. 9.2 that the wavelength of a photon follows $\lambda \sim R$. Therefore, the ratio of the scale factor R at a previous time to its value R_0 now is related to the redshift z through

$$\frac{R_0}{R} = \frac{\lambda_0}{\lambda} = 1 + z \,. \tag{11.3.2}$$

Furthermore, today's CMB temperature is measured to be $T_0 \approx 2.73$ K and it is known that $T \sim 1/R$. Therefore, one gets

$$1 + z = \frac{T}{T_0} \approx \frac{0.3 \, \text{eV}}{2.73 \, \text{K}} \times \frac{1}{8.617 \times 10^{-5} \, \text{eV K}^{-1}} \approx 1300 \,, \tag{11.3.3}$$

where Boltzmann's constant was inserted to convert temperature from units of eV to K. If one assumes that the scale factor follows $R \sim t^{2/3}$ from this point up to the present, then it is found that recombination was occurring at

$$t_{\text{rec}} = t_0 \left(\frac{R}{R_0} \right)^{3/2} = t_0 \left(\frac{T_0}{T_{\text{rec}}} \right)^{3/2} \doteq \frac{t_0}{(1 + z_{\text{rec}})^{3/2}}$$
$$\approx \frac{1.4 \times 10^{10} \, \text{years}}{(1300)^{3/2}} \approx 300\,000 \, \text{years} \,. \tag{11.3.4}$$

Shortly after recombination, the mean free path for a photon became so long that photons effectively decoupled from matter. While the universe was an ionized plasma, the photon scattering cross section was dominated by Thomson scattering, i.e., elastic scattering of a photon by an electron. The mean free path of a photon is determined by the number density of electrons, which can be predicted as a function of time and by the Thomson scattering cross section, which can be calculated. As the universe expands, the electron density decreases leading to a longer mean free path for the photons. This path length becomes longer than the horizon distance (the size of the observable universe at a given time) at a *decoupling temperature* of $T_{\text{dec}} \approx 0.26$ eV (3000 K) corresponding to a redshift of $1 + z \approx 1100$. This condition defines *decoupling* of photons from matter. The decoupling time is

$$t_{\text{dec}} = t_0 \left(\frac{T_0}{T_{\text{dec}}} \right)^{3/2} = \frac{t_0}{(1 + z_{\text{dec}})^{3/2}} \approx 380\,000 \, \text{years} \,. \tag{11.3.5}$$

Once the photons and matter decoupled, the photons simply continue unimpeded to the present day. One can define a *surface of last scattering* as the sphere centered about us with a radius equal to the mean distance to the last place, where the CMB photons scattered. To a good approximation this is equal to the distance, to where decoupling took place and the time of last scattering is essentially the same as t_{dec}. So, when the CMB is detected, one is probing the conditions in the universe at a time of approximately 380 000 years after the Big Bang.

11.4 CMB Anisotropies

> *As physics advances farther and farther every day and develops new axioms, it will require fresh assistance from mathematics.*
>
> Francis Bacon

The initial measurements of the CMB temperature by Penzias and Wilson indicated that the temperature was independent of direction, i.e., that the radiation was isotropic, to within an accuracy of around 10%. More precise measurements eventually revealed that the temperature is about one part in one thousand hotter in one particular direction of the sky than in the opposite. This is called the dipole anisotropy and is interpreted as being caused by the motion of the Earth through the CMB. Then in 1992 the COBE satellite found anisotropies at smaller angular separations at a level of one part in 10^5. These small variations in temperature have recently been measured down to angles of several tenths of a degree by several groups, including the WMAP[1] satellite, from which one can extract a wealth of information about the early universe. The Planck satellite was able to further improve the resolution limit, depending on the frequency, to about 4 arc minutes. An impression on the improvements of the three satellites COBE, WMAP, and Planck can be seen from Fig. 11.2.

These high-quality data allow to obtain a large variety of information on the early universe.

In order to study the CMB anisotropies one begins with a measurement of the CMB temperature as a function of direction, i.e., $T(\theta, \phi)$, where θ and ϕ are spherical coordinates, i.e., polar and azimuthal angle, respectively. As with any function of direction, it can be expanded in spherical harmonic functions $Y_{lm}(\theta, \phi)$ (a *Laplace series*),

$$T(\theta, \phi) = \sum_{l=0}^{\infty} \sum_{m=-l}^{l} a_{lm} Y_{lm}(\theta, \phi) . \tag{11.4.1}$$

This expansion is analogous to a Fourier series, where the higher-order terms correspond to higher frequencies. Here, terms at higher l correspond to structures at smaller angular scales. The same mathematical technique is used in the *multipole*

---

[1] WMAP—Wilkinson Microwave Anisotropy Probe.

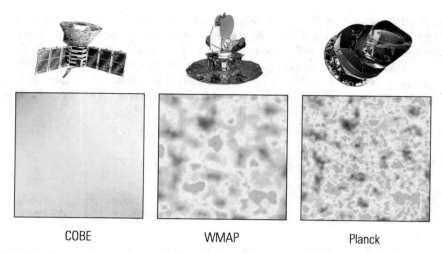

COBE   WMAP   Planck

Fig. 11.2 Comparison of the angular resolution and the related measurement accuracy of the cosmic background radiation of the COBE, WMAP, and Planck satellites [201]

expansion of the potential from an electric charge distribution. The terminology is borrowed from this example and the terms in the series are referred to as multipole moments. The $l = 0$ term is the monopole, $l = 1$ the dipole, etc.

Once one has estimates for the coefficients a_{lm}, the amplitude of regular variation with angle can be summarized by defining

$$C_l = \frac{1}{2l + 1} \sum_{m=-l}^{l} |a_{lm}|^2 . \tag{11.4.2}$$

The set of numbers C_l is called the *angular power spectrum*. The value of C_l represents the level of structure found at an angular separation

$$\Delta\theta \approx \frac{180°}{l} . \tag{11.4.3}$$

The measuring device will in general only be able to resolve angles down to some minimum value; this determines the maximum measurable l.

11.5   The Monopole and Dipole Terms

> *The universe contains the record of its past the way that sedimentary layers of rock contain the geological record of the Earth's past.*
>
> Heinz R. Pagels

The $l = 0$ term in the expansion of $T(\theta, \phi)$ gives the temperature averaged over all directions. The most accurate determination of this value comes from the COBE satellite,

$$\langle T \rangle = 2.725 \pm 0.0013 \, \text{K} \,. \tag{11.5.1}$$

In the 1970s it was discovered that the temperature of the CMB in a particular direction was around 0.1% hotter than in the opposite direction. This corresponds to a non-zero value of the $l = 1$ or *dipole* term in the Laplace expansion. The dipole anisotropy has recently been remeasured by the WMAP experiment, which finds a temperature difference of

$$\frac{\Delta T}{T} = 1.23 \times 10^{-3} \,. \tag{11.5.2}$$

This temperature variation has a simple explanation. It is caused by the movement of the Earth through the local, unique reference frame, in which the CMB has no dipole anisotropy. This frame is in some sense the (local) 'rest frame' of the universe. The solar system and with it the Earth are moving through it with a speed of $v = 371 \, \text{km/s}$ towards the constellation Crater (between Virgo and Hydra). The CMB is blueshifted to slightly higher temperature in the direction of motion and redshifted in the opposite direction.

11.6 Small-Angle Anisotropy

> *The universe is full of magical things, patiently waiting for our wits to grow sharper.*
>
> Eden Phillpotts

If small density fluctuations were to exist in the early universe, then one would expect these to be amplified by gravity, with more dense regions attracting even more matter, until the matter of the universe was separated into clumps. This is how galaxy formation is expected to have taken place. Given that one sees a certain amount of lumpiness today, one can predict what density variations must have existed at the time of last scattering. These variations would correspond to regions of different temperature, and so from the observed large-scale structure of the universe one expected to see anisotropies in the CMB temperature at a level of around one part in 10^5.

These small-angle anisotropies were finally observed by the COBE satellite in 1992. COBE had an angular resolution of around 7° and could therefore determine the power spectrum up to a multipole number of around $l = 20$. In the following years, balloon experiments were able to resolve much smaller angles but with limited sensitivity. In 2003 the WMAP project made very accurate measurements of the CMB temperature variations with an angular resolution of 0.2°. Finally, the Planck satellite managed to further improve the angular resolution to four arc minutes! In the following Figs. 11.3, 11.4, and 11.5 the results of the three satellites COBE, WMAP, and Planck are shown and clearly demonstrate the gain in information due to the improved angular resolution of the instruments on board of these satellites.

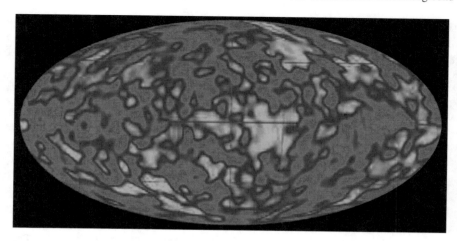

Fig. 11.3 Cosmographic map of the CMB temperature as measured by the COBE satellite after subtraction of the dipole component [202]

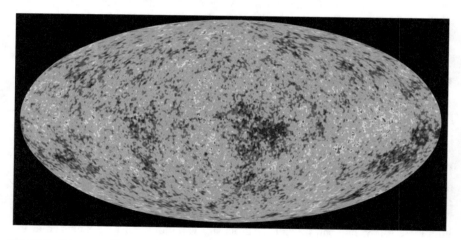

Fig. 11.4 Cosmographic map of the CMB temperature as measured by the WMAP satellite after subtraction of the dipole component, WMAP Science Team/NASA Goddard [203]

WMAP was able to measure the angular power spectrum up to $l \approx 1000$ and Planck managed to reach $l \approx 2500$.

The angular power spectrum of the Planck satellite is shown in Fig. 11.6. According to

$$\Delta\theta \approx \frac{180°}{l} \tag{11.6.1}$$

multipole moments have here also been converted to angular separations.

Fig. 11.5 Cosmographic map of the CMB temperature as measured by the Planck satellite after subtraction of the dipole component [157]

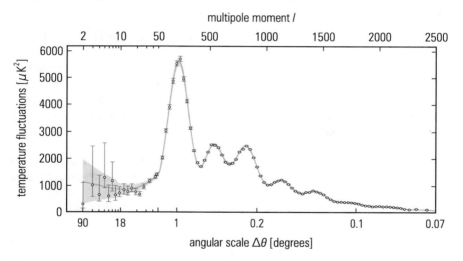

Fig. 11.6 The angular power spectrum of the Planck satellite. The main maxima are linked to dark energy, normal matter and dark matter (s. Chap. 12: 'Inflation') [204]

11.7 Determination of Cosmological Parameters

> *There are probably few features of theoretical cosmology that could not be completely upset and rendered useless by new observational discoveries.*
>
> Sir Hermann Bondi

The angular power spectrum of the CMB can be used to make accurate determinations of many of the most important cosmological parameters, including the Hubble constant H, the baryon-to-photon ratio η, the total energy density over the critical density, Ω, as well as the components of the energy density from baryons, Ω_b, and from all non-relativistic matter Ω_m.

As an example, in the following a rough idea will be given of how the angular power spectrum is sensitive to Ω. Consider the largest region that could be in causal contact at the time of last scattering $t_{ls} \approx t_{dec} \approx 380\,000$ years. This distance is called the *particle horizon* d_H. Naively, one would expect this to be $d_H = t$ (i.e., ct, but $c = 1$ has been assumed). This is not quite right because the universe is expanding. The correct formula for the particle-horizon distance at a time t in an isotropic and homogeneous universe is (see, e.g., [205]),

$$d_H(t) = R(t) \int_0^t \frac{dt'}{R(t')} .   (11.7.1)$$

If the time before matter–radiation equality is considered ($t_{mr} \approx 50\,000$ years), then $R \sim t^{1/2}$ and $d_H(t) = 2t$. For the matter-dominated era one has $R \sim t^{2/3}$ and $d_H = 3t$. If a sudden switch from $R \sim t^{1/2}$ to $R \sim t^{2/3}$ is assumed at t_{mr}, then one finds from integrating (11.7.1) a particle horizon at t_{ls} of [206]

$$d_H(t_{ls}) = 2t_{mr} + 3t_{ls} - 3t_{ls}^{2/3} t_{mr}^{1/3} \approx 660\,000 \text{ (light-)years} .   (11.7.2)$$

A detailed modeling of the density fluctuations in the early universe predicts a large level of structure on distance scales roughly up to the horizon distance. These fluctuations are essentially sound waves in the primordial plasma, i.e., regular pressure variations resulting from the infalling of matter into small initial density perturbations. These initial perturbations may have been created at a much earlier time, e.g., at the end of the inflationary epoch.

By looking at the angular separation of the temperature fluctuations, in effect one measures the distance between the density perturbations at the time when the photons were emitted.

To relate the angles to distances, one needs to review briefly the *proper distance* and *angular diameter distance*. The *proper distance* d_p at a time t is the length one would measure if one could somehow stop the Hubble expansion and lay meter sticks end to end between two points. In an expanding universe one finds that the current proper distance (i.e., at t_0) to the surface of last scattering can be approximated by $d_p(t_{ls}) \approx 3t_0$ if matter domination for this period is assumed.

Now, what one wants to know is the angle subtended by a temperature variation, which was separated by a distance perpendicular to our line of sight of $\delta = 3t_{ls}$ when the photons were emitted. To obtain this one needs to divide δ not by the current proper distance from us to the surface of last scattering, but rather by the distance that it *was* to us at the time when the photons started their journey. This location has been carried along with the Hubble expansion and is now further away by a factor equal to the ratio of the scale factors, $R(t_0)/R(t_{ls})$. Using (11.3.2), therefore, one finds

$$\Delta\theta = \frac{\delta}{d_p(t_{ls})} \frac{R(t_0)}{R(t_{ls})} = \frac{\delta}{d_p(t_{ls})}(1+z) ,   (11.7.3)$$

where $z \approx 1100$ is the redshift of the surface of last scattering. Thus, if a region is considered whose size was equal to the particle-horizon distance at the time of last scattering as given by (11.7.2) as viewed from today, then it will subtend an angle (compare Fig. 11.6)

$$\Delta\theta \approx \frac{d_H(t_{ls})}{3t_0}(1 + z)$$

(11.7.4)

$$\approx \frac{660\,000\,a}{3 \times 1.4 \times 10^{10}\,a} \times 1100 \times \frac{180°}{\pi} \approx 1°.$$

The structures at these angular scales correspond to the so-called 'acoustic peaks', which can be seen in the power spectrum from about $l \approx 200$ onwards.

The naming of the structures in the power spectrum as 'acoustic peaks' comes about for the following reason: as already mentioned, the density fluctuations in the early universe caused gravitational instabilities. When the matter fell into these gravitational potential wells, this matter was compressed, thereby getting heated up. This hot matter radiated photons causing the plasma of baryons to expand, thereby cooling down and producing less radiation as a consequence. With decreasing radiation pressure the irregularities reach a point, where gravity again took over initiating another compression phase. The competition between gravitational accretion and radiation pressure caused longitudinal acoustic oscillations in the baryon fluid. After decoupling of matter from radiation the pattern of acoustic oscillations became frozen into the CMB. CMB anisotropies therefore are a consequence of sound waves in the primordial proton fluid.

Table 11.1 Some of the cosmological parameters obtained by the Planck satellite from the angular power spectrum. The Hubble constant—using the 2015 Planck data—has been determined to $H_0 = (67.8 \pm 0.9)$ km/s/Mpc and the matter density is $\Omega_m = 0.308 \pm 0.012$ [208–211]

Parameter	Value and exp. error
Hubble constant H_0	(67.8 ± 0.9) km/s/Mpc
Ratio of the total energy density to ϱ_c, Ω	1.0002 ± 0.0026
Baryon-to-photon ratio η	$(6.19 \pm 0.14) \times 10^{-10}$
Baryon energy density to ϱ_c, Ω_b	0.0483 ± 0.0004
Matter density to ϱ_c, Ω_m	0.308 ± 0.012
Vacuum energy density to ϱ_c, Ω_Λ	0.694 ± 0.007
h parameter h	(0.697 ± 0.006)
Number of neutrinos N_{eff}	(3.15 ± 0.23)
Sum of neutrino masses Σm_ν	≤ 0.23 eV
Age of the universe t	$(13.799 \pm 0.021) \times 10^9$ years

The angle subtended by the horizon distance at the time of last scattering depends, however, on the geometry of the universe, and this is determined by Ω, the ratio of the energy density to the critical density. It can be shown, that the position of the first acoustic peak in the angular power spectrum is related to Ω in the following way [207]:

$$l_{peak} \approx \frac{220}{\sqrt{\Omega}} \qquad (11.7.5)$$

and can be approximated to $l_{peak} \approx 200$ for $\Omega \approx 1$.

The detailed structure of the peaks in the angular power spectrum depends not only on the total energy density but on many other cosmological parameters as well, such as the Hubble constant H_0, the baryon-to-photon ratio η, the energy-density contributions from matter, baryons, vacuum energy, etc.

Using the power-spectrum measurement shown in Fig. 11.6, the Planck team has determined many of these parameters to a precision of one percent or better. Some are shown in Table 11.1. Among the most important of these, one sees that the Hubble constant is now finally known to an accuracy of better than 2%, and the value obtained is in good agreement with the previous average ($h = 0.7 \pm 0.1$). Furthermore, the universe is flat, i.e., Ω is so close to unity that surely this cannot be a coincidence. The age of the universe has slightly 'grown' compared to earlier measurements, but it is still within the errors of, e.g., WMAP data. The Planck results provide a very precise value of $(13.799 \pm 0.021) \times 10^9$ years.

An independent determination of the Hubble constant using a distance-ladder argument based on properties of Cepheid stars and type Ia supernovae recently arrived at a value of $H_0 = (73.48 \pm 1.66)$ km/s/Mpc, which is more than three sigmas away from the value based on determinations from the cosmological microwave background observation [212]. Very recent measurements strengthen the discrep-

Fig. 11.7 Constraints on dark energy from Chandra X-ray observations of the largest galaxy clusters. The $1\,\sigma$ and $2\,\sigma$ confidence constraints in the $\Omega_{\mathrm{m}}, \Omega_{\Lambda}$ plane for the Chandra results on the gas fraction f_{gas} (*red contours*), CMB data (*blue contours*), and SNIa data (*green contours*) are given. A ΛCDM model is assumed [215–217]

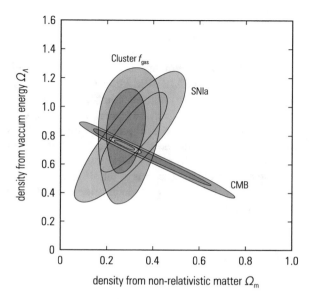

ancy between the determination of the Hubble constant from the early universe and the direct distance measurements. This deviation stands now at 4.4 standard deviations [213]. If confirmed with even higher precision, one might be forced to look for a physics solution: Additional relativistic particles, interacting dark matter, or dynamic dark energy would shrink the difference between the results. Whether systematic errors in the distance-ladder method or the interpretation of the baryonic acoustic oscillations in the primordial background radiation are responsible for the difference remains to be seen. It would be desirable to have a third independent method to determine the Hubble constant.

Data from the Cosmic Background Imager (CBI) [214]—an array of antennae operating at frequencies from 26 to 36 GHz in the Atacama Desert—confirm the picture of a flat universe, giving $\Omega = 0.99 \pm 0.12$ in agreement with results from the Boomerang[2] and Maxima[3] experiments and, of course, with WMAP and Planck.

In addition, WMAP has determined the baryon-to-photon ratio η to an accuracy of several percent. The value found is a bit higher than that obtained from the deuterium abundance, $\eta = (5.1 \pm 0.5) \times 10^{-10}$, but the latter measurement's error did not represent the entire systematic uncertainty and, in fact, the agreement between the two is quite reasonable. The fact that two completely independent measurements yield such close values is surely a good sign. The agreement of the Planck result with the WMAP value on the baryon-to-photon ratio $(6.1 \pm 0.25) \times 10^{-10}$ is perfect.

---

[2] Boomerang—Balloon Observations of Millimetric Extragalactic Radiation and Geophysics.

[3] Maxima—Millimeter Anisotropy Experiment Imaging Array.

The densities from non-relativistic matter Ω_m and from vacuum energy Ω_Λ are similarly well determined, and these measurements confirm earlier values based on completely different observables.

These older measurements based on the Chandra gas data [215], a compilation of SNIa [216], and the CMB-WMAP results [217] (see Fig. 11.7) show the constraint obtained from all three data sets combined with the result of $\Omega_m = 0.275 \pm 0.033$ and $\Omega_\Lambda = 0.735 \pm 0.023$ in agreement with the recent Planck data (see Table 11.1).

The satellite experiments represent a close bridge between cosmology and the particle physics of the early universe. In addition to helping pin down important cosmological parameters, the CMB may provide the most important clues needed to understand particle interactions at ultrahigh energy scales—energies that will never be attained by man-made particle accelerators. This theme will be further explored in the next chapter on *inflation*.

11.8   Summary

The cosmological background radiation was discovered more or less by chance in the microwave range by Penzias and Wilson in 1965. The left-over from the Big Bang was already predicted by George Gamow in the fifties of the last century. The satellites COBE, WMAP, and Planck have measured this radiation in great detail. The temperature of this radiation is not completely constant. The observed variations in the range of 10^{-5} are considered as seed for galaxy formation. The acoustic oscillations in the primordial quark soup and the angular power spectrum allow to determine very many cosmological parameters of the universe to high precision. The experimental error on the age of the universe (13.8 billion years) is only 0.2%.

11.9   Problems

1. What is the probability that a photon from the Big Bang undergoes a scattering after having passed the *surface of last scattering*?
2. Estimate a limit for the cosmological constant!
3. The Friedmann equation extended by the cosmological constant is nothing but a relation between different forms of energy:

$$\underbrace{\frac{m}{2}\dot{R}^2}_{\text{kinetic}} + \underbrace{\left(-\frac{GmM}{R} - \frac{1}{6}m\Lambda c^2 R^2\right)}_{\text{potential}} = \underbrace{-kc^2\frac{m}{2}}_{\text{total energy}}$$

(m is the mass of a galaxy at the edge of a galactic cluster of mass M).
Work out the pressure as a function of R related to the classical potential energy

in comparison to the pressure caused by the term containing the cosmological constant.

4. Work out the average energy of blackbody microwave photons! For the integration of the Planck distribution refer to Chap. 27 in [218] or Formulae 3.411 in [219] or [220]. For the Riemann zeta function look at [221].

5. Estimate the energy density of the cosmic microwave background radiation at present and at the time of last scattering. (In the limit of relativistic particles the energy density for a given bosonic particle type is $\varrho = 1/30 \cdot \pi^2 g T^4$, where g is the number of internal degrees of freedom for the particle.)

6. Why is the decoupling temperature for photons at last scattering ($0.3\,\text{eV}$) much lower compared to the ionization energy of hydrogen ($13.6\,\text{eV}$)?

7. Work out the spatial number density of radio photons emitted by the cosmological background!

To obtain this number one has to integrate the Planck distribution

$$B_\nu(T)\,d\nu = \frac{2h\nu^3}{c^2} \cdot \frac{1}{e^{h\nu/kT} - 1}\,d\nu. \tag{11.9.1}$$

The spectral energy density $u(T)$ is related to $B_\nu(T)$ by

$$u(T)\,d\nu = \frac{4\pi}{c} \cdot B_\nu(T)\,d\nu, \tag{11.9.2}$$

and the relevant number density is

$$n(T)\,d\nu = \frac{u(T)}{h\nu}\,d\nu = \frac{8\pi}{c^3} \cdot \frac{\nu^2}{e^{h\nu/kT} - 1}\,d\nu. \tag{11.9.3}$$

Hint: look at [222].

Chapter 12
Inflation

Inflation hasn't won the race, but so far it's the only horse.

Andrei Linde

The Standard Cosmological Model appears to be very successful in describing observational data, such as the abundances of light nuclei, the isotropic and homogeneous expansion of the universe, and the existence of the cosmic microwave background. The CMB's very high degree of isotropy and the fact that the total energy density is close to the critical density, however, pose problems in that they require a very specific and seemingly arbitrary choice of initial conditions for the universe. Furthermore, it turns out that Grand Unified Theories (GUTs) as well as other possible particle physics theories predict the existence of stable particles such as magnetic monopoles, which no one has yet succeeded in observing. These problems can be solved by assuming that at some very early time, the total energy density of the universe was dominated by vacuum energy. This leads to a rapid, accelerating increase in the scale factor called *inflation*.

This chapter takes a closer look at the problems mentioned above, how inflation solves them, what else inflation predicts, and how these predictions stand in comparison to observations. In doing this it will be necessary to make predictions for the expansion of the universe from very early times up to the present. For these purposes it will be sufficient to treat the universe as being matter dominated after a time of around 50 000 years, preceded by an era of radiation domination (except for the period of inflation itself). In fact it is now believed this picture is not quite true, and the current energy density is dominated again by a sort of vacuum energy. This fact will not alter the arguments relevant for the present chapter and it will be ignored here; the topic of vacuum energy in the present universe will be taken up again in the next chapter on dark energy and dark matter.

© Springer Nature Switzerland AG 2020
C. Grupen, *Astroparticle Physics*, Undergraduate Texts in Physics,
https://doi.org/10.1007/978-3-030-27339-2_12

Retrospective Consideration

12.1   The Horizon Problem

> *The existing universe is bounded in none of its dimensions; for then it must have an outside.*
>
> Lucretius

In the previous chapter it was shown that, after correcting for the dipole anisotropy, the CMB has the same temperature to within one part in 10^5 coming from all directions. This radiation was emitted from the surface of last scattering at a time of around $t_{ls} \approx 380\,000$ years. In Sect. 11.7 it was calculated that two places separated by the particle-horizon distance at t_{ls} are separated by an angle of around $1°$ as viewed today. This calculation assumed $R \sim t^{1/2}$ for the first $50\,000$ years during the radiation-dominated era, followed by a matter-dominated phase with $R \sim t^{2/3}$ from then up to t_{ls}. For any mixture of radiation and matter domination one would find values in the range of about one degree.

So, if regions of the sky separated in angle by more than, say, around $2°$ are considered, one would expect them not to have been in causal contact at the time of last scattering. Furthermore, the projected entire sky can be divided into more than 10^4 patches that should not have been in causal contact, and yet, they are all at almost the same temperature.

The unexplained uniform temperature in regions that appear to be causally disconnected is called the 'horizon problem'. It is not a problem in the sense that this model makes a prediction that is in contradiction with observation. The different temperatures could have perhaps all had the same temperature 'by chance'. This option is not taken seriously. The way that systems come to the same temperature is by interacting, and it is hard to believe that any mechanism other than this was responsible.

Horizon Problems

12.2 The Flatness Problem

> *This type of universe, however, seems to require a degree of fine-tuning of the initial conditions that is in apparent conflict with 'common wisdom'.*
>
> Idit Zehavi and Avishai Dekel

It was found in Sect. 11.7 that the total energy density of the universe is very close to the critical density or $\Omega \approx 1$. Although this is now known to hold to within less than 1%, it has been clear for many years that Ω is at least constrained to roughly $0.2 < \Omega < 2$. (A lower limit can be obtained from, e.g., the motions of galaxies in clusters, and an upper bound comes from the requirement that the universe is at least as old as the oldest observed stars.) The condition $\Omega = 1$ gives a flat universe, i.e., one with zero spatial curvature.

The problem with having an almost flat universe today, i.e., $\Omega \approx 1$, is that it then must have been very much closer to unity at earlier times. To see this, recall first from Chap. 8 the Friedmann equation,

$$H^2 + \frac{k}{R^2} = \frac{8\pi G}{3}\varrho , \qquad (12.2.1)$$

where as usual $H = \dot{R}/R$ and k is the curvature parameter. One obtains $k = 0$, i.e., a flat universe, if the density ϱ is equal to the critical density

$$\varrho_c = \frac{3H^2}{8\pi G} . \qquad (12.2.2)$$

By dividing both sides of the Friedmann equation by H^2, using (12.2.2) and $H = \dot{R}/R$, and then rearranging terms, one finds

$$\Omega - 1 = \frac{k}{\dot{R}^2} . \qquad (12.2.3)$$

Now, for the matter-dominated era one has $R \sim t^{2/3}$, which gives $\dot{R} \sim t^{-1/3}$ and thus $R\dot{R}^2$ is constant. Therefore, the difference between Ω and unity is found to be as follows:

$$\Omega - 1 \sim kR \sim kt^{2/3} . \qquad (12.2.4)$$

If matter domination from the time of matter–radiation equality, $t_{mr} \approx 50\,000$ years, to the present, $t_0 \approx 1.4 \times 10^{10}$ years, is assumed, then (12.2.4) implies

$$\frac{\Omega(t_{mr}) - 1}{\Omega(t_0) - 1} = \frac{R(t_{mr})}{R(t_0)} = \left(\frac{t_{mr}}{t_0}\right)^{2/3}$$

$$= \left(\frac{50\,000\,\text{a}}{1.4 \times 10^{10}\,\text{a}}\right)^{2/3} \approx 2 \times 10^{-4} . \qquad (12.2.5)$$

From the recent CMB data $\Omega(t_0) - 1$ is currently measured to be less than around 0.003. Using this with (12.2.5) implies that $\Omega - 1$ was less than 10^{-6} at $t = 50\,000$ years after the Big Bang.

ART MEETS COSMOLOGY

FLAT UNIVERSE

CHAOTIC UNIVERSE

Claus Grupen 2013

In the Art Museum

Going further back in time makes the problem more acute. Suppose that the universe was radiation dominated for times less than t_{mr} going all the way back to the Planck time, $t_{Pl} \approx 10^{-43}$ s. Then one should take $R \sim t^{1/2}$, which means $\dot{R} \sim t^{-1/2}$ and therefore[1]

$$\Omega - 1 \sim kR^2 \sim kt. \tag{12.2.6}$$

Using this dependence for times earlier than t_{mr}, one finds for $\Omega - 1$ at the Planck time relative to the value now:

$$\begin{aligned}
\frac{\Omega(t_{Pl}) - 1}{\Omega(t_0) - 1} &= \left(\frac{R(t_{Pl})}{R(t_{mr}))}\right)^2 \frac{R(t_{mr})}{R(t_0)} = \frac{t_{Pl}}{t_{mr}}\left(\frac{t_{mr}}{t_0}\right)^{2/3} \\
&\approx \frac{10^{-43}\,\text{s}}{50\,000\,\text{a} \times 3.2 \times 10^7\,\text{s/a}}\left(\frac{50\,000\,\text{a}}{1.4 \times 10^{10}\,\text{a}}\right)^{2/3} \\
&\approx 10^{-59}.
\end{aligned} \tag{12.2.7}$$

That is, to be able to find Ω of order unity today, the model requires it to be within 10^{-59} of unity at the Planck time.

---

[1] The Boltzmann constant and the curvature parameter are traditionally denoted with the same letter k. From the context it should always be clear, which parameter is meant.

As with the horizon problem, the issue is not one of a prediction that stands in contradiction with observation. There is nothing to prevent Ω from being arbitrarily close to unity at early times, but within the context of the cosmological model that has been described so far, it could have just as easily had some other value. And if other values are *a priori* just as likely, then it seems ridiculous to believe that nature would pick Ω 'by chance' to begin so close to unity. One feels that there must be some reason why Ω came out the way it did.

Cosmological Flatness Problem

12.3 The Monopole Problem

> *From the theoretical point of view one would think that monopoles should exist, because of the prettiness of the mathematics.*
> Paul Adrian Maurice Dirac

In order to understand the final 'problem' with the model that has been presented up to now, one needs to recall something about phase transitions in the early universe. In Chap. 9 it was remarked that a sort of phase transition took place at a critical temperature around the GUT energy scale, $T_c \approx E_{GUT} \approx 10^{16}\,\text{GeV}$.[2] As the temperature dropped below T_c, the Higgs field, which is responsible for the masses of the X and Y bosons, acquired a non-zero vacuum expectation value. There is an analogy between this process and the cooling of a ferromagnet, whereby the magnetic dipoles suddenly

---

[2] As usual, here and in the following the standard notation $c = 1$, $\hbar = 1$, and $k = 1$ (Boltzmann constant) is used.

line up parallel to their neighbours. Any direction is equally likely, but once a few of the dipoles have randomly chosen a particular direction, their neighbours follow along. The same is true for the more common transition of water when it freezes to snowflakes. In water the molecules move randomly in all directions. When the water is cooled below the freezing point, ice crystals or snowflakes form, where the water molecules are arranged in a regular pattern, which breaks the symmetry of the phase existing at temperatures above zero degree Celsius. In the case of the Higgs, the analogue of the dipole's direction or crystal orientation is not a direction in physical space, but rather in an abstract space, where the axes correspond to components of the Higgs field. As with the ferromagnet or the orientation axis of the ice crystals, the components of the field tend to give the same configuration the same way in a given local region.

If one considers two regions, which are far enough apart so they are not in causal contact, then the configuration acquired by the Higgs field in each will not in general be the same. At the boundary between the regions there will be what is called a 'topological defect', analogous to a dislocation in a ferromagnetic crystal. The simplest type of defect is the analogue of a point dislocation, and in typical Grand Unified Theories, these carry a magnetic charge: they are magnetic monopoles. Magnetic monopoles behave as particles with masses of roughly

$$m_{mon} \approx \frac{M_X}{\alpha_U} \approx 10^{17} \, \text{GeV} \,, \tag{12.3.1}$$

where the X boson's mass, $M_X \approx 10^{16} \, \text{GeV}$, is roughly the same as the GUT scale and the effective coupling strength is around $\alpha_U \approx 1/40$.

A further crucial prediction is that these monopoles are stable. Owing to their high mass they contribute essentially from the moment of their creation to the non-relativistic matter component of the universe's energy density. One expects monopoles to be produced with a number density of roughly one in every causally isolated region. The size of such a region is determined by the distance that light can travel from the beginning of the Big Bang up to the time of the phase transition at t_c. This distance is simply the particle horizon $d_H(t)$ [205, 223],

$$d_H(t) = R(t) \int_0^t \frac{dt'}{R(t')} \,, \tag{12.3.2}$$

at the time t_c. If a radiation-dominated phase for times earlier than t_c is assumed, then one gets $R \sim t^{1/2}$ and therefore a particle horizon of $2t_c$. The monopole number density is therefore predicted to be

$$n_{mon} \approx \frac{1}{(2t_c)^3} \,. \tag{12.3.3}$$

The critical temperature T_c is expected to be on the order of the GUT scale, $M_X \approx 10^{16}$ GeV, from which one finds the time of the phase transition to be $t_c \approx 10^{-39}$ s. As the monopoles are non-relativistic, their energy density is given by

$$\varrho_{mon} = n_{mon} m_{mon} \approx \frac{M_X}{\alpha_U} \frac{1}{(2t_c)^3} \approx 2 \times 10^{57}\,\text{GeV}^4 . \qquad (12.3.4)$$

This can compared with the energy density of photons at the same time, which is, see Sect. 9.2 and [224],

$$\varrho_\gamma = \frac{\pi^2}{15} T_{GUT}^4 \approx 2 \times 10^{63}\,\text{GeV}^4 . \qquad (12.3.5)$$

So, initially, the energy of photons still dominates over monopoles by a factor $\varrho_\gamma / \varrho_{mon} \approx 10^6$. But as photons are always relativistic, one has $\varrho_\gamma \sim 1/R^4$, whereas for the monopoles, which are non-relativistic, $\varrho_{mon} \sim 1/R^3$ holds. The two energy densities would become equal after R increased by 10^6, which is to say, after the temperature dropped by a factor of 10^6, because of $R \sim 1/T$. Since the time follows $t \sim T^{-2}$, equality of ϱ_γ and ϱ_{mon} occurs after the time increases by a factor of 10^{12}. So, starting at the GUT scale around $T_{GUT} \approx 10^{16}$ GeV or at a time $t_{GUT} \approx 10^{-39}$ s, one would predict $\varrho_\gamma = \varrho_{mon}$ at a temperature $T_{\gamma mon} \approx 10^{10}$ GeV or at a time $t_{\gamma mon} \approx 10^{-27}$ s.

This is clearly incompatible with what one observes today. Searches for magnetic monopoles have been carried out, and in a controversial experiment by Cabrera in 1982, evidence for a single magnetic monopole was reported [225]. This appears, however, to have been a one-time glitch, as no further monopoles were found in more sensitive experiments. Indeed, far more serious problems would arise from the predicted monopoles. Foremost among these is that the energy density would be so high that the universe would have recollapsed long ago. Given the currently observed expansion rate and photon density, the predicted contribution from monopoles would lead to a recollapse in a matter of days.

So, here is not merely a failure to explain the universe's initial conditions, but a real contradiction between a prediction and what one observes. One could always argue, of course, that Grand Unified Theories are not correct, and indeed there is no direct evidence that requires such a picture to be true. One can also try to arrange for types of GUTs, which do not produce monopoles, but these sorts of the theory are usually disfavoured for other reasons. Historically, the monopole problem was one of the factors that motivated Alan Guth to propose the mechanism of *inflation*, an accelerating phase in the early universe (Fig. 12.1). In the next sections it will first be defined more formally what inflation is and how it could come about, and then it will be shown how it solves not only the monopole problem but also the horizon and flatness problems as well.

Fig. 12.1 A bank note in the hyper-inflation period in Germany from 1914 to 1923 about 100 Billionen Mark: 10^{14} Mark. (recall: a German 'Billion' is 10^{12}, in contrast to an English or American billion, which is 10^{9})

12.4 How Inflation Works

No point is more central than this, that empty space is not empty. It is the seat of the most violent physics.

John Archibald Wheeler

Inflation is defined as meaning a period of accelerating expansion, i.e., where $\ddot{R} > 0$. In this section it will be investigated how this can arise. Recall first the Friedmann equation, which can be written as

$$\frac{\dot{R}^2}{R^2} + \frac{k}{R^2} = \frac{8\pi G}{3}\varrho, \qquad (12.4.1)$$

where here the term ϱ is understood to include all forms of energy including that of the vacuum, ϱ_{v}. It was shown in Chap. 8 that this arises if one considers a cosmological constant Λ. This gives rise to a constant contribution to the vacuum, which can be interpreted as vacuum energy density of the form

$$\varrho_{\mathrm{v}} = \frac{\Lambda}{8\pi G}. \qquad (12.4.2)$$

One can now ask what will happen if the vacuum energy or, in general, if any constant term dominates the total energy density. Suppose this is the case, i.e., $\varrho \approx \varrho_{\mathrm{v}}$, and assume as well that one can neglect the k/R^2 term; this should always be a good approximation for the early universe. The Friedmann equation then reads

$$\frac{\dot{R}^2}{R^2} = \frac{8\pi G}{3} \varrho_v . \tag{12.4.3}$$

Thus, one finds that the expansion rate H is a constant,

$$H = \frac{\dot{R}}{R} = \sqrt{\frac{8\pi G}{3} \varrho_v} . \tag{12.4.4}$$

The solution to (12.4.4) for $t > t_i$ is

$$R(t) = R(t_i) e^{H(t-t_i)} = R(t_i) \exp\left[\sqrt{\frac{8\pi G}{3} \varrho_v} (t - t_i)\right]$$

$$= R(t_i) \exp\left[\sqrt{\frac{\Lambda}{3}} (t - t_i)\right] . \tag{12.4.5}$$

That is, the scale factor increases exponentially in time.

More generally, the condition for a period of accelerating expansion can be seen by recalling the acceleration equation from Sect. 8.5,

$$\frac{\ddot{R}}{R} = -\frac{4\pi G}{3} (\varrho + 3P) . \tag{12.4.6}$$

This shows that one will have an accelerating expansion, i.e., $\ddot{R} > 0$, as long as the energy density and pressure satisfy

$$\varrho + 3P < 0 . \tag{12.4.7}$$

That is, if the equation of state is expressed as $P = w\varrho$, one has an accelerating expansion for $w < -1/3$. The pressure is related to the derivative of the total energy U of a system with respect to volume V at constant entropy S. If one takes $U/V = \varrho_v$ as constant, then the pressure is

$$P = -\left(\frac{\partial U}{\partial V}\right)_S = -\frac{U}{V} = -\varrho_v . \tag{12.4.8}$$

Thus, a vacuum energy density leads to a negative pressure and an equation-of-state parameter $w = -1$.

The determination of the w parameter can be achieved by combining various cosmological probes, which turns out to be very powerful for investigating the nature of dark energy. This is shown in Fig. 12.2, favouring $w = -1$.

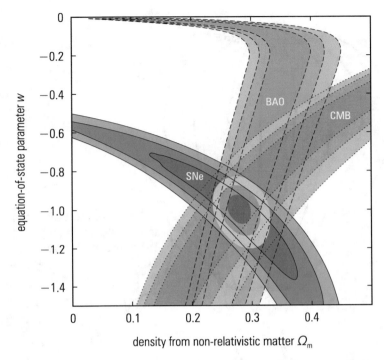

Fig. 12.2 wCDM model: 68.3, 95.4, and 99.7% confidence regions in the Ω_m, w plane using information from the cosmological microwave background (CMB), the baryon acoustic oscillations (BAO), and supernovae (SNe) with hydrogen emission in their spectra [226]

One now needs to ask several further questions: "What could cause such a vacuum energy density?", "How and why did inflation stop?", "How does this solve any of the previously mentioned problems?", and "What are the observational consequences of inflation?".

12.5 Mechanisms for Inflation

Using the forces we know now, you can't make the universe we know now.

George Smoot

Claus Grupen 2013

To predict an accelerating scale factor, an equation of state relating pressure P and energy density ϱ of the form $P = w\varrho$ with $w < -1/3$ was needed. Vacuum energy with $w = -1$ satisfies this requirement, but what makes us believe that it should exist?

The idea of vacuum energy arises naturally in a quantum field theory. A good analogy for a quantum field is a lattice of atoms occupying all space. The system of atoms behaves like a set of coupled quantum-mechanical oscillators. A particular mode of vibration, for example, will describe a plane wave of a certain frequency

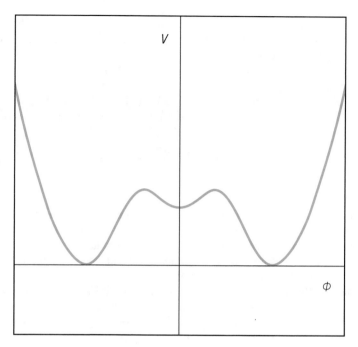

Fig. 12.3 Schematic representation of the potential $V(\phi)$, which has been proposed to cause inflation

and wavelength propagating in some direction through the lattice. Such a mode will carry a given energy and momentum, and in a quantum field theory this corresponds to a particle. This is in fact how *phonons* are described in a crystal lattice. They are quantized collective vibrations in a lattice that carry energy and momentum.

The total energy of the lattice includes the energies of all of the atoms. But a quantum-mechanical oscillator contributes an energy $\hbar\omega/2$ even in its lowest-energy state. This is the zero-point energy; it is the analogue of the vacuum energy in a quantum field theory.

In a quantum field theory of elementary particles, one must dispense with the atoms in the analogy and regard the 'interatomic' spacing as going to zero. In the Standard Model of particle physics, for example, there is an electron field, photon field, etc., and all of the electrons in the universe are simply an enormously complicated excitation of the electron field.

In 1981, Alan Guth [227] proposed that a scalar Higgs field associated with a Grand Unified Theory could be responsible for inflation. The potential of this field should have a dip around $\phi = 0$, as shown in Fig. 12.3.

As a consequence of the local minimum at $\phi = 0$, there should be a classically stable configuration of the field at this position. In this state, the energy density of the field is given by the height of the potential at $\phi = 0$. Suppose the field goes into this state at a time t_i. The scale factor then follows an exponential expansion,

$$R(t) = R(t_i) \exp\left(\sqrt{\frac{8\pi G \varrho}{3}} (t - t_i) \right), \qquad (12.5.1)$$

with $\varrho = V(0)$.

In a classical field theory, if the field were to settle into the local minimum at $\phi = 0$, then it would stay there forever. In a quantum-mechanical theory, however, it can tunnel from this 'false vacuum' to the true vacuum with $V = 0$. When this happens, the vacuum energy will no longer dominate and the expansion will be driven by other contributions to ϱ such as radiation.

In cosmology the expansion of the universe can be described by an equation of state or an interaction of a scalar field, the so-called *inflaton field* with negative pressure. In the quantum field theories such as the Standard Model of particle physics, different fields interact. That is, energy from one field can be transformed into excitations in another. This is how reactions are described, in which particles are created and destroyed. So, when the *inflaton field* (see below for more details) moves from the false to the true vacuum, one expects its energy to be transformed into other 'normal' particles such as photons, electrons, etc. Thus, the end of inflation simply matches onto the hot expanding universe of the existing Big Bang model. It may then appear that inflation has not made any predictions that one can verify by observation, and as described so far no verification has been given. Later in this chapter, however, it will be shown how inflation provides explanations for the initial conditions of the Big Bang, which otherwise would have to be imposed by hand.

Shortly after the original inflaton potential was proposed, it was realized by Guth and others that the model had important flaws. As the quantum-mechanical tunneling is a random process, inflation should end at slightly different times in different places. Therefore, some places will undergo inflation for longer than others. And because these regions continue to inflate, their contribution to the total volume of the universe remains significant. Effectively, inflation would never end. This has been called inflation's 'graceful exit problem'.

Graceful Exit Problem or "How can I exit this phase in a smart way?"

It was pointed out by Linde [228] and Albrecht and Steinhardt [229] that by a suitable modification of the potential $V(\phi)$, one could achieve a graceful ending to inflation. The potential needed for this is shown schematically in Fig. 12.4. The field quantum of the inflaton field, a supposed scalar boson, is called an *inflaton*. Its mass is difficult to predict. It could be somewhere between 10^{-10} and 10^{-5} Planck masses [230]. Its potential is not derived from particle physics considerations such as the Higgs mechanism, but rather its sole motivation is to provide for inflation. The local minimum in V at $\phi = 0$ is replaced by an almost flat plateau. The field can settle down into a metastable state near $\phi \approx 0$ and then effectively 'roll' down the plateau to the true vacuum. One can show that in this scenario, called *new inflation*, the exponential expansion ends gracefully everywhere.

As in Guth's original theory, however, new inflation does not end everywhere at the same time. But now this feature is turned to an advantage. It is used to explain the structure or lumpiness of the universe currently visible on distance scales less than

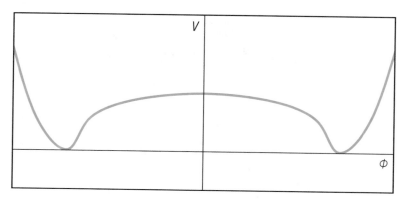

Fig. 12.4 Schematic representation of the potentials $V(\phi)$ for the new inflation

around 100 Mpc. Because of quantum fluctuations, the value of the field ϕ at the start of the inflationary phase will not be exactly the same at all places. As with the randomness of quantum-mechanical tunneling, these quantum fluctuations lead to differences, depending on position, in the time needed to move to the true vacuum state.

When a volume of space is undergoing inflation, the energy density $\varrho \approx V(0)$ is essentially constant. After inflation ends, the energy is transferred to particles such as photons, electrons, etc. Their energy density then decreases as the universe continues to expand, e.g., $\varrho \sim R^{-4}$ for relativistic particles. The onset of this decrease is therefore delayed in regions, where inflation goes on longer. Thus, the variation in the time of the end of inflation provides a natural mechanism to explain spatial variations in energy density. These density fluctuations are then amplified by gravity and finally result in the structures that one sees today, e.g., galaxies, clusters, and superclusters.

12.6   Solution to the Flatness Problem

I have just invented an anti-gravity machine. It's called a chair.

Richard P. Feynman

It will now be shown that an early period of inflationary expansion can explain the flatness problem, i.e., why the energy density today is so close to the critical density. Suppose that inflation starts at some initial time t_i and continues until a final time t_f, and suppose that during this time the energy density is dominated by vacuum energy ϱ_v, which could result from some inflaton field. The expansion rate H during this period is given by

$$H = \sqrt{\frac{8\pi G}{3}\varrho_v} \, . \tag{12.6.1}$$

So, from t_i to t_f, the scale factor increases by a ratio

$$\frac{R(t_f)}{R(t_i)} = e^{H(t_f - t_i)} \equiv e^N \, , \tag{12.6.2}$$

where $N=H(t_f-t_i)$ represents the number of e foldings of expansion during inflation.

Referring back to (12.2.3), it was shown that the Friedmann equation could be written as

$$\Omega - 1 = \frac{k}{\dot{R}^2} \,. \tag{12.6.3}$$

Now, during inflation, one has $R \sim e^{Ht}$, with $H = \dot{R}/R$ constant. Therefore, one finds

$$\Omega - 1 \sim e^{-2Ht} \tag{12.6.4}$$

during the inflationary phase. That is, during inflation, Ω is driven exponentially *towards* unity.

It is natural to assume that the inflaton field is related to the physics of grand unification (GUT theory). In that case one should expect that the inflational period extended from about 10^{-38} to 10^{-36} s. We now assume that the vacuum energy started to dominate at the beginning of the inflation period t_i at 10^{-38}, e.g., during inflation we had $H = 1/t_i$. So, the number of e foldings is basically determined by the duration of the inflationary period,

$$N = H(t_f - t_i) \approx \frac{t_f - t_i}{t_i} \,. \tag{12.6.5}$$

Supposing that the inflational period started at $t_i = 10^{-38}$ s and ended at $t_f = 10^{-36}$ s, we find $N \approx 100$, which means that $\Omega - 1$ exponentially approaches zero, in accordance of what we find today.

Simplified Universe

12.7   Solution to the Horizon Problem

> *There was no 'before' the beginning of our universe, because once upon a time there was no time.*
>
> John D. Barrow

It can also be shown that an early period of inflationary expansion can explain the horizon problem, i.e., why the entire sky appears at the same temperature. To be specific, suppose that inflation starts at $t_i = 10^{-38}$ s, ends at $t_f = 10^{-36}$ s, and has an expansion rate during inflation of $H = 1/t_i$. Assuming the universe was radiation dominated up to t_i, the particle-horizon distance d_H at this point was

$$d_H = 2ct_i \approx 2 \times 3 \times 10^8 \, \text{m/s} \times 10^{-38} \, \text{s} = 6 \times 10^{-30} \, \text{m} . \qquad (12.7.1)$$

This is the largest region, where one would expect to find the same temperature, since any region further away would not be in causal contact. Now, during inflation, a region of size d expands in proportion to the scale factor R by a factor e^N, where the number of e foldings is from (12.6.5) $N \approx 100$. After inflation ends, the region expands following $R \sim t^{1/2}$ up to the time of matter–radiation equality (50 000 a) and then in proportion to $t^{2/3}$ from then until the present, assuming matter domination. So, the current size of the region, which would have been in causal contact before inflation, is

$$d(t_0) = d(t_i) \, e^N \left(\frac{t_{mr}}{t_f} \right)^{1/2} \left(\frac{t_0}{t_{mr}} \right)^{2/3} \approx 10^{38} \, \text{m} . \qquad (12.7.2)$$

This distance can be compared to the size of the current Hubble distance, $c/H_0 \approx 10^{26}$ m. So, the currently visible universe, including the entire surface of last scattering, fits easily into the much larger region, which one can expect to be at the same temperature. With inflation, it is not true that opposite directions of the sky were never in causal contact. So, the very high degree of isotropy of the CMB can be understood.

12.8   Solution of the Monopole Problem

> *If you can't find them, dilute them.*
>
> Anonymous

"We are in a monopoly position: We are the only suppliers worldwide!"

Monopole Trade

The solution to the monopole problem is equally straightforward. One simply has to arrange for the monopoles to be produced before or during the inflationary period. This arises naturally in models, where inflation is related to the Higgs fields of a Grand Unified Theory and works, of course, equally well if inflation takes place after the GUT scale. The monopole density is then reduced by the inflationary expansion, leaving it with so few monopoles that one would not expect to see any of them.

To see this in numbers, let us suppose the monopoles are formed at a critical time $t_c = 10^{-39}$ s. Suppose, as in the previous example, the start and end times for inflation are $t_i = 10^{-38}$ s and $t_f = 10^{-36}$ s, and let us assume an expansion rate during inflation of $H = 1/t_i$. This gives $N \approx 100$ e foldings of exponential expansion. The volume containing a given number of monopoles increases in proportion to R^3, so that during inflation the density is reduced by a factor $e^{3N} \approx 10^{130}$. Putting together the entire time evolution of the monopole density, one takes $R \sim t^{1/2}$ during radiation domination from t_c to t_i, then $R \sim e^{Ht}$ during inflation, followed again by a period of radiation domination up to the time of matter–radiation equality at $t_{mr} = 50\,000$ a, followed by matter domination with $R \sim t^{2/3}$ up to the present, $t_0 = 14 \times 10^9$ a. Inserting explicitly the necessary factors of c, one would therefore expect today a monopole number density of

$$n_m(t_0) \approx \frac{1}{(2ct_c)^3} \left(\frac{t_i}{t_c}\right)^{-3/2} e^{-3(t_f-t_i)/t_i}$$

$$\times \left(\frac{t_{mr}}{t_f}\right)^{-3/2} \left(\frac{t_0}{t_{mr}}\right)^{-2}. \tag{12.8.1}$$

Using the relevant numbers one would get

$$n_{\rm m}(t_0) \approx 10^{-114}\,{\rm m}^{-3} \approx 3 \times 10^{-38}\,{\rm Gpc}^{-3} . \tag{12.8.2}$$

So, the number of monopoles one would see today would be suppressed by such a huge factor that one would not even expect a single monopole in the observable part of the universe.

12.9 Inflation and Growth of Structure

> *The universe is not made, but is being made continuously. It is growing, perhaps indefinitely.*
>
> Henri Bergson

The primary success of inflationary models is that they provide a dynamical explanation for specific initial conditions, which otherwise would need to be imposed by hand. Furthermore, inflation provides a natural mechanism to explain the density fluctuations, which grew into the structures such as galaxies and clusters as are seen today. But beyond this qualitative statement, one can ask whether the properties of the predicted structure indeed match what one observes.

The level of structure in the universe is usually quantified by considering the relative difference between the density at a given position, $\varrho(x)$, and the average density $\langle\varrho\rangle$,

$$\delta(x) = \frac{\varrho(x) - \langle\varrho\rangle}{\langle\varrho\rangle} . \tag{12.9.1}$$

The quantity $\delta(x)$ is called the *density contrast*. Suppose a cube of size L is considered, i.e., volume $V = L^3$, and the density contrast inside V is expanded into a Fourier series. Assuming periodic boundary conditions, this gives

$$\delta(x) = \sum \delta(k)\, e^{i k \cdot x} , \tag{12.9.2}$$

where the sum is taken over all values of $k = (k_x, k_y, k_z)$ that fit into the box, e.g., $k_x = 2\pi n_x/L$, with $n_x = 0, \pm 1, \pm 2, \ldots$, and similarly for k_y and k_z. Averaging over all directions provides the average magnitude of the Fourier coefficients as a function of $k = |k|$. Now the *power spectrum* is defined as

$$P(k) = \langle |\delta(k)|^2 \rangle . \tag{12.9.3}$$

Its interpretation is simply a measure of the level of structure present at a wavelength $\lambda = 2\pi/k$. Different observations can provide information about the power spectrum at different distance scales. For example, at distances up to around 100 Mpc, surveys of galaxies such as the Sloan Digital Sky Survey (SDSS) can be used to measure directly the galaxy density. At larger distance scales, i.e., smaller values of k, the

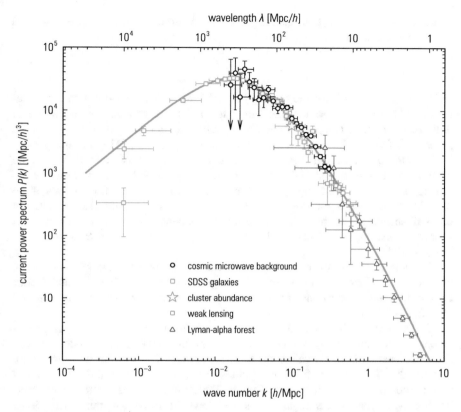

Fig. 12.5 Measurements of the power spectrum $P(k)$ at small and large structures by various observational methods. The part of the power spectrum at large structures is well described by the scale-invariant Harrison–Zel'dovich spectrum with $n = 1$. For the fit the h parameter was assumed to be 0.72 [235]

temperature variations in the CMB provide the most accurate information. Some recent measurements of $P(k)$ are shown in Fig. 12.5 [231]. The results on the power spectrum are based on measurements on the cosmological background radiation, the Sloan Digital Sky Survey (SDSS) [231], the observed frequency of galactic clusters, the growth of structure using the method of microgravitation, and the Lyman-alpha forest.[3]

---

[3]The Lyman-alpha forest describes the absorption lines in the spectrum of distant quasars. These absorption lines at different redshifts do not originate from the quasars themselves, but rather from hydrogen clouds, which are in the way between the quasars and Earth. The distribution of clouds at various redshifts allows to infer results on the structure in the universe.

In most cosmological theories of structure formation, one finds a power law valid for large distances (small k) of the form

$$P(k) \sim k^n .$$

(12.9.4)

Using the value $n = 1$ for the *scalar spectral index* gives what is called the scale-invariant Harrison–Zel'dovich spectrum. Most inflationary models predict $n \approx 1$ to within around 10%. The exact value for a specific model is related to the form of the potential $V(\phi)$, i.e., how long and flat its plateau is (see, e.g., [232] and Fig. 12.4).

The suppression of high wave-number modes on scales less than ≈ 10 Mpc at the present epoch is attributed to the so-called Silk damping. Silk damping describes an effect, which leads to a damping of matter condensation at small fluctuations of the radiation density during the very early phases of the universe. Since self-gravitation of small structures was insufficient, the matter condensation of such structures was suppressed leading eventually to a decrease of the power spectrum at small structures (large k) [233].

The details of the power spectrum for the full scope of wave numbers, resp. distances, depend on a number of cosmological parameters, like the total matter content Ω_m, the Hubble constant H_0, the baryon content Ω_b, and the vacuum energy density Ω_Λ [234].

The curve shown in Fig. 12.5 is based on a specific model with $n = 1$ for large distances, and this agrees well with the data. Depending on the specific model assumptions, the best determined values of the spectral index are equal to unity to within around 10%. To work out the spectral index of the power spectrum on small scales requires computationally intensive N-body simulations, which take into account non-linear gravitational effects in galaxies and a knowledge of dark-matter distributions. Recent estimates based on this kind of complication can motivate a power index at these small scales of n between -2 and -3 [236] in good agreement with the experimental results. The power spectrum is currently an area with a close interplay between theory and experiment, and future improvements in the measurement of the power spectrum should lead to increasingly tight constraints on models of inflation. Since there is no unique theory of inflation, the accurate determination of the power spectrum presumably allows to discriminate between different models of inflation.

12.10   Outlook on Inflation

> *The argument seemed sound enough, but when a theory collides with a fact, the result is tragedy.*
>
> Louis Nizer

So far, inflation seems to have passed several important tests. Its most generic predictions, namely, an energy density equal to the critical density, a uniform temperature for the observable universe, and a lack of relic particles created at any pre-inflationary time, are well confirmed by observation. Furthermore, there are no serious alternative theories that come close to this level of success. But inflation is not a single theory but rather a class of models that include a period of accelerating expansion. There are still many aspects of these models that remain poorly constrained, such as the nature of the energy density driving inflation and the time when it existed.

Concerning the nature of the energy density, it will be seen in Chap. 13 that around 70% of the current energy density in the universe is in fact 'dark energy', i.e., something with properties similar to vacuum energy. Data from type Ia supernovae and also information from CMB experiments such as WMAP and Planck indicate that for the last several billion years, the universe has been undergoing a renewed quasi-exponential expansion compatible with a value of $\Omega_{\Lambda,0} = \varrho_{v,0}/\varrho_{c,0} \approx 0.7$. (The subscript Λ refers to the relation between vacuum energy and a cosmological constant; the subscript 0 denotes as usual present-day values.) The critical density is $\varrho_{c,0} = 3H_0^2/8\pi G$, where $H_0 \approx 70 \, \text{km s}^{-1} \, \text{Mpc}^{-1}$ is the Hubble constant. This gives a current vacuum energy density of around ($\hbar c \approx 0.2 \, \text{GeV fm}$)

$$\varrho_{v,0} \approx 10^{-46} \, \text{GeV}^4 \,. \tag{12.10.1}$$

In the example of inflation considered above, however, the magnitude of the vacuum energy density is related to the expansion rate during inflation by (12.4.4). Furthermore, it was argued that the expansion rate is approximately related to the start time of inflation by $H \approx 1/t_i$. So, if one believes that inflation occurred at the GUT scale with, say, $t_i \approx 10^{-38}$ s, then one needs a vacuum energy density of, see (9.4.18),

$$\varrho_v = \frac{3m_{\text{Pl}}^2}{32\pi t_i^2} \approx 10^{64} \, \text{GeV}^4 \,. \tag{12.10.2}$$

The vacuum energies now and those, which existed during an earlier inflationary period, may well have a common explanation, but given their vast difference, it is by no means obvious what their relationship is.

One could try to argue that inflation took place at a later time, and therefore had a smaller expansion rate and correspondingly lower ϱ_v. The latest possible time for inflation would be just before the Big Bang nucleosynthesis era, around $t \approx 1$ s. With inflation any later than this, the predictions of BBN for abundances of light nuclei would be altered and would no longer stand in such good agreement with observation. Even at an inflation time of $t_i = 1$ s, one would need a vacuum energy density of $\varrho_v \approx 10^{-12} \, \text{GeV}^4$, still 34 orders of magnitude greater than what is observed today.

In addition to those observable features mentioned, inflation predicts the existence of gravitational waves. These would present a fossilized record of the first moments of time. In particular, the energy density of gravity waves is expected to be

$$\varrho_{\text{gravity waves}} = \frac{h^2 \omega^2}{32\pi G} \,, \tag{12.10.3}$$

which gives

$$\Omega_{\text{gravity waves}} = \frac{\varrho_{\text{gravity waves}}}{\varrho_c} = \frac{h^2 \omega^2}{12 H^2} . \qquad (12.10.4)$$

This would result in distortions of gravity-wave antennae on the order of $h = 10^{-27}$ for kHz gravity waves. This would correspond to a compression or elongation of a gravitational antenna of the size of the Earth orbit (150 million km) by 1/10 of the diameter of a hydrogen nucleus. This minute effect is not nearly in reach of presently operating gravitational-wave antennae, but the detection of gravitational waves due to gravitational background radiation would be as spectacular as the discovery of the cosmological 2.7-K background radiation. At the time of discovery of this echo of the Big Bang one also did not expect that the satellites COBE, WMAP, and Planck would reach a sensitivity of 10^{-5} in the temperature variations of the electromagnetic left-over from the Big Bang. If the sensitivity of gravitational-wave detectors could be increased by six orders of magnitude, the detection of signals at the level expected in inflationary models would result in a convincing credibility of inflation, even though this is still a long way to go.

However, it was a sensation when gravitational waves from the merger of two black holes were detected in 2015. This also indicated that spectacular improvements in gravitational-wave detection are achievable. The gravitational signal from this binary system consisting of two black holes was detected in coincidence of two Michelson interferometers at a distance of 3000 km [19]. The results of the two LIGO interferometers showed amplitudes of the gravitational wave of 10^{-21}. The signal was received from a distance of about 400 Mpc and originated from the inward spiral movement of two black holes (masses of 36 $M_\odot$ and 29 $M_\odot$), which merged to form a black hole of about 62 $M_\odot$. The signal was convincing and even textbook-like. The Nobel committee in Stockholm awarded the discoverers (Kip Thorne, Rainer Weiss, Barry Barish) the physics Nobel Prize in 2017 for this achievement.

In the subsequent time further signals (in total eleven—at the time of writing in 2018) were seen and in 2017 an event of the LIGO detectors was recorded, which was associated with a collision of two heavy neutron stars. VIRGO was operational at that time and confirmed a coincident signal. The afterglow of this spectacular event was also observed by the X-ray satellites Fermi (FGST) and INTEGRAL, and also in other spectral ranges.

12.11 Summary

The classical Big Bang model poses several questions: why is the Ω parameter so close to unity (flatness problem); why has the sky an almost identical temperature in all directions (horizon problem), why are there no magnetic monopoles? All these problems can be solved if one assumes that space has expanded exponentially with superluminal velocity shortly after the Big Bang. This rapid expansion made the universe flat, and also cared that regions of the universe, which were initially not causally connected, obtained the same temperature. One has to be careful, however, that there is no unique model of inflation. There are quite a number of variants, which differ in detail. Open is also the question how in detail inflation ended (graceful exit problem), i.e., what is the exact form of the inflation potential. The additional problem is to explain why the expansion is speeding up again since several billion years. In spite of this, there is almost general agreement that a rapid expansion in the early universe must have happened to understand why the present universe looks like the way it is now.

12.12   Problems

1. A value of the cosmological constant Λ can be estimated from the Friedmann equation extended by the cosmological term. It is often said that this value disagrees by about 120 orders of magnitude with the result from the expected vacuum energy in a unified supergravity theory. How can this factor be illustrated?
2. Work out the time dependence of the size of the universe for a flat universe with a cosmological constant Λ.
3. Work out the time evolution of the universe if Λ were a dynamical constant ($\Lambda = \Lambda_0(1 + \alpha t)$) for a Λ-dominated flat universe.
4. Estimate the size of the universe at the end of the inflation period ($\approx 10^{-36}$ s). Consider that for a matter-dominated universe its size scales as $t^{\frac{2}{3}}$, while for a radiation-dominated universe one has $R \sim t^{\frac{1}{2}}$.

Chapter 13
Dark Energy and Dark Matter

There is a theory, which states that if ever anyone discovers exactly what the universe is for and why it is here, it will instantly disappear and be replaced by something even more bizarre and inexplicable. There is another theory, which states that this has already happened.

Douglas Adams

About ninety years ago the general belief was that the universe was fairly understood. The general theory of relativity and the expansion of the universe successfully described the dynamics of stars and galaxies and even the large-scale structure of the universe. Questions, however, arose in the thirties of the last century, when Fritz Zwicky raised the question how the galaxies managed to remain stable, given the large rotational velocities of stars at the edges of the Milky Way. With the then known amount of gravitational matter stars could easily escape from the galaxy and the whole galaxy would disintegrate. Now we know: ordinary matter, humans, and the stars are made of material, which is relatively rare in the universe. Only somewhat less than 5% of the total mass, resp. energy, of the universe consists of normal, baryonic matter. The rest is something, which is essentially unknown. The dynamics of stars in galaxies and the dynamics of galaxies in galactic clusters require additional gravitational matter, which keeps the galaxies together. This additional matter is called *dark matter*.

The observation of accelerated expansion of the universe can only be understood if one assumes, in addition, a repelling force associated to *dark energy*, which might have its origin in empty space.

Finally, the dominance of matter over antimatter is still a mystery. In the Big Bang equal amounts of matter and antimatter have been created, but it is not understood why and how most of the antimatter has disappeared. The minute left-over from an annihilation process of matter with antimatter is what we call matter.

© Springer Nature Switzerland AG 2020
C. Grupen, *Astroparticle Physics*, Undergraduate Texts in Physics,
https://doi.org/10.1007/978-3-030-27339-2_13

13.1   Large-Scale Structure of the Universe

> *I sometimes think that the universe is a machine designed for the perpetual astonishment of astronomers.*
>
> Arthur C. Clarke

Originally it has been assumed that the universe is homogeneous and isotropic. However, all evidence speaks to the contrary. Nearly on all scales one observes inhomogeneities: stars form galaxies, galaxies form galactic clusters, and there are superclusters, filaments of galaxies, large voids, and great walls, just to name a few of them. The large-scale structure has been investigated up to distances of about 1 Gpc. On these large scales a surprising lumpiness was found. One has, however, to keep in mind that the spatial distribution of galaxies must not necessarily coincide with the distribution of matter in the universe.

The measurement of the COBE, WMAP, and Planck satellites on the inhomogeneities of the 2.7 K blackbody radiation has shown that the early universe was much more homogeneous. On the other hand, small temperature variations of the blackbody radiation have served as seeds for structures, which one now observes in the universe.

It is generally assumed that the large-scale structures have evolved from gravitational instabilities, which can be traced back to small primordial fluctuations in the energy density of the early universe. Small perturbations in the energy density are amplified due to the associated gravitational forces. In the course of time these gravitational aggregations collect more and more mass and thus lead to structure formation. The reason for the original microscopic small inhomogeneities is presumably to be found in quantum fluctuations. Cosmic inflation and the subsequent slow expansion have stretched these inhomogeneities to the presently observed size. One consequence of the idea of cosmic inflation is that the exponential growth has led to a smooth and flat universe, which means that the density parameter Ω should be very close to unity. To understand the formation of the large-scale structure of the universe and its dynamics in detail, a sufficient amount of mass is required because otherwise the original fluctuations could never have been transformed into distinct mass aggregations. It is, however, true that the amount of visible matter only does not support the critical density of $\Omega = 1$ as required by inflation.

To get a flat universe seems to require a second Copernican revolution. Copernicus had noticed that the Earth was not the center of the universe. Cosmologists now conjecture that the kind of matter, of which man and Earth are made, only plays a minor role compared to the dark non-baryonic matter, which is—in addition to the dark energy—needed to understand the dynamics of the universe and to reach the critical mass density.

13.2 Dark Energy

I find dark energy repelling!

Anonymous

Cosmic Menu

As has been described in the previous chapters, many measurements of cosmological parameters show that the universe is flat, i.e., that it has the critical density. It consists of dark energy, dark matter, and baryonic, normal matter. Dark energy is a hypothetical form of energy, which seems to be a property of empty space. The satellite experiments COBE, WMAP, and, in particular, the latest data from Planck show that the universe is dominated by dark energy. Its contribution to the critical density amounts to 68.3%. The remainder consists of dark matter (26.8%) and normal, baryonic matter (4.9%). The experimental evidence for dark energy is indirect. A comparison of distance measurements of far away supernovae with their redshift indicates that the universe undergoes again accelerated expansion since a few billion years. Apart from dark matter and normal matter an additional, special type of energy, which is repulsive, is required to understand the observed flatness of the universe. The structure of the universe on large scales can only be understood if a force exists, which is repelling. The critical density of the universe is quite low, about

10^{-29} g/cm^3. This corresponds to only several protons per cubic meter. According to common wisdom dark energy merely participates in gravitational interactions. To measure the effect of dark energy in laboratory measurements seems almost impossible. However, since dark energy appears to be a property of space and fills the whole universe it will influence the way how the universe expands. When Einstein developed the general theory of relativity he already introduced an interaction with repelling force: his famous cosmological constant Λ. Historically, the general theory of relativity evolved when one assumed that the universe was static and invariant in time. However, Einstein's field equations allowed only an expanding or collapsing dynamic universe. To prevent a collapse, Einstein *invented* a repelling interaction, represented by the cosmological constant, which stabilized the universe. After Hubble's discovery of the expanding universe, there seemed to be no longer a need for the cosmological constant, and Einstein discarded it, and he called his *invention* his biggest blunder. Only since the discovery of the accelerated expansion in the nineties of the last century, the cosmological constant experienced a renaissance. The cosmological constant corresponds to a negative pressure. This can already be inferred from classical thermodynamics: an expansion of the universe requires energy: A change dV of the volume is related to work, i.e., a change of energy according to $-P\,dV$. Here P is the pressure. An expansion naturally causes a volume increase, which means dV is positive. The energy content of an expanding volume, however, increases, since dark energy is a property of space. The energy gain is $\rho \cdot V$, if ρ is the energy density of dark energy. Therefore, the pressure must be negative, $P = -\rho$.

As has been mentioned earlier in the chapter on cosmology, a negative pressure and its cosmological effect can be illustrated as follows: Einstein has shown that all forms of energy are equivalent and all act gravitationally. But not only matter but also forces create gravitation. Even forces, which resist gravitation produce a certain gravitational effect. A celestial body resists a gravitative collapse by pressure forces. These pressure forces contribute energy, which in turn are equivalent to a gravitational effect. The harder the pressure forces resist gravity and try to push out, the more their own associated gravity pushes in [237]. This means that a positive pressure leads to a vicious circle: the pressure forces defeat themselves. The more they grow, the more their effect increases the gravitation. In contrast, a negative pressure is like a tension, which is a force that pulls things together. In the same way as a positive pressure leads to an increased attraction, the negative pressure starts a vicious circle in the opposite direction. A positive pressure, which gets out of hand, leads eventually to a compact object, like a neutron star or even a black hole, while a negative pressure results in an increasing expansion or even an inflationary scenario.

What kind of explanation can be offered for dark energy? The cosmological constant appears to be an elegant, economic solution. However, the experimentally observed expansion leads to an energy density corresponding to a density of roughly 10^{-29} g/cm^3, which is extremely small. In elementary particle physics also an energy density is known from vacuum fluctuations, which are firmly established, for example, by the observation of the Casimir effect describing the attractive interaction of two metal plates in vacuum: the number of virtual particles between the plates is smaller than on the outside, which leads to an attraction of the two plates, because the larger number of virtual particles outside the plates exerts a pressure on the plates. From these quantum-mechanical fluctuations one can derive a vacuum energy, which is larger by 120 orders of magnitude compared to the vacuum energy derived from the cosmological constant.

One has also to mention that the presently observed accelerated expansion of the universe seems to require again a repelling force, possibly described by a kind of cosmological constant, which could mean that the cosmological constant might vary in space and time.

It has also been proposed to describe the observed accelerated expansion in alternative approaches like in the quintessence model [238, 239]. Here the potential energy is characterized by a dynamical field. There is, however, no evidence to confirm this idea, but also no evidence to the contrary. A violation of Einstein's principle of equivalence or a variation of fundamental constants in space and time would support the idea of such a dynamical field. However, neither has been observed. An additional idea has been put forward assuming that the general theory of relativity must be modified at large cosmological scales. There is also no experimental evidence that would require such a modification of Newtonian or Einsteinian dynamics. Popular string theories suggest that our universe is embedded in a large number of universes, a multiverse. It has been discussed that there might be 10^{500} universes in a large multiverse! The vacuum energy could adopt different values in each different universe. If the vacuum energy in a universe were very large, no structures would develop. If it were very small, it would soon recollapse. The idea is that we happen to live in a special universe with a low, but appropriate cosmological constant, which allows to form stars, galaxies, and ourselves. This sounds very much like the anthropic principle.

According to present belief, the second accelerated expansion in our universe started about five billion years ago. If this accelerated expansion would have started much earlier, no structures like stars or galaxies could have formed, and there would be no life in our universe.

In this context it is also interesting that the latest Sloan Digital Sky Survey (SDSS-III) [240] has compiled a three-dimensional map of our universe for a volume with a transverse area of 6 times 4.5 billion light-years and a depth of 500 million light-years. The survey maps one quarter of the entire sky in detail. The clustering of galaxies and the occurrence of voids allows to derive results on dark energy and dark matter. In addition, the distribution of galaxies permits to determine the acoustic scale, which results from oscillations of the primordial baryon fluid. If dark energy drives the accelerated expansion this galaxy map, which has a depth of 500 million years, and using the acoustic scale, one can deduce that the expansion of the universe only proceeds at a relatively slow rate. It is estimated that in the time interval of

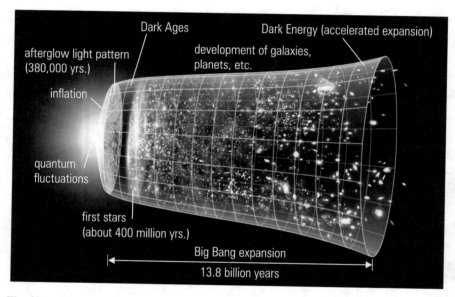

Fig. 13.1 This figure describes the time evolution of the universe. Right after the Big Bang, in the first fractions of a second, the universe expanded exponentially by a huge factor (inflation). When the universe cooled down due to the expansion, light elements could be formed, when the temperature dropped well below the binding energy of the deuteron, and the baryogenesis started. Within the next three minutes the light elements from hydrogen, over helium to beryllium and boron were synthesized. After that the temperature was insufficient to fuse heavier elements, which were then later made in supernova explosions. When the temperature had dropped so that hydrogen atoms could be formed (around 380 000 years), the universe became transparent. But then the blackbody radiation was redshifted out of the visible range and there was no other light, as the stars were not born yet. The first stars were then born about 400 million years after the Big Bang, and later on larger structures like galaxies and galaxy clusters formed. Since about five billion years the dark energy took over to start an accelerated expansion, which still persists [242]

seven billion years the change of the size of the universe due to expansion has increased by only 20%, if at all [241].

Completely open is the question why the primordial inflation in the very early universe in the first instants after the Big Bang has lead to a flat universe, and billions of years later the universe again accelerated. Possibly the cosmological constant is in fact a dynamical one.

The gross timeline of the evolution of the universe is sketched in Fig. 13.1 [242].

13.3 Dark Matter

The more we look, the more we know. The more we know, the greater the mystery.

Cees Nooteboom

The idea that our universe contains dark matter is not entirely new. Already in the thirties of the last century Zwicky [243] had argued that clusters of galaxies would not be gravitationally stable without additional invisible dark matter. Also, somewhat later, the astronomer Vera Rubin pioneered the work on galaxy rotation rates and provided evidence for the existence of dark matter [244].

Observations of high-z supernovae and detailed measurements of the cosmic microwave background radiation have now clearly demonstrated that large quantities of dark matter must exist, which fill up the universe. An argument for the existence of invisible dark matter can already be inferred from the Keplerian motion of stars in galaxies. Kepler had formulated his famous laws, based on the precision measurements of Tycho Brahe. The stability of orbits of planets in our solar system is obtained from the balance of centrifugal and the attractive gravitational force:

$$\frac{mv^2}{r} = G\frac{mM}{r^2} \tag{13.3.1}$$

(m is the mass of the planet, M is the mass of the Sun, r is the radius of the planet's orbit assumed to be circular). The resulting orbital velocity is calculated to be

$$v = \sqrt{GM/r}\,. \tag{13.3.2}$$

The radial dependence of the orbital velocity of $v \sim r^{-1/2}$ is perfectly verified in our solar system (Fig. 13.2).

The rotational curves of stars in galaxies, however, show a completely different pattern (Fig. 13.3). Since one assumes that the majority of the mass is concentrated at the center of a galaxy, one would at least expect for somewhat larger distances a Keplerian-like orbital velocity of $v \sim r^{-1/2}$. Instead, the rotational velocities of stars are almost constant even for large distances from the galactic center.

The flat rotational curves led to the conclusion that the galactic halo must contain nearly 90% of the mass of the galaxy. To obtain a constant orbital velocity, the mass

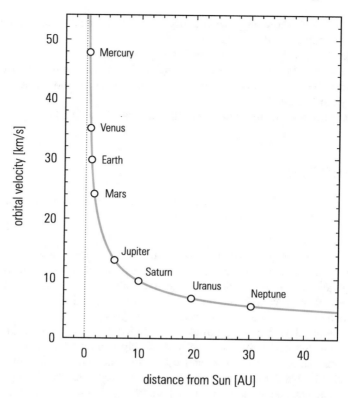

Fig. 13.2 Rotational curves of planets in the solar system (1 astronomical unit (AU) = distance Earth to Sun = 1.5×10^8 km)

of the galactic nucleus in (13.3.1) has to be replaced by the now dominant mass of invisible matter in the halo. This requirement leads to a radial dependence of the density of this mass of

$$\varrho \sim r^{-2} \,, \tag{13.3.3}$$

because

$$\frac{mv^2}{r} = G \, \frac{m \int_0^r \varrho \, dV}{r^2} \sim G \, \frac{m \, \varrho \, V}{r^2} \sim G \, \frac{mr^{-2} r^3}{r^2}$$
$$\Rightarrow v^2 = \text{const} \,. \tag{13.3.4}$$

Frequently the proportionality of (13.3.3) for larger r is parameterized in the following form:

$$\varrho(r) = \varrho_0 \, \frac{R_0^2 + a^2}{r^2 + a^2} \,, \tag{13.3.5}$$

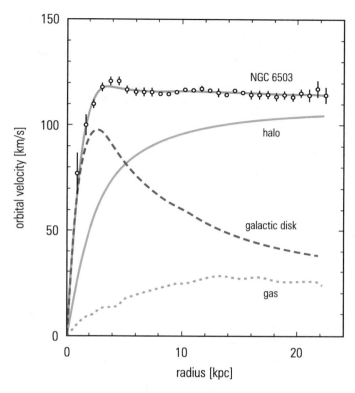

Fig. 13.3 Rotational curves of the spiral galaxy NGC 6503. The contributions of the galactic disk, the gas, and of the halo are shown *separately* [245]

where r is the galactocentric distance, $R_0 = 8.5\,\text{kpc}$ (for the Milky Way) is the galactocentric radius of the Sun, and $a = 5\,\text{kpc}$ is the radius of the halo nucleus. ϱ_0 is the local energy density in the solar system,

$$\varrho_0 = 0.3\,\text{GeV}/\text{cm}^3 . \tag{13.3.6}$$

If the mass density, like in (13.3.3) for flat rotational curves required, decreases like r^{-2}, the integral mass of a galaxy grows like $M(r) \sim r$, since the volume increases like r^3.

It should be mentioned that there are some results indicating that young galaxies from the very early period after the Big Bang seem to show star rotation curves, which fall at large distances from their galactic center instead of being flat. This could be a tentative evidence for the fact that massive galaxies about ten billion years ago might have contained less dark matter than today [246].

From elementary particle physics and nuclear physics it is well-known that the mass is essentially concentrated in atomic nuclei and thereby in baryons. In the framework of the primordial element synthesis the abundance of the light elements (D, ^{3}He, ^{4}He, ^{7}Li, ^{7}Be) formed by fusion can be determined. Based on these arguments and the findings on the cosmic microwave background radiation, the contribution of baryonic matter—expressed in terms of the critical density $\varrho_c = 3H^2/8\pi G$ and $\Omega_b = \varrho_b/\varrho_c$—is obtained to be

$$\Omega_b = 0.0486 \pm 0.0010 , \tag{13.3.7}$$

and the total matter density in the universe is

$$\Omega_m = 0.308 \pm 0.012 . \tag{13.3.8}$$

This corresponds to less than one baryon per m^3 and the visible luminous matter only provides a density of

$$0.003 \le \Omega_{lum} \le 0.007 , \tag{13.3.9}$$

that is, $\Omega_{lum} < 1\%$. That means, the fraction of luminous matter compared to the total matter density is very small.

From the fact that the rotational curves of all galaxies are essentially flat, the contribution of galactic halos to the mass density of the universe can be estimated to be

$$0.23 \le \Omega_{gal} \le 0.31. \tag{13.3.10}$$

The contribution of relativistic particles based on COBE, WMAP, and Planck data is

$$\Omega_{rel} = \Omega_\gamma + \Omega_{neutrinos} \approx 10^{-4} , \tag{13.3.11}$$

i.e., particularly small.

In total one gets for the total energy density of the universe, including the vacuum energy [210] using the most recent Planck results

$$\Omega_{total} = 1.0002 \pm 0.0026 . \tag{13.3.12}$$

The existence of large amounts of invisible non-luminous dark matter seems to be established in a convincing fashion. Consequently, the question arises what this matter is and how it is distributed.

13.3.1   Dark Stars

Everyone is a moon, and has a dark side which he never shows to anybody.

Mark Twain

"My wife prefers the black one; she finds dark matter so attractive!"

The rotational curves of stars in galaxies require a considerable fraction of gravitating matter, which is obviously invisible. The idea of primordial nucleosynthesis suggests that the amount of baryonic matter is larger than the visible matter. Therefore, it is obvious to assume that part of the galactic halos consists of baryonic matter. Because of the experimental result of $\Omega_{lum} < 0.007$, this matter cannot exist in form of luminous stars. Also hot and therefore luminous gas clouds, galactic dust, and cold galactic gas clouds are presumably unlikely candidates, too, because they would reveal themselves by their absorption. These arguments only leave room for special stellar objects, which are too small to shine sufficiently bright, or celestial objects, which just did not manage to initiate hydrogen burning. In addition, burnt-up stars in form of neutron stars, black holes, or dwarf stars could also be considered. Neutron stars and black holes are, however, unlikely candidates, because they would have emerged from supernova explosions. From the observed chemical composition of galaxies (dominance of hydrogen and helium) one can exclude that many supernova explosions have occurred, since these are a source of heavier elements.

Therefore, it is considered likely that galactic baryonic matter could be hidden in brown dwarves. Since the mass spectrum of stars increases to small masses, one would expect a significant fraction of small brown stars in our galaxy. The question is, whether one is able to find such massive, compact, non-luminous objects in galaxies (MACHOs).[1] As has already been shown in the introduction (Chap. 1), a dark star

---

[1]MACHO—MAssive Compact Halo Object, sometimes also called 'Massive Astrophysical Compact Halo Object'.

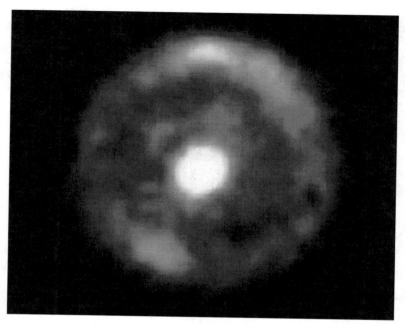

Fig. 13.4 Image of a background galaxy as Einstein ring (*Bull's-Eye Einstein Ring*), where a foreground galaxy *in the center of the image* acts as gravitational lens (deflector) [247]

can reveal itself by its gravitational effect on light. A point-like invisible deflector between a bright star and the observer produces two star images (see Fig. 1.7). If the deflecting brown star is directly on the line of sight between the star and the observer, a ring will be produced, the Einstein ring. The radius r_E of this ring depends on the mass M_d of the deflector like $r_E \sim \sqrt{M_d}$.

Such phenomena have frequently been observed. Figure 13.4 shows the famous Bull's-Eye Einstein Ring and Fig. 13.5 is a high-resolution image of a background galaxy, known as Cosmic Horseshoe (distance 1.6 Gpc), which is gravitationally lensed by a massive foreground galaxy. The deflector has a mass about hundred times of our galaxy. It is extraordinary luminous in the infrared spectral region.

If, however, the mass of the brown star is too small, the two star images or the Einstein ring cannot be resolved experimentally. Instead, one would observe an increase in brightness of the star, if the deflector passes through the line of sight between star and observer (*microlensing*). This brightness excursion is related to the fact that the light originally emitted into a larger solid angle (without the deflecting star) is focused by the deflector onto the observer. The brightness increase now depends on how close the dark star comes to the line of sight between observer and the bright star ('impact parameter' b). The expected apparent light curve is shown in Fig. 13.6 for various parameters. One assumes that b is the minimum distance of the dark star with respect to the line of sight between source and observer and that it passes this line of sight with the velocity v. A characteristic time for the brightness excursion is given by the time required to pass the Einstein ring ($t = r_E/v$).

Fig. 13.5 Hubble image of
the Cosmic Horseshoe as
Einstein ring. This blue
horseshoe is a distant galaxy,
which has been magnified
into a nearly complete
Einstein ring by the strong
gravitational pull of the
massive foreground galaxy,
which is visible *in the center
of the image* [248]

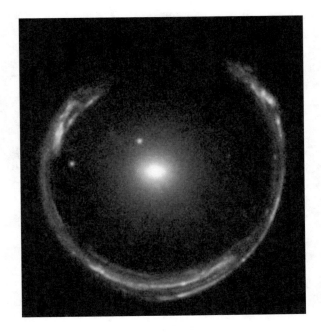

To be able to find non-luminous halo objects by gravitational lensing, a large
number of stars has to be observed over a long period of time to search for a brightness
excursion of individual stars. Excellent candidates for such a search are stars in the
Large Magellanic Cloud (LMC). The Large Magellanic Cloud is sufficiently distant
so that the light of its stars has to pass through a large region of the galactic halo and
therefore has a chance to interact gravitationally with many brown non-luminous star
candidates. Furthermore, the Large Magellanic Cloud is just above the galactic disk
so that the light of its stars really has to pass through the halo. From considerations of
the mass spectrum of brown stars ('MACHOs') and the size of the Einstein ring one
can conclude that a minimum of 10^6 stars has to be observed to have a fair chance
to find some MACHOs.

The experiments MACHO, EROS,[2] and OGLE[3] have found approximately a
dozen MACHOs in the halo of our Milky Way. Figure 13.7 shows the light curve
of the first candidate found by the MACHO experiment. The observed width of the
brightness excursion allows to determine the mass of the brown object.

---

[2]EROS—Expérience pour la Recherche d'Objets Sombres.

[3]OGLE—Optical Gravitational Lens Experiment.

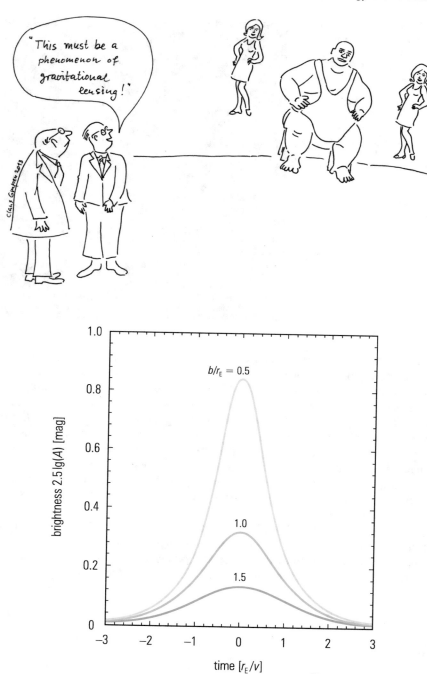

Fig. 13.6 Apparent light curve of a bright star produced by microlensing, when a brown dwarf star passes the line of sight between source and observer. The brightness excursion is given in terms of magnitudes generally used in astronomy [249]

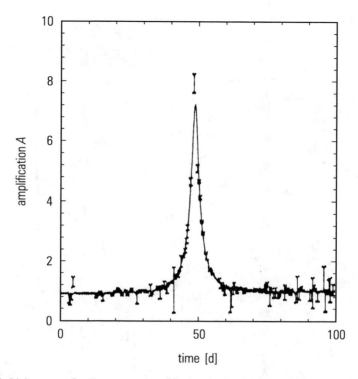

Fig. 13.7 Light curve of a distant star caused by gravitational lensing. Shown is the first brown object found by the MACHO experiment in the galactic halo [249]

If the deflector has a mass corresponding to one solar mass ($M_\odot$), one would expect an average brightness excursion of three months while for $10^{-6} M_\odot$ one would obtain only two hours. The measured brightness curves all lead to masses of approximately $0.5 M_\odot$. The non-observation of short brightness signals already excludes a large mass range of MACHOs as candidates for dark halo matter. If the few seen MACHOs in a limited solid angle are extrapolated to the whole galactic halo, one arrives at the conclusion that possibly 20% of the halo mass, which takes an influence on the dynamics of the Milky Way, could be hidden in dark stars. Because of the low number of observed MACHOs, this result, however, has a considerable uncertainty ($(20^{+30}_{-12})\%$).

The estimate of the fraction of lensing objects f_m has been derived to contribute

$$0.008\, f_m / h \tag{13.3.13}$$

to the critical density, where

$$f_m \geq 0.1 \tag{13.3.14}$$

and $h \approx 0.7$, i.e., $0.008\, f_m / h$ is on the order of 0.1% [250].

"Dark energy is
repelling!"

It has to be mentioned, however, that the lensing objects consist very likely of baryonic and not of *dark matter*, because they are considered to be star-like objects, which failed to grow into normal stars.

A remote possibility for additional non-luminous baryonic dark matter could be the existence of massive quark stars (several hundred solar masses). Because of their anticipated substantial mass the duration of brightness excursions would be so large that it would have escaped detection.

The Andromeda Galaxy with many target stars would be an ideal candidate for microlensing experiments. This galaxy is right above the galactic plane. Unfortunately, it is too distant that individual stars can be resolved. Still one could employ the 'pixel-lensing' technique by observing the apparent brightness excursions of individual pixels with a CCD camera. In such an experiment one pixel would cover several non-resolved star images. If, however, one of these stars would increase in brightness due to microlensing, this would be noticed by the change in brightness of the whole pixel.

One generally assumes that MACHOs constitute a certain fraction of gravitational invisible matter. However, it is not clear, which objects are concerned and where these gravitational lenses reside. That the MACHOs consist of dark, non-baryonic matter is considered unlikely.

A certain contribution to baryonic dark matter could also be provided by ultracold gas clouds (temperatures <10 K), which are very difficult to detect.

Recently a promising new technique of weak gravitational lensing has been developed for the determination of the density of dark matter in the universe. Weak gravitational lensing is based on the fact that images of distant galaxies will be distorted by dark matter between the observer and the galaxy. The particular pattern of the distortions mirrors the mass and its spatial distribution along the line of sight to the distant galaxy. Investigations on 145 000 galaxies in three different directions of observation have shown that dark matter is distributed in a manner consistent with either an open universe with $\Omega_{matter} = 0.45$ (mostly dark matter) or a flat universe that is dominated by a cosmological constant of $\Lambda = 0.67$ [251].

Results from strong gravitational lensing in galaxy clusters like Abell 1689 combined with cosmological constraints and data from X-ray clusters and WMAP data suggest a matter contribution of $\Omega_{matter} = 0.25 \pm 0.05$ [252].

Figure 13.8 shows an image of the Bullet Cluster, a galaxy cluster of about 40 galaxies at a distance of about 1.1 Gpc. The mass visible in the optical range is much

Fig. 13.8 Hubble image of the bullet cluster: The total projected matter reconstructed by weak gravitational lensing is shown *in blue*. Mass determinations based on X rays as measured by the X-ray satellite Chandra are superimposed *in red* [253]

smaller than the mass derived from the X-ray observations of the X-ray satellite Chandra. The two patches of X-ray emission are shown in red. The total mass of the bullet cluster is considerably larger than the sum of the visible mass plus the mass derived from the X-ray measurements. The technique of microlensing allows to infer the missing mass of the cluster. This missing mass, interpreted as dark matter, is shown in blue, superimposed on the combined optical and X-ray image.

There have been many exotic speculations what this enigmatic *dark matter* could be. There are discussions about X-matter, quintessence, NACHOs (Not Astrophysical Compact Halo Objects), Roll-ons, and possible other *smooth stuff* [254].

13.3.2   Neutrinos as Dark Matter

> *Neutrinos ... win the minimalist contest: zero charge, zero radius,*
> *and very possibly zero mass.*
>
> Leon M. Lederman

For a long time neutrinos were considered a good candidate for dark matter. A purely baryonic universe is in contradiction with the primordial nucleosynthesis of the Big Bang. Furthermore, baryonic matter is insufficient to explain the large-scale structure of the universe. The number of neutrinos approximately equals the number of blackbody photons. If, however, they had a small mass, they could provide a significant contribution to dark matter.

From direct mass determinations only limits for the neutrino masses can be obtained ($m_{\nu_e} < 2\,\mathrm{eV}, m_{\nu_\mu} < 170\,\mathrm{keV}, m_{\nu_\tau} < 15.5\,\mathrm{MeV}$). The deficit of atmospheric muon neutrinos, being interpreted as (ν_μ–ν_τ) oscillations, leads to a mass of the ν_τ neutrino of approximately $0.05\,\mathrm{eV}$. In recent reviews a limit from oscillations of $m_\nu \approx 0.01\,\mathrm{eV}$ is given [255].

Under the assumption of $\Omega = 1$ the expected number density of primordial neutrinos allows to derive an upper limit for the total mass that could be hidden in the three neutrino flavours. One expects that there are approximately equal numbers of blackbody neutrinos and blackbody photons. With $N \approx 330$ neutrinos/cm^3 and $\Omega = 1$ (corresponding to the critical density of $\varrho_c \approx 1 \times 10^{-29}\,\mathrm{g/cm^3}$ at an age of the universe of approximately 1.4×10^{10} years) one obtains[4]

$$N \sum m_\nu \le \varrho_c, \quad \sum m_\nu \le 20\,\mathrm{eV}. \tag{13.3.15}$$

The sum extends over all sequential neutrinos including their antiparticles. For the three known neutrino generations one has $\sum m_\nu = 2(m_{\nu_e} + m_{\nu_\mu} + m_{\nu_\tau})$. The consequence of (13.3.15) is that for each individual neutrino flavour a mass limit can be derived:

$$m_\nu \le 10\,\mathrm{eV}. \tag{13.3.16}$$

---

[4]For simplification $c = 1$ has been generally used. If numbers, however, have to be worked out, the correct numerical value for the velocity of light, $c \approx 3 \times 10^8\,\mathrm{m/s}$, must be used.

It is interesting to see that on the basis of these simple cosmological arguments the mass limit for the τ neutrino as obtained from accelerator experiments can be improved by about 6 orders of magnitude.

If the contribution of neutrino masses to dark matter is assumed to be $\Omega_\nu > 0.1$, a similar argument as before provides also a lower limit for neutrino masses. Under the assumption of $\Omega_\nu > 0.1$ (13.3.15) yields for the sum of masses of all neutrino flavours $\sum m_\nu > 2\,\mathrm{eV}$. If one assumes that in the neutrino sector the same mass hierarchy holds as with charged leptons ($m_e \ll m_\mu \ll m_\tau \to m_{\nu_e} \ll m_{\nu_\mu} \ll m_{\nu_\tau}$), the mass of the τ neutrino can be limited to the range

$$1\,\mathrm{eV} \leq m_{\nu_\tau} \leq 10\,\mathrm{eV}. \tag{13.3.17}$$

This conclusion, however, rests on the assumption of $\Omega_\nu > 0.1$. Recent estimates of Ω_ν indicate much smaller values ($\Omega_\nu < 10^{-4}$) and hence the ν_τ mass can be substantially smaller. If the argument given in (13.3.15) is applied to $\Omega_\nu < 10^{-4}$, one would even get

$$\sum m_\nu \leq 2\,\mathrm{meV}. \tag{13.3.18}$$

Neutrinos with low masses would be in thermal equilibrium in the early universe and, as relativistic particles, they would constitute so-called *hot dark matter*. With hot dark matter it would be difficult to understand the formation of structure on short scales. Also for this reason neutrinos alone could not be responsible for dark matter.

There is a possibility to further constrain the margin for neutrino masses in a completely different way. To contribute directly to the dark matter of a galaxy, neutrinos must be gravitationally bound to the galaxy, i.e., their velocity must be smaller than the escape velocity v_f. This allows to calculate a limit for the maximum momentum $p_{max} = m_\nu v_f$. If the neutrinos in a galaxy are treated as a free relativistic fermion gas in the lowest-energy state (at $T = 0$), one can derive from the Fermi energy

$$E_F = \hbar c (3\pi^2 n_{max})^{1/3} = p_{max} c \tag{13.3.19}$$

an estimation for the neutrino mass density (n_{max}—number density):

$$n_{max}\, m_\nu = \frac{m_\nu^4 v_f^3}{3\pi^2 \hbar^3}. \tag{13.3.20}$$

Since $n_{max}\, m_\nu$ must be at least on the order of magnitude of the typical density of dark matter in a galaxy, if one wants to explain its dynamics with neutrino masses, these arguments lead to a *lower* limit for the neutrino mass. Using $v_f = \sqrt{2GM/r}$, where M and r are galactic mass and radius, one obtains under plausible assumptions about the neutrino mass density and the size and structure of the galaxy

$$m_\nu > 1\,\mathrm{eV}. \tag{13.3.21}$$

Again, this argument is based on the assumption that neutrino masses might contribute substantially to the matter density of the universe. These cosmological arguments leave only a relatively narrow window for neutrino masses.

Cosmological arguments suggest that the sum of neutrino masses is less than 1 eV, therefore we have [256]

$$m_\nu \leq 1.0 \, \text{eV}. \tag{13.3.22}$$

Recent measurements of the Planck satellite limit the sum of the masses of the three neutrino flavours further to [257]

$$\Sigma m_{\nu_i} \leq 0.23 \, \text{eV}. \tag{13.3.23}$$

These considerations are not necessarily in contradiction to the interpretation of results on neutrino oscillations, because in that case one does not directly measure neutrino masses but rather the difference of their masses squared. From the deficit of atmospheric muon neutrinos one obtains

$$\delta m^2 = m_1^2 - m_2^2 = 3 \times 10^{-3} \, \text{eV}^2. \tag{13.3.24}$$

If $(\nu_\mu - \nu_\tau)$ oscillations are responsible for this effect, muon and tau neutrino masses could still be very close without getting into conflict with the cosmological mass limits. Only if the known mass hierarchy $(m_e \ll m_\mu \ll m_\tau)$ from the sector of charged leptons is transferred to the neutrino sector and if one further assumes $m_{\nu_\mu} \ll m_{\nu_\tau}$, the result $(m_{\nu_\tau} \approx 0.05 \, \text{eV})$ would be in conflict with certain cosmological limits (see, for example, (13.3.21)), but then the cosmological limits have been derived under the assumption that neutrino masses actually play an important role for the matter density in the universe, which is now known not to be the case.

Direct searches for neutrino masses or limits on neutrino masses (e.g., from tritium beta decay, like KATRIN [258]) presently aim at a mass sensitivity of 200 meV. Very recently KATRIN published a new result (September 2019) from the first science run with a limit of 1.1 eV for the antineutrino mass at 90% confidence level [259].

If light neutrinos would fill up the galactic halo, one would expect a narrow absorption line in the spectrum of high-energy neutrinos arriving at Earth. The observation of such an absorption line would be a direct proof of the existence of neutrino halos. Furthermore, one could directly infer the neutrino mass from the energetic position of this line. For a neutrino mass of 10 eV the position of the absorption line can be calculated from, see (3.1.5), $\nu + \bar{\nu} \to Z^0 \to$ to hadrons or leptons,

$$2m_\nu E_\nu = M_Z^2 \tag{13.3.25}$$

to be

$$E_\nu = \frac{M_Z^2}{2m_\nu} = 4.2 \times 10^{20} \, \text{eV}. \tag{13.3.26}$$

The verification of such an absorption line, which would result in a burst of hadrons or leptons from Z decay ('Z bursts') represents a substantial experimental challenge.

The fact that recent fits to cosmological data indicate that the contribution of neutrino masses to the total matter density in the universe is rather small ($\Omega_\nu \leq 10^{-4}$), makes the observation of such an absorption line rather unlikely.

13.3.3 Weakly Interacting Massive Particles (WIMPs)

In questions like this, truth is only to be had by laying together many variations of error.

Virginia Wolf

Baryonic matter and neutrino masses are insufficient to close the universe. A search for further candidates of dark matter must concentrate on particles, which are only subject to weak interactions apart from gravitation, otherwise one would have found them already. There are various scenarios, which allow the existence of weakly interacting massive particles (WIMPs). In principle, one could consider a fourth generation of leptons with heavy neutrinos. However, the LEP[5] measurements of the Z width have shown that the number of light neutrinos is exactly three (see Chap. 2, Fig. 2.1, and Sect. 10.6) so that for a possible fourth generation the mass limit

$$m_{\nu_x} \geq m(Z)/2 \approx 46\,\text{GeV} \tag{13.3.27}$$

holds. Such a mass, however, is considered to be too large to expect that a sizable amount of so heavy particles could have been created in the Big Bang.

An alternative to heavy neutrinos is given by WIMPs, which would couple even weaker to the Z than sequential neutrinos. Candidates for such particles are provided in supersymmetric extensions of the Standard Model. Supersymmetry is a symmetry between fundamental fermions (leptons and quarks) and gauge bosons (γ, W^+, W^-, Z, gluons, Higgs bosons, gravitons). In supersymmetric models all particles are arranged in supermultiplets and each fermion is associated with a bosonic partner and each boson gets a fermionic partner.

Bosonic quarks and leptons (called squarks and sleptons) are associated to the normal quarks and leptons. The counterparts of the usual gauge bosons are supersymmetric gauginos, where one has to distinguish between charginos (supersymmetric partners of charged gauge bosons: winos ($\widetilde{W}^+$, $\widetilde{W}^-$) and charged higgsinos ($\widetilde{H}^+$, $\widetilde{H}^-$)) and neutralinos (photino $\widetilde{\gamma}$, zino $\widetilde{Z}$, neutral higgsinos ($\widetilde{H}^0$, ...), gluinos $\widetilde{g}$, and gravitinos).

The non-observation of supersymmetric particles at accelerators means that supersymmetry must be broken and the superpartners obviously are heavier than known particles and not in the reach of present-day accelerators. The theory of supersymmetry appears to be at least for the theoreticians so aesthetic and simple that they expect it to be true. A new quantum number, the R parity, distinguishes normal particles from their supersymmetric partners. If the R parity is a conserved quantity—and this

---

[5]LEP—Large Electron–Positron Collider at CERN in Geneva.

is assumed to be the case in the most simple supersymmetric theories—the lightest supersymmetric particle (LSP) must be stable (it could only decay under violation of R parity). This lightest supersymmetric particle represents an ideal candidate for dark matter (Fig. 13.9).

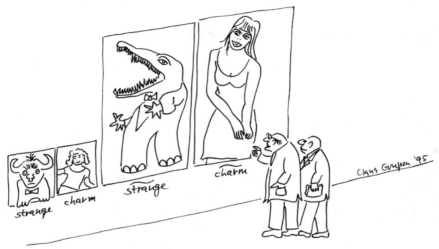

"These are the supersymmetric partners of *strange* and *charm*."

It was generally assumed that the Large Hadron Collider (LHC) at CERN had a fair chance to produce and to be able to successfully reconstruct the creation of supersymmetric particles. If plausible couplings of supersymmetric particles are assumed, and if it is further assumed that squarks and supersymmetric gluons have comparable masses, one can already exclude supersymmetric particles up to the mass range of about <1 TeV.

The low interaction strength of the LSP, however, is at the same time an obstacle for its detection. If, for example, supersymmetric particles would be created at an accelerator in proton–proton or electron–positron interactions, the final state would manifest itself by missing energy, because in decays of supersymmetric particles one lightest supersymmetric particle would always be created, which would leave the detector without significant interaction. Primordially produced supersymmetric particles would have decayed already a long time ago, apart from the lightest super-symmetric particles, which are expected to be stable if R parity is conserved. These would lose a certain amount of energy in collisions with normal matter, that could be used for their detection. Unfortunately, the recoil energy transferred to a target nucleus in a WIMP interaction (mass 10–100 GeV) is rather small, namely, in the range of about 10 keV. Still one tries to measure the ionization or scintillation pro-duced by the recoiling nucleus. Also a direct calorimetric measurement of the energy deposited in a bolometer is conceivable.

Fig. 13.9 Production and decay of supersymmetric particles as Feynman diagram (**a**) and in the detector (**b**)

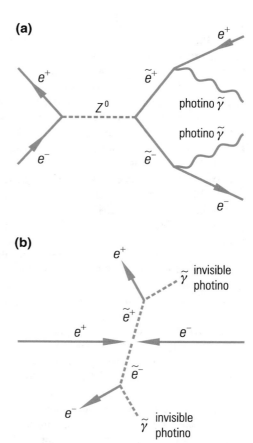

Because of the low energy ΔQ to be measured and the related minute temperature rise

$$\Delta T = \Delta Q / c_{sp} m \qquad (13.3.28)$$

(c_{sp} is the specific heat and m the mass of the calorimeter), these measurements can only be performed in ultrapure crystals (e.g., in sapphire) at extremely low temperatures (milli-kelvin range, $c_{sp} \sim T^3$). It has also been considered to use superconducting strips for the detection, which would change to a normal-conducting state upon energy absorption, thereby producing a usable signal.

Based on general assumptions on the number density of WIMPs one would expect a counting rate of at most one event per kilogram target per day. The main problem in these experiments is the background due to natural radioactivity and cosmic rays.

Due to their high anticipated mass WIMPs could also be gravitationally trapped by the Sun or the Earth. They would occasionally interact with the material of the Sun or Earth, lose thereby energy, and eventually obtain a velocity below the escape velocity of the Sun or the Earth, respectively. Since WIMPs and in the same way their antiparticles would be trapped, they could annihilate with each other and provide proton and antiproton or neutrino pairs. One would expect to obtain an equilibrium between trapping and annihilation rate.

The WIMP annihilation signal could be recorded in large existing neutrino detectors originally designed for neutrino astronomy, like ANTARES[6] or ICECUBE.

It is not totally inconceivable that particularly heavy WIMP particles could be represented by primordial black holes, which could have been formed in the Big Bang before the era of nucleosynthesis. They would provide an ideal candidate for cold dark matter. However, it is very difficult to imagine a mechanism by which such primordial black holes could have been produced in the Big Bang.

The Italian–Chinese DAMA[7] collaboration has published a result, which could hint at the existence of WIMPs. Similarly to our Sun, the Milky Way as a whole could also trap WIMPs gravitationally. During the rotation of the solar system around the galactic center the velocity of the Earth relative to the hypothetical WIMP halo would change depending on the season. In this way the Earth while orbiting the Sun would encounter different WIMP fluxes depending on the season. In the month of June the Earth moves against the halo ($\rightarrow$ large WIMP collision rate) and in December it moves parallel to the halo ($\rightarrow$ low WIMP collision rate).

The DAMA collaboration has measured a seasonal variation of the interaction rate in their 100-kg-heavy sodium-iodide crystal. The modulation of the interaction rate with an amplitude of 3% is interpreted as evidence for WIMPs. The results obtained in the Gran Sasso laboratory would favour a WIMP mass of 60 GeV. This claim, however, is in contradiction to the results of an American collaboration (CDMS[8]), which only observes a seasonal-independent background due to neutrons in their highly sensitive low-temperature calorimeter.

There have been many experiments over the years searching for heavy WIMP-like particles or dark-matter candidates. One of the most recent experiments on dark-matter detection with direct searches is the XENON1T detector installed in the Italian Gran Sasso laboratory. It consists of a ≈ 2000 kg liquid xenon time projection chamber (TPC). The recoil products of interactions of dark-matter particles in the liquid target are measured by scintillation and ionization in the liquid xenon TPC. The results so far have been negative. When no dark-matter particles are detected one can only give a limit on the mass-dependent cross section. This is shown in Fig. 13.10.

The results from this search show that the WIMP–nucleon cross section for the production of, say, 100 GeV WIMPs must be less than $\approx 10^{-46}$ cm^2.

---

[6]ANTARES—Astronomy with a Neutrino Telescope and Abyss environmental RESearch.
[7]DAMA—DArk MAtter search.
[8]CDMS—Cryogenic Dark Matter Search.

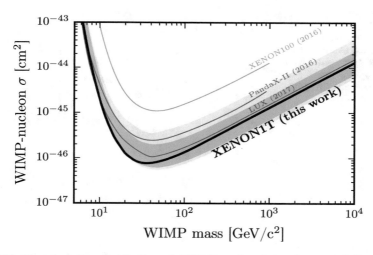

Fig. 13.10 The spin-independent limits on the WIMP–nucleon interaction cross section in their dependence on WIMP mass at 90% confidence level (*black*) for of XENON1T. In *green* and *yellow* are the 1σ and 2σ sensitivity bands. Results from other experiments like LUX (*red*), PandaX-II (*brown*), and XENON100 (*gray*) are also shown for reference [260]

13.3.4 Axions

For every complex natural phenomenon there is a simple, elegant, compelling, but wrong explanation.

Thomas Gold

Weak interactions not only violate parity P and charge conjugation C, but also the combined symmetry CP. The CP violation is impressively demonstrated by the decays of neutral kaons and B mesons. In the framework of quantum chromodynamics (QCD) describing strong interactions, CP-violating terms also originate in the theory. However, experimentally CP violation is not observed in strong interactions.

Based on theoretical considerations in the framework of QCD the electric dipole moment of the neutron should be on the same order of magnitude as its magnetic dipole moment. Experimentally one finds, however, that it is much smaller and even consistent with zero. This contradiction has been known as the so-called strong CP problem. The solution to this enigma presumably lies outside the Standard Model of elementary particles. A possible solution is offered by the introduction of additional fields and symmetries, which eventually require the existence of a pseudoscalar particle, called axion. The axion is supposed to have similar properties as the neutral pion. In the same way as the π^0 it would have a two-photon coupling and could be observed by its two-photon decay or via its conversion in an external electromagnetic field (Fig. 13.11).

Theoretical considerations appear to indicate that the axion mass should be somewhere between the μeV and meV range. To reach the critical density of the universe

Fig. 13.11 Coupling of an
axion to two photons via a
fermion loop (**a**); photons
could also be provided by an
electromagnetic field for
axion conversion (**b**) [261]

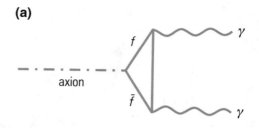

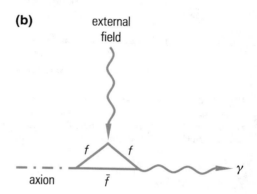

with axions only, the axion density—assuming a mass of 1 μeV—should be enormously high ($>10^{10}$ cm^{-3}). Since the conjectured masses of axions are very small, they must possess non-relativistic velocities to be gravitationally bound to a galaxy, because otherwise they would simply escape from it. For this reason axions are considered as cold dark matter.

As a consequence of the low masses the photons generated in the axion decay are generally of low energy. For axions in the preferred μeV range the photons produced by axion interactions in a magnetic field would lie in the microwave range. A possible detection of cosmological axions would therefore involve the measurement of a signal in a microwave cavity, which would significantly stand out above the thermal noise. Even though axions appear to be necessary for the solution of the strong *CP* problem and therefore are considered a good candidate for dark matter, all experiments on their search up to date gave negative results.

It is difficult to give a limit for a possible axion mass. To do this a number of assumptions on the coupling of the axions to photons has to be made. Also the supposed sources of the axions enter into this limit. For axions from the Sun the CAST experiment (CERN Axion Solar Telescope) at CERN gives a 95% limit for axion masses of $m_a \leq 0.02$ eV [262] for an axion–photon coupling of $g_{a\gamma} < 1.16 \times 10^{-10}$ GeV^{-1}.

13.3.5 The Role of Dark Matter and the Vacuum Energy Density

Sometimes I think that a vacuum is a hell of a lot better than some of the stuff that nature replaces it with.

Tennessee Williams

To obtain a flat universe a non-zero cosmological constant is required, which drives the exponential expansion of the universe. The cosmological constant is a consequence of the finite vacuum energy. This energy could have been stored originally in a false vacuum, i.e., a vacuum, which is not at the lowest-energy state. The energy of the false vacuum could be liberated in a transition to the true vacuum (see Sect. 12.5).

Paradoxically, a non-zero cosmological constant was introduced by Einstein as a parameter in the field equations of the theory of general relativity to describe a static universe, which was popular at that time and to prevent a dynamic behaviour, which followed from his theory. Now it appears that the dominant energy, which determines the dynamics of the universe, is stored in the empty space itself.

There is a fundamental difference between dark matter and the effect of the cosmological constant Λ.

The potential energy of a test mass m created by matter and the vacuum energy density can easily be written after Newton to be

$$
\begin{aligned}
E_{\text{pot}} &= -G\frac{m\,M_{\text{matter}}}{R} - G\frac{m\,M_{\text{vacuum energy}}}{R} \\
&\sim -\frac{\varrho_{\text{matter}}\,R^3}{R} - \frac{\varrho_{\text{vacuum}}\,R^3}{R} .
\end{aligned}
\tag{13.3.29}
$$

There is a fundamental difference between the matter density and the vacuum energy density during the expansion of the universe. For the vacuum-energy term in (13.3.29) the vacuum energy *density* remains constant, since this energy density is a property of the vacuum. In contrast to this the matter density does not remain constant during expansion, since only the *mass* is conserved leading to the dilution of the matter density. Therefore, the spatial dependence of the potential energy is given by

$$
E_{\text{pot}} \sim -\frac{M_{\text{matter}}}{R} - \varrho_{\text{vacuum}}\,R^2 .
\tag{13.3.30}
$$

Since $\varrho_{\text{vacuum}} \sim \Lambda$, (13.3.30) shows that the radial dependence of the mass term is fundamentally different from the term containing the cosmological constant. Therefore, the question of the existence of dark matter (M_{matter}) and a finite vacuum energy density ($\Lambda \neq 0$) are not trivially coupled. Furthermore, Λ could be a dynamical constant, which is not only of importance for the development of the early universe. Since the Λ term in the field equations appears to dominate also today, one would expect an accelerated expansion. The experimental determination of the acceleration parameter could provide evidence for the present effect of Λ. To do this, one would have to compare the expansion rate of the universe in earlier cosmological times with that of the present.

Such measurements have been performed with the Supernova Cosmology Project at Berkeley and with the High-z Supernova Search Team at Australia's Mount Stromlo Spring Observatory. The surprising results of these investigations was that the universe is actually expanding at a higher pace than expected (see also Sect. 8.7 and, in particular, Fig. 8.10). It is important to make sure that supernovae (SN Ia), upon which the conclusions depend, explode the same way now and at much earlier times so that these supernovae can be considered as dependable standard candles. This will be investigated by looking at older and more recent supernovae of type Ia. Involved in these projects surveying distant galaxies are the Cerro Tololo Interamerican Observatory (CTIO) in the Chilean Andes, the Keck Telescope in Hawaii, and for the more distant supernovae the Hubble Space Telescope (HST). Also the latest Sloan Digital Sky Survey (SDSS III) can throw light on the time dependence of the effect of dark energy.

13.3.6 Galaxy Formation

> There are at least as many galaxies in our observable universe as there are stars in our galaxy.
>
> Martin Rees

As already mentioned in the introduction to this chapter, the question of galaxy formation is closely related to the problem of dark matter. Already in the 18th-century philosophers like Immanuel Kant and Thomas Wright have speculated about the nature and the origin of galaxies. Today it seems to be clear that galaxies originated from quantum fluctuations, which have been formed right after the Big Bang.

With the Hubble telescope one can observe galaxies up to redshifts of $z = 12$ ($\lambda_{observed} = (1 + z)\lambda_{emitted}$); this corresponds to more than 90% of the total visible universe. The idea of cosmic inflation predicts that the universe is flat and expands forever, i.e., the Ω parameter is equal to unity.

The dynamics of stars in galaxies and of galaxies in galactic clusters suggests that less than $\approx$5% of matter is in form of baryons. Apart from the vacuum energy the main part of matter leading to $\Omega = 1$ has to be non-baryonic, which means it must consist of particles that do not occur in the Standard Model of elementary particle physics. The behaviour of this matter is completely different from normal matter. This dark matter interacts with other matter predominantly via gravitation. Therefore, collisions of dark-matter particles with known matter particles must be very rare. As a consequence of this, dark-matter particles lose only a small fraction of their energy when moving through the universe. This is an important fact when one discusses models of galaxy formation.

Candidates for dark matter are subdivided into 'cold' and 'hot' particles. The prototype of hot dark matter is the neutrino ($m_\nu \neq 0$), which comes in at least in three different flavour states. Low-mass neutrinos are certainly insufficient to close the universe. Under cold dark matter one normally subsumes heavy weakly interacting massive particles (WIMPs) or axions.

The models of galaxy formation depend very sensitively on whether the universe is dominated by hot or cold dark matter. Since in all models one assumes that galaxies have originated from quantum fluctuations, which have developed into larger gravitational instabilities, two different cases can be distinguished.

If the universe would have been dominated by the low-mass neutrinos, fluctuations below a certain critical mass would not have grown to galaxies because fast relativistic neutrinos could easily escape from these mass aggregations, thereby dispersing the 'protogalaxies'. Neutrinos as candidates for dark matter therefore would favour a scenario, in which first the large structures (superclusters), later clusters, and eventually galaxies would have been formed ('top–down' scenario). This would imply that galaxies have formed only for $z \leq 1$. However, there are indications that galaxy formation is a continuous process over cosmic time, even for larger values of the redshift [263]. This is also an argument to exclude a neutrino-dominated universe.

Massive, weakly interacting and mostly non-relativistic (i.e., slow) particles, however, will be bound gravitationally already to mass fluctuations of smaller size. If cold dark matter would dominate, initially small mass aggregations would collapse and grow by further mass attractions to form stars and galaxies. These galaxies would then develop galactic clusters and later superclusters. Cold dark matter therefore favours a scenario, in which smaller structures would be formed first and only later develop into larger structures ('bottom–up' scenario).

In particular, the COBE, WMAP, and Planck observations of the inhomogeneities of the 2.7 K radiation confirm the idea of structure formation by gravitational amplification of small primordial fluctuations. These observations therefore support a cosmogony driven by cold dark matter, in which smaller structures (galaxies) are formed first and later develop into galactic clusters (ΛCDM scenario—Cold Dark Matter and vacuum energy (Λ)).

The dominance of cold dark matter, however, does not exclude non-zero neutrino masses. The values favoured by the observed neutrino anomaly in the Super-Kamiokande and SNO experiments are compatible with a scenario of dominating cold dark matter. In this case one would have a cocktail—apart from baryonic matter—consisting predominantly of cold dark matter with a pinch of light neutrinos. Exact, very detailed computer-intensive simulations of structure formation in the early universe might be able to understand the formation of galaxies, galactic clusters, and voids. Generally the bottom–up scenario is favoured [264, 265].

13.3.7 Resume on Dark Matter and Dark Energy

> *Dark matter and dark energy are two things we measure in the universe that are making things happen, and we have no idea what the cause is.*
>
> Neil deGrasse Tyson

It is undisputed that large quantities of dark matter must be hidden somewhere in the universe. The dynamics of galaxies and the dynamics of galactic clusters can only be understood, if the dominating part of gravitation is caused by non-luminous matter. In theories of structure formation in the universe in addition to baryonic matter (Ω_b) also other forms of matter or energy are required. These could be represented by hot dark matter (e.g., light relativistic neutrinos: Ω_{hot}) or cold dark matter (e.g., WIMPs: Ω_{dark}). The effect of neutrinos and photons for structure formation can safely be ignored.

At the present time cosmologists prefer a mixture of these components. Recent measurements of distant supernovae and precise observations of inhomogeneities of the blackbody radiation lead to the conclusion that the dark energy, mostly interpreted in terms of the cosmological constant Λ, also has a very important impact on the structure of the universe. In general, the density parameter Ω can be presented as a sum of four contributions,

$$\Omega = \Omega_b + \Omega_{hot} + \Omega_{dark} + \Omega_\Lambda. \tag{13.3.31}$$

The present state of the art in cosmology gives $\Omega_b \approx 0.05$, $\Omega_{hot} \leq 10^{-4}$, $\Omega_{dark} \approx 0.26$, and $\Omega_\Lambda \approx 0.69$, ($\Omega_{matter} = \Omega_M = \Omega_{dark} + \Omega_b$).

Figure 13.12 shows a compilation of results in the Ω_M, Ω_Λ plane from a supernova Ia compilation, the cosmological background radiation, the large-scale galactic distribution (labeled BAO—baryon acoustic oscillations) under the assumption of a flat universe, where the error bands are just statistical. A cosmological constant (dark energy) with $w = -1$ has been assumed [266]. The overlap region for the different results confirms the values given above. If Ω_Λ were much larger than Ω_M there would have been no Big Bang, because the repulsive force would then be much larger than the attractive force.

As demonstrated by observations of distant supernovae, the presently observed expansion is accelerated. Therefore, it is clear that the cosmological constant plays a dominant role even today. Even if the repulsive force of the vacuum were small compared to the action of gravity, it will eventually win over gravity if there is enough empty space, see also (13.3.30). The present measurements (status 2018) indicate a value for Ω_Λ of ≈ 0.7. This means that the vacuum is filled with a weakly interacting hypothetical scalar field, which essentially manifests itself only via a repulsive gravitation and negative pressure. There are models (e.g., 'quintessence', see glossary), in which a field produces a dynamical energy density of the vacuum in such a way that the repulsive gravity plays an important role also today.

The temperature variations of the cosmological background radiation throw light on the structure of the universe about 380 000 years after the Big Bang when the universe became transparent. The universe before that time was too hot for atoms to form and photons were trapped in a sea of electrically charged particles. Pressure waves pulsated in this sea of particles like sound waves in water. The response of this sea to the gravitational potential fluctuations allows to measure the properties of this fluid in an expanding universe consisting of ordinary and dark matter. When the universe cooled down, protons and electrons recombined to form electrically neutral hydrogen atoms, thereby freeing the photons. These hot photons have cooled since then to a temperature of around 2.7 K at present. Temperature variations in

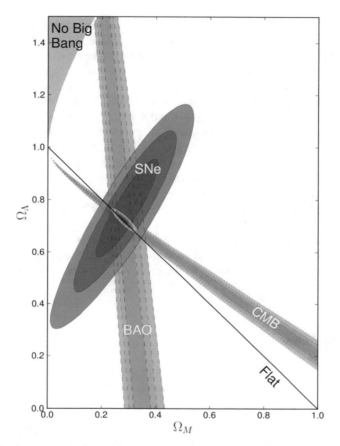

Fig. 13.12 68.3, 95.4, and 99.7% confidence regions in the Ω_M, Ω_Λ plane using a supernova Ia compilation, results from the cosmological background radiation, the large-scale galactic distribution (labeled BAO—baryon acoustic oscillations) under the assumption of a flat universe. A cosmological constant for dark energy with $w = -1$ has been assumed. The error bands are just statistical errors [266]

this radiation are fingerprints of the pattern of sound waves in the early universe, when it became transparent. The size of the spots in the thermal map is related to the curvature or the geometry of the early universe and gives information about the energy density. The satellite experiments COBE, WMAP, and Planck, as well as Boomerang and Maxima, observed temperature clusters of a size of about one degree across. This information is experimentally obtained from the power spectrum of the primordial sound waves in the dense fluid of particles (see Chap. 11). The observed power spectrum corresponding to characteristic cluster sizes of the temperature map of about 1° indicates—according to theory—that the universe is flat. If the results of the supernova projects (see Sect. 8.3), the COBE, WMAP, and Planck measurements, and the Boomerang and Maxima data are combined, a value for the cosmological constant corresponding to $\Omega_\Lambda \approx 0.7$ is favoured. Such a significant contribution would mean that the vacuum is filled with an incredibly weakly interaction substance, which reveals itself only via its repulsive gravitation (e.g., in terms of a quintessence model). The suggested large contribution of the vacuum energy to the Ω parameter would naturally raise the question, whether appreciable amounts of exotic dark matter (e.g., in the form of WIMPs) are required at all. However, it appears that for the understanding of the dynamics of galaxies and the interpretation of the fluctuations in the blackbody temperature a contribution of classical, non-baryonic dark *matter* corresponding to $\Omega_{\text{dark}} \approx 0.26$ is indispensable.

In this scenario the long-term fate of the universe is characterized by eternal accelerated expansion. A Big Crunch event as anticipated for Ω larger than unity is ruled out. Since the nature of the dark energy is essentially unknown, also its long-term properties are not understood. Therefore, one must be prepared for surprises. For example, it is possible that the dark energy becomes so powerful that the known forces are insufficient to preserve the universe. Galactic clusters, galaxies, planetary systems, and eventually atoms would be torn apart. In such theories one speculates about an apocalyptic Big Rip, which would shred the whole physical structure of the universe at the end of time.

13.4 Summary

For the moment we might very well call them DUNNOS (for Dark Unknown Nonreflective Nondetectable Objects Somewhere).

Bill Bryson

The analysis of the geometry of the universe suggests that it is flat, which means $\Omega = 1$. If the masses of the *Sun, the Moon, and the stars* and other baryonic matter is added one only finds a value of the mass density $\Omega_b \approx 0.05$; i.e., a relatively small fraction of the total mass/energy density. This had already been realized by the Swiss–American physicist Fritz Zwicky in the thirties of the last century, and later by the American astronomer Vera Rubin. They pioneered the work on galaxy rotation rates and provided evidence for the existence of dark matter. In addition, the observed large rotational velocities of stars at large distances from galactic centers also did not follow the rules found in other orbital systems with most of their mass at their center. The only way to prevent the stars in the cluster from flying apart was to assume that the cluster must contain large quantities of some invisible gravitational dark matter. The nature of this attractive dark matter, which keeps the stars bound to the galaxy, is unknown. In some models of particle physics supersymmetric particles are considered as candidates for this enigmatic dark matter, but there are no hints from accelerators whether such particles do exist. Even worse, dark energy, the energy of empty space with its repelling gravitation, which seems to be essential for the understanding of the expansion of the universe, leaves us completely in the dark.

13.5 Problems

1. In the vicinity of the galactic center of the Milky Way celestial objects have been identified in the infrared and radio band, which appear to rotate around an invisible center with high orbital velocities. One of these objects circles the galactic center with an orbital velocity of $v \approx 110\,\text{km/s}$ at a distance of approximately 2.5 pc. Work out the mass of the galactic center assuming Keplerian kinematics for a circular orbit!

2. What is the maximum energy that a WIMP ($m_W = 100\,\text{GeV}$) of 1 GeV kinetic energy can transfer to an electron at rest?

3. Derive the Fermi energy of a classical and relativistic gas of massive neutrinos. What is the Fermi energy of the cosmological neutrino gas?

4. If axions exist, they are believed to act as cold dark matter. To accomplish the formation of galaxies they must be gravitationally bound; i.e., their velocities should not be too high.
   Work out the escape velocity of a $1\,\mu\text{eV}$-mass axion from a protogalaxy with a nucleus of 10^{10} solar masses of radius 3 kpc!

5. Consider a (not highly singular) spherically symmetric mass density $\varrho(r)$, the total mass M_t of which is assumed to be finite.

   (a) Determine the potential energy of a test mass m in the gravitational field originating from $\varrho(r)$. The force is given by Newton's formula, in which only the mass $M(r)$ inside that sphere enters, whose radius is given by the position of m.

   (b) To calculate the potential energy of the full distribution, determine the potential energy of a test mass m at r for a distribution that is cut off outside the radius r and replace m by $dM = M'(r)\,dr$ to build the whole expression by r integration, i.e., by adding spherical shells successively. Show with an integration by parts that the total potential energy can be written as

   $$E_{\text{pot}} = -\frac{G}{2} \int_0^\infty \frac{M^2(r)}{r^2}\, dr \;.$$

   (c) Determine the potential energy of a massive spherical shell of radius R and mass M, i.e., $M(r) = 0$ for $r < R$ and $M(r) = M$ for $r > R$.

   (d) Work out the potential energy of a sphere of radius R and mass M with homogeneous mass density.

(This problem is a little difficult and tricky.)

6. Motivate the lower mass limit of neutrinos based on their maximum escape velocity from a galaxy (13.3.21), if it is assumed that they are responsible for the dynamics of the galaxy.

7. In the discussion on the search for MACHOs there is a statement that the radius of the Einstein ring varies as the square root of the mass of the deflector. Work out this dependence and determine the ring radius for

- distance star–Earth = 55 kpc (LMC),
- distance deflector–Earth = 10 kpc (halo),
- Schwarzschild radius of the deflector = 3 km (corresponding to the Schwarzschild radius of the Sun).

8. In a cryogenic argon calorimeter ($T = 1.1$ K, $m = 1$ g) a WIMP may deposit 100 keV. By how much does the temperature rise in the calorimeter (specific heat of argon $c_{Ar} = 8 \times 10^{-5}$ J/(g K))?

Chapter 14
Astrobiology

*I believe in aliens. I think it would be way too selfish of us as
mankind to believe we are the only lifeforms in the universe.*

Demi Lovato

Astrobiology is a truly interdisciplinary science, which resorts to many scientific
disciplines. In the early days of space travel this field was also named exobiology. The
advanced instrumentation of certain aspects of astrobiology (Kepler satellite finder
and the Transiting Exoplanet Survey Satellite TESS) are mainly based on techniques
used for astroparticle physics experiments. Also the interpretation of the results
on exoplanets relies on techniques known from classical astroparticle experiments.
Therefore, there is a substantial overlap of astrobiology with astroparticle physics.
There are also implications of the laws of particle physics on potential life in other
parts of the universe or in a possible multiverse. Because of these common aspects
of astrobiology and astroparticle physics, a short section on astrobiology seems to
be appropriate and justified to be included in the present book.

14.1 Extrasolar Planets

*To consider the Earth as the only populated world in infinite space
is as absurd as to assert that in an entire field of millet, only one
grain will grow.*

Metrodoros of Chios, fourth century before Christ.

Since the discovery of extrasolar planets in 1995 by Mayor and Queloz (Nobel Prize
2019) this discipline has created a lot of interest. With the help of new observational
methods and modern satellite techniques a large number of planets has been discovered in other stellar systems. Meanwhile about 4000 extrasolar planets are known,
where—in particular—the Kepler space telescope, which was launched in 2009 by

© Springer Nature Switzerland AG 2020

C. Grupen, *Astroparticle Physics*, Undergraduate Texts in Physics,
https://doi.org/10.1007/978-3-030-27339-2_14

NASA as planet hunter, was especially successful. The majority of these planets were gas giants, which are relatively easy to find, but there were also rocky planets like our Earth. Out of the so far found terrestrial-like planets, about 100 lie in habitable zones. Among those the exoplanet Proxima Centauri b is the closest at a distance of only 4.2 light-years. Proxima Centauri b orbits a red dwarf star in only 11.2 days and seems to lie in a habitable zone. This planet in our cosmic neighbourhood might also be reached by robot expeditions even within this century.

The exoplanet Kepler-452b, discovered in 2015, is one of the smallest planets to date orbiting the star Kepler-452 in a habitable zone. Kepler-452 is a G-type star, like our Sun, and its planet is at a similar distance from its star that Earth is from the Sun. Kepler-452b orbits its mother star in 385 days, also similar to the Earth. The distance to this stellar system is 430 pc, corresponding to 1400 light-years.

Exoplanets are searched for with different techniques. These methods include the measurement of the radial velocity using the Doppler effect, the transit method, which takes advantage of the partial eclipse of the star when the planet crossed its host star, microlensing, and the direct observation by observing of the light arrival times, and by the brightness variation along the orbit (see also Fig. 14.1).

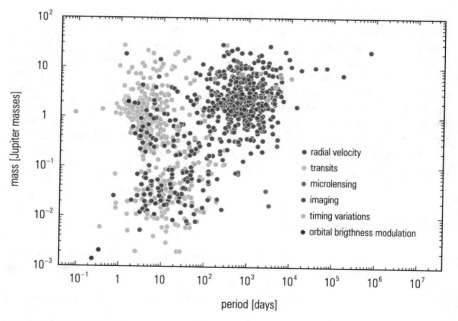

Fig. 14.1 Distribution of masses and orbital periods of planets found up to 2015. The different detections techniques used are (a) radial velocity (via the Doppler effect), (b) transit method, (c) microlensing, (d) imaging (direct observation), (e) variation of light arrival times (the center of mass of the star system with planets is shifted under the influence of gravitational forces), (f) brightness excursion along the orbit [267]

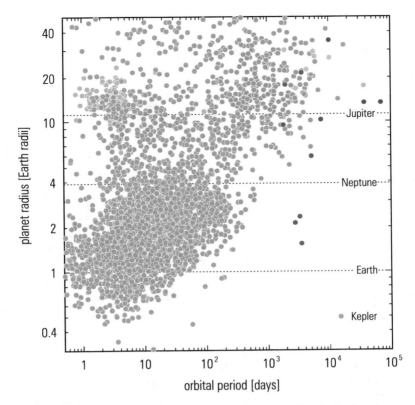

Fig. 14.2 Distribution of radii and orbital times of exoplanets found by the Kepler telescope until 2015. A comparison with Jupiter, Neptune, and the Earth is indicated. Among those exoplanets found quite a number are Earth-like [268]

While Fig. 14.1 includes predominantly heavy and large planets, the Kepler telescope (see Fig. 14.2) has found a multitude of low-mass, Earth-like planets.

The Kepler satellite was slowly running out of fuel and has been replaced by TESS (The Transiting Exoplanet Survey Satellite). TESS was launched in April 2018. The new satellite will cover 85% of the sky (Kepler only observed 0.25% of it), and it will monitor more than 500 000 stars looking for temporary drops in brightness caused by planetary transits.

Figure 14.3 shows the detection of an exoplanet using gravitational microlensing. If a foreground star, which serves as a deflector, passes the line of sight of a background star, the apparent brightness of the background star increases by gravitational microlensing. Depending on the mass of the foreground star this leads to a Gaussian brightness profile. If the foreground star has a planet, it also works as additional deflector and produces an extra, even small, intensity excursion of the background star. This extra intensity excursion is seen as spike on the Gaussian profile.

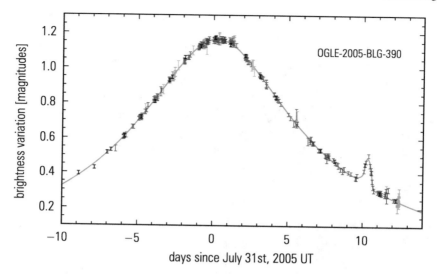

Fig. 14.3 Detection of an exoplanet by gravitational microlensing. The *full line* describes the microlensing event of OGLE-2005-BLG-390. The measurements have been obtained by six different experiments. The small brightness excursion around day 10 hints at an Earth-like planet of five Earth masses, where the mass was estimated from the intensity profile of the spike [269]

To estimate the Earth similarity index of an exoplanet, one has been guided by the properties of our Earth. What is needed is a main-sequence star like our Sun, which provides an energy flux comparable to our solar constant. For possible evolution of life one would like temperatures, which allow water to exist in liquid form. One believes to ask also for an atmosphere similar to ours. Also a local cosmic environment similar to our solar system seems to be important. Earth and Moon are a well-balanced pair. Without the stabilizing effect of the Moon it would be difficult to develop a constant climate on Earth. The large and massive nearby planet Jupiter prevents Earth from frequent impacts of comets and asteroids. Jupiter has banned these cosmic missiles into the asteroid belt beyond Mars. There are a number of reasons, which enabled life to be formed on Earth. Carl Sagan has argued that given the huge number of stars in our galaxy and beyond, there should be much room for the evolution of life, and life could be a common phenomenon in the universe.

Recently the detailed analysis of X rays emitted from a black hole has revealed the existence of even extragalactic planets in a galaxy 3.8 billion light-years away located between the black hole and Earth. The X rays from the black hole are characteristically distorted by massive objects in the distant galaxy. The microlensing pattern of X rays observed by the Chandra satellite is not uniform, and it can be interpreted as originating from massive objects like stars. Short time variations of the brightness excursions for particular objects seem to point also at the existence of small bodies of the size between the Moon and larger planets like Jupiter [270].

This result—even not unexpected—seems to enhance the probabilities of finding life in distant galaxies and planets in those galaxies.

With about 200 billion stars in our Milky Way and about 100 billion galaxies in our universe, there are about 10^{22} stars in total. Recent searches for exoplanets seem to suggest that every star has some planets, out of which there should be a reasonable number with habitable conditions. Opinions of what 'reasonable' means differ, however, substantially. Sagan assumed a probability of one per mill to obtain Earth-like living conditions on a planet. This would lead to about 10^{19} habitable planets. Possible visits to these planets would encounter substantial problems due to the large average distances, the limited lifetime of civilizations, and the impossibility to exceed the speed of light in space travel. Upon closer look one has realized that there are further particularities, which are essential for the evolution and development of life. Therefore, the estimates for the number of civilizations in our universe vary between the optimistic number of 10^{19} and zero.

However, a recent observation (September 2019) of water vapor on an habitable exoplanet orbiting a dim red-dwarf star K2-18 in the Leo constellation at a distance of 110 light-years indicates progress in the search for extraterrestrial life in our galaxy [271].

14.2 Extremophiles

Life has evolved to thrive in environments that are extreme only by our limited human standards: in the boiling battery acid of Yellowstone hot springs, in the cracks of permanent ice sheets, in the cooling waters of nuclear reactors, miles beneath the Earth's crust, in pure salt crystals, and inside the rocks of the dry valleys of Antarctica.

Jill Tarter

If people think of extraterrestrial life or aliens they usually image some humanoid beings. It is not even clear what actually 'life' is. There are life forms, which exist under extreme environments. Take the water bear, which belongs to the species of tardigrades. The water bear can live without water for months and even years, they can be frozen to low temperatures. They shut down their metabolism completely, and one would consider them according to human standards as dead. They are in a state of cryptobiosis. With just a little water they return to life, and it appears that they have not aged. Cryptobiosis is also known from nematodes and rotifers. The discovery of cryptobiosis has created some thought about the nature and definition of life. Apart from these amazing life forms there are also many other forms of so-called extremophiles. There are, for example, certain microorganisms, which can tolerate very high doses of ionizing radiation. Bacteria of the species Deinococcus radiodurans can withstand radiation doses up to 20 000 sievert (for humans 4.5 sievert are lethal). The bacterium Deinococcus radiophilus—as the name already indicates— even loves high doses of radiation and is able to repair up to 10 000 strand breaks of the DNA. In passing one should mention that these bacteria have gained some interest for data security. Obviously Deinococcus radiodurans is able to effectively cure radiation hazards. This is possible because there exists for each DNA strand a backup copy. Therefore, this bacterium is also of interest for IT industry. One takes advantage of the nucleic acids adenine, guanine, cytosine, and thymine (A, G, C, T) to transfer information into the genetic code of Deinococcus radiodurans. Different combinations of A, G, C, T will code a certain letter. Corresponding DNA parts are built up and are introduced into the genetic make-up of this bacterium. The encoded information will survive a number of generations in this bacterium, and should be able to be read out even after decades. To prevent mutations multiple copies are created. There is, however, the disadvantage that the readout of the biologically stored information erases the memory.

The bacterium Cryptococcus neoformans can even benefit from radiation energy for its growth by using its pigment melanin. This bacterium seems to be able—similar to plants—to exploit gamma-ray energies for their growth. Cryptococcus neoformans also prefers high radiation doses and is about to penetrate into the Chernobyl reactor, because it finds there tasty food.

Apart from these exotic radiophiles there are also life forms, which can withstand extremely low temperatures (cryophiles) or even prefer to live under high pressure, for example, in the Mariana Trench at 1000 atmospheres. Other life forms can also stand extreme heat, like in black smokers, or very poisonous environments. In addition, some organisms can survive in nutrient-poor surroundings, live without oxygen, or even live inside rocks.

All this indicates that there are life forms, which can live and survive in space for very long times. Frequently it is discussed that life on Earth could have been imported by comets from space (panspermia).

This raises the question of the conditions, which are necessary for the emergence of life in the universe, and which alternatives exist to the life forms that we know from Earth. In cosmology it is also discussed that our universe might be embedded in a multiverse. Probably our fantasy is insufficient to think of possible life forms in environments, which are completely different and alien from what we can imagine. Another possibility is that the laws of nature must not necessarily be the same in other universes.

14.3 Finely Tuned Parameters of Life

> *It is either coincidence piled on top of coincidence, or it is deliberate design.*
>
> Robert J. Sawyer

Apart from considering the physical development of our universe, it is interesting to speculate about alternatives to our universe, i.e., whether the evolution in a different universe might have led to different physical laws, other manifestations of matter, or to other forms of life. Einstein had already pondered about such ideas and had raised the question whether the necessity of logical simplicity leaves any freedom for the construction of the universe at all.

The numerous free parameters in the Standard Model of elementary particles appear to have values such that stable nuclei and atoms, in particular, carbon, which seems to be vital for the development of life, could be produced. α particles are formed in pp fusion processes. In $\alpha\alpha$ collisions ^{8}Be could be formed, but ^{8}Be is highly unstable and does not exist in nature. To reach ^{12}C, a triple collision of three α particles has to happen to get on with the production of elements. This appears at first sight very unlikely. Due to a curious mood of nature, however, this reaction has a resonance-like large cross section. Also the coincidence of time scales for stellar evolution and evolution of life on planets is strange. In the early universe, hydrogen and helium were mainly created, whereas biology, as we know it, requires all elements of the periodic table. These other elements are provided in the course of stellar evolution and in supernova explosions. The long-lived, primordial, radioactive elements (^{238}U, ^{232}Th, ^{40}K) that have half-lives on the order of giga-years, appear to be particularly important for the development of life. Furthermore, a sufficient amount of iron must be created in supernovae explosions such that habitable planets can develop a liquid iron core. This is essential in the generation of magnetic fields, which shield life on the planets against lethal radiation from solar and stellar winds and cosmic radiation.

The oldest sedimentary rocks on Earth (3.9×10^9 years old) contain fossils of cells. It is known that bacteria even existed 3.5×10^9 years ago. Since the biological evolution requires extremely long time scales, the development of the higher forms of life demands stars with long lifetimes, such that the fuel of stars typically lasts for 10^{10} years.

The origin of life is a question of much debate. In addition to the theory of a terrestrial origin also the idea of extraterrestrial delivery of organic matter to the early Earth has gained much recognition. These problems are investigated in the framework of bioastronomy.

The many free parameters in the Standard Model of electroweak and strong inter- actions can actually be reduced to a small number of parameters, whose values are of eminent importance for the astrobiological development of the universe. Among these important parameters are the masses of the u and d quarks and their mass difference. Experiments in elementary particle physics have established the fact that the mass of the u quark (≈ 5 MeV) is smaller than that of the d quark (≈ 10 MeV). The neutron, which has a quark content udd, is therefore heavier than the proton (uud), and can decay according to

$$n \rightarrow p + e^- + \bar{\nu}_e. \tag{14.3.1}$$

If, on the other hand, m_u were larger than m_d, the proton would be heavier than the neutron, and it would decay according to

$$p \rightarrow n + e^+ + \nu_e. \tag{14.3.2}$$

Stable elements would not exist, and there would be no chance for the development of life as we know it. On the other hand, if the d quark were much heavier than 10 MeV, deuterium ($d = {}^2$H) would be unstable and heavy elements could not be created. This is because the synthesis of elements in stars starts with the fusion of hydrogen to deuterium,

$$p + p \rightarrow d + e^+ + \nu_e, \tag{14.3.3}$$

and the formation of heavier elements requires stable deuterium. Life could also not develop under these circumstances. Also the value of the lifetime of the neutron takes an important influence on primordial chemistry and thereby on the chances to create life (see also Sect. 10.4). There are many other parameters, which appear to be fine-tuned to the development of life in the form that we know.

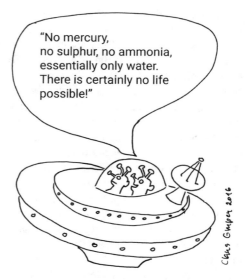

In his book 'Just six numbers' [272] the present Astronomer Royal, Martin Rees, mentions six parameters, which are crucial for the emergence and evolution of life. These parameters include the coupling constant of strong interactions, the Ω parameter, which represents the density parameter and determines the expansion of the universe, the cosmological constant Λ, and the number of spatial dimensions.

The fact that we live in a flat universe, i.e., the Ω parameter is very close to unity, is essential for our universe. Were Ω much larger than unity, the universe were collapsed long ago, and there would have been no time for the evolution of life. In this context the cosmological constant Λ describing a repulsive gravitation plays an important role. It is remarkable that Λ, just as attractive gravitation, is negligible on microscopic scales. The discrepancy between the low value of Λ in cosmology and the extremely large value of the vacuum energy in quantum field theories clearly shows, that essential ingredients are missing in the understanding in the unification of all interactions and the evolution of the universe.

An important problem in the framework of unified interactions is the low strength of gravitation. If, however, gravitation were much stronger, we would live—if at all— in a very short-lived universe. No creatures could grow larger than a few centimeters, and there were very little time for biological evolution.

Another critical parameter, also mentioned by Martin Rees, is the number of dimensions. It is possible to formulate superstring theories in eleven dimensions— out of which seven are rolled up to tiny dimensions, i.e., they are compacted, so that we live in a world of only three spatial and one time dimension. In two spatial dimensions no life in the form that we know is possible. Two-dimensional creatures would simply fall apart. What has singled out our case?

Also the efficiency of energy production in stars is important for the formation of chemical elements. If the effectiveness for mass-to-energy conversion in nuclear fusion were larger than 0.7%, stars would exhaust their hydrogen supply in a much shorter time compared to that needed for biological evolution.

The question whether the universe is a grand concept based on an intelligent design is discussed in a very controversial way. In the idea of dysteleology (this describes the non-purposeful, inexpedient, or even inappropriate creation) a number shortcomings in the design of creatures are listed. Already Oscar Wilde suspected "I think God, in creating man, somewhat overestimated his ability." Carl Sagan argued "The universe seems neither benign nor hostile, merely indifferent." and he also said "The universe is not required to be in perfect harmony with human ambition." One actually has to admit that the universe in many respects is inhospitable: it is practically empty, one is either irradiated by high-energy cosmic particles, frozen to very low temperatures, or burned in the vicinity of stars [273]. The physicist Robert L. Park has also criticized the theistic interpretation of the fine-tuning of the parameters [274]: "If the universe was designed for life, it must be said that it is a shockingly inefficient design. There are vast reaches of the universe in which life as we know it is clearly impossible: gravitational forces would be crushing, or radiation levels are too high for complex molecules to exist, or temperatures would make the formation of stable chemical bonds impossible. Fine-tuned for life? It would make more sense to ask why God designed a universe so inhospitable to life."

The physicist Victor Stenger also objected to the fine-tuning, and especially to theist use of fine-tuning arguments [275]. His numerous criticisms included what he called "the wholly unwarranted assumption that only carbon-based life is possible."

14.4   Multiverses and the Anthropic Principle

In an infinite multiverse, there is no such thing as fiction.

Scott Adsit

It has become clear that the fine-tuning of parameters in the Standard Model of elementary particle physics and cosmology is very important for the development of stars, galaxies, and life. If some of the parameters that describe our world were not finely tuned, then our universe would have grossly different properties. These different properties would have made unlikely that life—in the form we know it—would have developed. Consequently, physicists would not be around to ask the question why the parameters have exactly the values that they have. Our universe could have been the result of a selection effect on the plethora of universes in a multiverse. It is also conceivable that there could exist even a diversity of physical laws in other universes. Only in universes, where the conditions allow life to develop, questions about the specialness of the parameters can be posed. As a consequence of this *anthropic principle* there is nothing mysterious about our universe being special. It might just be that we are living in a most probable universe, which allows life to develop.

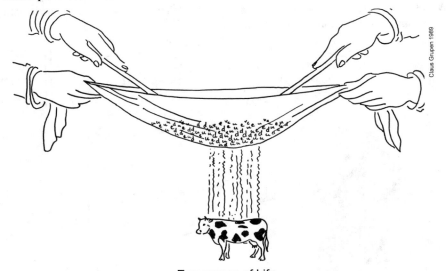

Emergence of Life

Nevertheless, it is the hope of particle theorists and cosmologists that a Theory of Everything can be found such that all sensitive parameters are uniquely fixed to the values that have been found experimentally. Such a theory might eventually also require a deeper understanding of time. To find such a theory and also experimental verifications of it, is the ultimate goal of cosmoparticle physics.

This is exactly what Einstein meant when he said: "What I am really interested in, is whether God could have made the world in a different way; that is, whether the necessity of logical simplicity leaves any freedom at all."

14.5 Summary

Man, who is he? Too bad, to be the work of God: Too good for the work of chance!

Doris Lessing

Astrobiology, in the past also called exobiology, is a new field that covers a broad domain of astroparticle physics. In particular, since the discovery of meanwhile about 4000 extrasolar planets this field has gained a lot of attention. Which conditions must be satisfied to allow the formation of life in general, or which special provisions must be fulfilled to enable life on other celestial bodies? Already on Earth we know extremophiles, which live in incredible environments, at high temperatures, under high pressures, and substantial radiation hazards, or even under conditions without oxygen. However, life, as we know it, requires finely tuned parameters in particle physics and gravitation. These constraints should also be valid for extrasolar planets. Whether or not these requirements also hold for other universes, is an unanswered or presumably even unanswerable question.

Open Question

14.6 Problems

1. Estimate the maximum stable size of a human in the Earth's gravitational field.
   What would happen to this result if the acceleration due to gravity would double?
   Hint: Consider that the weight of a human is proportional to the cube of a typical dimension, while its strength only varies as its cross section.
2. What are the conditions to synthesize carbon in stars?
   Hint: The detailed, quantitative solution of this problem is difficult. To achieve carbon fusion, one has to have three alpha particles almost at the same time in the same place. Consequently, high alpha-particle densities are required and high temperatures to overcome the Coulomb barriers. Also the cross section must be large. Check with [276] and [277].
3. What determines whether a planet has an atmosphere?

Hint: Gas atoms or molecules will be lost from an atmosphere if their average velocity exceeds the escape velocity from the planet.

4. Rotationally invariant long-range radial forces: (non-relativistic) two-body problem in n dimensions.[1]

   (a) How does such a force F between two bodies depend on the distance between two point-like bodies?
   Recall from three dimensions how the force depends on the surface or the solid angle of a sphere, respectively ('force flow'). The (hyper)surface $s_n(r)$ of the n-dimensional (hyper)sphere can be written as $s_n(r) = s_n r^{g(n)}$, where $g(n)$ needs to be fixed.

   (b) What is the corresponding potential energy and how does the effective potential (which includes the centrifugal barrier) look like? Find conditions for stable orbits with attractive forces.

   (c) Consider the introduction of a field-energy density $w \sim F^2$ (mediated by a field strength $\sim F$) and integrate w within the range of the two radii λ and Λ. The n-dimensional volume element is given by $dV_n = s_n(r) dr$, see also part (a). At what n does this expression diverge for $\lambda \to 0$ or $\Lambda \to \infty$? If quantum corrections are supposed to compensate the divergences, at what limit should they become relevant?

5. The Drake equation tries to estimate the number of 'intelligent' civilisations in our galaxy,

$$N = R_* \cdot f_p \cdot f_e \cdot f_l \cdot f_i \cdot f_t \cdot L,$$

where R_* is the star formation rate in our galaxy per year, f_p is the fraction of those stars that have planets, f_e is the fraction of habitable planets, f_l is the fraction of habitable planets that might develop life, f_i is the fraction of planets developing life that might develop intelligent life, f_t is the fraction of planets developing intelligent life with the ability to send signals, L is the lifetime of those civilisations that send suitable signals.
Estimate the number of 'intelligent' planets in our galaxy using reasonable inputs!

6. An astronaut is landed on an asteroid and left there. The asteroid's radius is r, and its density ρ is assumed to be equal to that of the Earth. What is the limit on the size of the asteroid such that the astronaut can reach the escape velocity from its surface to reach the spacecraft in its orbit, when he managed, as a former high jumper to jump on Earth, to lift his center of mass by 1.2 m against the pull of the Earth's gravitational field? Here any complications like the limitations of mobility in a spacesuit shall not be taken into account.

7. An extraterrestrial alien on a planet orbiting Proxima Centauri (distance 4.26 light-years) tries to reach a possible nearby civilization by using a 1-GW transmitter: He sends an isotropic signal with a bandwidth of 100 kHz. What kind of flux density would we receive on Earth using a circular radio antenna of 300 m diameter?

---

[1] This problem discusses unconventional dimensions. It is quite tricky and mathematically demanding.

8. Mass extinctions in Africa between the Pliocene and the Pleistocene might have indicated that a relatively nearby supernova had exploded at a distance of very approximately 150 light-years. The origin of a supernova for this mass extinction was suggested by an unusually high abundance of the isotope ^{60}Fe in sedimentary rocks based on geochronic geology. The annual radiation dose of a supernova explosion at a distance of 150 light-years due to the shower of cosmic rays caused by it in the atmosphere is estimated to be equivalent to the annual dose limit of radiation workers (20 mSv). This is considered to be not enough to completely kill most of the species. The estimated so-called 'kill-zone' for a supernova explosion is regarded to be about 25 light-years.

   It is assumed that only supernovae can synthesize the radioactive isotope ^{60}Fe in quantities that had been found earlier in rocks and also recently (2019) in the antarctic ice. It is considered that other sources of ^{60}Fe like nuclear interactions of cosmic rays can be excluded. How can one estimate when this nearby supernova explosion has occurred?

Chapter 15
Outlook

My goal is simple. It is complete understanding of the universe: why it is as it is and why it exists at all.

Stephen Hawking

Astroparticle physics has developed from the field of cosmic rays. As far as the particle physics aspect is concerned, accelerator experiments have taken over and played a leading role in this field since the sixties. However, physicists have realized that the energies in reach of terrestrial accelerators represent only a tiny fraction of nature's wide window, and it has become obvious that accelerator experiments will never match the energies required for the solution of important astrophysical and cosmological problems.

Since the last edition of this book in 2005 essential progress has been made. The missing particle of the Standard Model supposed to be giving masses to the elementary fermions, the Higgs particle, has been discovered at the Large Hadron Collider (LHC) at CERN by the ATLAS and CMS experiments in 2012. Its mass is $\approx 125\,\text{GeV}$, and it was just missed earlier at the Large Electron–Positron Collider LEP, because it was not in kinematical reach of that storage ring. Peter Higgs and François Englert, who had developed together with Robert Brout, Gerald Guralnik, Carl R. Hagen and Tom Kibble the Higgs mechanism, were awarded the Nobel Prize in 2013. Whether this newly discovered heavy particle is the only missing object of the Standard Model, or whether it is just a member of a whole new family, needs to be investigated in further experiments. It could be that these conjectured new particles could also be found at the LHC at higher energies and higher luminosities.

Detailed measurements of solar and atmospheric neutrinos and the study of neutrino interactions at accelerators have established the oscillation model in the neutrino sector. The neutrino eigenstates created in weak interactions ν_e, ν_μ, and ν_τ are mixtures of the mass eigenstates ν_1, ν_2, and ν_3. The mass of neutrinos is still an open question. There is only a powerful experimental mass limit on the electron-

© Springer Nature Switzerland AG 2020
C. Grupen, *Astroparticle Physics*, Undergraduate Texts in Physics,
https://doi.org/10.1007/978-3-030-27339-2_15

type antineutrino from the measurement of the tritium beta decay of $\leq 2\,\text{eV}$. The neutrino oscillation experiments measure only the squared mass difference and not absolute masses. Therefore, they only provide a limit like $\delta(m_{ik}^2) \lesssim 10^{-3}\,\text{eV}^2$ for all $i, k = 1, 2, 3$. This implies that—given the mass limit from the tritium beta decay, and information from the cosmological background radiation—that also the masses of ν_μ and ν_τ, must be low ($\leq 2\,\text{eV}$).

In 2015 the LIGO experiment presented the data on the discovery of gravitational waves, a major step into a new domain of astronomy. So far only several mergers (11 events as of 2019) of pairs of black holes have been seen, however, in coincidence over a distance of about 3000 km, which gives comforting confidence to the discovery. The detailed signatures of the signals are in excellent agreement with expectation. Meanwhile the Italian gravitational-wave detector VIRGO has also confirmed LIGO's findings.

The Standard Model of particle physics is strengthened by the discovery of the Higgs particle. The electromagnetic and weak interactions unify to a common electroweak force at center-of-mass energies of around 100 GeV, the scale of $W^\pm$ and Z masses. Already the next unification of strong and electroweak interactions at the GUT scale of 10^{16} GeV is beyond any possibility to reach in accelerator experiments even in the future. This is even more true for the unification of all interactions, which is supposed to happen at the Planck scale, where quantum effects of gravitation become important ($\approx 10^{19}$ GeV).

These energies will never be reached, not even in the form of energies of cosmic-ray particles. The highest energy observed so far of cosmic-ray particles, which has been measured in extensive-air-shower experiments (3×10^{20} eV), corresponds to a center-of-mass energy of about 800 TeV in proton–proton collisions, which is about 50-fold the energy that has presently been reached at the Large Hadron Collider LHC. It has, however, to be considered that the rate of cosmic-ray particles with these high energies is extremely low. In the early universe ($< 10^{-35}$ s), on the other hand, conditions have prevailed corresponding to GUT- and Planck-scale energies. The search for stable remnants of the GUT or Planck time probably allows to get information about models of the all-embracing theory (TOE—Theory of Everything). These relics could manifest themselves in exotic objects like heavy supersymmetric particles, magnetic monopoles, axions, or primordial black holes.

Apart from the unification of all interactions the problem of the origin of cosmic rays has still not been solved. There is a number of known cosmic accelerators (supernova explosions, pulsars, active galactic nuclei, M87(?), ...), but it is completely unclear how the highest energies ($> 10^{20}$ eV) are produced. Even the question of the identity of these particles (protons?, heavy nuclei?, photons?, neutrinos?, new particles?) has not been answered. It is conjectured that active galactic nuclei, in particular, those of blazars, are able to accelerate particles to such high energies. If, however, protons or gamma rays are produced in these sources, our field of view into the cosmos is rather limited due to the short mean free path of these particles at high energies ($\lambda_{\gamma p} \approx 50\,\text{Mpc}$, $\lambda_{\gamma\gamma} \approx 30\,\text{kpc}$). Therefore, the community of astroparticle physicists is optimistic to be able to explore the universe with neutrino astronomy, or even with gravitational astronomy, where cosmic neutrinos are produced in a similar way to γ rays in cosmic beam-dump experiments via pion decays, and spectacular cosmic events like mergers of black holes emit gravitational waves. Neutrinos directly point back to the sources, they are not subject to deflections by magnetic fields and they are barely attenuated or even absorbed by interactions. To determine the angular resolution of gravitational waves is much more complicated, and it also depends on details of the associated detector system.

The enigmatic neutrinos could also give in principle a small contribution to the dark matter, which constitutes a major ingredient of the universe. The presently favoured cosmological models, however, show that their contribution to the energy density can only be at most in the percent range or, according to the data from the Planck experiment, even less than 10^{-5}.

The investigations on the flavour composition of atmospheric neutrinos have shown that a deficit of muon neutrinos exists, which obviously can only be explained by oscillations. Such neutrino oscillations require a non-zero neutrino mass, which carries elementary particle physics already beyond the well-established Standard Model. It was clear already for a long time that the Standard Model of elementary particles with its large number of free parameters cannot be the final answer, however, first hints for a possible extension of the Standard Model do not originate from accelerator experiments but rather from investigations of cosmic rays.

Obviously, neutrinos alone are by far unable to solve the problem of dark matter. To which extent exotic particles (WIMPs, SUSY particles, axions, quark nuggets, ...) contribute to the invisible mass remains to be seen. In addition, there is also the cosmological constant, not really beloved by Einstein, which provides an important contribution to the structure and to the expansion of the universe via its associated vacuum energy density. In the standard scenario of the classical Big Bang model the presently observed expansion of the universe is expected to be decelerated due to the pull of gravity. Quite on the contrary, the observations of distant supernovae explosions indicate that the expansion has changed gear. It has turned out that the cosmological constant with its associated repulsive effect—even at present times—plays a dominant role. A precise time-dependent measurement of the acceleration parameter—via the investigation of the expansion velocity in different cosmological epochs, i.e., distances—has shown that the universe expands presently at a faster rate than at earlier times.

Another input to cosmology, which does not come from particle physics experiments, is due to precise observations of the cosmic microwave background radiation. The results of these experiments in conjunction with the findings of the distant supernova studies have shown that the universe is flat, i.e., the Ω parameter is equal to unity. Given the low amount of baryonic mass (about 5%) and that dark matter constitutes only about 26%, there is large room for dark energy (69%) to speculate about. These results from non-accelerator experiments have led to a major breakthrough in cosmology.

Eventually the problem of the existence of cosmic antimatter remains to be solved. From accelerator experiments and investigations in cosmic rays it is well-known that the baryon number is a sacred conserved quantity. If a baryon is produced, an antibaryon is always created at the same time. In the same way with each lepton an antilepton is produced. The few antiparticles ($\bar{p}$, e^+) measured in astroparticle physics are assumed to be dominated by secondary origin. Even the antiparticles measured by the satellite experiment PAMELA and the AMS experiment on board the International Space Station ISS, and, in particular, the measured positron excess at energies around 100 GeV could possibly be explained by nearby supernova explosions.

Since one has to assume that equal amounts of quarks and antiquarks had been produced in the Big Bang, a mechanism is urgently required, which acts in an asymmetric fashion on particles and antiparticles. Since it is known that weak interactions violate not only parity P and charge conjugation C but also CP, a CP-violating effect could produce an asymmetry in the decay of the parent particles of protons and antiprotons so that the numbers of protons and antiprotons developed in a slightly different way. In subsequent $p\bar{p}$ annihilations very few particles, which we now call protons, would be left over. Since in $p\bar{p}$ annihilations a substantial amount of energy is transformed into photons, the observed γ/p ratio (n_γ/n_p) of about 10^9 indicates that even a minute difference in the decay properties of the parent particles of protons and antiprotons has a significant effect on the matter dominance in the universe.

Whether the small CP violation, as known from kaon and B-meson decays, is sufficient to accomplish this remains to be seen. At the moment this is considered rather unlikely.

Such a matter–antimatter asymmetry could have its origin in Grand Unified Theories of electroweak and strong interactions. The matter–antimatter asymmetry could then be explained by different decay properties of heavy X and $\bar{X}$ particles, which are supposed to have existed in the early universe. Equal numbers of X and $\bar{X}$ particles decaying into quarks, antiquarks, and leptons could lead to different numbers of quarks and antiquarks if baryon-number and lepton-number conservation were violated, thereby leading to the observed matter dominance of the universe.

The non-observation of primordial antimatter is still not a conclusive proof that the universe is matter dominated. If galaxies made of matter and 'antigalaxies' made of antimatter would exist, one would expect that they would attract each other occasionally by gravitation. This should lead to a spectacular annihilation event with strong radiation. A clear signal for such a catastrophe would be the emission of the 511-keV line due to e^+e^- annihilation. It is true that such a 511-keV γ-ray line has been observed—also in our galaxy—but its intensity is insufficient to be able to understand it as an annihilation of large amounts of matter and antimatter. On the other hand, it is conceivable that the annihilation radiation produced in interactions of tails of galaxies with tails of antigalaxies would establish such a radiation pressure that the galaxies would be driven apart and the really giant spectacular radiation outburst would never happen.

Questions about a possible dominance of matter are unlikely to be answered in accelerator experiments alone. On the other hand, the early universe provides a laboratory, which might contain—at least in principle—the answers. However, the step to bridge the gap from the Planck era to the energies available from present-day accelerators and cosmic rays is still hidden in the mist of space and time.

The investigations in the framework of astroparticle physics represent an important tool for a deeper understanding of the universe, from its creation to the present and future state.

Finally, the search for extrasolar planets has gained a lot of attention. Which conditions must be satisfied to allow the formation of life in general, or which special provisions must be fulfilled to enable life on other celestial bodies? Life, as we know it, requires finely tuned parameters in particle physics and gravitation. Can the investigation of the early universe provide a clue in this matter?

15.1   Summary

Apart from the substantial progress in astroparticle physics, there are still many severe problems, where presently no solutions are imaginable. The degree of severity of the problems is to a certain extent subjective. In the following the main problems as considered by the author are listed:

- Quantum mechanics and the general theory of relativity are brilliantly confirmed in their area of application. However, nobody has really achieved to unify these two famous theories in a common framework. Some physicists consider string theories as promising candidates. A reasonable question is, how string theories can be tested experimentally. Can quantum gravity solve the Gordian knot?
- Dark energy appears to dominate the universe. We have absolutely no idea what dark energy could be.
- In addition to dark energy we also need dark matter. From gravitational lensing we know, where dark matter resides, and there are also ideas what dark matter could be made of. In contrast to dark energy there are a number of experiments searching directly for dark matter. Is supersymmetry a solution? At the LHC such particles could be found, but up to now there was no suitable candidate found.
- How do the cosmic accelerators manage to accelerate cosmic rays to incredible energies like 10^{20} eV?
- Where are these cosmic accelerators, and why can one not find cosmic-ray anisotropies clearly hinting to where these sources are? But then, it is difficult to do astronomy with a handful of events.
- What is the chemical composition of primary galactic and extragalactic cosmic rays at the highest energies?

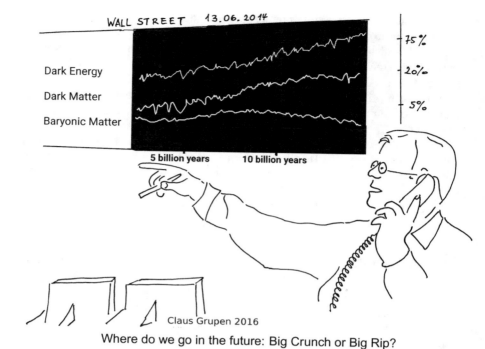

Where do we go in the future: Big Crunch or Big Rip?

- The theory of inflation can solve many cosmological puzzles. However, there is no decisive test to confirm the idea of inflation.
- How can one solve the enigma of the matter–antimatter asymmetry? Is the solution hidden in the field of elementary particles?
- There has been considerable progress in the neutrino sector. Their exact masses are still not known. And then: are neutrinos Dirac or Majorana particles? Are there also sterile neutrinos? How can one measure primordial blackbody neutrinos?
- The discovery of meanwhile about 4000 extrasolar planets has gained a lot of attention. Life obviously requires finely tuned parameters in particle physics and gravitation. Is the understanding of the prerequisites for the emergence of life hidden in the early universe?

The solution to all these problems is certainly a challenge to accelerator and astroparticle physics, and probably requires new detector concepts in particle physics and multi-messenger experiments in astroparticle physics.

15.2   Problems

1. It is generally believed that the cosmological constant represents the energy of the vacuum. On the other hand, quantum field theories lead to a vacuum energy,

which is larger by about 120 orders of magnitude. This discrepancy can be solved by an appropriate quantization of Einstein's theory of general relativity. Start from the Standard Model of particle physics (see Chap. 2) and use the Friedmann equation (8.3.7) extended by a term for the vacuum energy ϱ_v to solve the discrepancy between the value of the cosmological constant obtained from astroparticle physics measurements and the vacuum energy as obtained from quantum field theories!

Solutions

The precise statement of any problem is the most important step in its solution.

Edwin Bliss

Chapter 1

1. (a) The centrifugal force is balanced by the gravitational force between the Earth and the satellite:

$$\frac{mv^2}{R_\oplus} = G\frac{mM_\oplus}{R_\oplus^2} \approx mg \, ,$$

where it has been taken into account that the altitude is low and $R \approx R_\oplus$. As a result,

$$v \approx \sqrt{g \, R_\oplus} \approx 7.9 \, \text{km/s} \, .$$

(b) The escape velocity is found from the condition that at infinity the total energy equals zero. Since

$$E_{\text{tot}} = \frac{1}{2}mv^2 + E_{\text{pot}}$$

and the potential energy is given by

$$E_{\text{pot}} = -G\frac{mM_\oplus}{R_\oplus} \, ,$$

one obtains

$$\frac{1}{2}mv^2 = G\frac{mM_\oplus}{R_\oplus} \quad \Rightarrow \quad v = \sqrt{2M_\oplus G/R_\oplus} \approx 11.2 \, \text{km/s} \, .$$

© Springer Nature Switzerland AG 2020
C. Grupen, *Astroparticle Physics*, Undergraduate Texts in Physics,
https://doi.org/10.1007/978-3-030-27339-2

(c) The centrifugal force is balanced by the gravitational force between Earth and the satellite. From the equality of the magnitudes of these two forces one gets:

$$\frac{v^2}{r} = G\frac{M_\oplus}{r^2}.$$

From the relation between the velocity v and revolution frequency ω one obtains:

$$v^2 = r^2\omega^2 = G\frac{M_\oplus}{r} \quad \Rightarrow \quad r^3 = \frac{GM_\oplus}{\omega^2}.$$

Taking into account that for the geostationary satellite the revolution period $T_\oplus = 1$ sidereal day $\approx 86\,164$ s and $\omega_\oplus = 2\pi/T_\oplus$, one obtains

$$r = \sqrt[3]{\frac{GM_\oplus\, T_\oplus^2}{4\pi^2}} \approx 42\,164\,\text{km}.$$

The altitude above ground level therefore is $H = r - R_{\oplus,\text{Eq.}} \approx 35\,786\,\text{km}$. Geostationary satellites can only be positioned above the Equator because only there the direction of the centrifugal force can be balanced by the direction of the gravitational force. In other words, the centrifugal force for a geostationary object points outward perpendicular to the Earth's axis and only in the Equator plane the gravitational force is collinear to the former.

2. The centrifugal force is balanced by the Lorentz force, so

$$\frac{mv^2}{\varrho} = evB \quad \Rightarrow \quad p = eB\varrho \quad \Rightarrow \quad \varrho = \frac{p}{eB}.$$

Since the kinetic energy of the proton (1 MeV) is small compared to its mass (≈ 938 MeV), a classical treatment for the energy–momentum relation, $E = p^2/2m_0$, is appropriate, from which one obtains $p = \sqrt{2m_0 E_{\text{kin}}}$. Then the bending radius is given by

$$\varrho = \frac{p}{eB} = \frac{\sqrt{2m_0 E_{\text{kin}}}}{eB} \quad \text{or} \quad \varrho \approx 2888\,\text{m}.$$

Dimensional analysis:

$$p = eB\varrho, \quad p\left\{\frac{\text{kg m}}{\text{s}}\right\}\, c\left\{\frac{\text{m}}{\text{s}}\right\} = (pc)\,\{\text{J}\} = e\,\{\text{A s}\}\, B\,\{\text{T}\}\, \varrho\,\{\text{m}\}\, c\,\{\text{m s}^{-1}\},$$

$$(pc)\,[\text{J}] \times 1.6 \times 10^{-19}\,\text{J/eV} = 1.6 \times 10^{-19} \times B\,[\text{T}] \times \varrho\,[\text{m}] \times 3 \times 10^8,$$

$$(pc)[\text{eV}] = 3 \times 10^8\, B[\text{T}]\, \varrho[\text{m}] = 300\, B\,[\text{G}]\, \varrho\,[\text{cm}],$$

$$\varrho = \frac{(pc)\,[\text{eV}]}{300\,B\,[\text{G}]}\,\text{cm} = 2.888 \times 10^5\,\text{cm}\,.$$

3. From Fig. 4.4 one can read the average energy loss of muons in air-like materials to be $\approx 2\,\text{MeV}/(\text{g}/\text{cm}^2)$. This number can also be obtained from (4.2.1). The column density of the atmosphere from the production altitude is approximately $940\,\text{g}/\text{cm}^2$. Finally, the average energy loss in the atmosphere is

$$-\frac{\text{d}E}{\text{d}x}\,\Delta x \approx 2\,\text{MeV}/(\text{g}/\text{cm}^2) \times 940\,\text{g}/\text{cm}^2 = 1.88\,\text{GeV}\,.$$

4. By definition (see the Glossary) the following relation holds between the ratio of intensities I_1, I_2 and the difference of two star magnitudes m_1, m_2:

$$m_1 - m_2 = -2.5\log_{10}(I_1/I_2)\,,$$

so that

$$I_1/I_2 = 10^{-0.4(m_1 - m_2)}\,,$$

and from $\Delta m = 1$ one gets $I_1/I_2 \approx 0.398$ or $I_2 \approx 2.512\,I_1$.

5. Let N be the number of atoms making up the celestial body, μ the mass of the nucleon, and A the average atomic number of the elements constituting the celestial object. Gravitational binding dominates if

$$\frac{GM^2}{R} > N\,\varepsilon\,,$$

where ε is a typical binding energy for solid material ($\approx 1\,\text{eV}$ per atom). Here numerical factors of order unity are neglected; for a uniform mass distribution the numerical factor would be $3/5$, see Problem 13.5. The mass of the object is $M = N\,\mu\,A$, so that the condition above can be written as

$$\frac{GM^2}{R} = \frac{GM}{R}\,N\,\mu\,A > N\,\varepsilon \quad \text{or} \quad \frac{M}{R} > \frac{\varepsilon}{\mu A G}\,.$$

M/R can be rewritten as

$$\frac{M}{R} = \frac{4}{3}\pi R^2 \varrho = \left(\frac{4}{3}\pi\right)^{1/3}\underbrace{\left(\frac{4}{3}\pi\right)^{2/3} R^2\,\varrho^{2/3}\,\varrho^{1/3}}_{M^{2/3}} = \left(\frac{4}{3}\pi\right)^{1/3} M^{2/3}\,\varrho^{1/3}\,,$$

which leads to

$$\left(\frac{4}{3}\pi\right)^{1/3} M^{2/3}\varrho^{1/3} > \frac{\varepsilon}{\mu A G} \quad \text{or} \quad M > \left(\frac{\varepsilon}{\mu A G}\right)^{3/2}\frac{1}{\sqrt{\frac{4}{3}\pi\varrho}}\,.$$

The average density can be estimated to be (again neglecting numerical factors of order unity; for spherical molecules the optimal arrangement of spheres without overlapping leads to a packing fraction of 74%)

$$\varrho = \frac{\mu}{\frac{4}{3}\pi r_{\mathrm{B}}^3} \, ,$$

where $r_{\mathrm{B}} = r_e/\alpha^2$ is the Bohr radius ($r_{\mathrm{B}} = 0.529 \times 10^{-10}$ m), r_e the classical electron radius, and α the fine-structure constant. This leads to

$$M > \left(\frac{\varepsilon}{\mu A G}\right)^{3/2} \frac{1}{\sqrt{\frac{4}{3}\pi\mu/(\frac{4}{3}\pi r_{\mathrm{B}}^3)}} = \left(\frac{\varepsilon}{\mu A G}\right)^{3/2} \frac{r_{\mathrm{B}}^{3/2}}{\sqrt{\mu}} = \frac{1}{\mu^2}\left(\frac{\varepsilon\, r_{\mathrm{B}}}{A G}\right)^{3/2} \, .$$

With

$$\mu = 1.67 \times 10^{-27}\,\mathrm{kg}\,, \quad \varepsilon = 1\,\mathrm{eV} = 1.6 \times 10^{-19}\,\mathrm{J}\,,$$
$$G = 6.67 \times 10^{-11}\,\mathrm{m}^3\,\mathrm{kg}^{-1}\,\mathrm{s}^{-2}\,, \quad A = 50$$

one gets the condition

$$M > 4.58 \times 10^{22}\,\mathrm{kg}\,.$$

This means that our moon is gravitationally bound, while the moons of Mars (Phobos and Deimos) being much smaller are bound by solid-state effects, i.e., essentially by electromagnetic forces.

6. In special relativity the redshift is given by

$$z = \frac{\lambda - \lambda_0}{\lambda_0} = \sqrt{\frac{1+\beta}{1-\beta}} - 1 \, .$$

Let us assume that this relation can be applied even for the young universe, where a treatment on the basis of general relativity would be more appropriate. Solve $z(\beta)$ for β or v and apply $z = 7.085$:

$$\beta = \frac{(z+1)^2 - 1}{(z+1)^2 + 1} \quad \Rightarrow \quad v = \frac{(z+1)^2 - 1}{(z+1)^2 + 1}\, c = H d \, ,$$
$$t = \frac{d}{c} = \frac{1}{H}\frac{(z+1)^2 - 1}{(z+1)^2 + 1} \approx \frac{0.970}{H} \, .$$

Since $\frac{1}{H}$ is the age of the universe, we see the distant quasar when it had $\approx 3\%$ of its present age.

7. Diffraction patterns using a hole aperture show that two objects, which appear under an angle of δ, can be resolved if

$$\delta \geq 1.22 \, \frac{\lambda}{d} , \tag{1.1}$$

where λ is the used wavelength and d the diameter of the diaphragm.
This formula is a little difficult to arrive at, but it can be motivated by the following argument:
diffraction at a slit of width d leads to a first minimum given by

$$d \sin \delta = \lambda . \tag{1.2}$$

If one uses a circumscribed square of side length d this is just the same result. For the largest inscribed square inside the circle the side length is $\frac{d}{\sqrt{2}}$. In this case the first minimum is observed for

$$\sin \delta = \frac{\lambda}{\frac{d}{\sqrt{2}}} . \tag{1.3}$$

The above referred result can be interpreted as the approximate average of these two cases, if $\sin \delta \approx \delta$ is used for small angles.
Using $d = 535$ nm one gets

$$\sin \delta = 1.22 \, \frac{535 \times 10^{-9} \, \text{m}}{200 \times 2.54 \times 10^{-2} \, \text{m}} = 1.285 \times 10^{-7} . \tag{1.4}$$

Using the distance to the Sun, $s = 149\,600\,000$ km $= 1.496 \times 10^{11}$ m, one obtains for the minimum distance that can be resolved in this way

$$\Delta x = \delta \cdot s \approx 19.2 \, \text{km} . \tag{1.5}$$

Chapter 2

1. (a) lepton-number conservation is violated: not allowed,
   (b) possible,
   (c) possible, a so-called Dalitz decay of the π^0,
   (d) both charge and lepton-number conservation violated, not allowed,
   (e) kinematically not allowed ($m_{K^-} + m_p > m_\Lambda$),
   (f) possible,
   (g) possible,
   (h) possible.
2. For any unstable elementary particle like, e.g., a muon, the quantity 'lifetime' should be considered in the average sense only. In other words, it does not mean that a particle with lifetime τ will decay exactly the time τ after it was produced. Its actual lifetime t is a random number distributed with a probability density

function,

$$f(t; \tau)\, dt = \frac{1}{\tau}\, e^{-\frac{t}{\tau}}\, dt \, ,$$

giving a probability that the lifetime t lies between t and $t + dt$. It can easily be checked that the mean value of t equals τ. For an unstable relativistic particle, a mean range before it decays is given by the product of its velocity βc and lifetime $\gamma \tau$ to $\beta \gamma c \tau$. For muons $c \tau = 658.653$ m. Therefore, to survive to sea level from an altitude of 20 km, the average range should equal $l = 20$ km or $\beta \gamma c \tau_\mu = l = 20$ km and $\beta \gamma = l / c \tau_\mu$. From $\beta \gamma = \sqrt{\gamma^2 - 1}$ one gets $\gamma^2 = (l / c \tau_\mu)^2 + 1$. Since $l / c \tau \gg 1$, one finally obtains

$$\gamma \approx l / c \tau_\mu = \frac{20 \times 10^3 \text{ m}}{658.653 \text{ m}} \approx 30.4 \, .$$

Then the total energy is

$$E_\mu = \gamma m_\mu c^2 \approx 3.2 \text{ GeV}$$

and the kinetic energy is

$$E_\mu^{\text{kin}} = E_\mu - m_\mu c^2 \approx 3.1 \text{ GeV} \, .$$

3. The Coulomb force is

$$F_{\text{Coulomb}} = \frac{1}{4\pi \varepsilon_0} \frac{q_1 q_2}{r^2} \approx \frac{1}{4\pi \times 8.854 \times 10^{-12} \text{ F m}^{-1}} \frac{(1.602 \times 10^{-19} \text{ A s})^2}{(10^{-15} \text{ m})^2}$$
$$\approx 230.7 \text{ N}$$

and the gravitational force is

$$F_{\text{gravitation}} = G \frac{m_1 m_2}{r^2} \approx 6.674 \times 10^{-11} \frac{\text{m}^3}{\text{kg s}^2} \frac{(2.176 \times 10^{-8} \text{ kg})^2}{(10^{-15} \text{ m})^2} \approx 31\,600 \text{ N} \, .$$

4. Energy–momentum conservation requires

$$q_{e^+} + q_{e^-} = q_{\text{f}} \, ,$$

where q_{e^+}, q_{e^-}, and q_{f} are the four-momenta of the positron, electron, and final state, respectively. To produce a Z, the invariant mass of the initial state squared should be not less than the invariant mass of the required final state squared or

$$(q_{e^+} + q_{e^-})^2 \geq m_Z^2 \, .$$

Since $q_{e^+} = (E_{e^+}, \mathbf{p}_{e^+})$ and $q_{e^-} = (m_e, 0)$, one obtains

$$2m_e(E_{e^+} + m_e) \geq m_Z^2$$

or

$$E_{e^+} \geq \frac{m_Z^2}{2m_e} - m_e \approx 8.1 \times 10^{15} \text{ eV} = 8.1 \text{ PeV}.$$

Chapter 3

1. Similarly to Problem 2.4, for the reaction $\gamma + \gamma \to \mu^+ + \mu^-$ the threshold condition is

$$(q_{\gamma_1} + q_{\gamma_2})^2 \geq (2m_\mu)^2.$$

For photons $q_{\gamma_1}^2 = q_{\gamma_2}^2 = 0$. Then

$$(q_{\gamma_1} + q_{\gamma_2})^2 = 2(E_{\gamma_1} E_{\gamma_2} - \boldsymbol{p}_{\gamma_1} \cdot \boldsymbol{p}_{\gamma_2}) = 4E_{\gamma_1} E_{\gamma_2}$$

for the angle π between the photon 3-momenta. Finally,

$$E_{\gamma_1} \geq \frac{(m_\mu c^2)^2}{E_{\gamma_2}} \approx 1.1 \times 10^{19} \text{ eV}.$$

2. The number of collisions is obtained by subtracting the number of unaffected particles from the initial number of particles:

$$\Delta N = N_0 - N = N_0(1 - e^{-x/\lambda}).$$

For thin targets $x/\lambda \ll 1$ and the expansion in the Taylor series gives

$$\Delta N = N_0(1 - (1 - x/\lambda + \cdots)) = \frac{N_0 x}{\lambda} = N_0 N_A \sigma_N x.$$

For the numerical example one gets

$$\Delta N = 10^8 \times 6.022 \times 10^{23} \text{ g}^{-1} \times 10^{-24} \text{ cm}^2 \times 0.1 \text{ g/cm}^2 \approx 6 \times 10^6.$$

3. The threshold condition is

$$(q_{\bar{\nu}_e} + q_p)^2 \geq (m_n + m_e)^2.$$

Taking into account that $q_{\bar{\nu}_e}^2 = 0$ and $q_p = (m_p, 0)$, one obtains

$$m_p^2 + 2E_{\bar{\nu}_e} m_p \geq (m_n + m_e)^2$$

and

$$E_{\bar{\nu}_e} \geq \frac{(m_n + m_e)^2 - m_p^2}{2m_p} \approx 1.8\,\text{MeV}\,,$$

where the following values of the particles involved were used: $m_n = 939.565\,\text{MeV}$, $m_p = 938.272\,\text{MeV}$, $m_e = 0.511\,\text{MeV}$.

4.

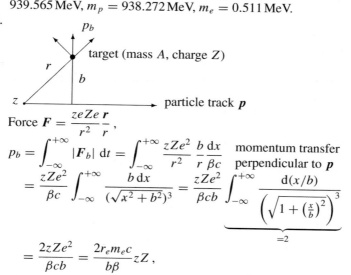

Force $\boldsymbol{F} = \dfrac{zeZe}{r^2}\dfrac{\boldsymbol{r}}{r}$,

$$p_b = \int_{-\infty}^{+\infty} |F_b|\,\mathrm{d}t = \int_{-\infty}^{+\infty} \frac{zZe^2}{r^2}\frac{b}{r}\frac{\mathrm{d}x}{\beta c} \quad \begin{matrix}\text{momentum transfer}\\ \text{perpendicular to } \boldsymbol{p}\end{matrix}$$

$$= \frac{zZe^2}{\beta c}\int_{-\infty}^{+\infty}\frac{b\,\mathrm{d}x}{(\sqrt{x^2+b^2})^3} = \frac{zZe^2}{\beta cb}\underbrace{\int_{-\infty}^{+\infty}\frac{\mathrm{d}(x/b)}{\left(\sqrt{1+\left(\frac{x}{b}\right)^2}\right)^3}}_{=2}$$

$$= \frac{2zZe^2}{\beta cb} = \frac{2r_e m_e c}{b\beta}zZ\,,$$

where the classical electron radius is $r_e = \dfrac{e^2}{m_e c^2}$.

5. Previously the transverse momentum transfer was obtained to be ($z = 1$)

$$p_b = 2\,Z\frac{r_e m_e c}{b\beta} = 2p\frac{r_e Z}{b\beta^2}\,,$$

where $p = m_e v$ was assumed (classical treatment). The transverse momentum transfer is given by $p_b = p\sin\vartheta$. Since

$$\sin 2\gamma = 2\sin\gamma\cos\gamma = 2\tan\gamma\cos^2\gamma = \frac{2\tan\gamma}{1+\tan^2\gamma}\,,$$

one gets, using Rutherford's scattering formula:

$$p_b = 2p\frac{r_e Z/b\beta^2}{1+(r_e Z/b\beta^2)^2}\,.$$

Chapter 4

1. $\phi = \dfrac{N_A}{A}\,\sigma_A = N_A\,\sigma_N\,, \left.\begin{array}{l} [N_A] = \mathrm{mol}^{-1} \\ [A] = \mathrm{g\,mol}^{-1} \\ [\sigma_A] = \mathrm{cm}^2 \end{array}\right\} \Rightarrow [\phi] = (\mathrm{g/cm}^2)^{-1}.$

2. The relative energy resolution is determined by the fluctuations of the number N of produced particles. If W is the energy required for the production of a particle (pair), the relative energy resolution is

$$\frac{\Delta E}{E} = \frac{\Delta N}{N} = \frac{\sqrt{N}}{N} = \frac{1}{\sqrt{N}} = \sqrt{\frac{W}{E}}\,,$$

since $N = E/W$. Here E and ΔE are the energy of the particle and its uncertainty,

$$\frac{\Delta E}{E} = \begin{cases} \text{(a) } 10\% \\ \text{(b) } 5.5\% \\ \text{(c) } 1.9\% \\ \text{(d) } 3.2 \times 10^{-4} \end{cases}.$$

3. $R = \displaystyle\int_E^0 \frac{dE}{dE/dx} = \int_0^E \frac{dE}{a+bE} = \frac{1}{b}\ln\left(1 + \frac{b}{a}E\right).$
   The numerical calculation gives

$$R \approx \frac{1}{4.4 \times 10^{-6}}\ln\left(1 + \frac{4.4 \times 10^{-6}}{2}\,10^5\right) \frac{\mathrm{g}}{\mathrm{cm}^2} \approx 45\,193\,\frac{\mathrm{g}}{\mathrm{cm}^2}$$

   or $R = 181\,\mathrm{m}$ of rock taking into account that $\varrho_{\mathrm{rock}} = 2.5\,\mathrm{g/cm}^3$.

4. The Cherenkov angle θ_C is given by the relation

$$\cos\theta_C = \frac{1}{n\beta}\,.$$

   From the relation between the momentum p and β:

$$p = \gamma m_0 \beta c \quad \Rightarrow \quad \beta = \frac{p}{\gamma m_0 c} \quad \text{and} \quad \cos\theta_C = \frac{\gamma m_0 c}{np}\,.$$

   Then

$$\frac{np\cos\theta_C}{m_0 c} = \frac{E}{m_0 c^2} = \frac{c\sqrt{p^2 + m_0^2 c^2}}{m_0 c^2}$$

   and

$$(np\cos\theta_C)^2 = p^2 + m_0^2 c^2 \quad \Rightarrow \quad p^2(n^2\cos^2\theta_C - 1) = m_0^2 c^2\,.$$

Finally,

$$m_0 = \frac{p}{c}\sqrt{n^2 \cos^2 \theta_C - 1} \, .$$

5. From the expression for the change of the thermal energy, $\Delta Q = c_{\mathrm{sp}} m \Delta T$, one can find the temperature rise:

$$\Delta T = \frac{\Delta Q}{c_{\mathrm{sp}} m} = \frac{10^4 \, \mathrm{eV} \times 1.6 \times 10^{-19} \, \mathrm{J/eV}}{8 \times 10^{-5} \, \mathrm{J/(g\,K)} \times 1\,\mathrm{g}} = 2 \times 10^{-11} \, \mathrm{K} \, .$$

6. $q_\gamma + q_e = q'_\gamma + q'_e \;\; \Rightarrow \;\; q_\gamma - q'_\gamma = q'_e - q_e \;\; \Rightarrow$
$q_\gamma^2 + q_{\gamma'}^2 - 2q_\gamma q'_\gamma = -2(E_\gamma E'_\gamma - \boldsymbol{p}_\gamma \boldsymbol{p}'_\gamma) = q'^2_e + q^2_e - 2q'_e q_e \;\; \Rightarrow$
$-2E_\gamma E'_\gamma(1 - \cos\theta) = 2m_e^2 - 2E'_e m_e \;\; ; \;\; (\boldsymbol{p}_e = 0)$
$$= -2m_e(E'_e - m_e) = -2m_e E_e^{\mathrm{kin}} = -2m_e(E_\gamma - E'_\gamma) \;\; \Rightarrow$$

$$\frac{E_\gamma - E'_\gamma}{E'_\gamma} = \frac{E_\gamma}{E'_\gamma} - 1 = \frac{E_\gamma}{m_e}(1 - \cos\theta) \;\; \Rightarrow$$

$$\frac{E'_\gamma}{E_\gamma} = \frac{1}{1 + \frac{E_\gamma}{m_e}(1 - \cos\theta)} = \frac{1}{1 + \varepsilon(1 - \cos\theta)} \, .$$

7. Start from (4.5.1). The maximum energy is transferred for backscattering, $\theta = \pi$;

$$\frac{E'_\gamma}{E_\gamma} = \frac{1}{1 + 2\varepsilon} \, ,$$

$$E_e^{\max} = E_\gamma - E'_\gamma = E_\gamma - \frac{E_\gamma}{1 + 2\varepsilon} = E_\gamma \frac{2\varepsilon}{1 + 2\varepsilon} = \frac{2\varepsilon^2}{1 + 2\varepsilon} m_e c^2 \, ; \quad (*)$$

with numbers: $E_e^{\max} = 478 \, \mathrm{keV}$ ('Compton edge').
For $\varepsilon \to \infty$ (*) yields $E_e^{\max} = E_\gamma$.
For $\theta_\gamma = \pi$ one has

$$E'_\gamma = E_\gamma \frac{1}{1 + 2\varepsilon}$$

and consequently

$$E'_\gamma = \frac{m_e c^2 \varepsilon}{1 + 2\varepsilon} = \frac{m_e c^2}{2 + 1/\varepsilon} \, .$$

For $\varepsilon \gg 1$ this fraction approaches $m_e c^2/2$.

8. Momentum: $p = m\,v; m = \gamma\,m_0$, where m_0 is the rest mass and $\gamma = \dfrac{1}{\sqrt{1 - \beta^2}}$:
$p = \gamma\,m_0\,\beta\,c \;\; \Rightarrow \;\; \gamma\beta = p/m_0 c \, .$

9. In the X-ray region the index of refraction is $n = 1$, therefore there is no dispersion and consequently no Cherenkov radiation.

10.

$$R = \int_E^0 \frac{dE}{-dE/dx} = \int_0^E \frac{E \, dE}{az^2 \ln(bE)} \approx \frac{1}{az^2} \int_0^E \frac{E \, dE}{(bE)^{1/4}}$$

$$= \frac{1}{a\sqrt[4]{b} \, z^2} \int_0^E E^{3/4} \, dE = \frac{4}{7a\sqrt[4]{b} \, z^2} E^{7/4} \sim E^{1.75} \, ;$$

experimentally the exponent is obtained to be 1.73 for protons and 1.65 for carbon ions.

Chapter 5

1. (a) In the classical non-relativistic case ($v \ll c$) the kinetic energy is just

$$E^{\text{kin}} = \frac{1}{2}m_0 v^2 = \frac{1}{2}m_0 \left(\frac{eRB}{m_0}\right)^2 = \frac{1}{2}\frac{e^2 R^2 B^2}{m_0} \, ,$$

where the velocity v was found from the usual requirement that the centrifugal force be balanced by the Lorentz force:

$$\frac{m_0 v^2}{R} = evB \quad \Rightarrow \quad v = \frac{eBR}{m_0} \, .$$

(b) In the relativistic case

$$E^{\text{kin}} = \gamma m_0 c^2 - m_0 c^2 = m_0 c^2 \left(\frac{1}{\sqrt{1 - \frac{v^2}{c^2}}} - 1\right) = c\sqrt{p^2 + m_0^2 c^2} - m_0 c^2$$

$$= c\sqrt{e^2 R^2 B^2 + m_0^2 c^2} - m_0 c^2 = ecRB\sqrt{1 + \frac{m_0^2 c^2}{e^2 R^2 B^2}} - m_0 c^2$$

$$\approx 5.95 \times 10^7 \, \text{eV} \, .$$

Alternatively, with an early relativistic approximation

$$E^{\text{kin}} = c\sqrt{p^2 + m_0^2 c^2} - m_0 c^2 \approx cp = ecRB \approx 6 \times 10^7 \, \text{eV} \, .$$

2. Let us first find the number of electrons N_e in the star:

$$N_e = M_{\text{star}}/m_p = 10 \, M_\odot/m_p \approx \frac{2 \times 10^{34} \, \text{g}}{1.67 \times 10^{-24} \, \text{g}} \approx 1.2 \times 10^{58} \, .$$

Here it is assumed that the star consists mainly of hydrogen ($m_p \approx 1.67 \times 10^{-24}$ g) and is electrically neutral, so that $N_e = N_p$.

The pulsar volume is

$$V = \frac{4}{3}\pi r^3 \approx 4.19 \times 10^{18}\,\text{cm}^3$$

and the electron density is $n = 0.5\,N_e/V \approx 1.43 \times 10^{39}/\text{cm}^3$. Then the Fermi energy is

$$E_F = \hbar c(3\pi^2 n)^{1/3}$$

$$\approx 6.582 \times 10^{-22}\,\text{MeV s} \times 3 \times 10^{10}\,\frac{\text{cm}}{\text{s}}\,(3 \times 3.1416^2 \times 1.43 \times 10^{39})^{1/3}\,\text{cm}^{-1}$$

$$\approx 688\,\text{MeV}\,.$$

Consequences:

The electrons are pressed into the protons,

$$e^- + p \rightarrow n + \nu_e\,.$$

The neutrons cannot decay, since in free neutron decay the maximum energy transfer to the electron is only $\approx 780\,\text{keV}$, and all energy levels in the Fermi gas are occupied, even if only 1% of the electrons are left: ($n^* = 0.01\,n \Rightarrow E_F \approx 148\,\text{MeV}$).

3. Event horizon of a black hole with mass $M = 10^6\,M_\odot$:

$$R_S = \frac{2GM}{c^2} \approx 2.96 \times 10^9\,\text{m}\,,$$

$$\Delta E = -\int_\infty^{R_S} G\frac{m_p M}{r^2}\,dr = G\frac{m_p M}{R_S} = \frac{1}{2}m_p c^2 \approx 469\,\text{MeV} \approx 7.5 \times 10^{-8}\,\text{J}\,.$$

The result is independent of the mass of the black hole. This classical calculation, however, is not suitable for this problem and just yields an idea of the energy gain. In addition, the value of the energy depends on the frame of reference. In the vicinity of black holes only formulae should be used that hold under general relativity.

4. (a) Conservation of angular momentum

$$\boldsymbol{L} = \boldsymbol{r} \times \boldsymbol{p} \quad \Rightarrow \quad L \approx mrv = mr^2\omega$$

requires $r^2\omega$ to be constant (no mass loss):

$$R_\odot^2\omega_\odot = R_{NS}^2\omega_{NS} \quad \Rightarrow \quad \omega_{NS} = \left(\frac{R_\odot}{R_{NS}}\right)^2\omega_\odot \approx 588\,\text{Hz}\,.$$

Then the rotational energy is

$$E_{\text{rot}} = \frac{1}{2}\frac{2}{5}M_{\text{NS}}R_{\text{NS}}^2\omega_{\text{NS}}^2 \approx 0.4 \times 10^{30} \text{ kg} \times (5 \times 10^4)^2 \text{ m}^2 \times 588^2 \text{ s}^{-2}$$
$$\approx 0.35 \times 10^{45} \text{ J}.$$

(b) Assume that the Sun consists of protons only (plus electrons, of course). Four protons each are fused to He with an energy release of 26 MeV corresponding to a mass–energy conversion efficiency of $\approx 0.7\%$:

$$E = M_\odot c^2 \times 7 \times 10^{-3} = 2 \times 10^{30} \text{ kg} \times (3 \times 10^8)^2 \text{ m}^2/\text{s}^2 \times 7 \times 10^{-3}$$
$$= 1.26 \times 10^{45} \text{ J},$$

which is comparable to the rotational energy of the neutron star.

5. A dipole field is needed to compensate the centrifugal force

$$\frac{mv^2}{R} = evB_{\text{guide}}.\qquad(*)$$

Equation (5.8.1) yields $p = mv = \frac{1}{2}eRB$. Comparison with $(*)$ gives

$$B_{\text{guide}} = \frac{1}{2}B \quad \text{(Wideroe condition)}.$$

6. The result of this estimate ($T = 1$ million years) shows that black holes of 100 solar masses might have been created indeed shortly after the Big Bang.

7. Wavelength of 400 nm converted to frequencies: $\nu = \frac{c}{\lambda} = 7.5 \times 10^{14}$ Hz; number of photons of a 1 PW laser: $N = \frac{P}{h\cdot\nu} = 2 \times 10^{33}$ per second; photon momentum $p = \frac{h}{\lambda} = \frac{h\cdot\nu}{c}$; momentum transfer $2 \cdot \frac{h\cdot\nu}{c}$; force $F = \frac{dp}{dt} = \frac{P}{h\cdot\nu} \cdot 2 \cdot \frac{h\cdot\nu}{c} = \frac{2\cdot P}{c} = 6.7 \times 10^6$ N.

To make such a propulsion work for lower laser powers, one has to shine this laser permanently onto the solar sail of the spacecraft to achieve permanent acceleration.

If $P = 1$ petawatt were used, one would achieve an acceleration of $a = \frac{F}{m} \approx 6.7 \times 10^3 \frac{m}{s^2}$ for a spacecraft mass of 1 ton. For an exposure of the laser on 1000 s, one would get a velocity of 2.2% of the velocity of light, where the relativistic increase of the spacecraft mass is neglected.

8. The centrifugal acceleration of the star is $\omega^2 \cdot r$. The gravitational (centripetal) pull must be larger than that to prevent the star from disintegration:

$$\frac{\gamma\frac{4}{3}\pi r^3\rho}{r^2} \geq \omega^2 \cdot r,\qquad(5.1)$$

where γ is the gravitational constant and ρ the density.
Solving for ρ gives

$$\rho \geq \frac{3\omega^2}{4\pi\gamma},$$ (5.2)

which gives

$$\rho = \frac{3\pi}{\gamma T^2},$$ (5.3)

where

$$T = \frac{2\pi}{\omega}$$ (5.4)

is the rotational period of the star.
For $T = 1$ ms one achieves

$$\rho \geq 1.4 \times 10^{17} \frac{\text{kg}}{\text{m}^3}.$$ (5.5)

Even though the assumptions on homogeneity of the star are rather strong, the obtained value is on the correct order of magnitude.

Chapter 6

Section 6.1

1. C, O, and Ne are even–even nuclei, oxygen is even doubly magic, while F as even–odd and N, Na as odd–odd configurations are less tightly bound. The pairing term δ in the Bethe–Weizsäcker formula gives the difference in the nuclear binding energies for even–even, even–odd, and odd–odd nuclei:

$$m(Z, A) = Zm_p + (A - Z)m_n - a_v A + a_s A^{2/3} + a_C \frac{Z^2}{A^{1/3}} + a_a \frac{(Z - A/2)^2}{A} + \delta$$

$a_v A$ – volume term,
$a_s A^{2/3}$ – surface term,
$a_C \frac{Z^2}{A^{1/3}}$ – Coulomb repulsion,
$a_a \frac{(Z-A/2)^2}{A}$ – asymmetry term,
$$\delta = \begin{cases} -a_p A^{-3/4} & \text{for even-even} \\ 0 & \text{for even-odd/odd-even} \\ +a_p A^{-3/4} & \text{for odd-odd} \end{cases}, \quad \delta \text{ - pairing term}.$$

2. From Fig. 7.9 the rate of primary particles can be estimated as

$$\Phi \approx 0.2 \, (\text{cm}^2 \, \text{s} \, \text{sr})^{-1}.$$

This particle flux is to be averaged over the zenith angle ϑ, where a factor of $\cos \vartheta$ for the projected surface element appears in the solid-angle integral over

dΩ with $\vartheta \geq 0$, leading to a factor $1/2$, thus yielding $\bar{\Phi} = \Phi/2$.
The surface of the Earth is

$$S = 4\pi R_{\oplus}^2 \approx 4\pi (6370 \times 10^5)^2 \, \text{cm}^2 \approx 5.10 \times 10^{18} \, \text{cm}^2 \, .$$

The age of the Earth is $T = 4.5 \times 10^9$ years or 1.42×10^{17} s. The total charge accumulated in the solid angle of 2π during the time T is

$$\int \bar{\Phi}(x, t) \, dx \, dt \approx 1.45 \times 10^{35} \times 2\pi/2$$

$$\approx 4.56 \times 10^{35} \text{ equivalent protons} \, \widehat{=} \, 7.3 \times 10^{16} \, \text{C} \, .$$

Still, there is no charge-up. Primary particles mentioned in Sect. 6.1 are mainly those of high energy (typically >1 GeV). In this high-energy domain positively charged particles actually dominate. If all energies are considered, there are equal amounts of positive and negative particles. Our Sun is also a source of large numbers of protons, electrons, and α particles. In total, there is no positive charge excess if particles of all energies are considered. This is not in contrast to the observation of a positive charge excess of sea-level muons because they are the result of cascade processes in the atmosphere initialized by energetic (i.e., mainly positively charged) primaries.

3. (a) The average column density traversed by primary cosmic rays is $\approx 6\text{g/cm}^2$ (see Sect. 6.1). The interaction rate Φ is related to the cross section by, see (3.6.5),

$$\Phi = \sigma \, N_A \, .$$

Because of $\Phi[(6\text{g/cm}^2)^{-1}] \approx 0.1$ and assuming that the collision partners are nucleons, one gets

$$\sigma \approx 0.1 \times (6\text{g/cm}^2)^{-1}/N_A \approx 1.66 \times 10^{-25} \, \text{cm}^2 = 166 \, \text{mbarn} \, .$$

(b) If the iron–proton fragmentation cross section is ≈ 170 mb, the cross section for iron–air collisions can be estimated to be $\sigma_{\text{frag}}(\text{iron–air}) \approx 170 \, \text{mb} \times A^\alpha$, where $A \approx 0.8 \, A_N + 0.2 \, A_O = 11.2 + 3.2 = 14.4$. With $\alpha \approx 0.75$ one gets $\sigma_{\text{frag}}(\text{iron–air}) \approx 1.26 \, \text{b}$. The probability to survive to sea level is $P = \exp(-\sigma N_A d/A) = 1.3 \times 10^{-23}$, where it has been used that the thickness of the atmosphere is $d \approx 1000 \, \text{g/cm}^2$.

Section 6.2

1. In principle, neutrons are excellent candidates. Because they are neutral, they would point back to the sources of cosmic rays. Their deflection by inhomogeneous magnetic fields is negligible. The only problem is their lifetime, $\tau_0 = 885.7$ s. At 10^{20} eV the Lorentz factor extends this lifetime considerably to

$$\tau = \gamma \tau_0 = \frac{10^{20}\,\text{eV}}{m_n c^2}\,\tau_0 = 9.4 \times 10^{13}\,\text{s} \, \hat{=} \, 2.99 \times 10^6\,\text{light-years} = 0.916\,\text{Mpc}\,.$$

Still the sources would have to be relatively near, and there is no evidence for point sources of this energy in the close vicinity of the Milky Way.

2. The relativistic mass is

$$m = \frac{m_0}{\sqrt{1 - \beta^2}}\,, \tag{6.1}$$

where $\beta = v/c$. Solved for β one gets

$$\beta = \frac{\sqrt{m^2 - m_0^2}}{m}\,. \tag{6.2}$$

Assuming for the rest mass of a proton 1 GeV, one gets

$$\beta = 1 - \frac{1}{1.8 \times 10^{23}} \tag{6.3}$$

leading to

$$\beta \approx 1 - 5.5 \times 10^{-24} \tag{6.4}$$

or a velocity of the Oh-My-God proton of

$$v = 0.999\,999\,999\,999\,999\,999\,999\,9945\,c\,. \tag{6.5}$$

3. In his reference frame the time would pass according to

$$t_0 = t \cdot \sqrt{1 - \beta^2}\,, \tag{6.6}$$

and since

$$\beta \approx 1 - 5.5 \times 10^{-24} \tag{6.7}$$

or, by using the ratio of the masses, one would get

$$\frac{t}{t_0} = \frac{m}{m_0} \approx 3 \times 10^{11}\,. \tag{6.8}$$

Therefore, in the rest frame of the alien traveler his time is slowed down by this huge factor, and it would take him only about 3.8 min to arrive at the experimental site at Dugway Proving Ground.

See also [278].

4. According to the reaction probability given by

$$W = N_A \sigma d\rho \qquad (6.9)$$

this would lead to

$$W = 6.022 \times 10^{23} \times 10^{-45} \times 10^5 \approx 6 \times 10^{-17} . \qquad (6.10)$$

For a dark-matter flux of $1\,\text{TeV}$ particles of $10\,\text{cm}^{-2}\,\text{s}^{-1}$ this could lead to one event in ICECUBE every second day.

5. The air-shower rate F in AGASA is

$$F = \frac{11\ \text{events}}{10^8\,\text{m}^2 \times 3.15 \times 10^7\,\text{s} \times 14 \times 2\pi} \approx 4 \times 10^{-17}\,\text{m}^{-2}\,\text{s}^{-1}\,\text{sr}^{-1} . \qquad (6.11)$$

The interaction probability of a crypton is

$$\Phi = N_A \sigma d\rho = 6.022 \times 10^{23}/\text{g} \times 10^{-44}\,\text{cm}^2 \times 1000\,\text{g/cm}^2 \approx 6 \times 10^{-18} . \qquad (6.12)$$

The crypton rate can be derived from

$$F = F_{\text{crypton}} \cdot \Phi \qquad (6.13)$$

to be

$$F_{\text{crypton}} = \frac{F}{\Phi} = \frac{4 \times 10^{-17}\,\text{m}^{-2}\,\text{s}^{-1}\,\text{sr}^{-1}}{6 \times 10^{-18}} \approx 7\,\text{m}^{-2}\,\text{s}^{-1}\,\text{sr}^{-1} . \qquad (6.14)$$

Of course, this estimate is only valid under the assumption of a cross section being $\approx 10^{-8}$ pb and the event rate as observed by the AGASA experiment being correct.

Section 6.3

1. The neutrino flux ϕ_ν is given by the number of fusion processes $4p \rightarrow {}^4\text{He} + 2e^+ + 2\nu_e$ times 2 neutrinos per reaction chain:

$$\phi_\nu = \frac{\text{solar constant}}{\text{energy gain per reaction chain}} \times 2$$

$$\approx \frac{1400\,\text{W/m}^2}{28.3\,\text{MeV} \times 1.6 \times 10^{-13}\,\text{J/MeV}} \times 2 \approx 6.2 \times 10^{10}\,\text{cm}^{-2}\,\text{s}^{-1} .$$

2. $(q_{\nu_\alpha} + q_{e^-})^2 = (m_\alpha + m_{\nu_e})^2$, $\alpha = \mu, \tau$;
   assuming m_{ν_α} to be small ($\ll m_e, m_\mu, m_\tau$), one gets

$$2E_{\nu_\alpha} m_e + m_e^2 = m_\alpha^2 \quad \Rightarrow \quad E_{\nu_\alpha} = \frac{m_\alpha^2 - m_e^2}{2m_e} \quad \Rightarrow$$

$\alpha = \mu : E_{v_\mu} = 10.92\,\text{GeV}$, $\alpha = \tau : E_{v_\tau} = 3.09\,\text{TeV}$;
since solar neutrinos cannot convert into such high-energy neutrinos, the proposed
reactions cannot be induced.

3. (a) The interaction rate is

$$R = \sigma_N N_A\, d\, A\, \phi_v,$$

where σ_N is the cross section per nucleon, $N_A = 6.022 \times 10^{23}\,\text{g}^{-1}$ is the
Avogadro constant, d the area density of the target, A the target area, and
ϕ_v the solar neutrino flux. With $d \approx 15\,\text{g\,cm}^{-2}$, $A = 180 \times 30\,\text{cm}^2$, $\phi_v \approx 7 \times 10^{10}\,\text{cm}^{-2}\,\text{s}^{-1}$, and $\sigma_N = 10^{-45}\,\text{cm}^2$ one gets $R = 3.41 \times 10^{-6}\,\text{s}^{-1} = 107\,\text{a}^{-1}$.

(b) A typical energy of solar neutrinos is $100\,\text{keV}$, i.e., $50\,\text{keV}$ are transferred to
the electron. Consequently, the total annual energy transfer to the electrons
is

$$\Delta E = 107 \times 50\,\text{keV} = 5.35\,\text{MeV} = 0.86 \times 10^{-12}\,\text{J}.$$

(c) With the numbers used so far the mass of the human is $81\,\text{kg}$. Therefore, the
equivalent annual dose comes out to be

$$H_v = \frac{\Delta E}{m}\, w_R = 1.06 \times 10^{-14}\,\text{Sv},$$

actually independent of the assumed human mass. The contribution of solar
neutrinos to the normal natural dose rate is negligible, since

$$H = \frac{H_v}{H_0} = 5.3 \times 10^{-12}.$$

4. (a) The time evolution of the electron neutrino is

$$|v_e; t\rangle = \cos\theta\, e^{-iE_{v_1}t}|v_1\rangle + \sin\theta\, e^{-iE_{v_2}t}|v_2\rangle.$$

This leads to

$$\langle v_\mu|v_e; t\rangle = \sin\theta\cos\theta\left(e^{-iE_{v_2}t} - e^{-iE_{v_1}t}\right),$$

because

$$(-\sin\theta|v_1\rangle + \cos\theta|v_2\rangle)^+ \left(\cos\theta\, e^{-iE_{v_1}t}|v_1\rangle + \sin\theta\, e^{-iE_{v_2}t}|v_2\rangle\right)$$

$$= -\sin\theta\cos\theta\, e^{-iE_{v_1}t} + \sin\theta\cos\theta\, e^{-iE_{v_2}t}$$

since $|v_1\rangle$; $|v_2\rangle$ are orthogonal states.
Squaring the time-dependent part of this relation, i.e., multiplying the expression by its complex conjugate, yields as an intermediate step

$$\left(e^{-iE_{v_2}t} - e^{-iE_{v_1}t}\right)\left(e^{+iE_{v_2}t} - e^{+iE_{v_1}t}\right) = 1 - e^{i(E_{v_2}-E_{v_1})t} - e^{i(E_{v_1}-E_{v_2})t} + 1$$
$$= 2 - (e^{i(E_{v_2}-E_{v_1})t} + e^{-i(E_{v_2}-E_{v_1})t}) = 2 - 2\cos[(E_{v_2} - E_{v_1})t]$$
$$= 2(1 - \cos[(E_{v_2} - E_{v_1})t]) = 4\sin^2[(E_{v_2} - E_{v_1})t/2].$$

Using $2\sin\theta\cos\theta = \sin 2\theta$ one finally gets

$$P_{v_e \to v_\mu}(t) = |\langle v_\mu | v_e; t \rangle|^2 = \sin^2 2\theta \sin^2[(E_{v_1} - E_{v_2})t/2].$$

Since the states $|v_e\rangle$ and $|v_\mu\rangle$ are orthogonal, one has $P_{v_e \to v_e}(t) = 1 - P_{v_e \to v_\mu}(t)$. In the approximation of small masses, $E_{v_i} = p + m_i^2/2p + O(m_i^4)$, one gets $E_{v_1} - E_{v_2} \approx (m_1^2 - m_2^2)/2p$.
Since $t = E_{v_i}x/p$ and because of $m_i \ll E_{v_i}$, the momentum p can be identified with the neutrino energy E_v. If, finally, also the correct powers of $\hbar$ and c are introduced, one obtains

$$P_{v_e \to v_e}(x) = 1 - \sin^2 2\theta \sin^2\left(\frac{1}{4\hbar c}\delta m^2\frac{x}{E_v}\right)$$

with $1/4\hbar c = 1.27 \times 10^9$ eV^{-1} km^{-1}, which is the desired relation.
(b) On both sides of the Earth a number of N muon neutrinos are created. The ones from above, being produced at an altitude of 20 km almost have no chance to oscillate, while the ones crossing the whole diameter of the Earth have passed a distance of more than 12 000 km, so that from below only $N P_{v_\mu \to v_\mu}(2R_E)$ muon neutrinos will arrive, leading to a ratio of

$$S = 0.54 = \frac{N P_{v_\mu \to v_\mu}(2R_E)}{N} = 1 - \sin^2(1.27 \times 12\,700 \times \delta m^2).$$

Solving for δm^2 one obtains $\delta m^2 = 4.6 \times 10^{-5}$ eV^2 and $m_{v_\tau} \approx 6.8 \times 10^{-3}$ eV.
(c) No. The masses of the lepton flavour eigenstates are quantum-mechanical expectation values of the mass operator $M = \sqrt{H^2 - p^2}$. For an assumed $(v_\mu \leftrightarrow v_\tau)$ mixing with a similar definition of the mixing angle as for the $(v_e \leftrightarrow v_\mu)$ mixing, one gets

$$m_{v_\mu} = \langle v_\mu | M | v_\mu \rangle = m_1 \cos^2\theta + m_2 \sin^2\theta,$$
$$m_{v_\tau} = \langle v_\tau | M | v_\tau \rangle = m_1 \sin^2\theta + m_2 \cos^2\theta.$$

For maximum mixing ($\theta = 45°$, $\cos^2\theta = \sin^2\theta = 1/2$) one obtains

$$m_{v_\mu} = m_{v_\tau} = (m_1 + m_2)/2.$$

5. The typical flux of geoneutrinos at the Earth's surface is $\Phi_v \approx 3 \times 10^6$ cm^{-2} s^{-1}. Most of the geoneutrinos are electron antineutrinos, and they react in the human

body by the reaction

$$\bar{\nu}_e + p \rightarrow e^+ + n \,. \tag{6.15}$$

The energy threshold for this reaction is 1.8 MeV, and the corresponding cross section is approximately given by $3 \times 10^{-46}\,\text{cm}^2$ for the energy range in question. The interaction rate in the human body is then

$$R = \sigma_N \cdot N_A \cdot d \cdot A \cdot \Phi_\nu \,, \tag{6.16}$$

where σ_N is the cross section per nucleon, N_A the Avogadro constant, d the area density of a human (assumed to be $15\,\frac{\text{g}}{\text{cm}^2}$), and $A = 180 \times 30\,\text{cm}^2$ the humans' projected surface. This leads to

$$R = 4.4 \times 10^{-11}\,\text{s}^{-1} \tag{6.17}$$

or 1.4×10^{-3} per year.

For an assumed energy transfer to the human body of 1 MeV, the energy deposit is

$$\Delta E = R \cdot 1\,\text{MeV} = 2.24 \times 10^{-16}\,\text{J} \,. \tag{6.18}$$

For the mass of the human—as already used for the interaction rate (81 kg)—and the radiation weighting factor for positrons ($w_R = 1$), one gets for the equivalent dose

$$H = 2.8 \times 10^{-18}\,\text{Sv} \,. \tag{6.19}$$

That is almost nothing compared to the natural exposure due to normal cosmic rays, terrestrial radiation, intake of food, and inhalation of radon, which is about 2.5 mSv per year.

6. The flux of geoneutrinos at the surface of the Earth is

$$\Phi_\nu \approx 3 \times 10^6\,\text{cm}^{-2}\,\text{s}^{-1} \,. \tag{6.20}$$

The surface of the Earth is

$$S = 4\pi R^2 = 5.1 \times 10^{18}\,\text{cm}^2 \,. \tag{6.21}$$

The cross section of electron antineutrinos is

$$\sigma_N = 3 \times 10^{-46}\,\text{cm}^2 \,. \tag{6.22}$$

The rate of neutrino interactions in the Earth can then be estimated to be

$$R = \sigma_N \cdot N_A \cdot d_{\text{eff}} \cdot \rho \cdot A \cdot \Phi_\nu \,, \tag{6.23}$$

where d_{eff} is the average effective distance of the interaction target, assumed to be the Earth's radius, ρ the average density of the Earth ($5.5\,g/cm^3$). This leads to

$$R \approx 10^{13}\,s^{-1}\,. \tag{6.24}$$

The energy transfer to the Earth material is estimated to be 3 MeV per interaction, if also the beta decays of the isotopes in the decay chain are considered. This leads to

$$\Delta E = 4.5\,W\,. \tag{6.25}$$

If the total heat production by the decay of radioisotopes (now including also α decays) is considered ($\Delta E_{total} \approx 20\,MeV$ per decay chain), one obtains

$$\Delta E^* = \Phi_\nu \cdot S \cdot \Delta E_{total} \approx 50\,TW\,. \tag{6.26}$$

This very crude estimation is in agreement with more elaborate calculations.

7. The energy-dependent small-value interaction probability of neutrinos is given in Eq. (6.3.47); $\rho = 3.34\,g/cm^3$ is the average density of the Moon, N_A is the Avogadro number, and d the diameter of the Moon. Then the interaction probability is

$$W = N_A\sigma d\rho = 6.022 \times 10^{23} \times 6.7 \times 10^{-34} \times 2 \times 1.737 \times 10^8 \times 3.34 \approx 0.23\,. \tag{6.27}$$

(For larger values one has $W = 1 - \exp(-N_A\sigma d\rho)$.) So, about 77% of the 100 TeV neutrinos will pass the Moon without interaction, However, already for PeV neutrinos the Moon represents an absorber.

Section 6.4

1. $I = I_0\,e^{-\mu x}$ photons survive, $I_0 - I = I_0(1 - e^{-\mu x})$ get absorbed; from Fig. 6.49 one reads $\mu = 0.2\,cm^{-1}$;

detection efficiency: $\quad \eta = \dfrac{I_0(1 - e^{-\mu x})}{I_0} = 1 - e^{-\mu x} \approx 0.45 = 45\%$.

2. The duration of the brightness excursion cannot be shorter than the time span for the light to cross the cosmological object. Figure 6.60 shows $\Delta t = 1\,s \Rightarrow$ size $\approx c\,\Delta t = 300\,000$ km.

3. $\cos\vartheta = \dfrac{1}{n\beta}$; threshold at $\beta > \dfrac{1}{n} \Rightarrow v > \dfrac{c}{n}$;

$$E_\mu = \gamma m_0 c^2 = \frac{1}{\sqrt{1 - \beta^2}}m_0 c^2 = \frac{1}{\sqrt{1 - \frac{1}{n^2}}}m_0 c^2 = \frac{n}{\sqrt{n^2 - 1}}\,m_0 c^2\,,$$

$$E_\mu \approx \begin{cases} 4.5\,GeV \text{ in air} \\ 160.3\,MeV \text{ in water} \end{cases}, \quad E_\mu^{kin} \approx \begin{cases} 4.4\,GeV \text{ in air} \\ 54.6\,MeV \text{ in water} \end{cases}.$$

4. Number of emitted photons:

$$N_E = \frac{P}{h\nu} \approx \frac{3 \times 10^{27}\,W}{10^{11}\,eV \times 1.602 \times 10^{-19}\,J/eV} \approx 1.873 \times 10^{35}\,s^{-1}\,.$$

Solid angle: $\Omega = \dfrac{A}{4\pi R^2} \approx 6.16 \times 10^{-41}$.

Number of recorded photons: $N_R = \Omega N_E \approx 1.15 \times 10^{-5}/\text{s} \approx 364/a$.

Minimum flux: assumed $10/a$, $P_{min} \approx P \dfrac{\Omega}{A} \dfrac{10}{364} \approx 6.35 \times 10^{-19}\,\text{J}/(\text{cm}^2\,\text{s})$.

5. **(a)** Assume isotropic emission, which leads to a total power of $P = 4\pi r^2 P_S$, where $r = 150 \times 10^6$ km is the astronomical unit (distance Sun–Earth). One gets

$$P \approx 3.96 \times 10^{26}\,\text{W}.$$

(b) In a period of 10^6 years the emitted energy is $E = P \times t \approx 1.25 \times 10^{40}$ J, which corresponds to a mass of $m = E/c^2 \approx 1.39 \times 10^{23}$ kg, which represents a relative fraction of the solar mass of

$$f = \frac{m}{M_\odot} \approx 6.9 \times 10^{-8}.$$

(c) The effective area of the Earth is $A = \pi R^2$, so that the daily energy transport to Earth is worked out to be $E = \pi R^2 P_S t$. This corresponds to a mass of

$$m = \frac{E}{c^2} \approx 1.71 \times 10^5\,\text{kg} = 171\,\text{tons}.$$

Section 6.5

1. The power emitted in the frequency interval $[\nu, \nu + d\nu]$ corresponds to the one emitted in the wavelength interval $[\lambda, \lambda + d\lambda]$, $\lambda = c/\nu$, i.e., $P(\nu)\,d\nu = P(\lambda)\,d\lambda$, or

$$P(\lambda) = P(\nu)\frac{d\nu}{d\lambda} \sim \frac{\nu^3}{e^{h\nu/kT}-1}\frac{d\nu}{d\lambda} \sim \frac{1}{\lambda^5(e^{hc/\lambda kT}-1)},$$

since $d\nu/d\lambda = -c/\lambda^2$.

2. The luminosity of a star is proportional to the integral over Planck's radiation formula:

$$L \sim \int_0^\infty \varrho(\nu, T)\,d\nu = \int_0^\infty \frac{8\pi h\nu^3}{c^3}\frac{1}{e^{h\nu/kT}-1}\,d\nu\,;$$

use $x = \frac{h\nu}{kT} \Rightarrow$

$$L \sim \frac{8\pi}{c^3}\left(\frac{kT}{h}\right)^3 h \int_0^\infty x^3 \frac{1}{e^x-1}\frac{kT}{h}\,dx = \frac{8\pi}{c^3}\frac{k^4 T^4}{h^3}\int_0^\infty \frac{x^3\,dx}{e^x-1}$$

$$= \frac{8\pi}{c^3}\frac{k^4}{h^3}\frac{\pi^4}{15}T^4 \sim T^4.$$

In addition, the luminosity varies with the size of the surface ($\sim R^2$). This gives the scaling law

$$\frac{L}{L_\odot} = \left(\frac{R}{R_\odot}\right)^2 \left(\frac{T}{T_\odot}\right)^4 \; ;$$

(a) $R = 10\, R_\odot$, $T = T_\odot$ $\Rightarrow$ $L = 100\, L_\odot$;
(b) $R = R_\odot$, $T = 10\, T_\odot$ $\Rightarrow$ $L = 10\,000\, L_\odot$.

"This is the simplified version
for our beginners!"

3. The motion of an electron in a transverse magnetic field is described by

$$\frac{mv^2}{\varrho} = evB \quad \Rightarrow \quad \frac{p}{\varrho} = eB.$$

Since at high energies $cp \approx E$, one gets

$$\frac{cp}{\varrho} = ceB \quad \Rightarrow \quad B = \frac{E}{ce\varrho},$$

where ϱ is the bending radius. This leads to

$$P = \frac{e^2 c^3}{2\pi} C_\gamma E^2 \frac{E^2}{c^2 e^2 \varrho^2} = \frac{c C_\gamma}{2\pi} \frac{E^4}{\varrho^2}.$$

The energy loss for one revolution around the pulsar is

$$\Delta E = \int_0^T P \, d\tau = \frac{cC_\gamma}{2\pi} \frac{E^4}{\varrho^2} 2\pi \varrho/c = C_\gamma \frac{E^4}{\varrho}$$

$$= 8.85 \times 10^{-5} \times \frac{10^{12} \, \text{GeV}^4}{10^6 \, \text{m}} \, \text{m} \, \text{GeV}^{-3} = 88.5 \, \text{GeV} \,.$$

The magnetic field is

$$B = \frac{E}{ce\varrho} = \frac{10^{12} \, \text{eV} \times 1.6 \times 10^{-19} \, \text{J/eV}}{3 \times 10^8 \, \text{m/s} \times 1.6 \times 10^{-19} \, \text{A s} \times 10^6 \, \text{m}} = 0.0033 \, \text{T} = 33 \, \text{G} \,.$$

4. The Lorentz force is related to the absolute value of the momentum change by $|\dot{p}| = F = evB$. The total radiated power is taken from the kinetic (or total) energy of the particle, i.e., $\dot{E}_{\text{kin}} = \dot{E} = -P$. From the centrifugal force the bending radius results to $\varrho = p/eB$, see also Problem 3 in this section. As differential equation for the energy one gets

$$\dot{E} = -P = -\frac{2}{3} \frac{e^4 B^2}{m_0^2 c^3} \gamma^2 v^2 \,.$$

(a) In general $E = \gamma m_0 c^2$ holds. The ultrarelativistic limit is given by $v \to c$, hence

$$\dot{E} = -\frac{2}{3} \frac{e^4 B^2 c^3}{(m_0 c^2)^4} E^2 = -\alpha E^2 \,, \quad \alpha = \frac{2}{3} \frac{e^4 B^2 c^3}{(m_0 c^2)^4} \,.$$

The solution of this differential equation is obtained by separation of variables,

$$\int_{E_0}^E \frac{dE'}{E'^2} = -\alpha \int_0^t dt' \quad \Rightarrow \quad \frac{1}{E_0} - \frac{1}{E} = -\alpha t \quad \Rightarrow \quad E = \frac{E_0}{1 + \alpha E_0 t} \,.$$

It is only valid for $E \gg m_0 c^2$. (This limit is not respected for longer times.) The bending radius follows with $E \to pc$ to

$$\varrho = \frac{E}{ceB} = \frac{1}{ceB} \frac{E_0}{1 + \alpha E_0 t} = \frac{\varrho_0}{1 + \alpha ce B \varrho_0 t} \,.$$

(b) In the general (relativistic) case there is $p^2 = \gamma^2 m_0^2 v^2$, i.e., $\gamma^2 v^2 = p^2/m_0^2 = (E^2 - m_0^2 c^4)/m_0^2 c^2$, thus

$$\dot{E} = -\frac{2}{3} \frac{e^4 B^2 c^3}{(m_0 c^2)^4} (E^2 - m_0^2 c^4) = -\alpha (E^2 - m_0^2 c^4) \,, \quad \alpha \text{ as in (a)} \,.$$

The solution is calculated as in (a):

$$\int_{E_0}^{E} \frac{dE'}{E'^2 - m_0^2 c^4} = -\int_0^t dt' = -\alpha t \quad \Rightarrow$$

$$\frac{1}{2m_0 c^2} \int_{E_0}^{E} \left(\frac{1}{E' - m_0 c^2} - \frac{1}{E' + m_0 c^2} \right) dE' = \frac{1}{2m_0 c^2} \ln \frac{E' - m_0 c^2}{E' + m_0 c^2} \Bigg|_{E'=E_0}^{E}$$

$$= -\frac{1}{m_0 c^2} \left(\frac{1}{2} \ln \frac{1 + \frac{m_0 c^2}{E}}{1 - \frac{m_0 c^2}{E}} - \frac{1}{2} \ln \frac{1 + \frac{m_0 c^2}{E_0}}{1 - \frac{m_0 c^2}{E_0}} \right)$$

$$= -\frac{1}{m_0 c^2} \left(\text{artanh} \frac{m_0 c^2}{E} - \text{artanh} \frac{m_0 c^2}{E_0} \right) = -\alpha t \, .$$

Solving this equation for E leads to

$$E = \frac{m_0 c^2}{\tanh \left(\alpha m_0 c^2 t + \text{artanh} \frac{m_0 c^2}{E_0} \right)} = m_0 c^2 \coth \left(\alpha m_0 c^2 t + \text{artanh} \frac{m_0 c^2}{E_0} \right) .$$

For larger times (in the non-relativistic regime) the rest energy is approached exponentially.

From this the bending radius results in

$$\varrho = \frac{p}{eB} = \frac{\sqrt{E^2 - m_0^2 c^4}}{ceB} = \frac{m_0 c}{eB} \sqrt{\coth^2 \left(\alpha m_0 c^2 t + \text{artanh} \frac{m_0 c^2}{E_0} \right) - 1}$$

$$= \frac{m_0 c / eB}{\sinh \left(\alpha m_0 c^2 t + \text{artanh} \frac{m_0 c^2}{E_0} \right)} = \frac{m_0 c / eB}{\sinh \left(\alpha m_0 c^2 t + \text{artanh} \frac{m_0 c}{\sqrt{\varrho_0^2 e^2 B^2 + m_0^2 c^2}} \right)} .$$

5. Measured flux R = source flux $R^* \times$ efficiency $\varepsilon \times$ solid angle Ω;

$$\varepsilon = 1 - e^{-\mu x} = 1 - e^{-125 \times 5.8 \times 10^{-3}} \approx 51.6\% \, , \quad \Omega = \frac{10^4 \, \text{cm}^2}{4\pi (55 \, \text{kpc})^2} \approx 2.76 \times 10^{-44} \, ,$$

$$R^* = \frac{R}{\varepsilon \, \Omega} \approx 1.95 \times 10^{40} / \text{s} \, .$$

6. In contrast to Compton scattering of photons on an electron target at rest, all three-momenta are different from zero in this case. For the four-momenta k_i, k_f (photon) and q_i, q_f (electron) one has $k_i - k_f = q_f - q_i$. Squaring this equation gives $k_i k_f = q_i q_f - m_e^2$. On the other hand, rewriting the four-momentum conservation as $q_f = q_i + k_i - k_f$ and multiplying with q_i leads to $q_i q_f - m_e^2 = q_i (k_i - k_f)$, yielding

$$k_i k_f = q_i (k_i - k_f) \, .$$

Let ϑ be the angle between k_i and k_f, which gives

$$\omega_i \omega_f (1 - \cos \vartheta) = E_i (\omega_i - \omega_f) - |\boldsymbol{q}_i| (\omega_i \cos \varphi_i - \omega_f \cos \varphi_f) \, .$$

Solving for ω_f leads to

$$\omega_f = \omega_i \frac{1 - \sqrt{1 - (m_e/E_i)^2} \cos \varphi_i}{1 - \sqrt{1 - (m_e/E_i)^2} \cos \varphi_f + \omega_i(1 - \cos \vartheta)/E_i} .$$

This expression is still exact. If the terms ω_i/E_i and m_e/E_i are neglected, the relation quoted in the problem is obtained.

7. Starting from the derivative $dP/d\nu$,

$$dP/d\nu \sim \frac{3\nu^2 (e^{h\nu/kT} - 1) - \nu^3 \frac{h}{kT} e^{h\nu/kT}}{(e^{h\nu/kT} - 1)^2} = \nu^2 \frac{3 (e^x - 1) - x e^x}{(e^x - 1)^2} ,$$

and the condition $dP/d\nu = 0$ one obtains the equation

$$e^{-x} = 1 - \frac{x}{3}, \quad x = \frac{h\nu}{kT} .$$

An approximated solution to this transcendental equation gives $x \approx 2.8$, leading to a frequency, resp. energy of the maximum of the Planck distribution of

$$h\nu_M \approx 2.8 \, kT .$$

This linear relation between the frequency in the maximum and the temperature is called *Wien's displacement law*. For $h\nu_M = 50 \, \text{keV}$ a temperature of

$$T \approx 2 \times 10^8 \, \text{K} .$$

is obtained.

Section 6.6

1. $\Delta E = \dfrac{GMm_\gamma}{R} - \dfrac{GMm_\gamma}{R + H} = GMm_\gamma \dfrac{H}{R (H + R)} ,$

$\dfrac{\Delta E}{E} = \dfrac{GM}{c^2} \dfrac{H}{R (H + R)} = \dfrac{GM}{R^2 c^2} \dfrac{H R}{H + R} = \dfrac{g_\odot}{c^2} \dfrac{H R}{H + R} \approx \dfrac{g_\odot}{c^2} H \approx 3 \times 10^{-12}$

for $H \ll R$.

2. (a) $P \approx \dfrac{G}{c^5} \omega^6 m^2 r^4 .$

For $\omega = 1/\text{year}$, $m = 5.97 \times 10^{24} \, \text{kg}$, $r = 1 \, \text{AU}$ one gets

$$P \approx 5 \times 10^{-4} \, \text{W} = 0.5 \, \text{mW} .$$

This is only a very small fraction of about 1.3×10^{-30} of the solar emission in the optical range.

(b) Assume $\omega = 10^3 \, \text{s}^{-1}$, $m = 10 \, \text{kg}$, $r = 1 \, \text{m}$, which leads to

$$P \approx 2.7 \times 10^{-33} \, \text{W} \, .$$

Chapter 7

1. $p = \dfrac{F}{A} = \dfrac{mg}{A} = 1.013 \times 10^5 \, \dfrac{\text{N}}{\text{m}^2} \, ,$

   $\dfrac{m}{A} = \dfrac{p}{g} = \dfrac{1.013 \times 10^5}{9.81} \, \dfrac{\text{kg}}{\text{m}^2} \approx 10\,326 \, \text{kg/m}^2 \approx 1.03 \, \text{kg/cm}^2 \, .$

2. $p = p_0 \, \text{e}^{-20/7.99} \approx 82.9 \, \text{hPa} \, ,$ $\dfrac{m}{A} \approx \dfrac{8.29 \times 10^3}{9.81} \, \dfrac{\text{kg}}{\text{m}^2} \approx 845 \dfrac{\text{kg}}{\text{m}^2} = 84.5 \, \text{g/cm}^2 \, .$

3. The differential sea-level muon spectrum can be parameterized by

   $$N(E)\,\text{d}E \sim E^{-\gamma}\,\text{d}E \quad \text{with} \quad \gamma = 3 \quad \Rightarrow \quad I(E) = \int N(E)\,\text{d}E \sim E^{-2} \, ,$$

   where $I(E) \to 0$ for $E \to \infty$. The thickness of the atmosphere varies with zenith angle like

   $$d(\theta) = \frac{d(0)}{\cos\theta} \, .$$

   For 'low' energies ($E <$ several $100 \, \text{GeV}$) the muon energy loss is constant $\left(\frac{\text{d}E}{\text{d}x} = a \right)$

   $\Rightarrow E \sim d \, , \quad I(\theta) \sim E^{-2} \sim d^{-2} = I(0) \cos^2\theta \, .$

4. $N(>E, R) = A(aR)^{-\gamma} \, , \quad \text{see (7.3.9)} \, ,$

   $\dfrac{\Delta N}{\Delta R} = -\gamma A a^{-\gamma} R^{-(\gamma+1)} \, , \quad \dfrac{\Delta N}{N} = -\gamma R^{-1} \Delta R = -\gamma \dfrac{\Delta R}{R} \, ,$

   traditionally $\quad \Delta R = 100 \, \text{g/cm}^2$.

   With $\gamma = 2$ (exponent of the integral sea-level spectrum):

   $\left| \dfrac{\Delta N}{N} \right| = \left| \dfrac{S}{N} \right| = \dfrac{200 \, \text{g/cm}^2}{R} = 2 \times 10^{-2} \, , \quad \text{e.g., for } 100 \, \text{m w.e.}$

5. $\vartheta_{\text{r.m.s.}} = \dfrac{13.6 \, \text{MeV}}{\beta c p} \sqrt{\dfrac{x}{X_0}} \left(1 + 0.038 \ln \dfrac{x}{X_0} \right) \qquad \text{(see [112])}$

   $\approx 4.87 \times 10^{-3}(1 + 0.27) \approx 6.19 \times 10^{-3} \approx 0.35° \, .$

6. The lateral separation is caused by
   (a) transverse momenta in the primary interaction,
   (b, c) multiple scattering in the air and rock.

   (a) $p_\text{T} \approx 350 \, \text{MeV}/c \, ,$

   $\vartheta = \dfrac{p_\text{T}}{p} \approx \dfrac{350 \, \text{MeV}/c}{100 \, \text{GeV}/c} = 3.5 \times 10^{-3} \, ,$

   average displacement: $\quad \Delta x_1 = \vartheta \, h$, where h is the production height $(20 \, \text{km})$,

$$\Delta x_1 \approx 70\,\text{m}\,.$$

(b) Multiple-scattering angle in air,

$$\vartheta_{\text{air}} \approx 6.90 \times 10^{-4}(1 + 0.123) \approx 7.75 \times 10^{-4} \approx 0.044°\,,$$

$$\Delta x_2 = 15.5\,\text{m}\,.$$

(c) Multiple scattering in rock,

$$\Delta x_3 \approx 0.8\,\text{m}\,.$$

$$\Rightarrow \Delta x = \sqrt{\Delta x_1^2 + \Delta x_2^2 + \Delta x_3^2} \approx 72\,\text{m}\,.$$

7. Essentially only those geomagnetic latitudes are affected, where the geomagnetic cutoff exceeds the atmospheric cutoff. This concerns latitudes between 0° and approximately 50°. The average latitude can be worked out from

$$\langle \lambda \rangle = \frac{\int_{0°}^{50°} \lambda \cos^4 \lambda \, d\lambda}{\int_{0°}^{50°} \cos^4 \lambda \, d\lambda} \approx 18.4°$$

corresponding to an average geomagnetic cutoff of $\langle E \rangle \approx 12\,\text{GeV}$. For zero field all particles with $E \geq 2\,\text{GeV}$ have a chance to reach sea level. Their intensity is $(\varepsilon = E/\text{GeV})$

$$N_1(>2\,\text{GeV}) = a \int_2^\infty \varepsilon^{-2.7}\, d\varepsilon = -(a/1.7)\,\varepsilon^{-1.7}\Big|_2^\infty = 0.181\,a\,.$$

With full field on, only particles with $E > \langle E \rangle = 12\,\text{GeV}$ will reach sea level for latitudes between 0° and 50°:

$$N_2(>12\,\text{GeV}) \approx 0.0086\,a\,.$$

These results have to be combined with the surface of the Earth that is affected. Assuming isotropic incidence and zero field, the rate of cosmic rays at sea level is proportional to the surface of the Earth,

$$\Phi_1 = A\,N_1 = 4\pi R^2\,N_1\,.$$

With full field on, only the part of the surface of Earth is affected, for which $0° \leq \lambda \leq 50°$. The relevant surfaces can be calculated from elementary geometry yielding a flux

$$\Phi_2 = A(50°\text{--}90°)\,N_1 + A(0°\text{--}50°)\,N_2\,,$$

where

$$A(50°\text{–}90°) = 2\pi \left[(R\cos 50°)^2 + (R(1 - \sin 50°))^2 \right]$$

and

$$A(0°\text{–}50°) = 4\pi R^2 - A(50°\text{–}90°) .$$

This crude estimate leads to $\Phi_1/\Phi_2 = 3.7$, i.e., in periods of transition when the magnetic field went through zero, the radiation load due to cosmic rays was higher by about a factor of 4.

8. A charged particle traversing a medium with refractive index n with a velocity v exceeding the velocity of light c/n in that medium, emits Cherenkov radiation. The threshold condition is given by

$$\beta_{\text{thres}} = \frac{v_{\text{thres}}}{c} \geq \frac{1}{n} . \tag{7.1}$$

The angle of emission increases with the velocity reaching a maximum value for $\beta = 1$, namely,

$$\Theta_{\mathrm{c}}^{\mathrm{max}} = \arccos \frac{1}{n}. \tag{7.2}$$

The threshold velocity translates into a threshold energy

$$E_{\mathrm{thres}} = \gamma_{\mathrm{thres}} m_0 c^2 \tag{7.3}$$

yielding

$$\gamma_{\mathrm{thres}} = \frac{1}{\sqrt{1 - \beta_{\mathrm{thres}}^2}} = \frac{n}{\sqrt{n^2 - 1}}. \tag{7.4}$$

The number of Cherenkov photons per path length $\mathrm{d}x$ is given by

$$\frac{\mathrm{d}N}{\mathrm{d}x} = 2\pi \alpha z^2 \int \left(1 - \frac{1}{n^2 \beta^2}\right) \frac{\mathrm{d}\lambda}{\lambda^2} \tag{7.5}$$

for $n(\lambda) > 1$, z—electric charge of the incident particle, λ—wavelength, and α—fine-structure constant. The yield of Cherenkov-radiation photons is proportional to $1/\lambda^2$, but only for those wavelengths, where the refractive index is larger than unity. Integrating Eq. (7.5) over the radio range from 40 to 80 MHz ($\lambda = 7.5$ m to $\lambda = 3.75$ m) gives

$$\frac{\mathrm{d}N}{\mathrm{d}x} = 2\pi \alpha z^2 \frac{\lambda_2 - \lambda_1}{\lambda_1 \lambda_2} \sin^2 \Theta_{\mathrm{c}} \approx \frac{0.006}{\mathrm{m}} \cdot z^2 \cdot \sin^2 \Theta_{\mathrm{c}}.$$

The Cherenkov angle in air is $n = 1.00029$, and therewith $\sin^2 \Theta_{\mathrm{c}} \approx 6.0 \times 10^{-4}$, which then gives for an electron ($z = 1$):

$$\frac{\mathrm{d}N}{\mathrm{d}x} \approx 3.5 \times 10^{-6}/\mathrm{m} \tag{7.6}$$

or, for a path length of 100 m, just 3.6×10^{-4}, which is a very small number compared to the number of geosynchrotron photons produced under otherwise identical conditions. Therefore, this effect can be safely neglected for the radio measurement of air showers.

9. The flux of microwaves is measured in W/m^2. If the bandwidth is considered, the flux density in radio astronomy is normally given in $\mathrm{W}/\mathrm{m}^2/\mathrm{Hz}$, where mostly the unit jansky is used, where $1\,\mathrm{Jy} = 10^{-26}\,\mathrm{W}/\mathrm{m}^2/\mathrm{Hz}$. A typical radio source has flux densities between $1\,\mathrm{Jy}$ and $1\,\mathrm{mJy}$.

$$P_\nu = \frac{1}{2} P_{\mathrm{oven}} \cdot \frac{1}{4\pi r^2} \cdot \frac{1}{\delta \nu_{\mathrm{oven}}}, \tag{7.7}$$

where P_{oven} is the power of the microwave oven, r the distance to the Moon, and $\delta\nu_{oven}$ the bandwidth centered on 2.7 GHz. The factor 1/2 takes into account that one observes normally in horizontal or vertical polarization only. This gives

$$P_\nu \approx 5 \times 10^{-26}\,\text{W/m}^2/\text{Hz}, \tag{7.8}$$

which is a radio power that could be seen on Earth by typical radio telescopes. See also [279].

10. In accelerator physics the energy loss by synchrotron radiation per turn in a circular synchrotron is usually given by

$$\Delta E[\text{keV}] = 88.5 \cdot \frac{(E[\text{GeV}])^4}{\rho[\text{m}]}, \tag{7.9}$$

where ρ is the bending radius. In the atmosphere the bending radius for relativistic 100 MeV electrons is

$$\rho[\text{m}] = \frac{p[\text{GeV/c}]}{0.3 \cdot B[\text{T}]}, \tag{7.10}$$

which leads to a bending radius of $\rho \approx 6700\,\text{m}$ in the 50 μT Earth magnetic field. The corresponding energy loss per turn (a 100 MeV electron would not make a complete turn in the atmosphere because of its ionization energy loss) is

$$\Delta E = 1.3 \times 10^{-6}\,\text{keV}. \tag{7.11}$$

If one assumes that the electron travels on its curved path about 100 m, the corresponding energy loss is only 3.1×10^{-9} keV.

How many synchrotron photons of 60 MHz will be created over the path length of 100 m?

$$E_{60\,\text{MHz}} = 2.5 \times 10^{-7}\,\text{eV}, \tag{7.12}$$

which leads to

$$N = 12.4, \tag{7.13}$$

i.e., about 12 photons on a path of 100 m length.

11. Using $T = 25°$ and $S = 35$ parts per thousand for shallow depths and $T = 4°$ and $z = 1000\,\text{m}$ for larger depths and constant salinity, one gets $c_{shallow} = 1547\,\text{m/s}$ and $c_{deep} = 1534\,\text{m/s}$. This is a 1% effect, which might lead to a systematic error for the determination of the angle of incidence.

12. The attenuation is really very small. For 100 Hz the attenuation over 1 km is only 1.00014. Therefore, huge effective volumes may be possible!

13. The ambient noise level has to be transformed into a pressure amplitude. The definition of decibel is

$$L \approx 20\lg\left(\frac{p}{p_0}\right)\,\text{dB}, \tag{*}$$

where p_0 is a reference pressure defined to be $20\,\mu\mathrm{Pa}$.
Equation (*) solved for p gives

$$p = p_0 \times 10^{L/20} = 2 \times 10^{-5} \times 10^{50/20} = 6.3\,\mathrm{mPa}\,. \tag{7.14}$$

This result, inserted into Eq. (7.9.4), leads to

$$E = \frac{p}{6 \times 10^{-21}} \approx 10^{18}\,\mathrm{eV}\,. \tag{7.15}$$

Chapter 8

1. Due to the relative motion of the galaxy away from the observer, the energy of the photon appears decreased. The energies of emitted photons and observed photons are related by the Lorentz transformation,

$$E_{\mathrm{em}} = \gamma E_{\mathrm{obs}} + \gamma \beta c p_{\|\mathrm{obs}} = \gamma(1 + \beta) E_{\mathrm{obs}}\,.$$

Since $E = h\nu = hc/\lambda$, one gets

$$\frac{1}{\lambda_{\mathrm{em}}} = \frac{1}{\lambda_{\mathrm{obs}}} \gamma(1 + \beta) \quad \Rightarrow \quad \lambda_{\mathrm{obs}} = \lambda_{\mathrm{em}} \gamma(1 + \beta)\,,$$

$$z = \frac{\lambda_{\mathrm{obs}} - \lambda_{\mathrm{em}}}{\lambda_{\mathrm{em}}} = \gamma(1 + \beta) - 1 = \frac{1}{\sqrt{1 - \beta^2}}(1 + \beta) - 1$$

$$= \frac{1 + \beta}{\sqrt{(1 + \beta)(1 - \beta)}} - 1 = \frac{\sqrt{1 + \beta}}{\sqrt{1 - \beta}} - 1\,,$$

which reduces for $\beta \ll 1$ to

$$z = \sqrt{(1 + \beta)(1 + \beta)} - 1 \approx \beta.$$

2. For an orbital motion in a gravitational potential one has

$$\frac{mv^2}{R} = G\frac{mM}{R^2} \quad \Rightarrow \quad \frac{1}{2}mv^2 = \frac{1}{2}G\frac{mM}{R}\,,$$

i.e., $E_{\mathrm{kin}} = \frac{1}{2}E_{\mathrm{pot}}$. The kinetic energy of the gas cloud is simply

$$E_{\mathrm{kin}} = \frac{3}{2}kT\frac{M}{\mu}\,.$$

The potential energy can be obtained by integration: the mass in a spherical subvolume of radius r is

$$m = \frac{4}{3}\pi r^3 \varrho .$$

A spherical shell surrounding this volume contains the mass

$$dm = 4\pi r^2 \varrho\, dr$$

leading to a potential energy of gravitation of

$$dE = \frac{Gm\, dm}{r} = \frac{G}{r}\frac{4}{3}\pi r^3 \varrho\, 4\pi r^2 \varrho\, dr = \frac{(4\pi)^2}{3}\varrho^2 r^4 G\, dr ,$$

so that the potential energy is

$$E_{\text{pot}} = \int_0^R dE = \frac{(4\pi)^2}{3}\varrho^2 G \frac{R^5}{5} .$$

Using $M = \frac{4}{3}\pi R^3 \varrho$, one obtains

$$E_{\text{pot}} = \frac{3}{5}G\frac{M^2}{R} .$$

From the relation between E_{kin} and E_{pot} above one gets

$$2E_{\text{kin}} = 3kT\frac{M}{\mu} = \frac{3}{5}G\frac{M^2}{R} \quad \text{and} \quad R = \frac{1}{5}\frac{GM\mu}{kT} .$$

Then

$$M = \frac{4\pi}{3}R^3 \varrho = \frac{4\pi}{3}\varrho\left(\frac{1}{5}\right)^3\frac{G^3 M^3 \mu^3}{(kT)^3}$$

and finally

$$M \approx \left(\frac{kT}{\mu G}\right)^{3/2}\frac{1}{\sqrt{\varrho}} \times 5.46 .$$

If $M_{\text{cloud}} > M$, the cloud gets unstable and collapses.

3. $\dfrac{mv^2}{R} \geq \dfrac{GMm}{R^2} \quad \Rightarrow \quad v^2 \geq \dfrac{GM}{R} .$

Since the density is constant, $M = \frac{4}{3}\pi\varrho R^3$, the revolution time will be

$$T_r = 2\pi R/v = 2\pi\frac{R\sqrt{R}}{\sqrt{GM}} = 2\pi\frac{R^{3/2}}{\sqrt{G}\sqrt{\varrho}R^{3/2}\sqrt{\frac{4\pi}{3}}} = \frac{\sqrt{3\pi}}{\sqrt{G\varrho}} ,$$

so that

$$v = \sqrt{\frac{4\pi}{3}} R \sqrt{G\varrho} \sim R$$

if ϱ = constant.

This behaviour is characteristic of the orbital velocities of stars in galaxies not too far away from the galactic center (see Figs. 1.26 or 13.3).

4. Photon mass $m = \dfrac{h\nu}{c^2} = \dfrac{E}{c^2}$;

energy loss of a photon in a gravitational potential ΔU: $\Delta E = m\Delta U$;

reduced photon energy: $E' = E - \Delta E = h\nu - \dfrac{h\nu}{c^2}\Delta U \Rightarrow$

$\nu' = \nu\left(1 - \dfrac{\Delta U}{c^2}\right)$, $\dfrac{\Delta\nu}{\nu} = \dfrac{\Delta U}{c^2}$;

gravitational potential: $\Delta U = \dfrac{GM}{R} \Rightarrow$

$\dfrac{\Delta\nu}{\nu} = \dfrac{GM}{Rc^2} \Rightarrow \dfrac{GM}{Rc^2}$ is dimensionless;

using the Schwarzschild radius $R_S = \dfrac{2GM}{c^2}$ one has $\dfrac{\Delta\nu}{\nu} = \dfrac{R_S}{2R}$.

This result is, however, only valid far away from the event horizon. The exact result from general relativity reads, see, e.g., [280]: $\Delta\nu/\nu = z = 1/\sqrt{1 - R_S/R} - 1$.

5.

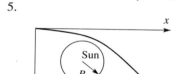

Acceleration due to gravity at the solar surface: $g = GM/R^2$.

Assumption: the deflection takes place essentially over the Sun's diameter $2R$.

The photons travel on a parabola:

$$y = \frac{g}{2}t^2, \quad x = ct \Rightarrow y(x) = \frac{GM}{2R^2}\frac{x^2}{c^2}.$$

The deflection δ corresponds to the increase of $y(x)$ at $x = 2R$,

$$\frac{dy}{dx} = \frac{GM}{R^2c^2}x \Rightarrow y'(2R) = \frac{2GM}{Rc^2} = \frac{R_S}{R} = \delta,$$

$R = 6.961 \times 10^5$ km, $M = 1.9884 \times 10^{30}$ kg $\Rightarrow \delta \approx 4.24 \times 10^{-6} \approx 0.87$ arcsec, 1 arcsec = $1''$.

This is the classical result using Newton's theory. The general theory of relativity gives $\delta^* = 2\delta = 1.75$ arcsec.

The solution to this problem may alternatively be calculated following Problem 13.7, see also Problem 3.4.

6. An observer in empty space measures times with an atomic clock of frequency ν_0. The signals emitted by an identical clock on the surface of the pulsar reach the observer in empty space with a frequency $\nu = \nu_0 - \Delta\nu$, where $\dfrac{\Delta\nu}{\nu_0} = \dfrac{\Delta U}{c^2} = \dfrac{GM}{Rc^2}$ (see Problem 4 in this chapter) $\Rightarrow$

$$\frac{f_{\text{pulsar}}}{f_{\text{empty space}}} = \frac{\nu_0 - \Delta\nu}{\nu_0} = 1 - \frac{GM}{Rc^2} = 1 - \varepsilon.$$

For the pulsar this gives $\varepsilon = 0.074$, i.e., the clocks in the gravitational potential on the surface of the pulsar are slow by 7.4%. For our Sun the relative slowing-down rate, e.g., with respect to Earth, is 2×10^{-6}. At the surface of the Earth clocks run slow by 1.06×10^{-8} with respect to clocks far away from any mass, from which just 7×10^{-10} results from the Earth's gravitation and the main contribution is caused by the gravitational potential of the Sun.

7. Gravitational force

$$\mathrm{d}F = -G\frac{M(r)\,\mathrm{d}m}{r^2} = -\frac{GM(r)}{r^2}\varrho(r)\,\mathrm{d}r\,4\pi r^2, \quad \text{inward force}$$

$$\mathrm{d}p = \frac{\mathrm{d}F}{4\pi r^2} = -\frac{GM(r)}{r^2}\varrho(r)\,\mathrm{d}r, \quad \frac{\mathrm{d}p}{\mathrm{d}r} = -\frac{GM(r)}{r^2}\varrho(r). \qquad (*)$$

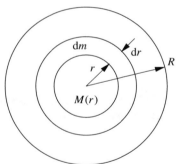

On the other hand $\dfrac{\mathrm{d}p}{\mathrm{d}r} \approx \dfrac{p(R) - p(r = 0)}{R} = -\dfrac{p}{R}$. Compare with $(*)$ and with the replacement $r \to R$ and $\varrho(r) \to$ average density ϱ one gets $\dfrac{p}{R} = \dfrac{GM}{R^2}\varrho$. If a uniform density $\varrho(r) = \varrho$ is assumed, $(*)$ can be integrated directly using $M(r) = \dfrac{4\pi}{3}\varrho r^3$ and thus $\dfrac{\mathrm{d}p}{\mathrm{d}r} = -\dfrac{4\pi}{3}G\varrho^2 r$,

$$p(0) = \int_R^0 \frac{\mathrm{d}p}{\mathrm{d}r}\,\mathrm{d}r = \frac{4\pi}{3}G\varrho^2 \int_0^R r\,\mathrm{d}r = \frac{4\pi}{3}G\varrho^2\frac{R^2}{2} = \frac{1}{2}\frac{GM}{R}\varrho.$$

This leads to $p = \dfrac{1}{2}\dfrac{GM}{R}\varrho \approx \begin{cases} 1.3 \times 10^{14}\,\mathrm{N/m^2} & \text{for the Sun} \\ 1.7 \times 10^{11}\,\mathrm{N/m^2} & \text{for the Earth} \end{cases}$.

8. Schwarzschild radius

$$\left.\begin{array}{l} R_S = \frac{2GM}{c^2} \\ M = \frac{4}{3}\pi R_S^3 \varrho \end{array}\right\} \;\Rightarrow\; \left(\frac{3}{4\pi\varrho}M\right)^{1/3} = \frac{2GM}{c^2} \;\Rightarrow\; \left(\frac{c^2}{2GM}\right)^3 \frac{3M}{4\pi} = \varrho .$$

$$\varrho = \frac{3c^6}{32\pi G^3 M^2} = \frac{3c^6}{32\pi G^3 M_\odot^2}\left(\frac{M_\odot}{M}\right)^2 \approx 1.8 \times 10^{19}\left(\frac{M_\odot}{M}\right)^2 \frac{\mathrm{kg}}{\mathrm{m}^3},$$

(a) $M \approx 10^{11} M_\odot \;\Rightarrow\; \varrho \approx 1.8 \times 10^{-3}\,\mathrm{kg/m^3}$
(b) $M = M_\odot \;\Rightarrow\; \varrho \approx 1.8 \times 10^{19}\,\mathrm{kg/m^3}$
(c) $M = 10^{15}\,\mathrm{kg} \;\Rightarrow\; \varrho \approx 7.3 \times 10^{49}\,\mathrm{kg/m^3}$

9. (a) $\dfrac{mv^2}{R} = \dfrac{GmM}{R^2} = \dfrac{Gm}{R^2}\dfrac{4}{3}\pi\varrho R^3 \;\Rightarrow\; v^2 = \dfrac{4}{3}\pi\varrho G R^2 \;\Rightarrow$

$$v = R\sqrt{\frac{4}{3}\pi\varrho G} = 245\,\mathrm{km/s} \quad \text{for } R = 20\,000 \text{ light-years}.$$

(b) For $R > 20\,000$ light-years the orbital velocities show Keplerian characteristics,

$$\frac{mv^2}{R} \approx G\frac{mM}{R^2} \;\Rightarrow\; v^2 \approx G\frac{M}{R}, \quad M \approx v^2\frac{R}{G} = 1.7 \times 10^{41}\,\mathrm{kg} = 8.6 \times 10^{10} M_\odot.$$

(c) The energy density of photons in the universe is approximately $0.3\,\mathrm{eV/cm^3}$. The critical density amounts to $\varrho_c = 0.945 \times 10^{-29}\,\mathrm{g/cm^3}$, which corresponds to an energy density of $\varrho_c c^2 = 5.3 \times 10^3\,\mathrm{eV/cm^3}$, which means $\varrho_{\mathrm{photons}}/\varrho_c c^2 \approx 5 \times 10^{-5}$; i.e., photons contribute only a small fraction to the total Ω parameter.

Chapter 9

1. For a gas of non-relativistic particles the ideal-gas law holds: $P = nT$. The density is given by $\varrho = nm$. Since $T \ll m$ for non-relativistic particles, one has $P \approx 0$. The fluid equation for $P = 0$ reads

$$\dot\varrho + \frac{3\dot R}{R}\varrho = 0 \;\Rightarrow\; \frac{1}{R^3}\frac{d}{dt}\left(\varrho R^3\right) = 0 \;\Rightarrow\; \varrho R^3 = \mathrm{const} \;\Rightarrow\; \varrho \sim \frac{1}{R^3}.$$

2. Assume $P = 0$, which gives $\varrho \sim \frac{1}{R^3}$ (see Problem 1 in this chapter). The last relation can be parameterized by

$$\varrho = \varrho_0 \left(\frac{R_0}{R}\right)^3.$$

The Friedmann equation can be approximated:

$$\frac{\dot R^2}{R^2} + \frac{k}{R^2} = \frac{8\pi}{3}G\varrho \;\Rightarrow\; \frac{\dot R^2}{R^2} = \frac{8\pi}{3}G\varrho,$$

since the second term on the left-hand side in the Friedmann equation is small compared to ϱ $(\sim \frac{1}{R^3})$ for the early universe.

With the ansatz

$$R = A\,t^p\,, \quad \dot{R} = p\,A\,t^{p-1}$$

one gets

$$\frac{\dot{R}^2}{R^2} = \frac{p^2 A^2 t^{2p} t^{-2}}{A^2 t^{2p}} = \frac{p^2}{t^2} = \frac{8\pi}{3} G\varrho_0 \left(\frac{R_0}{R}\right)^3 = \frac{8\pi}{3} G\varrho_0 R_0^3 A^{-3} t^{-3p}\,.$$

Comparing the t dependence on the right- and left-hand sides gives

$$p = \frac{2}{3} \quad \Rightarrow \quad R = A\,t^{2/3}\,.$$

In this procedure the constant of proportionality A is automatically fixed as

$$A = \sqrt[3]{6\pi\,G\varrho_0\,R_0}\,.$$

3. For the early universe one can approximate the Friedmann equation by

$$\frac{\dot{R}^2}{R^2} = \frac{8\pi}{3} G\varrho\,; \quad \text{with } \varrho = \frac{\pi^2}{30} g T^4 \quad \Rightarrow \quad \frac{\dot{R}^2}{R^2} = \frac{8\pi}{3} G \frac{\pi^2}{30} g T^4 = \frac{8\pi^3}{90} G g T^4\,.$$

Since $\dfrac{\dot{R}}{R} = H$ and $G = \dfrac{1}{m_{\mathrm{Pl}}^2}$, one has

$$H = \sqrt{\frac{8\pi^3 g}{90}\,\frac{T^2}{m_{\mathrm{Pl}}}}\,.$$

4. $[G] = \mathrm{m^3\,s^{-2}\,kg^{-1}}$, $[c] = \mathrm{m\,s^{-1}}$, $[\hbar] = \mathrm{J\,s} = \mathrm{kg\,m^2\,s^{-1}}$.

Try $\left[\dfrac{\hbar G}{c^3}\right] = \dfrac{\mathrm{kg\,m^2\,s^{-1}\,m^3\,s^{-2}\,kg^{-1}}}{\mathrm{m^3\,s^{-3}}} = \mathrm{m^2}$.

Therefore, $\sqrt{\dfrac{\hbar G}{c^3}}$ has the dimension of a length, and this is the Planck length.

5. The escape velocity from a massive object can be worked out from

$$\frac{1}{2}mv^2 = G\frac{m\,M}{R}\,,$$

where m is the mass of the escaping particle, and M and R are the mass and radius of the massive object. The mass of the escaping object does not enter into the escape velocity. If the escape velocity is equal to the velocity of light, even light cannot escape from the object. If v is replaced by c, this leads to

$$c^2 = \frac{2\,G\,M}{R} \quad \text{or} \quad R = \frac{2\,G\,M}{c^2},$$

which is the Schwarzschild radius. The result of this classical treatment of the problem (even though classical physics does not apply in this situation) accidentally agrees with the outcome of the correct derivation based on general relativity. By definition the Schwarzschild radius is $R_S = \dfrac{2GM}{c^2}$ and the calculation gives

$$R_S = \begin{cases} 8.9\,\text{mm for Earth} \\ 2.95\,\text{km for the Sun} \end{cases}.$$

6. $\dfrac{G M^2}{R} > \dfrac{3}{2}kT\,\dfrac{M}{\mu}$, μ—mass of a hydrogen atom,

$$M > \frac{3}{2}\frac{kT}{\mu G} \quad R = \frac{3}{2}\frac{kT}{\mu G}\left(\frac{3M}{4\pi\varrho}\right)^{1/3};$$

since $M = \dfrac{4}{3}\pi\varrho R^3$: $M^3 > \left(\dfrac{3}{2}\dfrac{kT}{\mu G}\right)^3 \dfrac{3M}{4\pi\varrho}$,

$$\varrho > \frac{3}{4\pi}\frac{1}{M^2}\left(\frac{3}{2}\frac{kT}{\mu G}\right)^3 \approx 3.9 \times 10^{-10}\,\text{kg/m}^3$$

$(k = 1.38 \times 10^{-23}\,\text{J/K},\ M = 2 \times 10^{30}\,\text{kg},\ \mu = 1.67 \times 10^{-27}\,\text{kg})$.

7.

The first observation is made at time t_0. The star is at a distance D from the observer. It moves at an angle γ to the line of sight. During Δt it has moved $d_1 = d\cos\gamma$ closer to the observer. The first light beam has further to travel. The light was emitted $\Delta t + \frac{d_1}{c}$ earlier compared to the second measurement. The apparent velocity is

$$v^* = \frac{d_2}{\Delta t} = \frac{d \sin \gamma}{\Delta t} = \frac{v \left(\Delta t + \frac{d_1}{c} \right) \sin \gamma}{\Delta t} = v \left(1 + \frac{d_1}{c \Delta t} \right) \sin \gamma$$

$$= v \left(1 + \frac{d_2}{c \Delta t} \cot \gamma \right) \sin \gamma = v \left(1 + \frac{v^*}{c} \cot \gamma \right) \sin \gamma = v \sin \gamma + \frac{v}{c} v^* \cos \gamma .$$

Solve for v^*: $v^* \left(1 - \frac{v}{c} \cos \gamma \right) = v \sin \gamma$, $v^* = \frac{v \sin \gamma}{1 - \frac{v}{c} \cos \gamma}$; if v approaches

c: $v^* = c \dfrac{\sin \gamma}{1 - \cos \gamma} = c \dfrac{2 \sin \frac{\gamma}{2} \cos \frac{\gamma}{2}}{2 \sin^2 \frac{\gamma}{2}} = c \cot \dfrac{\gamma}{2} > c$ for $0 < \gamma < \dfrac{\pi}{2}$.

Chapter 10

1. The number of degrees of freedom depends on the bosonic degrees of freedom and fermionic degrees of freedom. Due to the difference between Bose–Einstein and Fermi–Dirac statistics a factor of 7/8 must be introduced to give

$$g = 2 + \frac{7}{8}(4 + 2N_v) ;$$

for 3 neutrino generations this gives $g = 10.75$ [281]. Equation (10.1.2) related the time to the temperature and the Planck mass,

$$t = \frac{0.301 \, m_{\mathrm{Pl}}}{\sqrt{g} \, T^2} \quad \Rightarrow \quad t T^2 = 0.301 \frac{m_{\mathrm{Pl}}}{\sqrt{g}} .$$

To get a number, one has to obtain a dimension time $\times$ energy^2 for $t T^2$. Therefore, a factor $\hbar$ is missing:

$$t T^2 = 0.301 \times 6.582 \times 10^{-22} \, \mathrm{MeV \, s} \times 1.22 \times 10^{22} \, \mathrm{MeV} \times \frac{1}{\sqrt{g}} \approx 0.74 \, \mathrm{s \, MeV^2} .$$

2. The process $v_e n \rightarrow e^- p$ is governed by the weak interaction. Its coupling strength is given by $G_F = 10^{-5}/m_p^2$, where m_p is the proton mass. Because of $\lambda/2\pi = \hbar/p$, the length can be measured in units of energy^{-1} ($\hbar$ is usually set to 1). Therefore, one gets from

$$\sigma(v_e n \rightarrow e^- p) \sim G_F^2 \, f(s) ,$$

knowing that $G_F^2 \sim$ energy^{-4}:

$$f(s) \sim s ,$$

$$\sigma(v_e n \rightarrow e^- p) \left\{ \mathrm{length}^2 \hat{=} \mathrm{energy}^{-2} \right\} = \left(G_F^2 \, s \right) \left\{ \mathrm{energy}^{-4} \times \mathrm{energy}^2 \right\} .$$

Since for relativistic particles $v \approx c$ and $s = (\frac{3}{2}kT)^2 \sim T^2$, one gets

$$\sigma \sim G_F^2 T^2 .$$

3. The hadronic cross section $\sigma_{hadr}(e^+e^- \to Z \to \text{hadrons})$ at the Z resonance can be described by a Breit–Wigner distribution

$$\sigma_{hadr} = \sigma_{hadr}^0 \frac{s\Gamma_Z^2}{(s - M_Z^2)^2 + s^2\Gamma_Z^2/M_Z^2} (1 + \delta_{rad}(s)) ,$$

where

σ_{hadr}^0 – peak cross section,
$\sqrt{s}$ – center-of-mass energy,
Γ_Z – total width of the Z,
M_Z – mass of the Z,
δ_{rad} – radiative correction.

The peak cross section σ_{hadr}^0 for the process $e^+e^- \to Z \to$ hadrons is given by

$$\sigma_{hadr}^0 = \frac{12\pi}{M_Z^2} \frac{\Gamma_{e^+e^-}\Gamma_{hadr}}{\Gamma_Z^2} ,$$

where $\Gamma_{e^+e^-}$ is the partial e^+e^- width of the Z and Γ_{hadr} the partial width for $Z \to$ hadrons.

The total Z width can be written as

$$\Gamma_Z = \Gamma_{hadr} + \Gamma_{e^+e^-} + \Gamma_{\mu^+\mu^-} + \Gamma_{\tau^+\tau^-} + \Gamma_{inv} ,$$

where Γ_{inv} describes the invisible decay of the Z into neutrino pairs ($\nu_e \bar{\nu}_e$, $\nu_\mu \bar{\nu}_\mu$, $\nu_\tau \bar{\nu}_\tau$, and possibly other neutrino pairs from a suspected fourth neutrino family). Lepton universality is well established. Therefore,

$$\Gamma_Z = \Gamma_{hadr} + 3\Gamma_{\ell\bar{\ell}} + \Gamma_{inv} .$$

$$\frac{\Gamma_{inv}}{\Gamma_{\ell\bar{\ell}}} = \frac{\Gamma_Z}{\Gamma_{\ell\bar{\ell}}} - 3 - \frac{\Gamma_{hadr}}{\Gamma_{\ell\bar{\ell}}} = \frac{\Gamma_Z}{\Gamma_{\ell\bar{\ell}}} - 3 - R , \qquad (*)$$

where R is the usual ratio

$$R = \frac{\sigma(e^+e^- \to \text{hadrons})}{\sigma(e^+e^- \to \ell^+\ell^-)} .$$

The experiment provides σ_{hadr}^0, Γ_Z, and M_Z.
The measurement of the peak cross section σ_{hadr}^0 and Γ_Z together with M_Z fixes the product $\Gamma_{\ell\bar{\ell}} \Gamma_{hadr}$.

The measurement of the peak cross section for the process $e^+e^- \to Z \to \mu^+\mu^-$,

$$\sigma_{\mu\mu}^0 = \frac{12\pi}{M_Z^2} \frac{\Gamma_{e^+e^-} \, \Gamma_{\mu^+\mu^-}}{\Gamma_Z^2} \,,$$

determines $\Gamma_{\ell\bar\ell}$, if lepton universality is assumed. Therefore, also Γ_{hadr} is now known, so that Γ_{inv} can be obtained from (∗).
In the framework of the Standard Model the width of

$$Z \to \nu_x \bar\nu_x$$

is calculated to be ≈ 170 MeV. The LEP experiments resulted in

$$\Gamma_{inv} \approx 500 \text{ MeV}$$

indicating very clearly that there are only $500/170 \approx 3$ neutrino generations.

Chapter 11

1. This probability can be worked out from

$$\phi = \sigma_{Th} \, N \, d \,,$$

where σ_{Th} is the Thomson cross section (665 mb), N the number of target atoms per cm^3, and d the distance traveled.
Since the universe is flat, one has $\varrho = \varrho_{crit}$. Because of $\varrho_{crit} = 9.47 \times 10^{-30}$ g/cm^3 and $m_H = 1.67 \times 10^{-24}$ g, and the fact that only 4% of the total matter density is in the form of baryonic matter, one has $N = 2.27 \times 10^{-7}$ cm^{-3}. The distance traveled from the surface of last scattering corresponds to the age of the universe (14 billion years),

$$d = 14 \times 10^9 \times 3.156 \times 10^7 \text{ s} \times 2.998 \times 10^{10} \text{ cm/s} = 1.32 \times 10^{28} \text{ cm} \,,$$

resulting in
$$\phi = 1.99 \times 10^{-3} \approx 0.2\% \,.$$

2. The critical density $\varrho_c = 3H^2/8\pi G$ as obtained from the Friedmann equation has to be modified if the effect of the cosmological constant is taken into account:

$$\varrho_{c,\Lambda} = \frac{3H^2 - \Lambda c^2}{8\pi G} \,.$$

Since $\varrho_{c,\Lambda}$ cannot be negative, one can derive a limit from this equation:

$$3H^2 - \Lambda c^2 > 0 , \quad \Lambda < \frac{3H^2}{c^2} = 1.766 \times 10^{-56} \, \text{cm}^{-2} .$$

This leads to an energy density

$$\varrho \le \frac{c^4}{8\pi G} \Lambda = 8.51 \times 10^{-10} \, \text{J/m}^3 = 5.3 \, \text{GeV/m}^3 = 5.3 \, \text{keV/cm}^3 .$$

3. $p_1 = -\dfrac{\mathrm{d}E_{\text{class}}}{\mathrm{d}V} = -\dfrac{\mathrm{d}}{\mathrm{d}V}\left(-\dfrac{GmM}{R}\right) = GmM \dfrac{\mathrm{d}}{\mathrm{d}V}\left(\dfrac{1}{V^{1/3}}\right)\left(\dfrac{4\pi}{3}\right)^{1/3}$

$\sim \dfrac{\mathrm{d}}{\mathrm{d}V}\left(\dfrac{1}{V^{1/3}}\right) = -\dfrac{1}{3}V^{-4/3} \sim -\dfrac{1}{R^4} ,$

$p_2 = -\dfrac{\mathrm{d}E_\Lambda}{\mathrm{d}V} \sim -\dfrac{\mathrm{d}}{\mathrm{d}V}(-\Lambda R^2) \sim \dfrac{\mathrm{d}}{\mathrm{d}V}V^{2/3} = \dfrac{2}{3}V^{-1/3} \sim +\dfrac{1}{R} .$

p_1 is an inward pressure due to the normal gravitational pull, whereas p_2 is an outward pressure representing a repulsive gravity.

4. Planck distribution

$$\varrho(\nu, T) = \frac{8\pi h}{c^3} \nu^3 \frac{1}{e^{h\nu/kT} - 1} , \quad \langle \nu \rangle = \frac{\int_0^\infty \nu \, \varrho(\nu, T) \, \mathrm{d}\nu}{\int_0^\infty \varrho(\nu, T) \, \mathrm{d}\nu} ;$$

substitution $\frac{h\nu}{kT} = x$:

$$\int_0^\infty \varrho(\nu, T) \, \mathrm{d}\nu = \frac{8\pi h}{c^3}\left(\frac{kT}{h}\right)^3 \int_0^\infty \frac{x^3}{e^x - 1} \frac{kT}{h} \, \mathrm{d}x$$

$$= \frac{8\pi h}{c^3}\left(\frac{kT}{h}\right)^4 \int_0^\infty \frac{x^3}{e^x - 1} \, \mathrm{d}x ,$$

$$\int_0^\infty \nu \, \varrho(\nu, T) \, \mathrm{d}\nu = \frac{8\pi h}{c^3}\left(\frac{kT}{h}\right)^5 \int_0^\infty \frac{x^4}{e^x - 1} \, \mathrm{d}x ;$$

$$\int_0^\infty \frac{x^3}{e^x - 1} \, \mathrm{d}x = 3! \, \zeta(4) , \quad \zeta(4) = \pi^4/90 , \quad \zeta\text{—Riemann's zeta function,}$$

$$\int_0^\infty \frac{x^4}{e^x - 1} \, \mathrm{d}x = 4! \, \zeta(5) , \quad \zeta(5) = 1.036\,927\,7551\ldots ;$$

$$\langle h\nu \rangle = \frac{h \frac{8\pi h}{c^3}\left(\frac{kT}{h}\right)^5 \times 4! \, \zeta(5)}{\frac{8\pi h}{c^3}\left(\frac{kT}{h}\right)^4 \times 3! \, \zeta(4)} = h \frac{kT}{h} \times 4 \times \frac{\zeta(5)}{\zeta(4)}$$

$$= kT \times 4 \times \frac{\zeta(5)}{\pi^4} \times 90 = \frac{360}{\pi^4} kT \, \zeta(5) \approx 900 \, \mu\text{eV} .$$

5. (a) At present: number density of bb photons: $410/\text{cm}^3$, average energy $\langle E \rangle = 900\,\mu\text{eV}$ (from the previous problem). This leads to a first estimate of the present energy density of

$$\varrho_0 \approx 0.37\,\text{eV}/\text{cm}^3 \,.$$

Since, however, the temperature enters with the fourth power into the energy density, one has to use

$$\varrho_0 = \frac{\pi^2}{15} T^4$$

for photons (number of degrees of freedom $g = 2$, because for photons as massless particles there are right-handed and left-handed states, i.e., two spin states, see Table 9.1). One has to include the adequate factors of k and $\hbar c$ to get the correct numerical result:

$$\varrho_0 = \frac{\pi^2}{15} T^4 \frac{k^4}{(\hbar c)^3} \approx 0.26\,\text{eV}/\text{cm}^3$$

($k = 8.617 \times 10^{-5}\,\text{eV K}^{-1}$, $\hbar c = 0.197\,\text{GeV fm}$).

(b) At last scattering ($t = 380\,000\,\text{a}$, temperature at the time of last scattering: $T_{\text{dec}} = 0.3\,\text{eV}$, see Chap. 11):

$$T_{\text{dec}} = 0.3\,\text{eV}/k \approx 3500\,\text{K}\,,$$

$$\varrho_{\text{dec}} = \varrho_0 \left(\frac{T_{\text{dec}}}{T_0} \right)^4 \approx 0.26\,\text{eV}/\text{cm}^3 \times \left(\frac{3500}{2.725} \right)^4 \approx 0.7\,\text{TeV}/\text{cm}^3 \,.$$

6. Naively, one would expect the fraction of neutral hydrogen to become significant when the temperature drops below $13.6\,\text{eV}$. But this happens only at much lower temperatures because there are so many more photons than baryons, and the photon energy distribution, i.e., the Planck distribution, has a long tail towards high energies. The baryon-to-photon ratio, $\eta \approx 5 \times 10^{-10}$, is extremely small. Therefore, the temperature must be significantly lower than this before the number of photons with $E > 13.6\,\text{eV}$ is comparable to the number of baryons. Furthermore, interaction or ionization can take place in several steps via excited states of the hydrogen atom, the H_2 molecule, or the H_2^+ ion. One finds that the numbers of neutral and ionized atoms become equal at a *recombination temperature* of $T_{\text{rec}} \approx 0.3\,\text{eV}$ ($3500\,\text{K}$). At this point the universe transforms from an ionized plasma to an essentially neutral gas of hydrogen and helium.

7. Substitution of $\frac{h\nu}{kT} = x$ and replacing the differential correspondingly by $d\nu = \frac{kT}{h}\,dx$ leads to

$$\langle n \rangle = \frac{8\pi}{c^3} \frac{(kT)^3}{h^3} \cdot \int_0^\infty \frac{x^2}{e^x - 1} \cdot dx \,. \tag{11.1}$$

The integral extends over all frequencies from zero to infinity. It can be worked out numerically to be

$$\int_0^\infty \frac{x^2}{e^x - 1} \cdot dx \approx 2.404 .$$ (11.2)

This leads to

$$\langle n \rangle \approx \frac{8\pi}{c^3} \cdot 2.404 \cdot \frac{(kT)^3}{h^3} \approx 2.03 \times 10^7 \, (T/K)^3/m^3 .$$ (11.3)

For the background temperature of 2.725 K one arrives at a number density of

$$\langle n \rangle \approx 411 \, cm^{-3} .$$ (11.4)

Chapter 12

1. The Friedmann equation for $k = 0$, corresponding to the dominance of Λ, reads, see also Problem 11.2,

$$H^2 = \frac{8\pi G}{3}(\varrho + \varrho_v) .$$

With

$$\Lambda = \frac{8\pi G}{c^2}\varrho_v \quad \Rightarrow \quad H^2 - \frac{1}{3}\Lambda c^2 = \frac{8\pi G}{3}\varrho .$$

Since $\varrho > 0$, one has the inequality

$$H^2 - \frac{1}{3}\Lambda c^2 \geq 0 \quad \text{or} \quad \Lambda \leq \frac{3H^2}{c^2} \approx 2 \times 10^{-56} \, cm^{-2} .$$

This is just a reflection of the fact that in the visible universe there is no obvious effect of the curvature of space. The size of the visible flat universe being 10^{28} cm can be converted into

$$\Lambda_1 \leq 10^{-56} \, cm^{-2} .$$

If one assumes on the other hand that Einstein's theory of relativity is valid down to the Planck scale, then one would expect

$$\Lambda_2 \approx (\ell_{Pl}^2)^{-1} \approx 10^{66} \, cm^{-2} .$$

The difference between the two estimates is 122 orders of magnitude. The related mass densities for the two scenarios are estimated to be

$$\varrho_v^1 = \frac{\Lambda_1 c^2}{8\pi G} \approx \frac{10^{-52}\,\mathrm{m}^{-2} \times 9 \times 10^{16}\,\mathrm{m}^2/\mathrm{s}^2\,\mathrm{kg}\,\mathrm{s}^2}{8\pi \times 6.67 \times 10^{-11}\,\mathrm{m}^3}$$

$$\approx 5.37 \times 10^{-27}\,\mathrm{kg/m}^3 = 5.37 \times 10^{-30}\,\mathrm{g/cm}^3\,,$$

$$\varrho_v^2 = \frac{\Lambda_2 c^2}{8\pi G} \approx 5.37 \times 10^{95}\,\mathrm{kg/m}^3 = 5.37 \times 10^{92}\,\mathrm{g/cm}^3\,.$$

2. Starting from

$$H^2 = \frac{8\pi G}{3}(\varrho + \varrho_v) \quad \Rightarrow \quad H^2 - \frac{1}{3}\Lambda c^2 = \frac{8\pi G}{3}\varrho$$

and since $H = \dot{R}/R$, one obtains (for $\varrho = \mathrm{const}$)

$$\frac{\dot{R}}{R} = \sqrt{\frac{1}{3}\Lambda c^2 + \frac{8\pi G}{3}\varrho} \quad \Rightarrow \quad R = R_i \exp\left(\sqrt{\frac{1}{3}\Lambda c^2 + \frac{8\pi G}{3}\varrho}\, t\right),$$

which represents an expanding universe.

3. The Friedmann equation extended by the Λ term for a flat universe is

$$\frac{\dot{R}^2}{R^2} - \frac{1}{3}\Lambda c^2 = \frac{8\pi G}{3}\varrho\,.$$

For $\frac{1}{3}\Lambda c^2 \gg \frac{8\pi G}{3}\varrho$ this equation simplifies to

$$\frac{\dot{R}}{R} = \sqrt{\frac{1}{3}\Lambda c^2} = \sqrt{\frac{1}{3}\Lambda_0 c^2(1 + \alpha t)} = a\sqrt{1 + \alpha t} \quad \text{with} \quad a = \sqrt{\frac{1}{3}\Lambda_0 c^2}\,,$$

$$\ln R = \int a\sqrt{1 + \alpha t}\, dt + \mathrm{const} = a\frac{2}{3\alpha}(1 + \alpha t)^{3/2} + \mathrm{const}\,,$$

$$R = \exp\left(\frac{2a}{3\alpha}(1 + \alpha t)^{3/2} + \mathrm{const}\right);$$

boundary condition $R(t = 0) = R_i$

$$R(t = 0) = \exp\left(\frac{2a}{3\alpha} + \mathrm{const}\right) = R_i \quad \Rightarrow$$

$$\frac{2a}{3\alpha} + \mathrm{const} = \ln R_i \quad \Rightarrow \quad \mathrm{const} = \ln R_i - \frac{2a}{3\alpha}\,,$$

$$R = R_i \exp\left(\frac{2a}{3\alpha}(1 + \alpha t)^{3/2} - \frac{2a}{3\alpha}\right) = R_i \exp\left(\frac{2a}{3\alpha}[(1 + \alpha t)^{3/2} - 1]\right)$$

$$\text{with} \quad a = \sqrt{\frac{1}{3}\Lambda_0 c^2}\,.$$

For large t this result shows a dependence like

$$R \sim \exp\left(\beta t^{3/2}\right) \quad \text{with} \quad \beta = \frac{2a\sqrt{\alpha}}{3}.$$

4. $\dfrac{R_0}{R_{\mathrm{mr}}} = \left(\dfrac{t_0}{t_{\mathrm{mr}}}\right)^{\frac{2}{3}}$,

where R_{mr} is the size of the universe at the time of matter–radiation equality $t_{\mathrm{mr}} = 50\,000$ a. With $t_0 = 14$ billion years, one has

$$\frac{R_0}{R_{\mathrm{mr}}} \approx 4280.$$

The size of the universe at that time was

$$R_{\mathrm{mr}} \approx \frac{R_0}{4280} \approx 3.27 \times 10^6 \text{ light-years} \approx 3.09 \times 10^{22} \text{ m}.$$

Extrapolating to earlier times requires to assume $R \sim \sqrt{t}$, since for times $t < t_{\mathrm{mr}}$ the universe was radiation dominated:

$$\frac{R_{\mathrm{mr}}}{R_{\mathrm{infl}}} = \left(\frac{t_{\mathrm{mr}}}{t_{\mathrm{infl}}}\right)^{\frac{1}{2}} \approx \left(\frac{50\,000\,\mathrm{a}}{10^{-36}\,\mathrm{s}}\right)^{\frac{1}{2}} \approx 1.26 \times 10^{24}.$$

Correspondingly, the size of the universe at the end of inflation was

$$R_{\mathrm{infl}} \approx \frac{R_{\mathrm{mr}}}{1.26 \times 10^{24}} \approx 2.45 \text{ cm}.$$

If inflation had ended at 10^{-32} s, one would have obtained $R_{\mathrm{infl}} \approx 2.45$ m.

Chapter 13

1. $\dfrac{mv^2}{r} = G\dfrac{mM}{r^2} \quad\Rightarrow\quad M = \dfrac{v^2 r}{G}$,

$M \approx \dfrac{(1.10 \times 10^5 \text{ m/s})^2 \times 2.5 \times 3.086 \times 10^{16} \text{ m}}{6.67 \times 10^{-11} \text{ m}^3 \text{ s}^{-2}} \text{ kg} \approx 1.4 \times 10^{37} \text{ kg} \approx 7 \times$

$10^6\, M_\odot$.

2. Non-relativistic calculation: $\boldsymbol{p}_{\mathrm{W}} - \boldsymbol{p}'_{\mathrm{W}} = \boldsymbol{p}_e$;

$$\frac{p_{\mathrm{W}}^2}{2m_{\mathrm{W}}} = \frac{p_{\mathrm{W}}'^2}{2m_{\mathrm{W}}} + \frac{(\boldsymbol{p}_{\mathrm{W}} - \boldsymbol{p}'_{\mathrm{W}})^2}{2m_e} = \frac{p_{\mathrm{W}}^2}{2m_{\mathrm{W}}} + \frac{p_{\mathrm{W}}'^2}{2m_e} + \frac{p_{\mathrm{W}}'^2}{2m_e} - \frac{\boldsymbol{p}_{\mathrm{W}} \cdot \boldsymbol{p}'_{\mathrm{W}}}{m_e},$$

$$p_{\mathrm{W}}'^2\left(\frac{1}{m_{\mathrm{W}}} + \frac{1}{m_e}\right) - \frac{2\boldsymbol{p}_{\mathrm{W}} \cdot \boldsymbol{p}'_{\mathrm{W}}}{m_e} = p_{\mathrm{W}}^2\left(\frac{1}{m_{\mathrm{W}}} - \frac{1}{m_e}\right),$$

$$p_{\mathrm{W}}'^2 - 2\frac{m_{\mathrm{W}}}{m_{\mathrm{W}} + m_e}\boldsymbol{p}_{\mathrm{W}} \cdot \boldsymbol{p}'_{\mathrm{W}} = p_{\mathrm{W}}^2\frac{m_e - m_{\mathrm{W}}}{m_{\mathrm{W}} + m_e},$$

$$p_W'^2 - 2\frac{m_W}{m_W + m_e}p_W \cdot p_W' + \left(\frac{m_W}{m_W + m_e}\right)^2 p_W^2 = p_W^2\left(\frac{m_e - m_W}{m_W + m_e} + \frac{m_W^2}{(m_W + m_e)^2}\right),$$

$$\left(p_W' - \frac{m_W}{m_W + m_e}p_W\right)^2 = p_W^2\frac{m_e^2}{(m_W + m_e)^2}.$$

Central collision:

$$p_W' - \frac{m_W}{m_W + m_e}p_W = \pm p_W\frac{m_e}{m_W + m_e} \quad \Rightarrow \quad p_W' = \frac{m_W - m_e}{m_W + m_e}p_W,$$

since only the negative sign is physically meaningful.

$$\Delta E = \frac{1}{2}m_W(v_1^2 - v_1'^2) = E_W^{\text{kin}}\left(1 - \frac{v_1'^2}{v_1^2}\right) = E_W^{\text{kin}}\left(1 - \left(\frac{m_W - m_e}{m_W + m_e}\right)^2\right)$$

$$\approx 1\,\text{GeV} \times 2 \times 10^{-5} = 20\,\text{keV}.$$

3. Classical Fermi gas of neutrinos:

$$E_F = \frac{p^2}{2m_\nu} = \frac{\hbar^2 k^2}{2m_\nu}, \quad k - \text{wave vector}. \qquad (13.1)$$

In a quantized Fermi gas there is one k vector per $2\pi/L$ if one assumes that the neutrino gas is contained in a cube of side L. Number of states (at $T = 0$) for 2 spin states:

$$N = 2\frac{\frac{4}{3}\pi k_F^3}{(2\pi/L)^3} = V\frac{1}{3\pi^2}k_F^3 \quad \Rightarrow \quad k_F = (3\pi^2 n)^{1/3}$$

with $n = N/V$ particle density,

$$\Rightarrow E_F = \frac{\hbar^2}{2m}(3\pi^2 n)^{2/3}.$$

Relativistic Fermi gas:

$$E_F = p_F c = \hbar k_F c = \hbar c(3\pi^2 n)^{1/3};$$

assuming relativistic neutrinos one would get, e.g., for 300 neutrinos per cm^3:
$\hbar c = 197.3\,\text{MeV fm}$,
$E_F = \hbar c\,(3\pi^2 \times 300)^{1/3}\,\text{cm}^{-1} = 197.3 \times 10^6\,\text{eV} \times 10^{-13}\,\text{cm} \times 20.71\,\text{cm}^{-1}$
$\approx 409\,\mu\text{eV}.$

4. Since $v \ll c$ is expected, one can use classical kinematics:

$$\frac{1}{2}mv^2 = G\frac{mM}{R} \quad \Rightarrow \quad v = \sqrt{\frac{2GM}{R}} \quad \Rightarrow$$

$$v \approx \sqrt{\frac{2 \times 6.67 \times 10^{-11}\ \mathrm{m^3\ kg^{-1}\ s^{-2}} \times 10^{10} \times 2 \times 10^{30}\ \mathrm{kg}}{3 \times 10^3 \times 3.086 \times 10^{16}\ \mathrm{m}}} \approx 1.7 \times 10^5\ \mathrm{m/s},$$

$$\beta \approx 5.66 \times 10^{-4}.$$

The axion mass does not enter the calculation, i.e., the escape velocity does not depend on the mass.

A relativistic treatment, $(\gamma - 1)m_0 c^2 = G\gamma m_0 M/R$, leads to a similar result:

$$v = \beta c = \sqrt{\frac{2GM}{R}}\sqrt{1 - \frac{GM}{2Rc^2}}.$$

Since this expression is not based on general relativity, it is of limited use.

5. Spherically symmetric mass distribution.

(a) Mass inside a sphere: $M(r) = \int_0^r \varrho(r')\, dV' = 4\pi \int_0^r \varrho(r')r'^2\, dr'$;

potential energy of mass m: $E_{\mathrm{pot}}^{(m)}(r) = Gm \int_\infty^r \frac{M(r')}{r'^2}\, dr'$,

verification: $-\dfrac{\partial E_{\mathrm{pot}}^{(m)}(r)}{\partial r} = F = -G\dfrac{mM(r)}{r^2}$.

(b) Potential energy of mass m for $\varrho_{\mathrm{cut}}(r') = 0$, i.e., $M(r') = M(r)$ for $r' > r$:
Potential energy of mass shell $dM = M'(r)\, dr$: $E_{\mathrm{pot,cut}}^{(m)}(r) = GmM(r)\int_\infty^r \frac{1}{r'^2}\, dr' = -Gm\dfrac{M(r)}{r}$.

The incremental potential energy due to the additional mass shell $dM = M'(r)\, dr$ is: $dE_{\mathrm{pot}} = -GM'(r)\, dr\,\dfrac{M(r)}{r}$.

The total potential energy reads: $E_{\mathrm{pot}} = -G\int_0^\infty \frac{M'(r)M(r)}{r}\, dr$;

integration by parts: $E_{\mathrm{pot}} = -G\dfrac{M^2(r)}{2r}\Big|_{r=0}^\infty - \dfrac{G}{2}\int_0^\infty \frac{M^2(r)}{r^2}\, dr$;

the margin terms vanish if for $r \to 0$ one has $M(r) \to 0$ faster than $r^{1/2}$ and for finite masses at infinity, also causing the convergence of all the integrals:

$$E_{\mathrm{pot}} = -\frac{G}{2}\int_0^\infty \frac{M^2(r)}{r^2}\, dr.$$

By taking $\varrho_{\mathrm{cut}}(r') = 0$ for $r' > r$ double counting for the self-energy can be avoided.

(c) Mass M on a shell, i.e., $M(r < R) = 0$ and $M(r > R) = M$:

$$E_{\mathrm{pot}}^{(\mathrm{shell})} = -\frac{GM^2}{2}\int_R^\infty \frac{1}{r^2}\, dr = \frac{GM^2}{2}\frac{1}{r}\Big|_{r=R}^\infty = -\frac{GM^2}{2R}.$$

(d) Homogeneous mass distribution: $M(r < R) = M_t r^3/R^3$ and $M(r > R) = M_t$:

$$E_{\text{pot}}^{(\text{hom})} = -\frac{G}{2} \frac{M_t^2}{R^6} \int_0^R r^4 \, dr + E_{\text{pot}}^{(\text{shell})} = -\frac{G}{2} \frac{M_t^2}{R^6} \frac{r^5}{5}\bigg|_{r=0}^R - \frac{G}{2} \frac{M_t^2}{R} =$$

$$-\frac{3}{5}G\frac{M_t^2}{R}.$$

6. $n_{\text{max}}m_\nu = \dfrac{m_\nu^4 v_f^3}{3\pi^2\hbar^3}$. \hfill (*)

If $n_{\text{max}}m_\nu$ is assumed to be on the order of a typical dark-matter density ($\Omega = 0.3$), one can solve (*) for m_ν,

$$m_\nu = \left(\frac{3\pi^2\hbar^3\,n_{\text{max}}m_\nu}{v_f^3}\right)^{1/4}.$$

The mass of a typical galaxy can be estimated as $M = 10^{11}\,M_\odot = 2 \times 10^{41}$ kg. With an assumed radius of $r = 20$ kpc one gets

$$v_f = 2.11 \times 10^5\,\text{m/s}.$$

With $n_{\text{max}}m_\nu = 0.3\,\varrho_c \approx 3 \times 10^{-30}$ g/cm^3 one obtains

$$m_\nu \geq 1\,\text{eV}.$$

7.

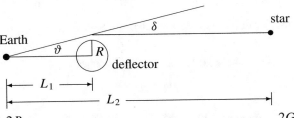

$\delta = \dfrac{2R_S}{R}$ as mentioned earlier (see Problem 8.5); $R_S = \dfrac{2GM_d}{c^2}$;

for $L_2 \gg L_1$: $\delta \approx \vartheta = \dfrac{R}{L_1} \Rightarrow R = L_1\delta$,

$\delta = \dfrac{2R_S}{R} = \dfrac{2R_S}{L_1\delta} \Rightarrow \delta = \sqrt{\dfrac{2R_S}{L_1}} = \sqrt{\dfrac{4GM_d}{L_1c^2}} \sim \sqrt{M_d}$.

Ring radius $R_E = L_1\,\delta = \sqrt{2L_1R_S}$, $L_1 = 10$ kpc $\approx 10 \times 10^3 \times 3.26 \times 3.15 \times 10^7$ s $\times 3 \times 10^8$ m s$^{-1} \approx 3.08 \times 10^{20}$ m: $R_E \approx \sqrt{2 \times 3.08 \times 10^{20}\,\text{m} \times 3000\,\text{m}} \approx 1.36 \times 10^{12}$ m;

opening angle for the Einstein ring: $\gamma = 2\delta = \dfrac{2R_E}{L_1} = 8.8 \times 10^{-9} \cong 0.0018$ arcsec

$\Rightarrow$ too small to be observable $\Rightarrow$ only brightness excursion visible.

8.

$$\Delta Q = c \cdot m\,\Delta T \implies$$

$$\Delta T = \frac{\Delta Q}{c \cdot m} = \frac{10^5\,\text{eV} \times 1.6 \times 10^{-19}\,\text{J/eV}}{8 \times 10^{-2}\,\text{J/(kg\,K)} \times 10^{-3}\,\text{kg}}$$

$$= 2 \times 10^{-10}\,\text{K}\,.$$

Chapter 14

1. The weight of a human is proportional to its volume, which in turn is proportional to the cube of its size;

$$W = W_0\,R^2\,20\,R\,,$$

if one assumes that the 'height' of a human is 20 times its 'radius'. The strength of a human is proportional to its cross section

$$S = S_0\,R^2\,.$$

For a mass of $100\,\text{kg}$, an assumed radius of $10\,\text{cm}$, a human has to carry its own weight plus, maybe, an additional $100\,\text{kg}$,

$$W_0 = \frac{100\,\text{kg}}{20\,R^3} = 5000\,\frac{\text{kg}}{\text{m}^3}\,,\quad S_0 = \frac{200\,\text{kg}}{R^2} = 20\,000\,\frac{\text{kg}}{\text{m}^2}\,.$$

From this the stability limit (weight $\,\widehat{=}\,$ strength) can be derived,

$$5000\,\frac{\text{kg}}{\text{m}^3} \times 20\,R^3 = 20\,000\,\frac{\text{kg}}{\text{m}^2} \times R^2\,,$$

which leads to

$$R_{\text{max}} = \frac{1}{5}\,\text{m} \quad \Rightarrow \quad H_{\text{max}} = 4\,\text{m}\,.$$

If the gravity were to double, the strength would be composed of $100\,\text{kg}$ and a body mass of $50\,\text{kg}$ (corresponding to $100\,\text{kg}$ 'weight', since $g^* = 2g$). Since

$$V = \frac{1}{2}V_0 \quad \Rightarrow \quad R_{\text{max}}(g^* = 2g) = \frac{1}{\sqrt[3]{2}}\,R_{\text{max}}^0\,,$$

$$R_{\text{max}}(g^* = 2g) = 0.1587\,\text{m} \quad \Rightarrow \quad H_{\text{max}}(g^* = 2g) = 3.17\,\text{m}\,.$$

2. Carbon is produced in the so-called triple-alpha process. In step one ^8Be is produced in $\alpha\alpha$ collisions

$$^4\text{He} + {}^4\text{He} \;\rightarrow\; {}^8\text{Be} + \gamma - 91.78\,\text{keV}\,.$$

The ^{8}Be produced in this step is unstable and decays back into helium nuclei in $2.6\,\mu s$. It therefore requires a high helium density to induce a reaction with ^{8}Be before it has decayed,

$$^8\text{Be} + {}^4\text{He} \rightarrow {}^{12}\text{C}^* + 7.367\,\text{MeV} .$$

The excited ^{12}C* state is unstable, but a few of these excited carbon nuclei emit a γ ray quickly enough to become stable before they disintegrate. The net energy release of the triple-alpha process is $7.275\,\text{MeV}$.

Only at extremely high temperatures ($\approx 10^8$ K) the bottleneck of carbon production from helium in a highly improbable reaction can be accomplished because of a lucky mood of nature: high temperatures are required to overcome the Coulomb barrier for helium fusion. Also high densities of helium nuclei are needed to make the triple-alpha process possible.

The reaction rate for the triple-alpha process depends on the number density N_α of α particles and the temperature T of the α plasma:

$$\sigma(3\alpha \rightarrow {}^{12}\text{C}) \sim N_\alpha^3 \, \frac{1}{T^3} \exp\left(-\frac{\varepsilon}{kT}\right) \Gamma_\gamma ,$$

where ε is a parameter ($\approx 0.4\,\text{MeV}$). For high temperatures up to $T \approx 1.5 \times 10^9$ K the exponential wins over the power T^3. The maximum cross section is derived to be in the following way:

$$0 = \frac{\text{d}}{\text{d}T}\left[\frac{1}{T^3}\exp\left(-\frac{\varepsilon}{kT}\right)\right] = \left(-\frac{3}{T^4} + \frac{\varepsilon}{kT^5}\right)\exp\left(-\frac{\varepsilon}{kT}\right) \Rightarrow$$

$$T_{\text{max}} = \frac{\varepsilon}{3k} \approx 1.5 \times 10^9\,\text{K} \Rightarrow \sigma_{\text{max}}(3\alpha \rightarrow {}^{12}\text{C}) \sim N_\alpha^3 \left(\frac{3k}{\varepsilon}\right)^3 \text{e}^{-3}\, \Gamma_\gamma .$$

Γ_γ is the electromagnetic decay width to the first excited state of ^{12}C. The conditions for the triple-alpha process can be summarized as follows:

- high plasma temperatures ($\approx 10^8$ K),
- high α-particle density,
- large, resonance-like cross section $\sigma(3\alpha)$.

3. Gravity sets the escape velocity (v_{esc}) and the temperature determines the mean speed, with which the molecules are moving (v_{mol}). The condition for a planet to retain an atmosphere is

$$v_{\text{esc}} \gg v_{\text{mol}} .$$

v_{esc} can be obtained from

$$\frac{1}{2}mv^2 = G\,\frac{m\,M_{\text{planet}}}{R_{\text{planet}}} \Rightarrow v_{\text{esc}} = \sqrt{2GM_{\text{planet}}/R_{\text{planet}}} \quad (= 11.2\,\text{km/s for Earth}) .$$

v_{mol} can be calculated from

$$\frac{1}{2}mv^2 = \frac{3}{2}kT \quad \Rightarrow \quad v_{\text{mol}} = \sqrt{3kT/m} \quad (= 517\,\text{m/s for N}_2 \text{ for Earth})\,.$$

For $v_{\text{esc}} = v_{\text{mol}}$ one obtains

$$2G\,\frac{M_{\text{planet}}}{R_{\text{planet}}} = \frac{3kT}{m}\,.$$

Let us assume Earth-like conditions ($T = 300\,\text{K}$, $m = m(^{14}\text{N}_2) = 4.65 \times 10^{-23}\,\text{g} = 4.65 \times 10^{-26}\,\text{kg}$, $k = 1.38 \times 10^{-23}\,\text{J K}^{-1}$):

$$2G\,\frac{M_{\text{planet}}}{R_{\text{planet}}} = 267\,097\,\text{J/kg} \quad \Rightarrow \quad \frac{M_{\text{planet}}}{R_{\text{planet}}} = 2 \times 10^{15}\,\text{kg/m}\,.$$

With $M = \frac{4}{3}\pi R^3 \varrho$, and assuming an Earth-like density $\varrho = 5.5\,\text{g/cm}^3 = 5500\,\text{kg/m}^3$, one gets

$$R = \sqrt[3]{\frac{3M}{4\pi\varrho}} \approx 0.035\,\text{m} \times \sqrt[3]{M/\text{kg}}\,,$$

$$\frac{M_{\text{planet}}}{\sqrt[3]{M_{\text{planet}}}} \approx 2 \times 10^{15}\,\frac{\text{kg}}{\text{m}} \times 0.035\,\frac{\text{m}}{\text{kg}^{1/3}} = 7.0 \times 10^{13}\,\text{kg}^{2/3}\,.$$

This results in

$$M_{\text{planet}} \approx 5.9 \times 10^{20}\,\text{kg}\,.$$

The result of this estimate depends very much on Earth-like properties and the assumption that other effects, like rotational velocity ($\rightarrow$ centrifugal force) are negligible.

4. (a) The long-range forces in three dimensions are the gravitational force and the electromagnetic force, the latter of which reduces in the non-relativistic (static) limit to the Coulomb force. The force is proportional to the masses (Newton's gravitation law) or to the charges (Coulomb's law), respectively. Both forces scale with $1/r^2$, where r is the distance between the two bodies. The surface of a three-dimensional sphere is $4\pi r^2$, and the forces are, roughly speaking, proportional to the solid angle, under which the bodies see each other. The surface of an n-dimensional sphere is (already from dimensional considerations) given by

$$s_n(r) = s_n\,r^{n-1}\,, \quad \text{i.e.,} \quad g_n = n - 1\,,$$

where the constant s_n is the surface of the corresponding unit sphere. Isotropic long-range two-body forces in n dimensions should therefore scale as

$$F(r) \sim \frac{1}{s_n(r)} \sim \frac{1}{r^{n-1}} \, .$$

(b) The potential of a radial force is given by ($\boldsymbol{F} \cdot \boldsymbol{r} < 0$ for attractive forces, $n > 2$)

$$V(r) = \int_{r_0}^r F(r') \, dr' \sim \int_{r_0}^r \frac{dr'}{r'^{n-1}} = -\frac{1}{n-2} \left(\frac{1}{r^{n-2}} - \frac{1}{r_0^{n-2}} \right)$$

$$\sim -\frac{1}{r^{n-2}} + \text{const} \, .$$

A particle of mass m, at distance r from a center, and having velocity v on a circular orbit has an angular momentum of $L = rp = mvr$. (In general, the velocity component perpendicular to $\boldsymbol{r}$ is considered.) The centrifugal force is then given by

$$F_{\text{c}}(r) = \frac{mv^2}{r} = \frac{L^2}{mr^3} \, ,$$

where the last expression is also valid for general motion and L is a constant for central forces. The corresponding centrifugal potential therefore reads

$$V_{\text{c}}(r) = -\int_{r_0}^r F_{\text{c}}(r') \, dr' = \frac{L^2}{2mr^2} + \text{const} \, .$$

The effective potential is the sum of V and V_{c},

$$V_{\text{eff}}(r) = V(r) + V_{\text{c}}(r) = -\frac{C_n}{r^{n-2}} + \frac{L^2}{2mr^2} + \text{const} \, .$$

Circular orbits take place for a vanishing effective radial force,

$$0 = -F_{\text{eff}}(r) = \frac{dV_{\text{eff}}(r)}{dr} = \frac{(n-2)C_n}{r^{n-1}} - \frac{L^2}{mr^3} \quad \Rightarrow \quad r_{\text{orb}} = \left(\frac{m\tilde{C}_n}{L^2} \right)^{\frac{1}{n-4}}$$

with $\tilde{C}_n = (n-2)C_n$, $n \neq 4$, also applicable for $n = 2$. Stable orbits are given for potential minima:

$$0 < \left. \frac{d^2 V_{\text{eff}}(r)}{dr^2} \right|_{r_{\text{orb}}} = \left. \left(-\frac{(n-1)\tilde{C}_n}{r^n} + \frac{3L^2}{mr^4} \right) \right|_{r_{\text{orb}}} = (4-n)\frac{L^2}{m} \left(\frac{L^2}{m\tilde{C}_n} \right)^{\frac{4}{n-4}} \, .$$

Thus, stable motion is only possible for $n < 4$. For $n \geq 4$ the motion is either unbounded in space or eventually leads to a collision of the bodies after finite time.

As discussed in books of classical mechanics, also the spatial direction of non-circular motion is conserved for Newton's law, described by the Runge–Lenz vector, leading to 'true' ellipses. This is a special feature of $V \sim 1/r$, so for $n = 2$ with a logarithmic potential the motion in contrast is ergodic, i.e., every energetically reachable space point eventually lies in the vicinity of a trajectory point.

(c) The field energy of the force in the radial interval $[\lambda, \Lambda]$ reads

$$
W = \int_{\lambda \leq r \leq \Lambda} w(r)\,dV_n = \int_\lambda^\Lambda w(r) s_n(r)\,dr \sim s_n \int_\lambda^\Lambda F^2(r) r^{n-1}\,dr \,.
$$

With the expression $F(r) \sim 1/r^{n-1}$ one yields

$$
W \sim \int_\lambda^\Lambda \frac{dr}{r^{n-1}} = \begin{cases} \ln(\Lambda/\lambda) \,, & n = 2 \\ (\lambda^{-(n-2)} - \Lambda^{-(n-2)})/(n-2) \,, & n > 2 \end{cases} \,,
$$

which is similar to the expression of the potential energy. For $n > 2$ the limit $\Lambda \to \infty$ leads to a finite result, whereas W diverges for $\lambda \to 0$. In the case $n = 2$ both limits are divergent. Quantum corrections, here for the so-called self-energy corrections, are supposed to become significant for small distances. Therefore, the limit $\lambda \to 0$ may diverge, whereas the limit $\Lambda \to \infty$ should exist. Thus, $n > 2$ dimensions are considered valid from this aspect. The degree of divergence is then smallest for $n = 3$.

5. Assuming wild guesses, $R_* = 10$ per year, $f_p = 1$, $f_e = 0.1$, $f_l = 0.01$, $f_i = 0.01$, $f_t = 0.01$, and $L = 1\,000\,000$ years, one arrives at exactly 1, and that is our Earth.

Estimates for the probability of 'intelligent' life in our galaxy have been estimated to vary over a wide range. Given that there are about 100 billion galaxies in our universe, one might expect that life in our universe must be very common, not to speak of multiverses.

6. The escape velocity v from the asteroid can be worked out from

$$
\frac{1}{2}mv^2 \geq mg_A h \,, \tag{14.1}
$$

where g_A is the acceleration due to gravity on the asteroid. In case of jumping more than 2 m on Earth, i.e., lifting the center of mass by about 1.2 m, g_A has to be replaced by the acceleration due to gravity of the Earth g_E. The gravitational potential of an asteroid of mass M is

$$
\gamma \frac{M \cdot m}{r} \,, \tag{14.2}
$$

which is

$$\frac{4}{3}\pi\gamma\rho m\frac{r^3}{r} \tag{14.3}$$

or, correspondingly,

$$\frac{4}{3}\pi\gamma\rho m r^2. \tag{14.4}$$

Leaving the asteroid is equivalent to

$$mg_{\mathrm{E}}h \geq \frac{4}{3}\pi\gamma\rho m\frac{r^3}{r}, \tag{14.5}$$

where

$$g_{\mathrm{E}} = \gamma\frac{M_{\mathrm{Earth}}}{R^2} = \gamma\frac{4}{3}\pi R\gamma\rho. \tag{14.6}$$

In this formula R is the Earth's radius. This leads to

$$R \cdot h \geq r^2 \tag{14.7}$$

or

$$r^2 \leq R \cdot h. \tag{14.8}$$

Using the Earth's radius (6.4×10^6 m) and $h = 1.2$ m, one gets $r \leq 2.77$ km.

7. The flux density can be worked out using the power 1 GW of the transmitter, the distance to Proxima Centauri of $r = 4.26 \times 0.9461 \times 10^{16}$ m $= 4.03 \times 10^{16}$ m, the isotropic emission into the full solid angle $\Omega = 4\pi$, and the bandwidth of 100 kHz to get $P = 4.9 \times 10^{-26}$ W/m^2.

Using the area of a 300 m-diameter dish and the fact that usually only one po-larization is measured (factor $1/2$), one obtains $P = 1.73 \times 10^{-21}$ W. This is a signal, which one should be able to record. In terms of the usual unit used in radio astronomy the signal in jansky (1 Jy $= 10^{-26}$ W/m^2/Hz), using again the factor of $1/2$, would correspond to 2.45×10^{-31} W/m^2/Hz $= 2.45 \times 10^{-5}$ Jy.

In modern radio astronomy a signal strength of micro-jansky can be detected. Our galaxy has a radio strength of one million Jy at 20 MHz, resp. the Orion Nebula (distance 1.3 light-years) as radio source has 100 Jy.

8. The geological epochs of Pliocene and the Pleistocene vary between 0.5 and 2.5 million years. Therefore, the age of the supernova explosion must also be on the order of one million years. In addition, ^{60}Fe is a beta-ray emitter with a half-life of 2.6 million years. To measure a reasonable rate of ^{60}Fe atoms, the supernova cannot be much older than a few half-lives. Therefore, an estimate of the age of this supernova of about 2.5 million years seems plausible.

Models of supernova explosions also predict the abundance of this isotope. The relative abundance of ^{60}Fe with respect to all iron isotopes produced in a supernova explosion is relatively small (10^{-15}), but in principle it is sufficient to also permit a rough distance estimate. This information and the flux dilution due to the solid-angle effect could also help to derive an estimate on the distance of this supernova, which is considered to be around 150 light-years.

Chapter 15

1. There is not yet a solution. If you have solved the problem successfully, you should book a flight to Stockholm because you will be the next laureate for the Nobel Prize in physics.

Glossary

A

absolute brightness The total luminosity emitted from an astrophysical object. A star with an apparent brightness m at a distance d (in parsec) is assigned an absolute brightness $M = m - 5(\lg d - 1) = m - 5(\lg(d/10\,\text{pc}))$; e.g., Rigel at a distance of $d = 860$ light-years ($1\,\text{pc} = 3.26\,\text{ly}$) has $m = 0.12$, corresponding to an absolute brightness of $M = 0.12 - 5(\lg(860/3.26) - 1) \approx -7$. The absolute brightness of a supernova of type Ia is $M = -17$. Supernovae of type Ia are assumed to have identical absolute brightness and are considered as standard candles.

absolute zero Lowest possible temperature, $0\,\text{K}$, at which all motion comes to rest except for quantum effects ($0\,\text{K} = -273.15\,^\circ\text{C}$).

abundance Percentage of an element occurring on Earth, in the solar system, or in primary cosmic rays in a stable isotopic form.

acceleration equation It describes the acceleration of the universe in its dependence on the energy density and negative pressure (i.e., due to the cosmological constant).

acceleration parameter A measurement for the change of the expansion rate with time. For an increased expansion rate the acceleration parameter is positive. It was generally expected that the present expansion rate is reduced by the attractive gravitation (negative acceleration parameter = deceleration). The most recent measurements of distant supernovae, however, contradict this expectation by finding an accelerated expansion.

accelerator A machine used to accelerate charged particles to high speeds. There are, of course, also cosmic accelerators.

accretion disk Accumulation of dust and gas into a disk, normally rotating around a compact object.

© Springer Nature Switzerland AG 2020
C. Grupen, *Astroparticle Physics*, Undergraduate Texts in Physics,
https://doi.org/10.1007/978-3-030-27339-2

acoustic peaks After decoupling of matter from radiation the pattern of acoustic oscillations of the primordial baryon fluid became frozen as structures in the power spectrum of the CMB.

active galaxy Galaxy with a bright central region, an active galactic nucleus. Seyfert galaxies, quasars, and blazars are active galaxies, which are presumably powered by a black hole.

AGASA AGASA (Akeno Giant Air Shower Array) was a large-area ($100\,\text{km}^2$) extensive-air-shower detector in Japan with the ability to measure electron and muon components.

age of the universe In the Big Bang model the inverse of the Hubble constant is identified with the age of the universe. The present estimate of the age of the universe is about 14 billion years.

AGN Active Galactic Nucleus. If one assumes that black holes powered by infalling matter reside at centers of AGNs and if these AGNs emit polar jets, then the different AGN types (Seyfert galaxies, BL-Lacertae objects, radioquasars, radiogalaxies, quasars) can be understood as a consequence of the random direction of view from Earth.

air-shower Cherenkov technique Measurement of extensive air showers via their produced Cherenkov light in the atmosphere.

alpha decay Nuclear decay consisting of the emission of an α particle (helium-4 nucleus).

AMS experiment Alpha Magnetic Spectrometer on board the International Space Station ISS for the measurement of primary cosmic rays.

AMANDA Antarctic Muon And Neutrino Detector Array for the measurement of high-energy cosmic neutrinos.

Andromeda Nebula M31, one of the main galaxies of the local group; distance from the Milky Way: 700 kpc.

Ångström The unit of length equal to 10^{-10} m.

annihilation A process, in which particles and antiparticles disintegrate, e.g., $e^+e^- \rightarrow \gamma\gamma$.

ANTARES Astronomy with a Neutrino Telescope and Abyss Environmental Research; large-volume water Cherenkov detector in the Mediterranean for the measurement of cosmic neutrinos.

anthropic principle One might ask why the universe is as it is. If some of the physics constants that describe our universe were not finely tuned, then it would have grossly different properties. It is considered highly likely that life forms as we know them would not develop under an even slightly different set of basic parameters. Therefore, if the universe were different, nobody would be around to ask why the universe is as it is.

antigravitation Repulsive gravitation caused by a negative pressure as a consequence of a finite non-zero cosmological constant.

antimatter Matter consisting only of antiparticles, like positrons, antiprotons, or antineutrinos. When antimatter particles and ordinary matter particles meet, they annihilate mostly into γ rays (e.g., $e^+e^- \rightarrow \gamma\gamma$).

antiparticles For each particle there exists a different particle type of exactly the same mass but opposite values for all other quantum numbers. This state is called antiparticle. For example, the antiparticle of an electron is a particle with positive electric unit charge, which is called positron. Also bosons have antiparticles apart from those, for which all charge quantum numbers vanish. An example for this is the photon or a composite boson consisting of a quark and its corresponding antiquark. In this particular case there is no distinction between particle and antiparticle, they are one and the same object.

antiquark The antiparticle of a quark.

apastron The point of greatest separation of two stars as in a binary star orbit.

aphelion The point in the orbit of a planet, which is farthest from the Sun.

apogee The point in the orbit of an Earth satellite, which is farthest from Earth.

apparent magnitude The brightness of a star as it appears to the observer. It is measured in units of magnitude. See magnitude.

asteroid belt Circumstellar disc of asteroids and dwarf planets in the solar system between Mars and Jupiter.

astrobiology Considerations on the formation and evolution of life in universes with different fundamental parameters (different physical laws, quark masses, etc.)

astronomical unit (AU) The average distance from Earth to Sun. $1\,\text{AU} \approx 149\,597\,870\,\text{km}$.

archeoastronomy Archeoastronomy is a branch of archaeology, which describes the use of early astronomical techniques by prehistoric civilizations and how they have understood the phenomena in the sky.

asymptotic freedom At large momenta quarks will be eventually de-confined and become asymptotically free.

atomic mass The mass of a neutral atom or a nuclide. The atomic weight of an atom is the weight of the atom based on the scale, where the mass of the carbon-12 nucleus is equal to 12. For natural elements with more than one isotope, the atomic weight is the weighted average for the mixture of isotopes.

atomic number The number of protons in the nucleus.

Auger experiment Pierre Auger Observatory; air-shower experiment in Argentina for the measurement of the highest-energy cosmic rays.

aurora The northern (aurora borealis) or southern (aurora australis) lights are bright emissions of atoms and molecules in the polar upper atmosphere around the south and north geomagnetic poles. The aurora is associated with luminous processes caused by energetic solar cosmic-ray particles incident on the upper atmosphere of the Earth.

AXAF Advanced X-ray Astrophysics Facility; now named Chandra.

axion Hypothetical pseudoscalar particle, which was introduced as quantum of a field to explain the non-observation of CP violation in strong interactions.

B

background radiation See cosmic background radiation.

baryogenesis Formation of baryons out of the primordial quark soup in the very early universe.

baryon–antibaryon asymmetry An asymmetry created by some, so far unknown baryon-number-violating processes in the early universe (see Sakharov conditions).

baryon number The baryon number B is unity for all strongly interacting particles consisting of three quarks. Quarks themselves carry baryon number $\frac{1}{3}$. For all other particles $B = 0$.

baryon acoustic oscillations Baryon acoustic oscillations (BAO) are periodic fluctuations of the baryonic matter density in the universe. BAO are frozen relics left over from the early universe.

baryons Elementary particles consisting of three quarks like, e.g., protons and neutrons.

BATSE Burst And Transient Source Experiment on board the CGRO satellite; lifetime 1991–2000.

beam-dump experiment If a high-energy particle beam is stopped in a sufficiently thick target, all strongly and electromagnetically interacting particles are absorbed. Only neutral weakly interacting particles like neutrinos can escape.

beta decay In the nuclear β decay a neutron of the nucleus is transformed into a proton under the emission of an electron and an electron antineutrino. The transition of the proton in the nucleus into a neutron under emission of a positron and an electron neutrino is called β^+ decay, contrary to the aforementioned β^- decay. The electron capture, which is the reaction of proton and electron into neutron and an electron neutrino ($p + e^- \rightarrow n + \nu_e$), is also considered as beta decay.

Bethe–Bloch formula Describes the energy loss of charged particles by ionization and excitation when passing through matter.

Bethe–Weizsäcker cycle Carbon–nitrogen–oxygen cycle: nuclear fusion process in massive stars, in which hydrogen is burnt to helium with carbon as catalyzer (CNO cycle).

Bethe–Weizsäcker formula Describes the nuclear binding energy in the framework of the liquid-drop model.

Big Bang Beginning of the universe when all matter and radiation emerged from a singularity.

Big Bang theory The theory of an expanding universe that begins from an infinitely dense and hot primordial soup. The initial instant is called the Big Bang. The theory says nothing at all about time zero itself.

Big Crunch If the matter density in the universe is larger than the critical density, the presently observed expansion phase will turn over into a contraction, which eventually will end in a singularity.

Big Rip The Big Rip is a cosmological hypothesis about the ultimate fate of the universe. The universe has a large amount of dark energy corresponding to a repulsive gravity, and in the long run it will win over the classical attractive gravity anyhow. Therefore, all matter will be finally pulled apart. Galaxies in galactic clusters will be separated, the gravity in galaxies will become too weak to hold the stars together, and also the solar system will become gravitationally unbound. Eventually also the Earth and the atoms themselves will be destroyed with the consequence that the Big Rip will shred the whole physical structure of the universe.

binary stars Binary stars are two stars that orbit around their common center of mass. An X-ray binary is the special case, where one of the stars is a collapsed object such as a neutron star or black hole. Matter is stripped from the normal star and falls onto the collapsed star producing X rays or also gamma rays.

binding energy The energy that has to be invested to disintegrate a nucleus into its single constituents (protons and neutrons, nuclear binding energy).

bioastronomy Scientific branch of astronomy that deals with the question of the origin of life, e.g., whether organic material has been delivered to Earth from exraterrestrial sources. The techniques are to look for biomolecules with spectroscopic methods used in astronomy.

blackbody A hypothetical body that absorbs all the radiation falling on it. The emissivity of a blackbody depends only on its temperature.

blackbody radiation The radiation produced by a blackbody. The blackbody is a perfect radiator and absorber of heat or radiation, see also Planck distribution.

black hole A massive star that has used up all its hydrogen can collapse under its own gravity to a mathematical singularity. The size of a black hole is characterized by the event horizon. The event horizon is the radius of that region, in which gravity is so strong that even light cannot escape (see also Schwarzschild radius).

blazars Short for variable active galactic nuclei, which are similar to BL-Lacertae objects and quasars except that the optical spectrum is almost featureless.

blueshift Reduction of the wavelength of electromagnetic radiation by the Doppler effect as observed, e.g., during the contraction of the universe.

BL-Lacertae objects Variable extragalactic objects in the nuclei of some galaxies, which outshine the whole galaxy. BL stands for radio loud emission in the B band ($\approx$100 MHz); lacerta = lizard.

bolometer Sensitive resistance thermometer for the measurement of small energy depositions.

Boltzmann constant A constant of nature, which describes the relation between the temperature and kinetic energy for atoms or molecules in an ideal gas.

Bose–Einstein distribution The energy distribution of bosons in its dependence on the temperature.

boson A particle with integer spin. The spin is measured in units of $\hbar$ (spin $s = 0, 1, 2, \ldots$). All particles are either fermions or bosons. Gauge bosons are associated with fundamental interactions or forces, which act between the fermions. Composite particles with an even number of fermion constituents (e.g., consisting of a quark and antiquark) are also bosons.

bottom quark b; the fifth quark flavour (if the quarks are ordered with increasing mass). The b quark has the electric charge $-\frac{1}{3}e$.

Brahe Tycho Brahe, a Danish astronomer, whose accurate astronomical observations form the basis for Johannes Kepler's laws of planetary motion. The supernova remnant SNR 1572 (Tycho) is named after Brahe.

brane Branes are higher-dimensional generalisations of strings. A p brane is a space-time object with p spatial dimensions.

brane cosmology Brane cosmology is related to superstring theory in the attempt to find a theory of everything. The concept is applied to cosmology in the framework of M-theory.

bremsstrahlung Emission of electromagnetic radiation when a charged particle is decelerated in the Coulomb field of a nucleus. Bremsstrahlung can also be emitted during the deceleration of a charged particle in the Coulomb field of an electron or any other charge.

brown dwarves Low-mass stars (<0.08 solar masses), in which thermonuclear reactions do not ignite, which, however, still shine because gravitational energy is transformed into electromagnetic radiation during a slow shrinking process of the objects.

C

Calabi–Yau space Complex space of higher dimension, which has become popular for the compactification of extraspatial dimensions in the framework of string theories.

carbon cycle See Bethe–Weizsäcker cycle.

cascade See shower.

Casimir effect A reduction in the number of virtual particles between two flat parallel metal plates in vacuum compared to the surrounding leads to a measurable attractive force between the plates.

cataclysmic variables Stars that have rapid and unpredictable changes in brightness, mostly associated with a binary star system.

CBR See cosmic background radiation.

CCD Charge-Coupled Device, a solid-state camera.

Centaurus A Strong galactic radio source in the constellation Centaurus.

Cepheid variables Type of pulsating variable stars, where the period of oscillation is proportional to the average absolute magnitude; often used as 'standard' candles.

CERN Centre Européen pour la Recherche Nucléaire. The major European center for particle physics located near Geneva in Switzerland.

CGRO Compton Gamma-Ray Observatory. Satellite with four experiments on board for the measurement of galactic and extragalactic gamma rays; lifetime 1991–2000.

Chandra X-ray satellite of the NASA (original name: AXAF) started in July 1999; named after Subrahmanyan Chandrasekhar.

Chandrasekhar mass Limiting mass for white dwarves (1.4 solar masses). If a star exceeds this mass, gravity will eventually defeat the pressure of the degenerate electron gas leading to a compact neutron star.

charge Quantum number carried by a particle. The charge quantum number determines whether a particle can participate in a special interaction process. Particles with electric charge participate in electromagnetic interactions, such with strong charge undergo strong interactions, and those with weak charge are subject to weak interactions.

charge conjugation The principle of charge invariance claims that all processes again constitute a physical reality if particles are exchanged by their antiparticles. This principle is violated in weak interactions.

charge conservation The observation that the charge of a system of particles remains unchanged in an interaction or transformation. In this context charge stands for electric charge, strong charge, or also weak charge.

charged current Interaction process mediated by the exchange of a virtual charged gauge boson.

charm quark c; the fourth quark flavour (if quarks are ordered with increasing mass). The c quark has the electric charge $+\frac{2}{3}e$.

chemical potential For a given component in a particle mixture the chemical potential describes the change in free energy with respect to a change in the amount of the component at fixed pressure and temperature.

Cherenkov effect Cherenkov radiation occurs, if the velocity of a charged particle in a medium with index of refraction n exceeds the velocity of light c/n in that medium.

CKM matrix The Cabibbo–Kobayashi–Maskawa matrix is a mixing matrix describing the mixing of the three quark flavour generations in the framework of the electroweak theory. It is a unitary matrix, which contains information on the strength of flavour-changing weak decays and takes account of the difference of the eigenstates of mass and the eigenstates of weak interactions.

closed universe A Friedmann–Lemaître model of the universe with positive curvature of space. Such a universe will eventually contract leading to a Big Crunch.

cluster of galaxies A set of galaxies containing from a few tens to several thousand member galaxies, which are all gravitationally bound together. The Virgo Cluster includes 2500 galaxies.

CMB Cosmic microwave background, see cosmic background radiation.

COBE Cosmic Background Explorer. Satellite, with which the minute temperature inhomogeneities ($\frac{\Delta T}{T} \approx 10^{-5}$) of the cosmic background radiation were first discovered.

cold dark matter Type of dark matter that was moving at much less than the velocity of light some time after the Big Bang. It could consist of Weakly Interacting Massive Particles (WIMPs, such as supersymmetric particles) or axions.

collider An accelerator, in which two counter-rotating beams are steered together to provide a high-energy collision between the particles from one beam with those of the other.

colour The strong 'charge' of quarks and gluons is called colour.

colour charge The quantum number that determines the participation of hadrons in strong interactions. Quarks and gluons carry non-zero colour charges.

compactification The universe may have extra dimensions. In string theories there are 10 or more dimensions. Since our world appears to have only three plus one dimension, it is assumed that the extra dimensions are curled up into sizes so small that one can hardly detect them directly, which means they are compactified.

COMPTEL Compton telescope on board the CGRO satellite.

Compton effect Compton effect is the scattering of a photon off a free electron. The scattering off atomic electrons can be considered as pure Compton effect, if the binding energy of the electrons is small compared to the energy of the incident photon.

confinement The property of strong interactions, which says that quarks and gluons can never be observed as single free objects, but only inside colour-neutral objects.

conserved quantity A quantity is conserved, if it does not change in the course of a reaction between particles. Conserved quantities are, for example, electric charge, energy, and momentum.

constellation A group of stars that produce in projection a shape often named after mythological characters or animals.

conversion electron An alternative process to X-ray emission during the de-excitation of an excited atom or an excited nucleus. If the excitation energy of the atomic shell is transferred to an atomic electron and if this electron is liberated from the atom, this electron is called an Auger electron. If the excitation energy of a nucleus is transferred to an atomic electron of the K, L, or M shell, this process is called internal conversion. Auger electron emission or emission of characteristic X rays is often a consequence of internal conversion.

corona A very hot outer layer of the Sun's atmosphere consisting of highly diffused ionized gas and extending into interplanetary space. The hot gas in the solar corona forms the solar wind.

CORSIKA COsmic Ray SImulation for KAscade; it is a detailed and frequently used Monte Carlo program for the simulation of high-energy extensive air showers.

COS-B European gamma-ray satellite started in 1975.

cosmic abundance Standard composition of elements in the universe as determined from terrestrial, solar, and extrasolar matter.

cosmic background radiation CBR; nearly isotropic blackbody radiation originating from the Big Bang ('echo of the Big Bang'). The cosmic background radiation has now a temperature of 2.7 K.

cosmic-ray sources Sources of cosmic rays are stars, supernovae, neutron stars, pulsars, active galactic nuclei, and black holes. The sources of the highest-energy cosmic rays are still to be identified.

cosmic rays Nuclear and subatomic particles moving through space at velocities close to the speed of light. Cosmic rays originate from stars and, in particular, from supernova explosions. The origin of the highest-energy cosmic rays is an unsolved problem.

cosmic strings One-dimensional defects, which might have been created in the early universe.

cosmic textures Topological defect, which involves a kind of twisting of the fabric of space-time.

cosmoarcheology The branch of particle physics and astronomy that tries to dig out results on elementary particles and cosmology from experimental evidence obtained from data about the early universe.

cosmogony The science of the origins of galaxies, stars, planets, and satellites, and, in particular, of the universe as a whole.

cosmographic map Two-dimensinonal representation of the universe in the light of the 2.7 K blackbody radiation showing temperature variations as seed for galaxy formation.

cosmological constant (Λ) The cosmological constant was introduced by hand into the Einstein field equations with the intention to allow an aesthetic cosmological solution, i.e., a steady-state universe, which was believed to represent the correct picture of the universe at the time. The cosmological constant is time independent and spatially homogeneous. It is physically equivalent to a non-zero vacuum energy. In an expanding universe the repulsive gravity due to Λ will eventually win over the classical attractive gravity.

cosmological principle Hypothesis that the universe at large scales is homogeneous and isotropic.

cosmological redshift Light from a distant galaxy appears redshifted by the expansion of the universe.

cosmology Science of the structure and development of the universe.

cosmoparticle physics The study of elementary particle physics using information from the early universe.

***CP* invariance** Nearly all interactions are invariant under simultaneous interchange of particles with antiparticles and space inversion. The *CP* invariance, however, is violated in weak interactions (e.g., in the decay of the neutral kaon and *B* meson).

***CPT* invariance** All interactions lead again to a physically real process, if particles are replaced by antiparticles and if space and time are reversed. It is generally assumed that *CPT* invariance is an absolutely conserved symmetry.

Crab Nebula Crab; supernova explosion of a star in our Milky Way (observed by Chinese astronomers in 1054). The masses ejected from the star form the Crab Nebula in whose center the supernova remnant resides.

critical density Cosmic matter density ϱ_c leading to a flat universe. In a flat universe without cosmological constant ($\Lambda = 0$) the presently observed expansion rate will asymptotically tend to zero. For $\varrho > \varrho_c$ the expansion will turn into a contraction (see Big Crunch); for $\varrho < \varrho_c$ one would expect eternal expansion.

cross section The cross section is that area of an atomic nucleus or particle, which has to be hit to induce a certain reaction.

CTA The Cherenkov Telescope Array is a ground-based observatory for high-energy gamma astronomy based on the Cherenkov technique.

curvature The curvature of space-time determines the evolution of the universe. A universe with zero curvature is Euclidean. Positive curvature is the characteristic of a closed universe, while an open universe has negative curvature.

cyclotron mechanism Acceleration of charged particles on circular orbits in a transverse magnetic field.

Cygnus X1 X-ray binary consisting of a blue supergiant and a compact object, which is considered to be a black hole.

Cygnus X3 X-ray and gamma-ray binary system consisting of a pulsar and a companion. The pulsar appears to be able to emit occasionally gamma rays with energies up to 10^{16} eV.

D

dark age In the early universe the end of the photon epoch was followed by the dark age, where the universe was transparent but no large-scale structures had yet been formed, which could have given new light. Also the blackbody radiation was shifted into the infrared.

dark energy A special form of energy that creates a negative pressure and is gravitationally repulsive. It appears to be a property of empty space. Dark energy appears to contribute about 70% to the energy density of the universe. The vacuum energy, the cosmological constant, and light scalar fields (like quintessence) are particular forms of dark-energy candidates.

dark matter Unobserved non-luminous matter, whose existence has been inferred from the dynamics of the universe. The nature of the dark matter is an unsolved problem.

dark-matter detectors During the past decade, the sensitivity of experiments trying to detect dark-matter particles has improved significantly. Dark-matter detectors include crystals, bolometers, and superconducting devices all operated at cryogenic temperatures to separate the low signals from the interactions of dark-matter particles from noise and background.

Davis experiment Historically first experiment for the measurement of solar neutrinos.

de Broglie wavelength Quantum-mechanical wavelength λ of a particle: $\lambda = h/p$ (p—momentum, h—Planck's constant).

decay A process, in which a particle disappears and in its place different particles appear. The sum of the masses of the produced particles is always less than the mass of the original decaying particle.

declination A measure of how far an object is above or below the celestial equator (in degrees); similar to latitude on Earth.

degeneracy pressure The pressure in a degenerate gas of fermions caused by the Pauli exclusion principle and the Heisenberg uncertainty principle.

degenerate matter Fermi gas (electrons, neutrons), whose stability is guaranteed by the Pauli principle. In a gas of degenerate particles quantum effects become important.

degrees of freedom The number of values of a system that are free to vary.

deleptonization Process during supernova explosions, where electrons and protons merge to form neutrons and neutrinos ($e^- + p \rightarrow n + \nu_e$).

density fluctuations Local increase or decrease of the mass or radiation density in the early universe leading to galaxy formation.

deuterium Isotope of hydrogen with one additional neutron in the nucleus.

differential gravitation Different forces of gravity acting on two different points of a body in a strong gravitational field lead to a stretching of the body.

dipole anisotropy The apparent change in the cosmic-background-radiation temperature caused by the motion of the Earth through the CBR.

disk The visible surface of any heavenly body projected against the sky.

distance ladder A number of techniques (e.g., redshift, standard candles, ...) used by astronomers to obtain the distances to progressively more distant astronomical objects.

DNA Deoxyribonucleic acid is a complicated biomolecule, which carries the genetic information.

domain wall Topological defect that might have been created in the early universe. Domain walls are two-dimensional defects.

Doppler effect Change of the wavelength of light caused by a relative motion between source and observer.

double pulsar Predictions of the general theory of relativity about the periastron rotation and the energy loss by emission of gravitational waves have been confirmed for the binary pulsar system PSR1913+16 with extreme precision.

down quark d; the second quark flavour, (if quarks are ordered according to increasing mass). The d quark has the electric charge $-\frac{1}{3}e$.

Drake equation The Drake equation is an attempt to estimate the number of extraterrestrial civilizations in the Milky Way. It represents a very general assumption about the existence and the probability of extraterrestrial life.

E

east–west effect Geomagnetically caused asymmetry in the arrival direction of primary cosmic rays, which is related to the fact that most primary cosmic-ray particles carry positive charge.

EGRET Energetic Gamma Ray Experiment Telescope on board the CGRO satellite.

Einstein ring An Einstein ring is the result of gravitational lensing of a source by a massive object between the source and the observer forming a ring-like structure around the massive deflector.

electromagnetic interactions The interactions of particles due to their electric charge. This type includes also magnetic interactions.

electron e; the lightest electrically charged particle. As a consequence of this it is absolutely stable, because there are no lighter electrically charged particles, in which it could decay. The electron is the most frequent lepton with the electric charge -1 in units of the elementary charge.

electron capture Nuclear decay by capture of an atomic electron. If the decay energy is larger than twice the electron mass, positron emission can also occur in competition with electron capture.

electron number The electron and its associated electron neutrino are assigned the electronic lepton number $+1$, its associated antiparticles the electronic lepton number -1. All other elementary particles have the electron number 0. The electron number is generally a conserved quantity. Only in neutrino oscillations it is violated.

electron volt eV; unit of energy and also of mass of particles. One eV is the energy that an electron (or, more generally, a singly charged particle) gets, if it is accelerated in a potential difference of one volt.

electroweak theory Standard Model of elementary particles, in which the electromagnetic and weak interactions are unified.

elemental abundance Fraction of an element on Earth, in the solar system, or in primary cosmic rays.

elliptical galaxy Galaxy of smooth elliptical structure without spiral arms.

energy density Radiation density in joule/cm^3 or eV/cm^3.

EROS Expérience pour la Recherche d'Objets Sombres. Experiment searching for dark objects; see also MACHO.

ESA European Space Agency.

escape velocity The minimum velocity of a body to escape a gravitational field caused by a mass M from a distance R from its center is $v = \sqrt{\frac{2GM}{R}}$. For the Earth the escape velocity is $v_E = 11.2\,\text{km/s}$.

event horizon Surface of a black hole. Particles or light originating from inside the event horizon cannot escape from the black hole.

exclusion principle The exclusion principle, also called Pauli exclusion principle, states that no two electrons or, generally, no two fermions can have identical quantum numbers in an atom or molecule.

exoplanet An exoplanet or extrasolar planet is a planet outside our solar system that orbits a star. Queloz and Mayor discovered the first exoplanet around a main-sequence star in 1995.

expansion model Theory, in which all galaxies fly apart from each other. The expansion behaviour looks the same from every galaxy.

extended dimension Our universe has three spatial and one time dimension. In string theories and supersymmetric theories there are in general p extended spatial dimensions, some of which may be compactified.

extensive air shower (EAS) Large particle cascade in the atmosphere initiated by a primary cosmic-ray particle of high energy.

extra dimensions Extension of the space-time continuum by additional spatial dimensions.

extragalactic radiation Radiation originating from outside our galaxy.

extrasolar planets Extrasolar planets (exoplanets) are not part of our solar system. They are rather planets orbiting a different star (so far mostly in our Milky Way).

extremophiles Organisms that can live or exist under extreme environmental conditions (high pressure, high temperature, strong ionizing radiation, ...).

F

false vacuum A metastable state describing a quantum field that is zero, even though its corresponding vacuum energy density does not vanish. A decaying false vacuum can in principle liberate the stored energy.

families Organization of matter particles into three groups with each group being known as a family. There is an electron, a muon, and a tau family on the lepton side and, equivalently, there are three families of quarks. LEP has shown that there are only three families (generations) of leptons with light neutrinos.

Fermi–Dirac distribution The energy distribution of fermions in its dependence on the temperature.

Fermi energy Energy of the highest occupied electron level at absolute zero for a free Fermi gas (e.g., electron, neutron gas).

Fermi mechanism Acceleration mechanism for charged particles by shock waves (the 1st-order process) or extended magnetic clouds (the 2nd-order process).

fermion A particle with half-integer spin ($\frac{1}{2}$, $\frac{3}{2}$, etc.) if the spin is measured in units of $\hbar$. An important consequence of this particular angular momentum is that fermions are subject to the Pauli principle. The Pauli principle says that no two fermions can exist in the same quantum-mechanical state at the same time. Many properties of ordinary matter originate from this principle. Electrons, protons, and neutrons—all of them are fermions—just as the fundamental constituents of matter, i.e., quarks and leptons.

Fermi satellite γ satellite, Fermi Gamma-ray Space Telescope (FGST), started in 2008.

Feynman diagrams Feynman diagrams are pictorial shorthands to describe the interaction processes in space and time. With the necessary theoretical tools Feynman diagrams can be translated into cross sections. In this book the agreement is such that time is plotted on the horizontal and spatial coordinates on the vertical axis. Particles move forward in space-time, while antiparticles move backwards.

fission The splitting of heavier atomic nuclei into lighter ones. Fission is the way how nuclear power plants produce energy.

fixed-target experiment An experiment, in which the beam of particles from an accelerator is aimed at a stationary target.

flare Short-duration outburst from stars in various spectral ranges.

flatness problem In classical cosmology any value of Ω in the early universe even slightly different from one would be amplified with time and driven away from unity. In contrast, in inflationary cosmology Ω is approaching exponentially a value of one, thus explaining naturally the presently observed value of $\Omega = 1.02 \pm 0.02$. In classical cosmology such a value would have required an unbelievably careful fine-tuning.

flavour Characterizes the assignment of a fermion (lepton or quark) to a particle family or a particle generation.

fluid equation It describes the evolution of the energy density in its dependence on the expansion rate, pressure, and density. It can be derived using classical thermodynamics.

Fly's Eye Particle detector for the measurement of large extensive air showers based on the measurement of the emitted scintillation light in the atmosphere.

Forbush effect The Forbush effect or Forbush decrease is an occasional decrease in the intensity of cosmic rays as observed on Earth. It is caused by the solar wind or sudden solar flares and their magnetic effect on the propagation of galactic cosmic rays.

four-momentum Vector with four components comprising energy and the three momentum components.

four-vector Vector with four components mostly comprising time in the first component and three additional spatial components.

fragmentation Break-up of a heavy nucleus into a number of lighter nuclei in the course of a collision.

Friedmann equation A differential equation that expresses the evolution of the universe depending on its energy density. This equation can be derived from Einstein's field equations of general relativity. Surprisingly, a classical treatment leads to the same result.

Friedmann–Lemaître universes Standard Big Bang models with negative ($\Omega < 1$), positive ($\Omega > 1$), or flat ($\Omega = 1$) space curvature.

fundamental particle A particle with no observable inner structure. In the Standard Model of elementary particles quarks, leptons, photons, gluons, W^+, W^- bosons, and Z bosons are fundamental. All other objects are composites of fundamental particles.

fusion In fusion lighter elements are combined into heavier ones. Fusion is the way how stars produce energy.

G

galactic cluster Aggregation of galaxies in a spatially limited region.

galactic halo A spherical region mainly consisting of old stars surrounding the center of a galaxy.

galactic magnetic field Magnetic field in our Milky Way with an average strength of $3\,\mu G = 3 \times 10^{-10}$ T.

Galilei transformation Coordinate transformation for frames of reference, in which force-free bodies move in straight lines with constant speed without consideration of the fact that there exists a limiting velocity (velocity of light). This theorem of the addition of velocities leads to conflicts with experience, in particular, for velocities close to the velocity of light.

GALLEX experiment Gallium experiment in the Gran Sasso laboratory for the detection of solar neutrinos from the pp cycle.

gamma astronomy Astronomy in the gamma energy range (>0.1 MeV).

gamma burster Extragalactic gamma-ray sources flaring up only once. Spectacular supernova explosions (hypernova) or colliding neutron stars are possible candidates for producing gamma-ray bursts.

gamma–gamma interactions Gamma–gamma interactions describe the interactions of photons among each other, which are only possible in quantum electrodynamics. In classical physics this type of interactions does not exist. In astroparticle physics this interaction limits the range of high-energy cosmic-ray photons through interactions with the blackbody radiation.

gamma rays Short-wavelength electromagnetic radiation corresponding to energies ≥ 0.1 MeV.

gamma-ray sources Gamma rays originate from cosmic-ray sources, mostly emitted from very hot plasmas, due to bremsstrahlung or synchrotron radiation from relativistic electrons or by inverse Compton scattering. Another possibility are gamma rays from π^0 decays produced in potential hadronic accelerators.

general principle of relativity The principle of relativity argues that accelerated motion and being exposed to acceleration by a gravitational field is equivalent.

general relativity Einstein's general formulation of gravity, which shows that space and time communicate the gravitational force through their curvature.

general theory of relativity Generalization of Newton's theory of gravity to systems with relative acceleration with respect to each other. In this theory gravitational mass (caused by gravitation) and inertial mass (mass resisting acceleration) are equivalent.

generation A set of two quarks and two leptons that form a family. The order parameter for families is the mass of the family members. The first generation (family) contains the up and down quark, the electron and the electron neutrino. The second generation contains the charm and strange quark as well as the muon and its associated neutrino. The third generation comprises the top and bottom quark and the tau with its neutrino.

GLAST Gamma-ray Large Area Space Telescope, launched in 2008; γ energy range from 8 keV to more than 300 GeV; renamed Fermi Gamma-ray Space Telescope, see Fermi satellite.

gluino Supersymmetric partner of the gluon.

gluon g; the gluon is the carrier of strong interactions. There are altogether eight different gluons, which differ from each other by their colour quantum numbers.

grand unification A theory, which combines the strong, electromagnetic, and weak interactions into one unified theory (GUT).

gravitational collapse If the gas and radiation pressure of a star can no longer resist the inwardly directed gravitational pressure, the star will collapse under its own gravity.

gravitational instability Process by which density fluctuations of a certain size grow by self-gravitation.

gravitational interaction The interaction of particles due to their mass or energy. Also particles without rest mass are subject to gravitational interactions if they have energy, since energy corresponds to mass according to $m = E/c^2$.

gravitational lens A large mass causes a strong curvature of space thereby deflecting the light of a distant radiation source. Depending on the relative position of source, deflector, and observer, multiple images, arcs, or rings are produced as images of the source.

gravitational redshift Increase of the wavelength of electromagnetic radiation in the course of emission against a gravitational field.

gravitino Supersymmetric partner of the graviton.

graviton Massless boson with spin $2\hbar$ mediating the interaction between masses.

gravity Attractive force between two bodies caused by their mass. The gravitation between two elementary particles is negligibly small due to their low masses.

gravity waves In the same way as an accelerated electrical charge emits electromagnetic radiation (photons), accelerated masses emit gravitational waves. The quantum of the gravity wave is called the graviton.

Great Attractor The Great Attractor is a cluster of galaxies in intergalactic space at the center of the local Laniakea Supercluster at a distance of several million light-years. It has a mass corresponding to $>10^4$ galaxies and it is observable by its gravitational effect on nearby galaxies.

Great Wall A 100 Mpc structure seen in the distribution of galaxies.

Greisen–Zatsepin–Kuzmin cutoff (GZK) Threshold energy of energetic protons for pion production off photons of the cosmic background radiation via the Δ resonance.

GUT (Grand Unified Theory) Unified theory of strong, electromagnetic, and weak interactions.

GUT epoch The very early universe, when the strong, weak, and electromagnetic forces were unified and of equal importance.

GUT scale Energy scale, at which strong and electroweak interactions merge into one common force, $E \approx 10^{16}$ GeV.

gyroradius A charged particle moving in a magnetic field will orbit around the magnetic field lines. The radius of this orbit is called the gyroradius or sometimes also the Larmor radius.

H

habitable zone A habitable zone corresponds to the range of orbits around a star, in which life can exist, i.e., where liquid water and a sufficient atmospheric pressure might provide the existence of life.

hadron A particle consisting of strongly interacting constituents (quarks or gluons). Mesons and baryons are hadrons. All these particles participate in strong interactions.

halo Spherical cloud mostly of old stars, which surrounds a galaxy, like our Milky Way.

Hawking radiation The gravitational field energy of a black hole allows to create particle–antiparticle pairs of which, for example, one can be absorbed by the black hole while the other can escape if the process occurs outside the event horizon. By this quantum process black holes can evaporate because the escaping particle reduces the energy of the system. The time scales for the evaporation of large black holes exceeds, however, the age of the universe considerably.

Hawking temperature The temperature of a black hole manifested by the emission of Hawking radiation.

HEAO 1 High Energy Astronomy Observatory 1, X-ray satellite, energy range 0.2 keV–10 MeV, lifetime 1977–1979.

HEAO 2 High Energy Astronomy Observatory 2, X-ray satellite, energy range 0.2–20 keV, lifetime 1978–1981.

HEAO 3 High Energy Astronomy Observatory 3, X-, γ-, and cosmic-ray satellite, energy range 60 keV–20 MeV for X and γ rays, lifetime 1979–1981.

Heisenberg's uncertainty principle Position and momentum of a particle cannot be determined simultaneously with arbitrary precision, $\Delta x \, \Delta p \geq \hbar/2$. This uncertainty relation refers to all complementary quantities, e.g., also to the energy and time uncertainty.

heliocentric Centered on the Sun.

heliosphere The large region starting at the Sun's surface and extending to the limits of the solar system.

Hertzsprung–Russell diagram Representation of stars in a colour–luminosity diagram. Stars with large luminosity are characterized by a high colour temperature, i.e., shine strongly in the blue spectral range.

H.E.S.S. The High Energy Stereoscopic System is a ground-based Cherenkov telescope system in Namibia for the measurement of high-energy primary cosmic gamma rays.

Higgs boson Hypothetical particle as a quantum of the Higgs field predicted in the framework of the Standard Model to give masses to fermions via the mechanism of spontaneous symmetry breaking. Named after the Scottish physicist Peter Higgs. Discovered 2012 at the LHC at CERN.

Higgs field A hypothetical scalar field, which is assumed to be responsible for the generation of masses in a process called spontaneous symmetry breaking.

horizon The observable range of our universe. The radius of the observable universe corresponds to the distance that light has traveled since the Big Bang.

horizon problem In classical cosmology regions in the sky separated by more than $\approx 2°$ are causally disconnected. Experimentally it is found that they still have the same temperature. This can naturally be explained by an exponential expansion in the early universe, which has smoothed out existing temperature fluctuations.

hot dark matter Relativistic dark matter. Neutrinos are hot-dark-matter candidates.

HRI High Resolution Imager, focal detector of the X-ray satellite ROSAT.

Hubble constant H Constant of proportionality between the receding velocity v of galaxies and their distance r, $v = H \, r$. According to recent measurements the value of the Hubble constant H is about $70 \, (\text{km/s})/\text{Mpc}$. The inverse Hubble constant corresponds to the age of the universe.

Hubble law The receding velocity of galaxies is proportional to their distance.

Hubble telescope Space telescope with a mirror diameter of 2.2 m (HST—Hubble Space Telescope); launched 1990; to be replaced by the James Webb telescope 2021.

hydrogen burning Hydrogen burning is the fusion of four hydrogen nuclei into a single helium nucleus. It starts with the fusion of two protons into a deuterium, positron, and electron neutrino ($p + p \rightarrow d + e^+ + \nu_e$).

Hyper-Kamiokande detector Detector planned to consist of an order of magnitude larger tank than the predecessor experiment Super-Kamiokande. Commissioning foreseen for the second half of 2020.

I

ICECUBE Large-volume neutrino detector ($1\,\text{km}^3$) in the antarctic ice.

IMB Irvine–Michigan–Brookhaven collaboration for the search of nucleon decay and for neutrino astronomy.

implosion A violent inward-bound collapse.

inertia Property of matter that requires a force to act on it to change the way it is moving.

inflation Hypothetical exponential phase of expansion in the early universe starting at $10^{-38}\,\text{s}$ after the Big Bang and extending up to $10^{-36}\,\text{s}$. The presently observed isotropy and uniformity of the cosmic background radiation can be understood by cosmic inflation.

infrared slavery At low momenta quarks are confined in hadrons. They cannot escape their hadronic prison.

infrared astronomy Astronomy in the light of infrared radiation emitted from celestial objects.

INTEGRAL International Gamma-Ray Astrophysics Laboratory. Space observatory for the measurement of primary gamma rays from 15 keV to 10 MeV. Launched 2002.

interaction A process, in which a particle decays or responses to a force due to the presence of another particle as in a collision.

interaction length λ Characteristic collision length for strongly interacting particles in matter.

interstellar medium The gas and dust that exist in the space between the stars.

interferometer Interferometers are mostly optical systems, superimposing usually light beams, whereby phenomena of interference are studied to extract information on the emitting light source or detailed information on changes of an optical light path. Interferometry is an important tool in astronomy.

inverse Compton scattering Energy transfer by an energetic electron to a low-energy photon.

ionization Liberation of atomic electrons by photons or charged particles, namely, photo ionization and ionization by collision.

isomers Long-lived excited states of a nucleus.

isospin Hadrons of equal mass (apart from electromagnetic effects) are grouped into isospin multipletts, in analogy to the spin. Individual members of the multipletts are considered to be presented by the third component of the isospin. Nucleons form an isospin doublet ($I = \frac{1}{2}$, $I_3(p) = +\frac{1}{2}$, $I_3(n) = -\frac{1}{2}$) and pions an isospin triplet ($I = 1$, $I_3(\pi^+) = +1$, $I_3(\pi^-) = -1$, $I_3(\pi^0) = 0$).

isotope Nuclei of fixed charge representing the same chemical element albeit with different masses. The chemical element is characterized by the atomic number. Therefore, isotopes are nuclei of fixed proton number, but variable neutron number.

ISS International Space Station; manned space station at an altitude of about 400 km; operational since 2000.

J

Jeans mass An inhomogeneity in a matter distribution will grow due to gravitational attraction and contribute to the formation of structure in the universe (e.g., galaxy formation), if it exceeds a certain critical size (Jeans mass).

JEM-EUSO JEM-EUSO is a planned air-shower experiment (Extreme Universe Space Observatory) to be installed on the Japanese experiment module (JEM) on the International Space Station ISS; planned after 2020.

jets Long narrow streams of matter emerging from radiogalaxies and quasars, or bundles of particles in particle–antiparticle annihilations.

K

Kamiokande Nucleon Decay Experiment in the Japanese Kamioka mine for the measurement of cosmic and terrestrial neutrinos.

kaon K; a meson consisting of a strange quark and an anti-up or anti-down quark or, correspondingly, an anti-strange quark and an up or down quark.

KASCADE The KArlsruhe Shower Core and Array DEtector with the extension of KASCADE-Grande was an air-shower experiment at the Karlsruhe Institute for Technology (KIT) for the measurement of primary cosmic rays in the energy range up to 10^{18} eV.

Keplerian orbit A Keplerian orbit is the motion of, e.g., a planet by its motion around a massive object like the sun as an ellipse, parabola, or hyperbola.

Kerr black hole A Kerr black hole is a special type of black hole, which has only mass and angular momentum. A black hole might also have electric charge. Such a type of black hole could also exist in Einstein's theory of gravitation.

kpc Kilo parsec (see parsec).

Kuiper belt Circumstellar disc in the outer solar system, consisting of asteroids, comets, and dwarf planets, extending from Neptune to approximately 50 AU from the Sun. Pluto is one of the largest objects in the Kuiper belt.

L

large attractor Hypothetical supercluster complex with a mass of $> 10^4$ galaxies at a distance of several 100 million light-years, which impresses an oriented proper motion on the local group of galaxies.

latitude effect Increase of the cosmic-ray intensity to the geomagnetic poles caused by the interaction of charged primary cosmic rays with the Earth's magnetic field.

LEP Short for Large Electron–Positron Collider, the e^+e^- storage ring at CERN with a circumference of 27 km.

leptogenesis The production of leptons in the early universe.

lepton A fundamental fermion that does not participate in strong interactions. The electrically charged leptons are the electron (e), the muon (μ), and the tau (τ), and their antiparticles. Electrically neutral leptons are the associated neutrinos (ν_e, ν_μ, ν_τ).

lepton number Quantum number describing the lepton flavour. The three generations of charged leptons (e^-, μ^-, τ^-) have different individually conserved lepton numbers, while their neutrinos are allowed to oscillate, thereby violating lepton-number conservation.

LHC Large Hadron Collider at CERN in Geneva. In the LHC protons collide with a center-of-mass energy of up to 14 TeV. LHC startet operation in 2010. LHC is presently the accelerator with the highest center-of-mass energy in the world. It is generally hoped that the physics of the Large Hadron Collider will help to clarify some open questions of particle physics, for example, to better understand the role of the Higgs boson (discovered in 2012), which is considered to be responsible for the generation of masses of fundamental fermions.

light-year The distance traveled by light in vacuum during one year; $1 \, \text{ly} = 9.46 \times 10^{15} \, \text{m} = 0.307 \, \text{pc}$.

lighthouse model The model that explains pulsars as flashes of light from a rotating pulsar.

LIGO Laser Interferometer Gravitational-wave Observatory; Michelson interferometer as gravitational observatory consisting of two independent detectors at about 3000 km distance in the USA. LIGO discovered in 2015 gravitational waves from the merger of two black holes.

LISA Laser Interferometric Space Array; planned space observatory for the measurement of gravitational waves consisting of three satellites forming an equilateral triangle of 5 km edge length. Meanwhile LISA has been replaced by eLISA (Evolved Laser Interferometer Space Antenna), which is supposed to be financed by European countries alone. Its new name is NGO (New Gravitational wave Observatory); planned launch 2034.

LMC—Large Magellanic Cloud Irregular galaxy in the southern sky at a distance of 170 000 light-years ($\hat{=}$52 kpc).

local group A system of galaxies comprising among others our Milky Way, the Andromeda Nebula, and the Magellanic Clouds. The diameter of the local group amounts to approximately 5 million light-years.

local supercluster Virgo Supercluster; 'Milky Way' of galaxies, to which also the local group and the Virgo Cluster belong. Its extension is approximately 30 Mpc.

LOFAR Low-Frequency Array; it is a radio interferometer consisting of a large number of radio detectors (about 10 000) sensitive in the frequency range 10–80 MHz and 110–240 MHz. LOFAR covers an extension of 1000 km and achieves an angular resolution in the range of arc seconds.

LOPES LOPES (LOFAR PrototypE Station) was a digital radio aerial array for the measurement of cosmic-ray air showers at the site of the KASCADE-Grande experiment in Karlsruhe.

Lorentz contraction Feature emerging from special relativity, in which a moving object appears shortened along its direction of motion.

Lorentz transformation Transformation of kinematical variables for frames of reference, which move at linear uniform velocity with respect to each other under the consideration that the speed of light in vacuum is a maximum velocity.

luminosity Total light emission of a star or a galaxy in all ranges of wavelengths. The luminosity therefore corresponds to the absolute brightness.

luminosity distance The distance to an astronomical object derived by comparing its apparent brightness to its total known or presumed luminosity by assuming isotropic emission. If the luminosity is known, the source can be used as standard candle.

Lyman-alpha forest The Lyman-alpha forest in the spectra of distant galaxies or quasars are absorption lines caused by Lyman-alpha electron transitions of neutral hydrogen in various gas clouds between a source and the observer. These clouds, depending on their distance, form a forest of spectral absorption lines thereby allowing to determine their redshifts.

M

M87 Galaxy in the Virgo Cluster at a distance of 11 Mpc. M87 is considered an excellent candidate for a source of high-energy cosmic rays.

MACHO Massive Compact Halo Object. Experiment for the search for dark compact objects in the halo of our Milky Way, see also EROS.

Magellanic Cloud Nearest galaxy neighbour to our Milky Way consisting of the Small (SMC) and the Large Magellanic Cloud (LMC); distance 52 kpc.

MAGIC MAGIC (Major Atmospheric Gamma Imaging Cherenkov Telescopes) has two large Cherenkov telescopes on La Palma for the measurement of high-energy cosmic gamma rays.

magic numbers Nuclei whose atomic number Z or neutron number N is one of the magic numbers 2, 8, 20, 28, 50, or 126. In the framework of the shell model of the nucleus these magic nucleon numbers form closed shells. Nuclei whose proton and also neutron number are magic are named doubly magic.

magnetar Special class of neutron stars having a superstrong magnetic field (up to 10^{11} T). Magnetars emit sporadic γ-ray bursts.

magnetic monopole Hypothetical particle that constitutes separate sources and charges of the magnetic field. Magnetic monopoles are predicted by Grand Unified Theories.

magnitude A measure of the relative brightness of a star or some other celestial object. The brightness differences between two stars with intensities I_1 and I_2 are fixed through the magnitude $m_1 - m_2 = -2.5 \log_{10} (I_1/I_2)$. The zero of this scale is defined in such a way that the apparent brightness of the polestar—a magnitude 2 star—comes out to be $m = 2.12$. An apparent brightness difference of $\Delta m = 2.5$ corresponds to an intensity ratio of $10 : 1$. On this scale the planet Venus has the magnitude $m = -4.4$. With the naked eye one can see stars down to the magnitude $m = 6$. The strongest telescopes are able to observe stars down to the 31st magnitude (Hubble).

main-sequence star Star of average age lying on the main sequence in the Hertzsprung–Russell diagram.

Majorana neutrinos Majorana neutrinos (named after Ettore Majorana) or, more generally, Majorana leptons are electrically neutral fermions, which are identical to their own antiparticles. Majorana neutrinos would allow neutrinoless double beta decay.

Markarian galaxies Distant galaxies with a bright active galactic nucleus emitting in the blue and UV range. The Markarian galaxies are also sources of $\geq$TeV gamma rays.

mass Rest mass; the rest mass m_0 of a particle is that mass, which one obtains, if the energy of an isolated free particle in the state of rest is divided by the square of the velocity of light. When particle physicists use the name mass, they always refer to the rest mass of a particle.

mass number The sum (A) of the number of neutrons (N) and protons (Z) in a nucleus.

matter–antimatter asymmetry An asymmetry created by some, so far unknown baryon- and lepton-number-violating process in the early universe (see Sakharov conditions).

matter oscillation Flavour oscillations that can be amplified by matter effects in a resonance-like fashion; e.g., suppression of solar neutrinos in $\nu_e e^-$ interactions in the Sun.

matter–radiation equality Moment when the energy density of radiation dropped below that of matter at around 50 000 years after the Big Bang.

meson A hadron consisting of a quark and an antiquark.

microlensing A small dark object in the line of sight to a bright background star can give rise to a brightness excursion of the background star due to a bending of the light rays by the dark body.

mini black holes Extremely small ($\approx \mu$g) black holes could have been formed in the early universe. These primordial black holes are not final states of collapsing or evaporating stars.

missing mass Mass or energy in cosmology that must be present because of its gravitational force it exerts on other matter. The missing mass does not emit detectable electromagnetic radiation (see dark matter). Missing mass is also encountered in relativistic kinematics, where in many experiments, in which the total energy is fully constrained by the center-of-mass energy, the mass, energy, and momentum of particles escaping from the detector without interaction (like neutrinos or SUSY particles), can be reconstructed due to the detection of all other particles.

models of the universe See Friedmann–Lemaître universes.

monopole problem All grand unified theories predict large numbers of massive magnetic monopoles in contrast to observation. Inflation during the era of monopole production would have diluted the monopole density to a negligible level, thereby solving the problem.

Mpc Megaparsec (see parsec).

MSW effect Matter oscillations first proposed by Mikheyev, Smirnov, and Wolfenstein; see matter oscillation.

M-theory Supersymmetric string theory ('superstring theory') unifying all interactions. In particular, the M-theory also contains a quantum theory of gravitation. The smallest objects of the M-theory are p-dimensional membranes. It appears now that all string theories can be embedded into an 11-dimensional theory, called M-theory. Out of the 10 spatial dimensions 7 are compactified into a Calabi–Yau space.

multiplicity Number of secondary particles produced in an interaction.

multiverse Hypothetical enlargement of the universe, in which our universe is only one of an enormous number of separate and distinct universes.

muon μ; the muon belongs to the second family of charged leptons. It has the electric charge -1.

muon number The muon μ^- and its associated muon neutrino have the muonic lepton number $+1$, their antiparticles the muonic lepton number -1. All other elementary particles have the muon number 0. The muon number—except for neutrino oscillations—is a conserved quantity.

N

NASA Independent US-American civilian space agency.

negative pressure In the same way as a positive pressure increases gravitation through its field, negative pressure (like, for example, in a spring) leads to a repulsive gravity. Dark energy represents a form of energy that is gravitationally repulsive due to its negative effective pressure.

neutral current Interaction mechanism mediated by the exchange of a virtual neutral gauge boson.

neutralino Candidate for a non-baryonic dark-matter particle. In the framework of supersymmetry fermions and bosons come in supermultiplets. Neutralinos is a collective term for the supersymmetric partners of the neutral standard gauge bosons, namely, the supersymmetric partners of the photon, the gluons, the Z, the graviton, and the neutral Higgs particles, called photino, gluinos, zino, gravitino, and higgsinos. The lightest neutralino is usually the lightest supersymmetric partner and it is in most models assumed to be stable and would therefore be a good dark-matter candidate.

neutrino A lepton without electric charge. Neutrinos only participate in weak and gravitational interactions and therefore are very difficult to detect. There are three known types of neutrinos, which are all very light. These leptons are ν_e, ν_μ, and ν_τ for the electron, muon, and tau family. At the LEP experiments it could be shown that apart from the three already known neutrino generations there is no further generation with light neutrinos ($m_\nu < 45\,\text{GeV}/c^2$).

neutrino oscillations Transmutation of a neutrino flavour into another by vacuum or matter oscillations.

neutrino vacuum oscillation Neutrino flavour oscillation in vacuum that can occur if neutrinos have mass and if the weak eigenstates are superpositions of different mass eigenstates.

neutron n; a baryon with electric charge zero. A neutron is a fermion with an internal structure consisting of two down quarks and one up quark, which are bound together by gluons. Neutrons constitute the neutral component of atomic nuclei. Different isotopes of the same chemical element differ from each other only by a different number of neutrons in the nucleus.

neutron decay Neutron decay process into a proton, an electron, and an electron antineutrino.

neutron evaporation Nuclear decay by emission of a neutron.

neutron star Star of extremely high density consisting predominantly of neutrons. Neutron stars are remnants of supernova explosions, where the gravitational pressure in the remnant star is so large that the electrons and protons are merged to neutrons and neutrinos ($e^- + p \to n + \nu_e$). Neutron stars have a diameter of typically 20 km. If the gravitational pressure surmounts the degeneracy pressure of neutrons, the neutron star will collapse to a black hole.

Newtonian gravitation Based on the mathematical techniques of calculus of Newton and Leibniz, the motion of planets orbiting the Sun due to its gravitational effect can be described in mathematical form. The Newtonian mechanics and gravitation is formulated in the three Newton's laws and the law of gravity.

'no hair' theorem For an external observer the black hole has only three properties: mass, electric charge, and angular momentum. If two black holes agree in these properties they are indistinguishable.

Northern lights Aurora borealis: solar particles entering the Earth's atmosphere at polar regions can penetrate deeply into the atmosphere while traveling parallel to the magnetic field lines. Due to interactions of electrons and protons with the molecules in the atmosphere they cause a natural light display in the Earth's sky. Polar lights occur always at the same time at the North and South Pole.

nova Star with a sudden increase in luminosity ($\approx 10^6$-fold). The star explosion is not as violent as in supernova explosions. A nova can also occur several times on the same star.

nuclear binding energy The energy, which is required to disintegrate an atomic nucleus into its constituents; $\approx 8\,\mathrm{MeV}/$nucleon.

nuclear fission Fission of a nucleus in two fragments of approximately equal size. In most cases fission leads to asymmetric fragments. It can occur spontaneously or be induced by nuclear reactions.

nuclear fusion Fusion of light elements to heavier elements. In a fusion reactor protons are combined to produce helium via deuterium. Our Sun is a fusion reactor.

O

OGLE Optical Gravitational Lens Experiment looking for dark stars in the galactic halo of the Milky Way.

Olbers' paradox If the universe were infinite, uniform, and unchanging, the sky at night would be bright since in whatever direction one looked, one would eventually see a star. The number of stars would increase in proportion to the square of the distance from Earth, while its intensity is inversely proportional to the square of the distance. Consequently, the whole sky should be about as bright as the Sun. The paradox is resolved by the fact that the universe is not infinite, not uniform, and not unchanging. Also light from distant stars and galaxies displays an extreme redshift and sometimes ceases to be visible. Moreover, the lifetime of stars is not unlimited.

Oort cloud A supposed cloud of icy astronomical objects assumed to surround the Sun to as far as somewhere between 50 000 and 200 000 AU; it is a cloud with a diameter of about three light-years.

open universe A Friedmann–Lemaître model of an infinite, permanently expanding universe.

orbit The path of an object (e.g., satellite) that is moving around a second object or point.

OSSE Oriented Scintillation Spectrometer Experiment on board the CGRO satellite.

P

pair production Creation of fermion–antifermion pairs by photons in the Coulomb field of atomic nuclei. Pair production—in most cases electron–positron pair production—can also occur in the Coulomb field of an electron or any other charged particle.

PAMELA Payload for Antimatter Matter Exploration and Light-nuclei Astrophysics; satellite experiment for the measurement of primary cosmic rays. PAMELA discovered an unexpected excess of positrons in the high-energy range. Launched 2006.

parallax An apparent displacement of a distant object with respect to a more distant background when viewed from two different positions.

parity The property of a wave function that determines its behaviour, when all its spatial coordinates are reversed. If a wave function ψ satisfies the equation $\psi(r) = +\psi(-r)$, it is said to have even parity. If, however, $\psi(r) = -\psi(-r)$, the parity of the wave function is odd. Experimentally only the square of the absolute value of the wave function is observable. If parity were conserved, there would be no fundamental way of distinguishing between left and right. Parity is conserved in electromagnetic and strong interactions, but it is violated in weak interactions.

parsec Parallax second; the unit of length used to express astronomical distances. It is the distance, at which the mean radius of the Earth's orbit around the Sun appears under an angle of 1 arc second. 1 pc = 3.086×10^{16} m = 3.26 light-years.

particle horizon Largest region that can be in causal contact.

Pauli principle See fermion.

periastron The point in an orbit of a star in a double-star system, in which the body describing the orbit is nearest to the star.

perigee The point in the orbit of the Moon or an artificial Earth satellite, at which it is closest to Earth.

perihelion The point in the orbit of a planet, comet, or artificial satellite in a solar orbit, at which it is nearest to the Sun.

PETRA Positron Electron Tandem Ring Accelerator, an electron–positron storage ring at DESY in Hamburg. Built from 1975–1978.

Pfotzer maximum Intensity maximum of secondary cosmic rays at an altitude of approximately 15 km produced by interactions of primary cosmic rays in the atmosphere.

photino Supersymmetric partner of the photon.

photoelectric effect Liberation of atomic electrons by photons.

photon The gauge boson of electromagnetic interactions.

photon decoupling Photon decoupling occurred during the recombination epoch, where electrons combined with protons to form neutral hydrogen. After this has happened (around 380 000 years after the Big Bang) the universe became transparent.

pion π; the lightest meson; pions constitute an isospin triplet, the members of which have electric charges of $+1$, -1, or 0.

Planck distribution Intensity distribution of blackbody radiation of a blackbody of temperature T as a function of the wavelength following Planck's radiation law.

Planck length Scale, at which the quantum nature of gravitation becomes visible, $L_{Pl} = \sqrt{G\hbar/c^3} \approx 1.62 \times 10^{-35}$ m.

Planck mass Energy scale, at which all forces including gravity can be described by a unified theory; $m_{Pl} = \sqrt{\hbar c/G} \approx 1.22 \times 10^{19}$ GeV/c^2.

Planck's radiation law A law giving the distribution of energy radiated by a blackbody. The frequency-dependent radiation density depends on the temperature of the blackbody. See also Planck distribution.

Planck Surveyor Satellite for the measurement of primordial cosmic rays (2009–2013).

Planck tension The tension of a typical string in string theories.

Planck time The time taken for a photon to pass the distance equal to the Planck length: $t_{Pl} = \sqrt{G\hbar/c^5} \approx 5.39 \times 10^{-44}$ s.

planetary nebula A shell of gas ejected from and expanding about a certain kind of extremely hot star.

PMNS matrix Neutrino-oscillation mixing matrix named after Pontecorvo, Maki, Nakagawa, and Sakata.

polar lights See northern lights or aurora.

positron e^+; antiparticle of the electron.

positron annihilation Positron decay in matter by annihilation with an electron ($e^+ e^- \rightarrow \gamma\gamma$).

power spectrum Describes a measure of the level of structure related to the density difference with respect to the average density in the early universe, which became frozen during the growth of structure.

primary cosmic rays Radiation of particles originating from our galaxy and beyond. It consists mainly of protons and α particles, but also elements up to uranium are present in primary cosmic rays.

primordial black holes See Mini Black Holes.

primordial nucleosynthesis Production of atomic nuclei occurring during the first three minutes after the Big Bang.

primordial particles Particles from the sources.

proton p; the most commonly known hadron, a baryon of electric charge $+1$. Protons are made up of two up quarks and one down quark bound together by gluons. The nucleus of a hydrogen atom is a proton. A nucleus with electric charge Z contains Z protons. Therefore, the number of protons determines the chemical properties of the elements.

proton–proton chain Nuclear reaction, in which hydrogen is eventually fused to helium. The pp cycle is the main energy source of our Sun.

protostar Very dense regions or aggregations of gas clouds, from which stars are formed.

Proxima Centauri Nearest neighbour of our Sun at a distance of 4.24 light-years.

pseudoscalar Scalar quantity, which changes sign under spatial inversion.

PSPC Position Sensitive Proportional Chamber on board of ROSAT.

pulsar Rotating neutron star with characteristic, pulsed emission in different spectral ranges (radio, optical, X-ray, gamma-ray emission—'pulsating radiostar'); discovered by Jocelyn Bell 1967.

Q

quantum A quantum is the minimum discrete amount, by which certain properties such as energy or angular momentum of a system can change. Planck's constant is the smallest quantity of a physical action. The elementary charge is the smallest charge of freely observable particles.

quantum anomalies Quantum anomalies can arise if a classical symmetry is broken in the process of quantization and renormalization.

quantum chromodynamics QCD; theory of strong interactions of quarks and gluons.

quantum field theory Quantum-mechanical theory applied to systems that have an infinite number of degrees of freedom. Quantum field theory also describes processes, in which particles can be created or annihilated.

quantum foam Frothy character of the space-time fabrique on ultramicroscopic scales in theories of quantum gravity.

quantum gravitation Quantum theory of gravitation aiming at the unification with the other types of interactions.

quantum mechanics Quantum mechanics describes the laws of physics that hold at very small distances. The main feature of quantum mechanics is that, for example, the electric charge, the energies, and the angular momenta come in discrete amounts, which are called quanta.

quark q; a fundamental fermion subject to strong interactions. Quarks carry the electric charge of either $+\frac{2}{3}$ (up, charm, top) or $-\frac{1}{3}$ (down, strange, bottom) in units, in which the electric charge of a proton is $+1$.

quasars Quasistellar radio sources; galaxies at large redshifts with an active nucleus, which outshines the whole galaxy and therefore makes the quasar appear like a bright star.

quintessence A scalar field model for the dark energy. In contrast to the cosmological constant the energy density in the quintessence field represents a time-varying inhomogeneous component with a negative pressure that satisfies $-1 < w < 0$, where $w \equiv P/\varrho$ and P and ϱ are the pressure and energy density of the quintessence field. The decay of the quintessence field could liberate energy into space-time, which could drive the expansion of the universe.

R

radiation belt Solar-wind particles can be trapped in the Earth's magnetic field. They form the Van Allen belts, in which charged particles spiral back and forth. There are separate radiation belts for electrons and protons. See also Van Allen belt.

radiation era Era up to 50 000 years after the Big Bang when radiation dominated the universe.

radiation length X_0 Characteristic attenuation length for high-energy electrons and gamma rays.

radiation pressure Force exerted by photons if they are absorbed or scattered on small dust or matter particles or absorbed by atoms.

radio astronomy Astronomy in the radio band; it led to the discovery of the cosmic blackbody radiation.

radio galaxy A galaxy with high luminosity in the radio band.

recombination Capture of an electron by a positive ion, frequently in connection with radiation emission.

recombination temperature The number of neutral and ionized atoms become equal at the recombination temperature (3500 K).

red giant If a main-sequence star has used up its hydrogen supply, its nucleus will contract leading to a temperature increase so that helium burning can set in. This causes the star to expand associated with an increase in luminosity. The diameter of a red giant is large compared to the size of the original star.

redshift Increase of the wavelength of electromagnetic radiation by the Doppler effect, the expansion of the universe, or by strong gravitational fields:

$$z = \frac{\Delta\lambda}{\lambda_0} = \frac{\lambda - \lambda_0}{\lambda_0}$$

(λ_0—emitted wavelength, λ—observed wavelength),

$$z = \begin{cases} v/c & \text{, classical} \\ \sqrt{\frac{c+v}{c-v}} - 1 & \text{, relativistic} \end{cases}.$$

re-ionization When the first stars and galaxies formed after the dark ages they released large amounts of ultraviolet light, which was energetic enough to ionize (hydrogen) atoms, a process called cosmic re-ionization.

relativity principle Equivalence of observation of the same event from different frames of reference having different velocities and acceleration. There is no frame of reference that is better or qualitatively different from any other.

residual interaction Interaction between objects that do not carry a charge but do contain constituents that have that charge. The residual strong interaction between protons and neutrons due to the strong charges of their quark constituents is responsible for the binding of the nucleus. In the thirties it was believed that the binding of protons and neutrons in a nucleus was mediated by the exchange of pions.

rest mass The rest mass of a particle is the mass defined by the energy of an isolated free particle at rest divided by the speed of light squared.

right ascension A coordinate that along with declination may be used to locate any object in the sky. Right ascension is the angular distance measured eastwards along the celestial equator from the vernal equinox to the intersection of the hour circle passing through the body. It is the celestial equivalent to longitude.

Robertson–Walker Metric The metric describing an isotropic and homogeneous space-time of the universe.

ROSAT German–British–American Roentgen satellite (launched 1990).

Rosetta European robotic spacecraft visiting the comet Churyumov–Gerasimenko and deploying the lander Philae on it; launched 2004.

R parity Quantum number that distinguishes supersymmetric particles from normal particles.

S

SAGE experiment Soviet–American Gallium Experiment for the measurement of solar neutrinos from the pp cycle.

Sakharov conditions Necessary conditions first formulated by Sakharov, which are required to create a matter-dominated universe: a baryon-number-violating process, violation of C or CP symmetry, and departure from thermal equilibrium.

Sanduleak Sanduleak was a blue supergiant star, located in the Tarantula Nebula in the Large Magellanic Cloud, whose explosion was observed in 1987 as SN 1987A.

SAS-2, SAS-3 Small Astronomy Satellite; gamma-ray satellites launched by NASA, 1972 (SAS-2) resp. 1975 (SAS-3).

scale factor The scale factor denotes an arbitrary distance, which can be used to describe the expansion of the universe. The ratio of scale factors between two different epochs indicates by how much the size of the universe has grown.

Schwarzschild radius Event horizon of a spherical black hole, $R = 2GM/c^2$.

scintillation Excitation of atoms and molecules by the energy loss of charged particles with subsequent light emission.

secondary cosmic rays Secondary cosmic rays are a complex mixture of elementary particles, which are produced in interactions of primary cosmic rays with the atomic nuclei of the atmosphere.

selectron Supersymmetric partner of the electron.

SETI The Search for Extraterrestrial Intelligence is the search for extraterrestrial civilisations using radio signals. SETI mainly concentrates on finding and listening for possible signals of extraterrestrial, technically advanced civilisations.

Seyfert galaxy Member of a small class of galaxies with brilliant nuclei and inconspicuous spiral arms. Seyfert galaxies are strong emitters in the infrared and are also detectable as radio and X-ray sources. The nuclei of Seyfert galaxies possess many features of quasars.

SGR objects Soft Gamma(-ray) Repeaters are objects with repeated emission of γ bursts. Presumably soft gamma-ray repeaters are neutron stars with extraordinarily strong magnetic fields (see magnetar).

shock front A sudden pressure, density, and temperature gradient.

shock wave A very narrow region of high pressure and temperature. Particles passing through a shock front can be effectively accelerated, if the velocity of the shock front and that of the particle have opposite direction.

shower Or cascade. High-energy elementary particles can generate numerous new particles in interactions, which in turn can produce particles in further interactions. The particle cascade generated in this way can be absorbed in matter. The energy of the primary initiating particle can be derived from the number of particles observed in the particle cascade. One distinguishes electromagnetic (initiated by electrons and photons) and hadronic showers (initiated by strongly interacting particles, e.g., p, α, Fe, $\pi^{\pm}$, ...).

singularity A space-time region with infinitely large curvature—a space-time point.

sky surveys A sky survey maps the whole sky in various ranges of electromagnetic radiation (gamma-ray sky, X-ray sky, radio survey of the sky, ...) or the density of stars or galaxies (Sloan Digital Sky Survey).

slepton Supersymmetric partner of a lepton.

SMC Small Magellanic Cloud. Galaxy in the immediate vicinity of the Large Magellanic Cloud (LMC).

SN 1987A Supernova explosion in the Large Magellanic Cloud in 1987. The progenitor star was Sanduleak.

SNAP SuperNova/Acceleration Probe. A proposed space-based experiment to measure the properties of the accelerating universe, which depend on the amounts of dark energy and dark matter. If selected by NASA, SNAP will be launched before 2020.

SNO Sudbury Neutrino Observatory.

SNR Remnant after a supernova explosion; mostly a rotating neutron star (pulsar).

solar flare Violent eruption of gas from the Sun's surface.

solar neutrinos In the different nuclear fusion processes in the Sun electron neutrinos are produced. Solar neutrinos constitute a huge flux of neutrinos from natural sources.

solar wind Flux of solar particles (electrons and protons) streaming away from the Sun.

spaghettification A body falling into a black hole will be stretched because of the differential gravitation.

spallation Nuclear transmutation by high-energy particles, in which—contrary to fission—a large number of nuclear fragments, α particles, and neutrons are produced.

special theory of relativity This theory refers to inertial non-accelerated frames of reference. It assumes that physical laws are identical in all frames of reference and that the speed of light in vacuum c is constant throughout the universe and independent of the speed of the observer. The theorem of the addition of velocities is modified to account for the deviation from the Galilei transformation.

spin Intrinsic angular momentum of a particle quantized in units of $\hbar$, where $\hbar = \frac{h}{2\pi}$ and h is Planck's constant.

spontaneous symmetry breaking Emergence of different properties of a system at low energies (e.g., weak and electromagnetic interaction), which do not exist at high energies, where these interactions are described by the Unified Theory.

squark Supersymmetric partner of a quark.

standard candle An astronomical object with well-known absolute luminosity that can be used to determine distances.

Standard Model Theory of fundamental particles and their interactions. Originally the Standard Model described the unification of weak and electromagnetic interactions, the electroweak interaction. In a more general sense it is used for the common description of weak, electromagnetic, and strong interactions.

stationary universe Older model of the world, in which matter is continuously created, which fills space that is produced by the expansion of the universe to provide a constant density.

starburst galaxies Galaxies with a high star-formation rate.

star cluster A bunch of stars, which are bound to each other by their mutual gravitational attraction.

star quake A quake in the crust of a neutron star.

steady-state universe Older model of the universe, in which matter is continuously produced to fill up the empty space created by expansion, thereby maintaining a constant density ('steady-state universe').

Stefan–Boltzmann law This law states that the amount of energy radiated by a blackbody is proportional to the fourth power of its temperature.

stellar wind The ejection of gas off the surface of a star.

sterile neutrino Hypothetical elementary particle presumably with only gravitational interactions that is considered as possible dark-matter candidate.

storage ring Synchrotron, in which counter-rotating particles and antiparticles are stored in a vacuum pipe. The particles usually stored in bunches collide in interaction points, where the center-of-mass energy is equal to twice the beam energy for a head-on collision. Storage rings are used in particle physics and for experiments with synchrotron radiation.

strangeness The strangeness of an s quark is -1. Strangeness is conserved in strong and electromagnetic interactions, however, violated in weak interactions. In weak decays or weak interactions the strangeness changes by one unit.

strange quark s; the third quark flavour (if quarks are ordered with increasing mass). The strange quark has the electric charge $-\frac{1}{3}e$.

string In the framework of string theories the known elementary particles are different excitations of elementary strings. The length of strings is given by the Planck scale.

string theory Theory, which unifies the general theory of relativity with quantum mechanics by introducing a microscopic theory of gravitation.

strong interaction Interaction responsible for the binding between quarks, antiquarks, and gluons, the constituents of hadrons. The residual interaction of strong interactions is responsible for nuclear binding.

sunspot cycle The recurring 11-year rise and fall in the number of sunspots.

sunspots A disturbance of the solar surface, which appears as a relatively dark center surrounded by less dark area. Sunspots appear dark because part of the thermal energy is transformed into magnetic field energy.

supercluster See galactic cluster.

supergalactic plane The supergalactic plane is a major sheet-like structure in the local universe, which contains the Local Supercluster, the Coma Supercluster, the Great Attractor, and other clusters and voids. Its radius is on the order of 50 Mpc.

super-Earth A super-Earth is an extrasolar planet with a mass higher than the Earth's, but substantially below the masses of the gas planets Jupiter or Neptune. There is no conjecture whether there could be life on such a super-Earth.

Super-Kamiokande detector Successor experiment of the Kamiokande detector (see Kamiokande).

superluminal speed Phenomenon of apparent superluminal speed caused by a geometrical effect related to the finite propagation time of light.

supermassive black hole Black hole at the center of a galaxy containing about 10^9 solar masses.

supernova Star explosion initiated by a gravitational collapse, if a star has exhausted its hydrogen and helium supply and collapses under its own gravity. Neutrinos are assumed to play a decisive role in triggering the explosion. In a supernova explosion the luminosity of a star is increased by a factor of 10^9 for a short period of time. The remnant star of a supernova explosion is a neutron star or pulsar.

supernova remnant Remnant of a supernova explosion; mostly a neutron star or pulsar.

superpartners Particles whose spin differs by $1/2$ unit from normal particles. Superpartners are paired by supersymmetry.

supersymmetry (SUSY) In supersymmetric theories each fermion is associated with a bosonic partner and each boson has a fermionic partner. In this way the number of elementary particles is doubled. The bosonic partners of leptons and quarks are sleptons and squarks. The fermionic partners, for example, of the photon, gluon, Z, and W are photino, gluino, zino, and wino. Up to now no supersymmetric particles have been found.

symmetry breaking A reduction in the amount of symmetry of a system, usually associated with a phase transition.

synchrotron Circular accelerator, in which charged particles travel in bunches at a fixed radius. The orbit is stabilized by synchronizing an external magnetic guiding field with the increasing momentum of the accelerated particles.

synchrotron radiation Electromagnetic radiation emitted by an accelerated charged particle in a magnetic field.

T

tachyon Particle that moves faster than the speed of light. Its mass squared is negative. Its presence in a theory generally yields inconsistencies.

tau τ; the third flavour of charged leptons (if the leptons are arranged according to increasing mass). The tau carries the electric charge -1.

tau-lepton number τ^- and its associated tau neutrino have the tau-lepton number $+1$, the antiparticles τ^+ and $\bar{\nu}_\tau$ the tau-lepton number -1. All other elementary particles have the tau-lepton number 0. The tau-lepton number is a conserved quantity except for neutrino oscillations.

Theory of Everything (TOE) The ultimate theory, in which the different phenomena of electroweak, strong, and gravitational interactions are unified.

thermonuclear reaction Nuclear fusion of light elements to heavier elements at high temperatures (e.g., pp fusion at $T \approx 10^7$ K).

tidal force The tidal force stretches a body towards the center of mass of another body in its differential gravitational field. The tidal force is responsible for tides, and it can break apart celestial bodies, like comets in a strong differential gravitational field.

time dilatation Stretching of time explained by special relativity; also called time dilation.

time-reversal invariance T invariance or simply time reversal. The operation of replacing time t by time $-t$. As with CP violation, T violation occurs in weak interactions of kaon decays. The CPT operation, which is a succession of charge conjugation, parity transformation, and time reversal, is regarded as a conserved quantity.

topological defect Topological defects like magnetic monopoles, cosmic strings, domain walls, or cosmic textures might have been created in the early universe.

top quark t; the sixth quark flavour (if quarks are arranged according to increasing mass). The top quark carries the electric charge $+\frac{2}{3}e$. The mass of the top quark is comparable to the mass of a gold nucleus ($\approx 175\,\text{GeV}/c^2$).

triple-alpha process Reaction, in which three α particles are fused to carbon. Such a process requires high α-particle densities and a large resonance-like cross section.

tritium Hydrogen isotope with two additional neutrons in the nucleus.

U

uncertainty relation The quantum principle first formulated by Heisenberg, which states that it is not possible to know exactly both the position x and the momentum p of an object at the same time. The same is true for the complementary quantities energy and time.

Unified Theory Any theory that describes all four forces and all of matter within a single all-encompassing framework (see also Grand Unified Theory, GUT).

up quark u; the lightest quark flavour with electric charge $+\frac{2}{3}e$.

V

vacuum energy density Quantum fields in the lowest-energy state describing the vacuum, which must not necessarily have the energy zero.

Van Allen belts Low-energy particles of the solar wind are trapped in certain regions of the Earth's magnetic field and stored.

Vela Pulsar Vela X1, supernova remnant in the constellation Vela at a distance of about 1500 light-years; the Vela supernova explosion was observed by the Sumerians 6000 years ago.

VERITAS VERITAS (Very Energetic Radiation Imaging Telescope Array System) is a Cherenkov telescope in Arizona for the measurement of high-energy cosmic gamma rays; started 2005.

velocity of light (c) The value of the velocity of light forms the basis for the definition of the length unit meter. It takes $1/299\,792\,458$ seconds for light in vacuum to pass 1 meter. The velocity of light in vacuum is the same in all frames of reference.

VIRGO Italian Michelson interferometer to measure gravitational waves. The Advanced Virgo detector is operational since 2017.

Virgo Cluster Concentration of galaxies in the direction of the constellation Virgin.

virtual particle A particle that exists only for an extremely brief instant of time in an intermediate process. For virtual particles the Lorentz-invariant mass does not coincide with the rest mass. Virtual particles with negative mass squared are called space-like, those with a positive mass squared time-like. Virtual particles can exist for times allowed by the Heisenberg's uncertainty principle.

W

W^+, W^- **boson** Charged gauge quanta of weak interactions. These quanta participate in so-called charged-current processes. These are interactions, in which the electric charge of the participating particles changes.

wave function Probability waves, upon which quantum mechanics is founded.

weak interaction The weak interaction acts on all fermions, e.g., in the decay of hadrons. It is responsible for the beta decay of particles and nuclei. In charged-current interactions the quark flavour is changed, while in neutral-current interactions the quark flavour remains the same.

weak gravitational lensing Image distortions due to weak gravitational lensing allow to quantify cosmic structures in the universe. They can also help to map the distribution of dark and visible matter in the sky.

Whipple Whipple was a Cherenkov telescope for the measurement of high-energy cosmic gamma rays (1997–2006).

white dwarf A star that has exhausted most or all of its nuclear fuel and has collapsed to a very small size. The stability of a white dwarf is not maintained by the radiation or gas pressure as with normal stars but rather by the pressure of the degenerate electron gas. If the white dwarf has a mass larger than 1.4 times the solar mass, also this pressure is overcome and the star will collapse to a neutron star.

WIMPs Weakly Interacting Massive Particles are candidates for dark matter.

WMAP Wilkinson Microwave Anisotropy Probe launched in 2001 to measure the fine structure of the cosmic background radiation.

world models See Friedmann–Lemaître Universes.

wormhole A proposed channel of space-time connecting distant regions of the universe. It is not totally inconceivable that wormholes might provide the possibility of time travel.

w parameter The w parameter is defined as the ratio of pressure over energy density ($w \equiv P/\varrho$). In models with a cosmological constant the pressure of the vacuum density P equals exactly the negative of the energy density ϱ; i.e., $w = -1$, in contrast to quintessence models, where $-1 < w < 0$.

X

X **boson** See *Y* boson.

XMM-Newton European X-ray space observatory mission, with its X-ray Multi-Mirror design using three telescopes; named in honour of Sir Isaac Newton; launched 1999.

X-ray astronomy Astronomy in the X-ray range (0.1–100 keV).

X-ray burster X-ray source with irregular sudden outbursts of X rays.

X rays X rays are a form of electromagnetic radiation with wavelengths ranging from 0.01 to 10 nm, discovered 1895 by W. C. Röntgen.

Y

Y **boson** The observed baryon–antibaryon asymmetry in the universe requires, among others, a baryon-number-violating process. Hypothetical heavy *X* and *Y* bosons, existing at around the GUT scale with masses comparable to the GUT energy scale ($\approx 10^{16}$ GeV), could have produced this matter–antimatter asymmetry in their decay to quarks and antiquarks, resp. baryons and antibaryons. The couplings of the *X* and *Y* bosons to fermion species is presently the simplest idea, for which baryon-number violation appears possible.

ylem Name for the state of matter before the Big Bang. Gamow proposed that the matter of the universe originally existed in a promordial state, which he coined 'ylem' (from the Greek ὕλη, which stands for 'matter', 'wood' via the medieval latin *hylem* meaning *the primordial elements of life*). According to his idea all elements were formed shortly after the Big Bang from this primary substance.

Yukawa particle Yukawa predicted a particle that should mediate the binding between protons and neutrons. After its discovery the muon was initially mistaken to be this particle. The situation was resolved with the discovery of the pion as Yukawa particle.

Z

Z boson Neutral boson of weak interactions. It mediates weak interactions in all those processes, where electric charge and flavour do not change.

zero-point energy The Heisenberg uncertainty relation does not allow a finite quantum-mechanical system to have a definite position and definite momentum at the same time. Thus any system even in the lowest-energy state must have a non-zero energy. Zero energy would lead to zero momentum and thereby to an infinite position uncertainty.

Other useful glossaries on astroparticle physics and related fields:

Glossary of the Pacific Northwest National Laboratory: https://npac.pnnl.gov/glossary. Accessed: April 1st, 2019

Sigl, G. Astroparticle Physics: Theory and Phenomenology, Springer, Heidelberg (2016). Glossary pages 801–809

Sarkar, U.: Particle and Astroparticle Physics—CRC Press Book, Taylor and Francis, New York, London (2008)

Glossary and Acronyms | Astronomy ESFRI and Research Astronomy ESFRI and Research Infrastructure Cluster, https://www.asterics2020.eu/glossary. Accessed: April 1st, 2019. ASTERICS is a project supported by the European Commission Framework Programme Horizon 2020; Rob van der Meer et al.

Unified Astronomy Thesaurus: http://astrothesaurus.org/glossary/. Accessed: April 1st, 2019. Frey, K., Harvard–Smithsonian Center for Astrophysics, Cambridge MA

Hildebrand, R., Cosmology Glossary—Kavli Institute for Cosmological Physics: https://kicp.uchicago.edu/research/cosmology_glossary.html. Accessed: April 1st, 2019. Kavli Foundation (2011)

Spiering, C. & Weinheimer, C.: Astroparticle Physics, John Wiley and Sons, Indianapolis (2016)

NASA's Cosmicopia Glossary: https://helios.gsfc.nasa.gov/glossary.html. Accessed: April 1st, 2019

Hubble Legacy Archive Glossary: https://hla.stsci.edu/hla_glossary.html. Accessed: April 1st, 2019. Hubble Legacy Archive Team (2018)

Astronomical Glossary: https://ned.ipac.caltech.edu/level5/Glossary/frames.html. Accessed: April 1st, 2019. A Knowledgebase for Extragalactic Astronomy and Cosmology, Madore, B.F., Pasadena, California U.S.A. (2002)

Glossary of Astronomy Terms: http://www.seasky.org/astronomy/astronomy-glossary.html. Accessed: April 1st, 2019. Knight, J.D., creator and Webmaster of Sea and Sky (2016)

Glossary of astronomy, Wikipedia: https://en.wikipedia.org/wiki/Glossary_of_astronomy. Accessed: April 1st, 2019. Wikipedia contributors (2018)

Astronomy Dictionary: https://www.enchantedlearning.com/subjects/astronomy/glossary/Astronomers.shtml. Accessed: April 1st, 2019

Astronomy Glossary: http://www.observatorysciencecentre.co.uk/astronomy-glossary. Accessed: April 1st, 2019. The Observatory Science Centre, Barry Howse (2018)

Astronomy glossary: http://m.espacepourlavie.ca/en/astronomy-glossary. Accessed: April 1st, 2019. (2018)

Hubblesite: http://hubblesite.org/reference_desk/glossary/. Accessed: April 1st, 2019. Public Outreach Space Telescope Science Institute, Villard, R. et al.

ESO Astronomical Glossary: https://www.eso.org/public/outreach/glossary/glossary_a. Accessed: April 1st, 2019. Barcons, X. (2017)

Photo Credits and References

1. Credner, T., Kohle, S., Bonn University, Calar Alto Astronomical Observatory
2. M1: Filaments of the Crab Nebula. With courtesy of S. Kohle, T. Credner et al., Astronomical Institute of the University of Berne. ◇ http://apod.nasa.gov/apod/ap980208.html. Accessed: March 18th, 2019
3. Mammana, D.: https://melitatrips.com/destinations/2016/alaska/alaska.html. Accessed: March 18th, 2019. Private communication by D. Mammana, 2017
4. Kuhn, D., Innsbruck University. Private communication, 2017
5. Hess, V.F.: Über Beobachtungen der durchdringenden Strahlung bei sieben Freiballonfahrten. Phys. Z. **13**, 1084 (1912)
6. Kohlhörster, W.: Messungen der durchdringenden Strahlung im Freiballon in größeren Höhen. Phys. Z. **14**, 1153 (1913)
7. NASA and ESA: The Gravitational Lens G2237 + 0305; NASA and ESA (HubbleSite; September 13th, 1990) http://hubblesite.org/image/22/news_release/1990-20. Accessed: March 18th, 2019. ◇ https://en.wikipedia.org/wiki/Einstein_Cross. Accessed: March 18th, 2019
8. East–West Effect after de Clercq, C., Vrije Universiteit Brussel. http://w3.iihe.ac.be/~cdeclerc/astroparticles/figures/. Accessed: March 18th, 2019. http://hep.bu.edu/~superk/ew-effect.html. Accessed: March 18th, 2019
9. Anderson, C.D., Neddermeyer, S.H.: Cloud Chamber Observations of Cosmic Rays at 4300 Meters Elevation and Near Sea-Level. Phys. Rev. **50**, 263 (1936)
10. Powell, C.F., Fowler, P.H., Perkins, D.H.: The Study of Elementary Particles by the Photographic Method. Pergamon Press, New York (1959)
11. Perkins, D.H., University of Oxford. Private communication, 2017
12. Rochester, G.D., Butler, C.C. (Manchester U.): Evidence for the Existence of New Unstable Elementary Particles. Nat. **160**, 855–857 (1947)
13. Eisenberg, Y.: Possible Existence of a new Hyperon. Phys. Rev. **96**, 541 (1954)
14. Oh, S.Y., Yi, Y.: A Simultaneous Forbush Decrease Associated with an Earthward Coronal Mass Ejection Observed by STEREO. Solar Phys. **280**, 197–204 (2012)
15. Very Large Telescope (VLT) detects farthest known galaxy. CERN Courier, 13 (May 2004). ◇ Pelló, R. et al.: Properties of faint distant galaxies as seen through gravitational telescopes. Astron. Astrophys. **416**, L35 (2004)
16. Penzias, A.A., Bell Labs, USA. With courtesy of A.A. Penzias
17. Penzias, A.A., Wilson R.W.: A Measurement Of Excess Antenna Temperature At 4080 Mc/s. Astrophys. J. Lett. **142**, 419–421 (July 1965). ◇ A.A. Penzias and R.W. Wilson, Bell Labs, Holmdel, NJ

© Springer Nature Switzerland AG 2020
C. Grupen, *Astroparticle Physics*, Undergraduate Texts in Physics,
https://doi.org/10.1007/978-3-030-27339-2

18. Niu, K., Mikumo, E., Maeda, Y.: Possible Decay in Flight of a New Type Particle. Prog. Theor. Phys. **46**, 1644–1646 (1971). ◇ Niu, K., Mikumo, E., Maeda, Y.: A Possible Decay in Flight of a New Type Particle. Conf. Paper, 12th Int. Cos. Ray Conf. (Hobart), 2792–2798 (1971). ◇ See also Niu, K.: Discovery of naked charm particles and lifetime differences among charm species using nuclear emulsion techniques innovated in Japan. Proc. Jpn. Acad. Ser. B **84(1)**, 1–16 (2008)

19. LIGO Collabotation. https://www.ligo.caltech.edu/. Accessed: March 18th, 2019. ◇ Kashlinsky, A. (NASA, Goddard and SSAI, Lanham): Astrophys. J. **823(2)**, L25 (2016). ◇ Abbott, B.P. et al. (LIGO Scientific Collaboration and Virgo Collaboration): Observation of Gravitational Waves from a Binary Black Hole Merger. Phys. Rev. Lett. **116**, 061102 (11 February 2016)

20. Grupen, C.: Cosmic Cartoon Collection; Siegen University Press, Siegen, Germany (2014); https://www.hep.physik.uni-siegen.de/~grupen/. Accessed: March 18th, 2019

21. Timm, U., PLUTO Collaboration. ◇ Stella, B.R. (Rome III U. & INFN, Rome3), Meyer, H.-J. (Siegen U.): Y(9.46 GeV) and the gluon discovery (a critical recollection of PLUTO results). (Aug 2010. 37 pp.). Eur. Phys. J. H **36**, 203–243 (2011)

22. ATLAS Collaboration, Higgs boson decaying into two photons. http://atlasexperiment.org/HiggsResources/. Accessed: March 18th, 2019

23. Malin, D., Anglo-Australian Observatory. With courtesy of D. Malin

24. H.E.S.S. Collaboration, The H.E.S.S. II Telescope Array. https://commons.wikimedia.org/wiki/File:H.E.S.S._II_Telescope_Array.jpg. Accessed: March 18th, 2019

25. T2K (Tokai to Kamioka) Experiment, http://t2k-experiment.org/. Accessed: March 18th, 2019

26. Drevermann, H.—DALI: The Aleph Offline Event Display; http://aleph.web.cern.ch/aleph/dali/. Accessed: March 18th, 2019. See therein http://aleph.web.cern.ch/aleph/dali/Z0_examples.html. Accessed: March 18th, 2019. See also https://www.hep.physik.uni-siegen.de/~grupen/mona.pdf. Accessed: March 18th, 2019

27. Aguilar, J.A.: Gamma-Ray Astronomy. Université Libre de Bruxelles (2012). http://w3.iihe.ac.be/~aguilar/PHYS-467/PA5_notes.pdf. Accessed: March 18th, 2019

28. Riegler, W. (CERN). http://images.slideplayer.com/23/6855375/slides/slide_25.jpg. Accessed: March 18th, 2019

29. Meurer, C., Blümer, J., Engel, R., Haungs, A., Roth, M. and HARP Collaboration: The muon component in extensive air showers and new $p + C$ data in fixed target experiments. AIP Conf. Proc. **896**, 158–167 (2007)

30. Meurer, C., now Assmus, C.; HSS 08.09.2006 Fermilab/USA. http://slideplayer.com/slide/8816088/. Accessed: March 18th, 2019. ◇ Meurer, C.: Muon production in extensive air showers and fixed target accelerator data. Ph.D. thesis, Univ. of Karlsruhe (2007)

31. Azmi, M.D., Cleymans, J.: Transverse Momentum Distributions at the LHC and Tsallis Thermodynamics. Acta Phys. Polon. B Proc. Suppl. **7** no. 1, 9 (2014); arXiv:1310.0217 [hep-ph]. https://inspirehep.net/record/1256351/plots. Accessed: March 18th, 2019

32. Grupen, C.: Particle Detectors. Cambridge University Press, Cambridge (1996)

33. Yu, W. (ALICE TPC Collaboration): Particle identification of the ALICE TPC via dE/dx. Nucl. Instr. Meth. Phys. Res. A **706**, 55–58 (2013). ◇ ALICE Collaboration, http://aliceinfo.cern.ch/Public/Welcome.html. Accessed: March 18th, 2019

34. Hashim, N.O.: Measurement of the Momentum Spectrum of Cosmic Ray Muons at a Depth of 320 m w.e. Ph.D. Thesis, Siegen University (2007)

35. Hettl, P. et al.: High-resolution x-ray spectroscopy with superconducting tunnel junctions. In: Fernandez, J.E., Tartari, A. (eds.): Proceedings of the European Conference of Energy Dispersive X-Ray Spectrometry 1998, EDXRS-98, Bologna (1999). ◇ Angelini, F., Bellazzini, R. et al.: The microstrip gas chamber. Nucl. Phys. B Proc. Suppl. **23A**, 254–260 (1991). And private communication by Prof. Bellazzini, 2013

36. Hochauflösende Spektroskopie mit Kryodetektoren: http://www.anjarohde.de/csp/news/reports/laborpraxisfeb00.htm. Accessed: March 18th, 2019. And private communication by G. Angloher, 2013. ◇ Grupen, C.: Radiation and Particle Detectors. In Seidel, P. (ed.): Applied Superconductivity: Handbook on Devices and Applications. Wiley-VCH, Weinheim, Germany, **2**, 843–859 (2015)

37. Pretzl, K.: Superconducting granule detectors. J. Low Temp. Phys. **93(3–4)**, 439–448 (1993)
38. Bavykina, I. et al.: Investigation of $ZnWO_4$ Crystals as Scintillating Absorbers for Direct Dark Matter Search Experiments. IEEE Trans. Nucl. Sci. **55(3:2)**, 1449–1452 (2008)
39. Pretzl, K.P.: Superconducting granule detectors. Particle World **1**, 153–162 (1990)
40. Pretzl, K.: Cryogenic calorimeters in astro and particle physics, Nucl. Instr. Meth. Phys. Res. A **454**, 114–27 (2000)
41. Bavykina, I.: Investigation of $ZnWO_4$ crystals as an absorber in the CRESST dark matter search. Master thesis, University of Siegen (2006)
42. Fiorini, E.: Introduction or "Low-temperature detectors: yesterday, today and tomorrow". Nucl. Instr. Meth. Phys. Res. A **520**, 1–3 (2004)
43. Alessandrello, A.: A massive thermal detector for alpha and gamma spectroscopy. Nucl. Instr. Meth. Phys. Res. A **440**, 397–402 (2000)
44. Fiorini, E.: Underground Cryogenic Detectors. Europhys. News **23**, 207–209 (1992)
45. Rando, N. et al.: Space science applications of cryogenic detectors. Nucl. Instr. Meth. Phys. Res. A **522**, 62–68 (2004). ◇ Collaudin, B., Rando, N.: Cryogenics in space: a review of the missions and of the technologies. Cryogenics **40(12)**, 797–819 (2000). ◇ Seidel, P. (ed.): Applied Superconductivity: Handbook on Devices and Applications, Vol. 2, Wiley-VCH, Weinheim, Germnay (2015)
46. Meunier, P. et al.: Discrimination Between Nuclear Recoils and Electron Recoils by Simultaneous Detection of Phonons and Scintillation Light. Appl. Phys. Lett. **75(9)**, (1999)
47. Risse, M., Homola, P.: Search for ultra-high energy photons using air showers. Mod. Phys. Lett. A **22**, 749–766 (2007)
48. Bustamante, M. et al.: High-energy cosmic-ray acceleration; https://cds.cern.ch/record/1249755/files/p533.pdf. Accessed: March 18th, 2019. Acceleration of cosmic-rays, Chapter 3; http://www.apc.univ-paris7.fr/Downloads/auger/Cours/Lectures2and3/Chapter3.pdf. Accessed: March 18th, 2019. ◇ Allard, D. (2016): High Energy Astrophysics: courses #2 and 3, Acceleration of high energy cosmic-rays. http://www.apc.univ-paris7.fr/Downloads/auger/Cours/Lectures2and3/NPAC_cours2_2016.pdf. Accessed: March 18th, 2019. ◇ Gaisser, T.K., Engel, R., Resconi, E.: Cosmic Rays and Particle Physics. 2nd ed., Cambridge University Press, Cambridge UK (2nd June 2016)
49. TRACE (Transition Region And Coronal Explorer) satellite (photo by NASA), NASA—http://trace.lmsal.com/POD/images/arcade_9_nov_2000.gif. Accessed: March 18th, 2019. Nemiroff, R. (MTU) & Bonnell, J. (USRA); TRACE Project, NASA, 2000, https://science.nasa.gov/missions/trace. Accessed: March 18th, 2019. http://www.spacesafetymagazine.com/space-hazards/space-weather/. Accessed: March 18th, 2019
50. De Rujula, A.: A Cannonball Model of Cosmic Rays. Nucl. Phys. Proc. Suppl. **151**, 23–32 (2006). arXiv:hep-ph/0412094
51. Dorman, L.: Cosmic Ray Interactions, Propagation, and Acceleration in Space Plasmas. Springer, Dordrecht, Netherlands (2016)
52. Aguilar, J.A.: Major components of the primary cosmic radiation. In: Particle Astrophysics, Lecture 3: Université Libre de Bruxelles. http://w3.iihe.ac.be/~aguilar/PHYS-467/PA3.pdf. Accessed: March 18th, 2019. ◇ *This figure originates from:* Boyle, P., Müller, D.: The Review of Particle Physics. Fig. 28.1. In Olive, K.A. et al. (Particle Data Group): Chin. Phys. C **38**, 090001 (2014) *and an update from 2015*
53. Blasi, P.: Recent results in cosmic ray physics and their interpretation. Braz. J. Phys. **44**, 426–440 (2014). arXiv:1312.1590 [astro-ph.HE]
54. Hu, H.: Status of the EAS studies of cosmic rays with energy below 10^{16} eV. arXiv:0911.3034 [astro-ph.HE]; http://inspirehep.net/record/837035/plots. Accessed: March 18th, 2019
55. Adriani, O. et al. (CALET Collaboration): Energy Spectrum of Cosmic-Ray Electron and Positron from 10 GeV to 3 TeV Observed with the Calorimetric Electron Telescope on the International Space Station. Phys. Rev. Lett. **119**, 181101 (2017)
56. H.E.S.S. Collaboration: Probing Local Sources with High Energy Cosmic Ray Electrons; https://www.mpi-hd.mpg.de/hfm/HESS/pages/home/som/2017/09/. Accessed: March 18th,

2019. Kerszberg, D., H.E.S.S. Collaboration (2017): The cosmic-ray electron spectrum measured up to $\sim 20\,$TeV with H.E.S.S.; https://indico.in2p3.fr/event/15018/. Accessed: March 18th, 2019. Kerszberg, D. et al. for the H.E.S.S. Collaboration: The High-Energy End of the Cosmic-Ray Electron Spectrum (TeVPA 2018); https://indico.desy.de/indico/event/18204/session/15/contribution/249/material/slides/0.pdf. Accessed: March 18th, 2019

57. High Resolution Fly's Eye Experiment (HiRes), http://www.cosmic-ray.org. Accessed: March 18th, 2019. ◇ HiRes Collaboration: First Observation of the Greisen–Zatsepin–Kuzmin Suppression. arXiv.astro-ph/0703099. ◇ Abbasi, R.U. et al. (High Resolution Fly's Eye Collaboration): First Observation of the Greisen–Zatsepin–Kuzmin Suppression. Phys. Rev. Lett. **100**, 101101 (2008)

58. Energy spectrum as measured at the Pierre Auger Observatory. https://www.auger.org/. Accessed: March 18th, 2019. Verzi, V. et al.: Measurement of Energy Spectrum of Ultra-High Energy Cosmic Rays. Prog. Theor. Exp. Phys. **2017(12)**, (2017) (31 pages). https://doi.org/10.1093/ptep/ptx082. ◇ Valiño, I., for the Pierre Auger Collaboration, in Proc. 34th ICRC 2015, The Hague, The Netherlands, PoS (ICRC2015) 271

59. Haungs, A. et al.: KASCADE-Grande: Composition studies in the view of the post-LHC hadronic interaction models. EPJ Web of Conferences 145, 13001 (2017)

60. Choutko, V., AMS-Collaboration: The Latest Results from the Alpha Magnetic Spectrometer (AMS) on the International Space Station. 26th E+CRS workshop, Barnaul, Russia (2018). https://ecrs18.asu.ru/event/1/contributions/271/attachments/22/34/ECRS-LR-july7.pdf. Accessed: March 18th, 2019

61. Aguilar, J.A.: Energy Dependence of the Position of the Shower Maximum and Width of High Energy Air Showers by the Auger and HiRes Experiments. In: Particle Astrophysics Lecture 3: Cosmic Rays; http://w3.iihe.ac.be/~aguilar/PHYS-467/PA3.pdf. Accessed: March 18th, 2019

62. Schilling, G.; http://news.sciencemag.org/2004/06/bloated-blackhole-babies. Accessed: March 18th, 2019. http://www.gothosenterprises.com/black_holes/images/agn_galaxyzoo.jpg. Accessed: March 18th, 2019. ◇ The SNO Detector: https://sno.phy.queensu.ca/sno/sno2.html. Accessed: April 1st, 2019

63. Super-Kamiokande detector; http://www-sk.icrr.u-tokyo.ac.jp/sk/index-e.html. Accessed: March 18th, 2019; Kamioka Observatory, ICRR (Institute for Cosmic Ray Research), The University of Tokyo. Ashie, Y. et al. (Super-Kamiokande Collaboration): Measurement of atmospheric neutrino oscillation parameters by Super-Kamiokande I. Phys. Rev. D **71**, 112005 (2005). DOI: https://doi.org/10.1103/PhysRevD.71.112005

64. Pontecorvo, B.: Inverse beta processes and nonconservation of lepton charge. Zhurnal Eksperimental'noi i Teoreticheskoi Fiziki **34**, 247 (1957). Reproduced and translated in Sov. Phys. JETP **7**, 172 (1958)

65. Maki, Z., Nakagawa, M., Sakata, S.: Remarks on the Unified Model of Elementary Particles. Prog. Theor. Phys. **28**, 870 (1962)

66. Totsuka, Y.; IEEE conference, Portland (2003). And: Experimental Highlights from Super-Kamiokande. Subnucl. Ser. **40**, 348–383 (2003)

67. Bahcall, J.N.: Neutrinos from the Sun. Sci. Am. **221(1)**, 28–37 (July 1969). https://pdfs.semanticscholar.org/7c87/dc28ea37df1261d56caca14edfa892c01df0.pdf. Accessed: March 18th, 2019

68. The Sudbury Neutrino Observatory, Kingston, Ontario, Canada: http://www.sno.phy.queensu.ca/. Accessed: March 18th, 2019. The SNO Detector: https://sno.phy.queensu.ca/sno/sno2.html. Accessed: March 18th, 2019

69. Nakamura, K., Petcov, S.T.: Neutrino mass, mixing, and oscillations. In: Beringer, J. et al. (Partcle Data Group), Phys. Rev. D **86**, 010001 (2012). http://pdg.lbl.gov. Accessed: March 18th, 2019

70. Nakamura, K., Petcov, S.T.: Neutrino mass, mixing, and oscillations – Particle Data Group, http://pdg.lbl.gov/2017/reviews/rpp2017-rev-neutrino-mixing.pdf. Accessed: March 18th, 2019. ◇ Patrignani, C. et al. (Particle Data Group), Chin. Phys. C **40**, 100001 (2016) and 2017 update; http://pdg.lbl.gov/. Accessed: March 18th, 2019

71. Copyright: Australian Astronomical Observatory, photo taken by D. Malin from CCDs of the AAT telescope, Anglo-Australian Observatory, Copyright Australian Astronomical Observatory, image by David Malin. ◇ http://www.messier.seds.org/xtra/ngc/lmc_sn1987A.html. Accessed: March 18th, 2019

72. Krist, J.: Deconvolving WFPC2 Images with Tiny Tim Models. Jet Propulsion Laboratory. http://www.stsci.edu/software/tinytim/deconwfpc2.html. Accessed: February 5th, 2019

73. Nakahata, M., Kamioka Observatory, ICRR and IPMU; Institute for Cosmic Ray Reseach (ICRR); University of Tokyo; Kavli Institute for the Physics and Mathematics of the Universe (IPMU), Kashiwa, Japan: Supernova Neutrino Detection Overview. 8th Symposium on Large TPCs for Low Energy Rare Event Detection, Paris Diderot University (2016). https://indico.cern.ch/event/473362/contributions/2317627/attachments/1384385/2107214/Nakahata-SNdetection-overview-161207_v2.pdf. Accessed: March 18th, 2019

74. https://icecube.wisc.edu/. Accessed: March 18th, 2019. https://masterclass.icecube.wisc.edu/de/learn-ger/neutrinonachweis-mit-icecube. Accessed: March 18th, 2019. ◇ Siegel, E.: Cosmic Neutrinos Detected, confirming the Big Bang's last great prediction. Forbes (Sep 9, 2016). ◇ Image credit: IceCube collaboration / NSF / University of Wisconsin, via https://icecube.wisc.edu/masterclass/neutrinos. Accessed: March 18th, 2019

75. Stenger, V.J.: High Energy Neutrino Astrophysics, Potential Sources and their Underwater Detection; DUMAND Note 9-92 (1992). Stenger, V.J., 2nd NESTOR International Workshop, Pylos, Greece (1992)

76. ICECUBE Detector. https://icecube.wisc.edu/. Accessed: June 14th, 2019. And John Felde, University of Maryland. With courtesy of F. Halzen

77. The ICECUBE Collaboration, Fermi-LAT, MAGIC, AGILE, ASAS-SN, HAWC, H.E.S.S., INTEGRAL, Kanata, Kiso, Kapteyn, Liverpool Telescope, Subaru, Swift/NuSTAR, VERITAS, VLA/17B-403 teams; Multimessenger observations of a flaring blazar coincident with high-energy neutrino IceCube-170922A. Science 361(6398), eaat1378 (13 Jul 2018). https://doi.org/10.1126/science.aat1378

78. Mimouni, J.: The Promises of Geoneutrinos. J. Phys.: Conf. Ser. 593, 012003 (2015)

79. Kamioka Liquid Scintillator Antineutrino Detector, https://en.wikipedia.org/wiki/Kamioka_Liquid_Scintillator_Antineutrino_Detector. Accessed: March 18th, 2019. The Kamioka Liquid-scintillator Anti-Neutrino Detector (KamLAND), http://kamland.stanford.edu/. Accessed: March 18th, 2019

80. Miramonti, L. et al., Geoneutrinos from 1353 days with the Borexino detector. Phys. Procedia 61, 340–344 (2015). Bellini, G. et al.: Measurement of geo-neutrinos from 1353 days of Borexino. Phys. Lett. B 722, 295–300 (2013). Spectroscopy of geoneutrinos from 2056 days of Borexino data, https://bxopen.lngs.infn.it/geoneutrinos/. Accessed: March 18th, 2019

81. Rybicki, G.B., Lightman, A.P., Radiative Processes in Astrophysics. Wiley, Weinheim (1991)

82. Tolstoy, E.: Radiative Processes in Astrophysics, Lecture 9: Synchrotron Radiation 2007; https://www.astro.rug.nl/~etolstoy/astroa07/. Accessed: March 18th, 2019

83. Mavromatos, N.: High-Energy Gamma-Ray Astronomy and String Theory. E. J. Phys. Conf. Ser. 174, 012016 (2009). ◇ De Naurois, M.: Workshop of HSSHEP, April 2008, Olympia (Greece). https://inspirehep.net/record/814530/plots?ln=de. Accessed: March 18th, 2019

84. CORSIKA – an Air Shower Simulation Program. Knapp, J. (University of Leeds, now DESY): https://www.ikp.kit.edu/corsika/. Accessed: March 18th, 2019

85. Hüntemeyer, P. et al. (2019), The HAWC Collaboration: https://www.hawc-observatory.org/collaboration/. Accessed: March 18th, 2019. https://www.hawc-observatory.org/. Accessed: March 18th, 2019

86. Colossal Bubbles at Milky Way's Plane – "May Be the Annihilation of Dark Matter"; Fermi Gamma-ray Space Telescope; http://fermi.gsfc.nasa.gov/. Accessed: March 18th, 2019. The Fermi Bubbles and the Galactic Center; http://www.stsci.edu/~afox/group/fb.html. Accessed: March 18th, 2019

87. The MAGIC telescopes; https://magic.mpp.mpg.de/. Accessed: March 18th, 2019. MAGIC Telescopes Observe Black Hole Gamma-Ray Lightning; https://scitechdaily.com/magic-

telescopes-observe-black-hole-gamma-ray-lightning/. Accessed: March 18th, 2019. Aleksić, J. et al., Black hole lightning due to particle acceleration at subhorizon scales. Science (2014); https://doi.org/10.1126/science.1256183

88. https://heasarc.gsfc.nasa.gov/docs/sas2/sas2.html. Accessed: March 18th, 2019. ◇ Hillier, R.: Gamma Ray Astronomy. Oxford University Press, Oxford, UK (1985). ◇ Hartman, R.C. et al.: Galactic plane gamma-radiation. Astrophys. J., Part 1, **230**, 597–606 (1979)

89. The Energetic Gamma Ray Experiment Telescope (EGRET), NASA: EGRET Data; CGRO EGRET Team, Drake, S. for the HEASARC: http://heasarc.gsfc.nasa.gov/docs/cgro/images/egret/EGRET_All_Sky.jpg. Accessed: March 18th, 2019. ◇ Hartman, R.C. et al.: The Third EGRET Catalog of High-Energy Gamma-Ray Sources. Astrophys. J. Suppl. Ser. **123**, 79–202 (1999)

90. Samorski, M., Stamm, W.: Detection of 2×10^{15} eV to 2×10^{16} eV gamma-rays from Cygnus X-3. Astrophys. J. **268**, L17–L21 (1983)

91. Suntzeff, N.B. et al.: The Bolometric Light Curve of SN 1987 A. Astrophys. J. Lett. **384**, L33 (1992). Suntzeff, N.B., Bouchet, P.: The bolometric light curve of SN 1987A. I – Results from ESO and CTIO U to Q0 photometry. Astron. J. **99**, 650–663 (1990). ◇ http://mitchell.tamu.edu/people/nicholas-suntzeff/. Accessed: March 18th, 2019. ◇ Pettini, M. (University of Cambridge): Structure and Evolution of Stars, Lecture 16 (2011): https://www.ast.cam.ac.uk/~pettini/STARS/. Accessed: March 18th, 2019

92. NASA Goddard Space Flight Center: Burst and Transient Source Experiment on board of the Gamma Ray Observatory (BATSE – GRO). NASA: What we know about gamma-ray bursts; https://imagine.gsfc.nasa.gov/science/objects/know_bursts.html. Accessed: April 1st, 2019

93. Fishman, G. et al.: BATSE Gamma-Ray-Burst Final Sky Map. http://apod.nasa.gov/apod/ap000628.html. Accessed: March 18th, 2019

94. Short- and Long-Duration Gamma-Ray-Bursts. https://imagine.gsfc.nasa.gov/Images/science/burst_durations_labelled.jpg. Accessed: March 18th, 2019

95. Hurley, K. et al.: Detection of a gamma-ray burst of very long duration and very high energy. Nat. **372**, 652–654 (1994)

96. LIGO Scientific Collaboration, Virgo Collaboration, Fermi GBM, INTEGRAL, ...: Multi-messenger Observations of a Binary Neutron Star Merger. arXiv:1710.05833 [astro-ph.HE]. Astrophys. J. Lett. **848(2)**, L12 (2017). ◇ https://en.wikipedia.org/wiki/GW170817. Accessed: April 1st, 2019

97. Castelvecchi, D.: Holy Cow! Astronomers Agog at Mysterious New Supernova. Nat. **563**, 168–169 (2018); and correction from 30 November 2018 and references therein

98. Gorenstein, P., Giacconi, R., Gursky H.: The spectra of several X-ray sources in Cygnus and Scorpio. Astrophys. J. **150**, L 85 (1967); DOI: https://doi.org/10.1086/180098

99. Max Planck Institute for Extraterrestrial Physics: The ROSAT Satellite. http://www.mpe.mpg.de/xray/wave/rosat/mission/rosat/index.php. Accessed: March 18th, 2019

100. Chandra X-ray Observatory. http://chandra.si.edu/. Accessed: March 18th, 2019

101. Artist's impression of XMM-Newton: XMM-Newton Science Operations Centre – ESA. http://xmm.esac.esa.int/. Accessed: March 18th, 2019. http://www.esa.int/spaceinimages/Images/2000/09/Artist_s_impression_of_XMM-Newton. Accessed: March 18th, 2019. ©: ESA – David Ducros

102. https://heasarc.gsfc.nasa.gov/docs/nustar/nustar_tech_desc.html. Accessed: March 18th, 2019. Harrison, F.A. et al.: The Nuclear Spectroscopic Telescope Array (NuSTAR) High-energy X-Ray Mission. Astrophys. J. **770(2)**, 103H (2013)

103. XMM-Newton Science Operations Centre – ESA: XMM-Newton observation of XMMU J2235.3-2557. http://sci.esa.int/xmm-newton/. Accessed: March 18th, 2019. And http://xmm.esac.esa.int/. Accessed: March 18th, 2019. http://sci.esa.int/xmm-newton/36649-xmm-newton-observation-of-xmmu-j2235-3-2557/. Accessed: March 18th, 2019. ©: ESA/ESO

104. https://www.jpl.nasa.gov/spaceimages/details.php?id=PIA20061. Accessed: March 18th, 2019. https://www.jpl.nasa.gov/news/news.php?feature=4811. Accessed: March 18th, 2019. Image credit: NASA/JPL-Caltech/GSFC, 2016; http://www.nasa.gov/nustar. Accessed: March 18th, 2019; Brian Dunbar, 2017

105. Hobbs, G.: The Parkes Pulsar Timing Array. arXiv:1307.2629 (2013). ⋄ PPTA Parkes Pulsar Timing Array: http://www.atnf.csiro.au/research/pulsar/ppta/Main.PPTA.html. Accessed: March 18th, 2019. ⋄ Babak, S. et al.: European Pulsar Timing Array Limits on Continuous Gravitational Waves from Individual Supermassive Black Hole Binaries. arXiv:1509.02165 (2015). See also Günther Hasinger, ESA Director of Science 2017: http://sci.esa.int/lisa-pathfinder/. Accessed: March 18th, 2019

106. Weisberg, J.M., Taylor, J.H.; Institute of Astronomy, University of Cambridge, UK (2005). ⋄ Hulse, R.A., Taylor, J.H.: A High-Sensitivity Pulsar Survey. Astrophys. J. **191**, L 59–61 (1974). ⋄ https://en.wikipedia.org/wiki/Hulse%E2%80%93Taylor_binary. Accessed: April 15th, 2019

107. Distortion of space from a gravitational wave. Unit 3: Gravity. In: http://www.learner.org/courses/physics/unit/text.html?unit=3&secNum=7. Accessed: March 18th, 2019. With courtesy of B. Heckel (2016)

108. Sabia, S., Tyler, P. (2019): https://lisa.nasa.gov/. Accessed: March 18th, 2019

109. Overbye, D.: Black Hole Picture Revealed for the First Time. New York Times, April 10th, 2019. https://www.nytimes.com/2019/04/10/science/black-hole-picture.html. Accessed: April 10th, 2019. ⋄ Akiyama 1, K. et al. (The Event Horizon Telescope Collaboration): First M87 Event Horizon Telescope Results. I. The Shadow of the Supermassive Black Hole. Astrophys. J. Lett. **875**, L1 (2019). ⋄ https://iopscience.iop.org/journal/2041-8205/page/Focus_on_EHT. Accessed: April 10th, 2019. ⋄ With kind permission of Nancy Wolfe Kotary, MIT Haystack Observatory, Westford, Massachusetts 01886, USA

110. IMAX, Isotope Matter Antimatter Experiment. http://ida1.physik.uni-siegen.de/imax.html. Accessed: March 18th, 2019. ⋄ Private communication by W. Menn, 2018

111. Grashorn, E.W. et al.: The atmospheric charged kaon/pion ratio using seasonal variation methods. Astropart. Phys. **33**, 140–145 (2010)

112. Altitude Variation of the main cosmic ray components. In: Allkofer, O.C., Grieder, P.K.F.: Cosmic Rays on Earth – Physik Daten. Fachinformationszentrum Karlsruhe, Karlsruhe, ISSN 0344-8401 (1984). ⋄ Particle Data Group, 2018: Review of Particle Physics Big Bang Cosmology. In: Tanabashi, M. et al. (Particle Data Group): Phys. Rev. D **98**, 030001 (2018); Cosmic Rays. http://pdg.lbl.gov/2017/reviews/rpp2017-rev-cosmic-rays.pdf. Accessed: April 11th, 2019

113. IMAX, Isotope Matter Antimatter Experiment. Private communication by M. Simon and W. Menn, Siegen University, 2018

114. Kaftanov, V.S., Liubimov, V.A.: Spark Chamber Use in High Energy Physics. Nucl. Instr. Meth. **20**, 195–197 (1963)

115. Engel, R., Karlsruhe Institute of Technology. Private communication, 2018

116. Schmelling, M., Grupen, C. et al.: Spectrum and charge ratio of vertical cosmic ray muons up to momenta of 2.5 TeV/c. Astropart. Phys. **49**, 1–5 (2013)

117. Allkofer, O.C., Grupen, C., Stamm, W.: Electromagnetic Interactions of High-Energy Cosmic-Ray Muons. Phys. Rev. D **4**, 638 (Published 1 August 1971)

118. Particle Data Group, http://pdg.lbl.gov/2018/reviews/rpp2018-rev-cosmic-rays.pdf. Accessed: March 18th, 2019. Figure 29.7

119. Krishnaswamy, M.R., Menon, M.G.K., Narasimham, V.S., Wolfendale, A.W.: The Kolar Gold Fields Neutrino Experiment. II. Atmospheric Muons at a Depth of 7000 hg cm^{-2} (Kolar). Proc. R. Soc. A **323(1555)**, 511–522 (July 1971). ⋄ Grieder, P.K.F.: Cosmics Rays at Earth, Elsevier (2001). ⋄ Adarkar, H. et al.: Study of prompt muon production by angular distribution of muons recorded in KGF nucleon decay experiment. Proc. Int. Symp. on Underground Physics Experiments (1990)

120. Crookes, J.N., Rustin, R.C.: An investigation of the absolute intensity of muons at sea-level. Nucl. Phys. B **39**, 493–508 (1972). ⋄ Grieder, P.K.F.: Cosmics Rays at Earth, Elsevier (2001)

121. The CosmoALEPH Experiment: http://aleph.web.cern.ch/aleph/cosmolep/. Accessed: February 9th, 2019. ⋄ Grupen, C.: Cosmic Ray Results from CosmoALEPH; https://www.hep.physik.uni-siegen.de/~grupen/dali.jpg. Accessed: March 18th, 2019

122. Maciuc, F., Grupen, C., Schmelling, M. et al.: Muon-Pair Production by Atmospheric Muons in CosmoALEPH. Phys. Rev. Lett. **96**, 021801 (2006)

123. Aartsen, M.G. et al. (ICECUBE Collaboration): Observation of the cosmic-ray shadow of the Moon with IceCube. Phys. Rev. D **89** 10, 102004 (2014). With courtesy of F. Halzen

124. Sinnis, G.: Air shower detectors in gamma-ray astronomy. New J. Phys. **11**, 055007 (2009)

125. Shellard, R.C.: First results from the Pierre Auger Observatory. Braz. J. Phys. **36(4a)**, 1184–1193 (2006). ◇ Knapp, J., Heck, D. et al.: Extensive Air Shower Simulations at the Highest Energies. Astropart. Phys. **19**, 77–99 (2003). ◇ Image initially from J. Knapp, D. Heck and T. Merz (KIT Karlsruhe, DESY)

126. Engel, R.: Indirect Detection of Cosmic Rays. In Grupen, C., Buvat, I. (eds.): Handbook of Particle Detection and Imaging. Springer, Berlin, **2**, 594 (2012). ◇ Engel, R., Heck, D., Pierog, T.: Extensive Air Showers and Hadronic Interactions at High Energy. Annu. Rev. Nucl. Part. Sci. **61**, 467–489 (2011). ◇ Heck, D., private communication, 2016

127. Pierre Auger Observatory, Dept. of Physics; Univ. of Roma – Tor Vergata Homepage: http://research.roma2.infn.it/~auger/integtest.html. Accessed: March 18th, 2019. ◇ Kampert, K.-H., private communication, 2016

128. http://www.telescopearray.org/. Accessed: September 25th, 2019

129. Image credit: Pierre Auger Collaboration. www.auger.org. Accessed: March 18th, 2019. ◇ http://research.roma2.infn.it/~auger/welcome.html. Accessed: March 18th, 2019

130. JEM-EUSO; Extreme Universe Space Observatory on Board the Japanese Experiment Module. https://en.wikipedia.org/wiki/JEM-EUSO. Accessed: March 18th, 2019. ◇ http://uhecr.sinp.msu.ru/en/jem-euso1.html. Accessed: March 18th, 2019. ◇ http://jem-euso.roma2.infn.it/. Accessed: March 18th, 2019

131. Haungs, A.: The air-shower experiment KASCADE-Grande. Searching for the Origins of Cosmic Rays, Trondheim; June 2009. http://web.phys.ntnu.no/~mika/haungs.pdf. Accessed: March 18th, 2019. Experimental high-energy astroparticle physics. Talk: Universidad Michoacana de San Nicolas de Hidalgo, Morelia, MEX, 11.November 2009

132. Ogio, S. et al.: The energy spectrum and the chemical composition of primary cosmic rays with energies from 10^{14} to 10^{18} eV. Astrophys. J. **612**, 268–275 (2004). ◇ KASCADE-Grande Collaboration; Haungs, A., Karlsruhe Institute of Technology. Private communication, 2018. ◇ KASCADE-Grande Collaboration, https://web.ikp.kit.edu/KASCADE/. Accessed: April 1st, 2019. ◇ Apel, W.D. et al.: KASCADE-Grande measurements of energy spectra for elemental groups of cosmic rays. Astropart. Phys. **47**, 54–66 (2013). https://doi.org/10.1016/j.astropartphys.2013.06.004

133. The Auger experiment: https://www.auger.org/. Accessed: March 18th, 2019. ◇ Abraham, J. et al.: Correlation of the highest-energy cosmic rays with the positions of nearby active galactic nuclei. Astropart. Phys. **29**, 188–204 (2008). http://inspirehep.net/record/767631. Accessed: March 18th, 2019

134. Pierre Auger Collaboration – Aab, A. et al.: Observation of a Large-scale Anisotropy in the Arrival Directions of Cosmic Rays above 8×10^{18} eV. Science **357(6537)**, 1266–1270 (2017). DOI: https://doi.org/10.1126/science.aan4338. arXiv:1709.07321 [astro-ph.HE]. FERMILAB-PUB-17-354, 2017

135. Allan, H.R.: Chapter III. In: Wilson, J.G., Wouthuysen, S.A. (eds.): Progress in Elementary Particle and Cosmic Ray Physics, Volume X, North-Holland Publishing Company (1971)

136. Falcke, H. et al. (LOPES Collaboration): Detection and imaging of atmospheric radio flashes from cosmic ray air showers. Nat. **435**, 313–316 (2005)

137. LOPES and KASCADE-Grande Collaboration, KIT Karlsruhe: http://www.timhuege.de/experiments/lopes. Accessed: March 18th, 2019. ◇ Grupen, C. et al.: Radio detection of cosmic rays with LOPES. Braz. J. Phys. **36(4a)**, 1157–1164 (2006)

138. Link, K. et al.: The LOPES experiment. Nucl. Phys. Proc. Suppl. **212–213**, 323–328 (2011)

139. Schröder, F.G.: Radio detection of Cosmic-Ray Air Showers and High-Energy Neutrinos. Prog. Part. Nucl. Phys. **93**, 1–68 (2017)

140. Learned, J.G.: Acoustic Detection of Charged Atomic Particles in Liquids. Phys. Rev. D **19**, 3293 (1979). ◇ Barret, W.L.: Acoustic Detection of Cosmic Ray Air Showers. Science

202(4369), 749–752 (1978). ◇ SOSUS – Sound Surveillance System; Pike, J. (2011): https://www.globalsecurity.org/intell/systems/sosus.htm. Accessed: March 18th, 2019. ◇ Sulak, L. et al.: Experimental Studies of the Acoustic Detection of Particle Showers. Proc. 15th ICRC **11**, 420 (1977). ◇ And Sulak, L. et al.: Experimental studies of the acoustic signature of proton beams traversing fluid media. Nucl. Instr. Meth. **161(2)**, 203–217 (1979)

141. Budnev, N. et al. (Baikal Collaboration) 1998: High energy neutrino acoustic detection activities in Lake Baikal: status and plans. https://slideplayer.com/slide/10793852/. Accessed: April 1st, 2019. ◇ Aynutdinov, V. et al. (Baikal Collaboration): ARENA 2008: High energy neutrino acoustic detection activities in Lake Baikal: Status and results. Nucl. Instr. Meth. **604(1–2) Suppl. 1**, S130–S135 (2009). ◇ Thompson, L.F.: Acoustic detection of ultra-high-energy neutrinos. Nucl. Instr. Meth. Phys. Res. A **588(1–2)**, 155–161 (2008)

142. Niess, V., Bertin, V.: Underwater Acoustic Detection of Ultra High Energy Neutrinos. Astropart. Phys. **26(4-5)**, 243–256 (2006)

143. Bevan, S. et al.: Simulation of Ultra High Energy Neutrino Interactions in Ice and Water. Astropart. Phys. **28(3)**, 366–379 (2007)

144. Thompson, L.F.: The Acoustic Detection of Ultra-High Energy Neutrinos – a Status Report. J. Phys.: Conf. Ser. **60**, 52 (2007). ◇ Thompson, L.: Acoustic Detection of Ultra High Energy Neutrinos. Rome International Conference on Astroparticle Physics; RICAP 2007. https://indico.cern.ch/event/14299/contributions/162083/attachments/130495/185273/LeeThompsonRICAP22Jun2007sito.pdf. Accessed: March 18th, 2019

145. Risse, M., private communication, 2018. ◇ Aab, A. et al. (Pierre Auger Collaboration): Highlights from the Pierre Auger Observatory. Proc. 35th ICRC2017, 1102 (2018). arXiv:1710.09478

146. Kampert, K-H., private communication, 2019

147. Jeong, Y.S. et al.: Quark mass effects in high energy neutrino nucleon scattering. Phys. Rev. D **81**, 114012 (2010). arXiv:1001.4175 [hep-ph]

148. Farrar G. R., Piran, T.: Violation of the Greisen-Zatsepin-Kuzmin Cutoff: A Tempest in a (Magnetic) Teapot? Why Cosmic Ray Energies above 10^{20} eV May Not Require New Physics. Phys. Rev. Lett. **84**, 3527 (2000). GZK Violation – a Tempest in a (Magnetic) Teapot?; see also https://arxiv.org/pdf/astro-ph/9906431.pdf. Accessed: April 1st, 2019

149. Huang, K.-H. et al.: Detection of Lyman-Alpha Emission from a Triply Imaged z = 6.85 Galaxy Behind MACS J2129.4-0741. Astrophys. J. Lett. **823**, L14 (2016). arXiv:1605.05771v1

150. Penrose, R., Floyd, R.M.: Extraction of Rotational Energy from a Black Hole. Nat. Phys. Sci. **229**, 177 (1971)

151. Zhao, Z.: Synchrotron Light Sources. In: Fan, C. and Zhao, Z. (eds.): Synchrotron Radiation in Materials Science: Light Sources, Techniques, and Applications. Wiley-VCH, Weinheim (2018). https://application.wiley-vch.de/books/sample/3527339868_c01.pdf. Accessed: March 18th, 2019. ◇ Kunz, C. (ed.): Synchrotron Radiation: Techniques and Applications. Springer, Berlin, Heidelberg, New York (1979)

152. https://en.wikipedia.org/wiki/Speed_of_sound. Accessed: March 18th, 2019. ◇ Wilcock, W.S.D. et al.: Sounds in the Ocean at 1–100 Hz. Annu. Rev. Mar. Sci. **6**, 117–40 (2014). http://faculty.washington.edu/wilcock/files/PaperPDFs/wilcocketal_annrev_2014.pdf. Accessed: March 18th, 2019

153. Riccobene, G. (2008): https://www.slac.stanford.edu/econf/C0805263/Slides/Riccobene.pdf. Accessed: March 18th, 2019

154. Turner, M.S., Tyson, J.A.: Cosmology at the millennium. Rev. Mod. Phys. **71** (1999) 145

155. Bahcall, N.A. (Department of Astrophysical Sciences, Princeton University, Princeton, NJ 08544): Hubble's Law and the expanding universe. Proc. Natl. Acad. Sci. USA **112(11)**, 3173–3175 (2015). https://www.pnas.org/content/pnas/112/11/3173.full.pdf. Accessed: March 18th, 2019. ◇ Jha, S.: Ph.D. thesis. Harvard Univ., Cambridge, MA (2002). ◇ Jha, S. et al.: The Type Ia Supernova 1998bu in M96 and the Hubble Constant. Astrophys. J. Suppl. Ser. **125(1)**, 73–97 (1999)

156. Particle Data Group, 2018 Review of Particle Physics, Big Bang Cosmology, Tanabashi, M. et al. (Particle Data Group). Phys. Rev. D **98**, 030001 (2018). ◇ Kirshner, R.P.: Hubble's

diagram and cosmic expansion. Proc. Natl. Acad. Sci. U.S.A. **101(1)**, 8–13 (Jan 6, 2004). https://doi.org/10.1073/pnas.2536799100. Accessed: March 18th, 2019

157. Planck Experiment: http://www.esa.int/Our_Activities/Space_Science/Planck. Accessed: March 18th, 2019

158. Particle Data Group, 2018 Review of Particle Physics, Big Bang Cosmology, Tanabashi, M. et al. (Particle Data Group). Phys. Rev. D **98**, 030001 (2018)

159. Liddle, A.: An Introduction to Modern Cosmology. Wiley, Chichester, 5th ed. (2015)

160. Huchra, J.P.; https://www.cfa.harvard.edu/~dfabricant/huchra/hubble/. Accessed: March 18th, 2019

161. Koberlein, B., 2015, Nothing But Net: https://briankoberlein.com/2015/03/06/nothing-but-net/. Accessed: March 18th, 2019. ◊ Davis, E.W. et al.: Review of Experimental Concepts for Studying the Quantum Vacuum Field. AIP Conf. Proc. **813**, 1390 (2006). https://doi.org/10.1063/1.2169324. Accessed: March 18th, 2019. https://www.researchgate.net/publication/228353872_Review_of_Experimental_Concepts_for_Studying_the_Quantum_Vacuum_Field/figures?lo=1. Accessed: March 18th, 2019

162. Nielsen, J.T., Guffanti, A., Sarkar, S.: Marginal evidence for cosmic acceleration from Type Ia supernovae. Sci. Rep. **6**, 35596 (2016) (Open Access). https://doi.org/10.1038/srep35596 (2016). Accessed: March 18th, 2019. arXiv:1506.01354 [astro-ph.CO]

163. Helligkeit – die Größe der Sterne (brightness of stars): http://www.br-online.de/wissen-bildung/spacenight/sterngucker/info/helligkeit.html. Accessed: March 18th, 2019

164. Apparent magnitude: https://en.wikipedia.org/wiki/Apparent_magnitude. Accessed: April 1st, 2019

165. The James Webb Telescope. Gardner, J.P.: https://www.jwst.nasa.gov/resources/JWST_SSR_JPG.pdf. Accessed: March 18th, 2019. ◊ Absolute magnitude limit in the infrared. https://en.wikipedia.org/wiki/Limiting_magnitude. Accessed: March 18th, 2019

166. Perlmutter, S.: Supernovae, Dark Energy, and the Accelerating Universe. Phys. Today, 53–60 (April 2003). https://physicsforme.files.wordpress.com/2011/10/perlmutter3.jpg. Accessed: April 1st, 2019

167. Accelerating universe and dark energy. http://www.physicsoftheuniverse.com/topics_bigbang_accelerating.html. Accessed: March 18th, 2019 ◊ Variation of the size of the universe. http://www.jaymaron.com/particles/friedmann.png. Accessed: March 18th, 2019. ◊ NASA / WMAP Science Team 2015. Expansion of the universe. http://map.gsfc.nasa.gov/media/990350/990350s.jpg. Accessed: March 18th, 2019. NASA Official: Dr. E. J. Wollack, Webmaster: B. Griswold

168. How Big is the Universe? https://www.space.com/24073-how-big-is-the-universe.html. Accessed: March 18th, 2019

169. Siegel, E. (2018): How Many Fundamental Constants Does It Take To Explain The Universe? https://ep-news.web.cern.ch/content/how-many-fundamental-constants-does-it-take-explain-universe. Accessed: March 18th, 2019. ◊ Kane, G.: The Mystery of Mass. Sci. Am. **293(1)**, 40–48 (2005). https://craftx.org/sites/all/themes/craft_blue/pdf/Interrogating_the_Text_Kane_Mysteries_of_Mass_p5.pdf. Accessed: March 18th, 2019

170. See the web site of the Relativistic Heavy Ion Collider, www.bnl.gov/RHIC/. Accessed: March 18th, 2019

171. Adriani, O. et al.: PAMELA Results on the Cosmic-Ray Antiproton Flux from 60 MeV to 180 GeV in Kinetic Energy. Phys. Rev. Lett. **105**, 121101 (2010). ◊ PAMELA Collaboration: http://pamela.roma2.infn.it/index.php. Accessed: March 18th, 2019. ◊ With courtesy of: M. Simon. ◊ http://inspirehep.net/record/860709. Accessed: April 1st, 2019

172. STAR Collaboration 2011, https://www.star.bnl.gov/. Accessed: April 1st, 2019

173. Max Planck Institute for Astrophysics, Garching; Annihilation of positrons in the Galaxy, https://wwwmpa.mpa-garching.mpg.de/mpa/research/current_research/hl2005-5/spi_spec-l.gif. Accessed: March 18th, 2019; https://wwwmpa.mpa-garching.mpg.de/mpa/research/current_research/hl2005-5/hl2005-5-en.html. Accessed: April 1st, 2019

174. Annihilation of positrons in the Galaxy: https://link.springer.com/chapter/10.1007/978-1-4020-5304-7_2. Accessed: March 18th, 2019. ◊ Churazov, E., Sunyaev, R., Sazonov, S.,

Revnivtsev, M., Varshalovich, D.: Positron annihilation spectrum from the Galactic Centre region observed by SPI/INTEGRAL. Mon. Notices Royal Astron. Soc. **357**, 1377–1386 (2005)

175. O. Adriani et al.: Cosmic-Ray Positron Energy Spectrum Measured by PAMELA. Phys. Rev. Lett. **111**, 081102 (2013). And a modification to the spectrum by M. Boezio; private communication, 2017

176. Adriani, O. et al.: An anomalous positron abundance in cosmic rays with energies 1.5–100 GeV. Nat. **458**, 607–609 (2 April 2009). ◇ http://www.nature.com/nature/journal/v458/n7238/fig_tab/nature07942_F2.html. Accessed: March 18th, 2019. ◇ Aguilar, M. et al. (AMS Collaboration): First Result from the Alpha Magnetic Spectrometer on the International Space Station: Precision Measurement of the Positron Fraction in Primary Cosmic Rays of 0.5–350 GeV. Phys. Rev. Lett. **110(14)** 141102 (2013). ◇ New results from the Alpha Magnetic Spectrometer on the International Space Station: http://www.ams02.org/2014/09/new-results-from-the-alpha-magnetic-spectrometer-on-the-international-space-station/. Accessed: March 18th, 2019. ◇ Aguilar, M. et al. (AMS Collaboration): Observation of Complex Time Structures in the Cosmic-Ray Electron and Positron Fluxes with the Alpha Magnetic Spectrometer on the International Space Station. Phys. Rev. Lett. **121(5)**, 051102 (2018)

177. Von Ballmoos, P.: Antimatter in the Universe: constraints from gamma-ray astronomy. Hyperfine Interact. **228(1–3)**, 91–100 (2014). ◇ CGRO/COMPTEL, arXiv:1401.7258 [astro-ph.HE] (2014)

178. Sakharov, A.D.: Violation of CP invariance, C asymmetry, and baryon asymmetry of the universe. JETP Letters **5**, 24–27 (1967)

179. Boyle, L. et al.: CPT-Symmetric Universe. Phys. Rev. Lett. **121**, 251301 (2018). https://journals.aps.org/prl/abstract/10.1103/PhysRevLett.121.251301. Accessed: March 18th, 2019

180. Aver, E. et al.: The primordial helium abundance from updated emissivities. J. Cosmol. Astropart. Phys. **2013(11)**, 017 (2013)

181. Hirata, C., Center for Cosmology and Astroparticle Physics at the Ohio State University. Big Bang Nucleosynthesis: Overview and Modern Cosmology by Dodelson, S. (2003). http://home.fnal.gov/~dodelson/book.html. Accessed: April 1st, 2019

182. The BBNcode can be downladed from www-thphys.physics.ox.ac.uk/users/SubirSarkar/bbn.html. Accessed: March 18th, 2019

183. Rieke, M. (Department of Astronomy and Steward Observatory): The relative amounts of deuterium, various isotopes of He, Be, and Li formed in the Big Bang depend on the temperature at a given pressure and hence size of the Universe. http://ircamera.as.arizona.edu/astr_250/Lectures/Lecture_27.htm. Accessed: March 18th, 2019. ◇ Astronomy and Cosmology: The description, origin, and development of the universe: http://burro.case.edu/Academics/Astr222/Cosmo/Early/bbn.html. Accessed: March 18th, 2019

184. Burles, S., Tytler, D.: The Matter Composition of the Universe. Astrophys. J. **499**, 699 (1998); **507** 732 (1998). ◇ https://ned.ipac.caltech.edu/level5/March03/Freedman/Freedman2_3.html. Accessed: March 18th, 2019. ◇ Pastor, S.: Neutrinos in Cosmology. IS-APP 2017, https://agenda.infn.it/event/11926/contributions/11326/attachments/8378/9402/pastor_ISAPP2017_1.pdf. Accessed: April 15th, 2019

185. The Dark and the Visible Side of the Universe, Texel, Netherlands (2017), https://indico.cern.ch/event/568904/. Accessed: April 15th, 2019

186. Berger, A. (ed.): The Big Bang and Georges Lemaître. D. Reidel Publishing Company, Dordrecht, Holland (1983); and Springer, Netherlands (1984)

187. Matthews, G.J. et al.: Introduction to Big Bang Nucleosynthesis and Modern Cosmology. Int. J. Mod. Phys. E **26(08)**, 1741001 (2017). See also https://doi.org/10.1142/S0218301317410014. Accessed: March 29th, 2019. And https://arxiv.org/abs/1706.03138. Accessed: March 29th, 2019

188. Steigman, G.: Abundances of primordial elements in their dependence on the baryon density: http://i.stack.imgur.com/h8KGc.jpg. Accessed: March 18th, 2019. ◇ *Big Bang Nucleosynthesis:* Steigman, G.: Probing the First 20 Minutes. Nat. **415**, 27–29 (3 January 2002). https://ned.ipac.caltech.edu/level5/Sept03/Trodden/Trodden4_5.html. Accessed: March 18th, 2019

189. Schramm, D.N., Turner, M.S.: Big-Bang Nucleosynthesis Enters the Precision Era. FERMILAB–Pub–97/186-A; arXiv:astro-ph/9706069v1 (1997); Rev. Mod. Phys. **70**, 303–318 (1998) and references therein

190. Burles, S. el al.: Sharpening the Predictions of Big-Bang Nucleosynthesis. Phys. Rev. Lett. **82**, 4176 (1999). ◇ Pastor, S.: Light Neutrinos in Cosmology; http://isapp2011.mi.infn.it/Slides/Pastor_Set1.pdf. Accessed: April 15h, 2019. ◇ *Image originally from* Burles, S. et al.: Big-Bang Nucleosynthesis: Linking Inner Space and Outer Space. arXiv:astro-ph/9903300v1 (1999). http://arxiv.org/abs/astro-ph/9903300. Accessed: March 18th, 2019

191. Cooke, R.J.: Big Bang Nucleosynthesis and the Helium Isotope Ratio. Astrophys. J. Lett. **812**, L12 (2015); https://doi.org/10.1088/2041-8205/812/1/L12. Accessed: March 18th, 2019; Creative Commons licences AAS. Reproduced with courtesy of Tyan Cooke. http://www.colorado.edu/aps/ryan-cooke-aps-colloquium. Accessed: March 18th, 2019

192. Vogt, S.S. et al.: HIRES: the high-resolution echelle spectrometer on the Keck 10-m Telescope. Proc. SPIE Instrumentation in Astronomy VIII, Crawford, D.L. and Craine, E.R. (eds.), **2198**, 362 (1994). https://doi.org/10.1117/12.176725. Accessed: March 18th, 2019. https://www2.keck.hawaii.edu/inst/hires/. Accessed: March 18th, 2019

193. Cooke R.J.: The primordial deuterium abundance of the most metal-poor damped Lyman-system (2016); https://arxiv.org/pdf/1607.03900.pdf. Accessed: March 18th, 2019. Cooke, R.J. et al.: One Percent Determination of the Primordial Deuterium Abundance. Astrophys. J. **855**, 102 (2018). arXiv:1710.11129 [astro-ph.CO]

194. Mangano, G. et al.: Relic neutrino decoupling including flavour oscillations. Nucl. Phys. B **729**, 221–234 (2005). See also https://doi.org/10.1016/j.nuclphysb.2005.09.041. Accessed: March 29th, 2019

195. The LEP and SLD experiments, A Combination of Preliminary Electroweak Measurements and Constraints on the Standard Model, CERN EP/2000-016 (2000)

196. Gamow, G.: The Origin of Elements and the Separation of Galaxies. Phys. Rev. **74**, 505 (1948)

197. Alpher, R.A., Herman, R.C.: Remarks on the Evolution of the Expanding Universe. Phys. Rev. **75**, 1089 (1949)

198. Penzias, A.A., Wilson, R.W.: A Measurement of Excess Antenna Temperature at 4080 Mc/s. Astrophys. J. **142**, 419 (1965)

199. Dicke, R.H., Peebles, P.J.E., Roll, P.G., Wilkinson, D.T.: Measurements of Cosmic Microwave Background Radiation. Astrophys. J. **142**, 414 (1965)

200. Fixsen, D.J. et al.: The Cosmic Microwave Background Spectrum from the Full COBE/FIRAS Data Set. Astrophys. J. **473**, 576 (1996). ◇ http://map.gsfc.nasa.gov/. Accessed: March 18th, 2019. ◇ http://lambda.gsfc.nasa.gov/product/cobe/about_firas.cfm. Accessed: March 18th, 2019

201. Cosmic Background Explorer: http://aether.lbl.gov/www/projects/cobe/. Accessed: March 18th, 2019. ◇ Wilkinson Microwave Anisotropy Probe: http://map.gsfc.nasa.gov/. Accessed: March 18th, 2019. ◇ Planck: http://www.esa.int/Our_Activities/Space_Science/Planck. Accessed: March 18th, 2019

202. Mather, J.C. (NASA Science and Exploration Directorate): Sky map of the cosmic microwave background radiation by COBE. http://science.nasa.gov/missions/cobe/. Accessed: March 18th, 2019

203. Goddard Space Flight Center and Princeton University: A full-sky map produced by the Wilkinson Microwave Anisotropy Probe. http://map.gsfc.nasa.gov. Accessed: March 18th, 2019. WMAP Science Team/NASA Goddard

204. Planck: http://www.esa.int/Our_Activities/Space_Science/Planck; http://sci.esa.int/planck/51555-planck-power-spectrum-of-temperature-fluctuations-in-the-cosmic-microwave-background/; http://sci.esa.int/science-e-media/img/63/Planck_power_spectrum_orig.jpg. Accessed: March 18th, 2019. ESA and the Planck Collaboration (2013)

205. Ryden, B.: Introduction to Cosmology. Addison-Wesley, San Francisco (2003)

206. Gumjudpai, B., National Astronomical Research Institute of Thailand. Private communication, 2017

207. Hu, W.: Cosmic Microwave Background Anisotropies. https://arxiv.org/abs/astro-ph/0110414. Accessed: March 18th, 2019. ◇ Hu, W., Dodelson, S.: Cosmic Microwave Background Anisotropies. Ann. Rev. Astron. Astrophys. **40**, 171–216 (2002); astro-ph/0110414. ◇ Melchiorri, A., Griffiths, L.M.: From Anisotropy to Omega. arXiv:astro-ph/0011147; New Astron. Rev. **45**, 321–328 (2001)

208. Planck 2013 Results, XVI. Cosmological Parameters, arXiv.org: astro-ph: arXiv:1303.5076, arXiv:1303.5076v3 (astro-ph.CO) Astronomy and Astrophysics (2014)

209. Planck Legacy Archive, ESA, http://pla.esac.esa.int/pla/#cosmology. Accessed: March 18th, 2019

210. Ade, P.A.R. et al.: Planck 2015 results. XIII. Cosmological parameters Planck Collaboration. arXiv:1502.01589 [astro-ph.CO] (2016)

211. Bartelmann, M.: Planck und wie er die Welt sah (Planck and how he viewed the world). Physik-Journal, 35–41 (2017)

212. Riess, A. et al.: New Parallaxes of Galactic Cepheids from Spatially Scanning the Hubble Space Telescope: Implications for the Hubble Constant. Astrophys. J. **855**, 136 (2018)

213. Riess, A.G. et al.: Large Magellanic Cloud Cepheid Standards Provide a 1% Foundation for the Determination of the Hubble Constant and Stronger Evidence for Physics Beyond ΛCDM. arXiv:1903.07603v2 [astro-ph.CO] (2019)

214. Readhead, A.C.S. et al.: CERN Cour. (July/August 2002); astro-ph/0402359 (2004). ◇ Readhead, A.C.S. et al.: Extended mosaic observations with the Cosmic Background Imager. Astrophys. J. **609**, 498–512 (2004)

215. Allen, S.W. et al.: Improved constraints on dark energy from Chandra X-ray observations of the largest relaxed galaxy clusters. Mon. Notices Royal Astron. Soc. **383(3)**, 879–896 (2008). arXiv:0706.0033

216. Davis, T.M. et al.: Scrutinizing Exotic Cosmological Models Using ESSENCE Supernova Data Combined with Other Cosmological Probes. Astrophys. J. **666**, 716 (2007)

217. Spergel, D.N. et al.: Three-Year Wilkinson Microwave Anisotropy Probe (WMAP) Observations: Implications for Cosmology. Astrophys. J. Suppl. **170**, 370 (2007)

218. Abramowitz, M. and Stegun, I.A.: Handbook of Mathematical Functions. Dover Publ., New York (1970)

219. Gradshteyn, I.S. and Ryzhik, I.M.: Tables of Integrals, Series, and Products. Harri Deutsch, Thun (1981)

220. Krisciunas, K.: How to Integrate Planck's Function: https://people.physics.tamu.edu/krisciunas/planck.pdf. Accessed: March 18th, 2019

221. http://mathworld.wolfram.com/RiemannZetaFunction.html. Accessed: March 18th, 2019

222. Wilson, T.L. and Hüttemeister, S.: Tools of Radio Astronomy – Problems and Solutions. 2nd ed., Springer, Cham, Switzerland (2018)

223. Peacock, J.A.: Cosmological Physics. Cambridge University Press (2003). ◇ Primack, J. (2012): Origin and Evolution of the Universe. http://physics.ucsc.edu/~joel/12Phys224/12Phys224-Wk2-GR-Dist.pdf. Accessed: March 18th, 2019

224. Hannu Kurki-Suonio, Department of Physics, Division of Particle Physics and Astrophysics, University of Helsinki, Thermal history of the Early Universe; (2017): http://www.helsinki.fi/~hkurkisu/cpt/Cosmo6.pdf. Accessed: March 18th, 2019

225. Cabrera, B.: First Results from a Superconductive Detector for Moving Magnetic Monopoles. Phys. Rev. Lett. **48**, 1378 (1982)

226. Suzuki, N. et al.: The Hubble Space Telescope Cluster Supernova Survey. V. Improving the Dark-Energy Constraints Above $z > 1$ and Building an Early-Type-Hosted Supernova Sample. Astrophys. J. **746**, 85 (2012). ◇ The Supernova Cosmology Project: http://supernova.lbl.gov/Union/. Accessed: March 18th, 2019

227. Guth, A.H.: Inflationary universe: A possible solution to the horizon and flatness problems. Phys. Rev. D **23**, 347 (1981)

228. Linde, A.D.: A New Inflationary Universe Scenario: A Possible Solution of the Horizon, Flatness, Homogeneity, Isotropy and Primordial Monopole Problems. Phys. Lett. B **108**, 389–393 (1982)

229. Albrecht A., Steinhardt, P.J.: Cosmology for Grand Unified Theories with Radiatively Induced Symmetry Breaking. Phys. Rev. Lett. **48**, 1220 (1982)

230. Brout, R.: The Inflaton and its Mass. arXiv:gr-qc/0201060, e-Print: gr-qc/0201060, https://arxiv.org/pdf/gr-qc/0201060.pdf. Accessed: April 1st, 2019

231. The Sloan Digital Sky Survey: Mapping the Universe: http://www.sdss.org/. Accessed: March 18th, 2019

232. Guth, A.H.: Inflation. In: Freedman, W.L. (ed.): Measuring and Modeling the Universe. Carnegie Observatories Astrophysics Series, Vol. 2, Cambridge University Press (2004). astro-ph/0404546

233. Liddle, A., Loveday, J.: The Oxford Companion to Cosmology. Oxford University Press, Oxford (2008)

234. Abdalla, F.B. (2017): Observational Cosmology: Inflation; http://zuserver2.star.ucl.ac.uk/~hiranya/PHAS3136/PHAS3136/PHAS3136_files/Cosmo4_78_inflation.pdf. Accessed: March 18th, 2019. ◇ Norman, M.L.: Cosmological Framework and Perturbation Growth in the Linear Regime; https://ned.ipac.caltech.edu/level5/Sept11/Norman/Norman2.html. Accessed: March 18th, 2019. ◇ Tegmark, M. et al.: Cosmological parameters from SDSS and WMAP. Phys. Rev. D **69(10)** 103501 (2004)

235. Tegmark, M. et al.: The 3-D power spectrum of galaxies from the SDSS. Astrophys. J. **606**, 702–740 (2004)

236. Widrow, L.M. et al.: Power spectrum for the small-scale Universe. Mon. Notices Royal Astron. Soc. **397(3)**, 1275–1285 (2009)

237. Hogan, C.J.: The Little Book of the Big Bang. Copernicus, Springer, New York (1998)

238. Ostriker, J.P., Steinhardt, P.: The Quintessential Universe. Sci. Am. **284(1)**, 46–53 (2001)

239. Tsujikawa, S.: Quintessence: A Review. arXiv:1304.1961 [gr-qc] (2013). https://arxiv.org/pdf/1304.1961.pdf. Accessed: April 1st, 2019

240. Sloan Digital Sky Survey III. A Slide Through a Map of our Universe. New 3-D Map of Massive Galaxies and Distant Black Holes Offers Clues to Dark Matter and Dark Energy. http://www.sdss3.org/press/dr9.php. Accessed: March 18th, 2019

241. Eisenstein, D. and the SDSS-III collaboration 2016: Astronomers map a record-breaking 1.2 million galaxies to study the properties of dark energy; July 2016; Tojeiro, R., Tinker, J. et al., https://www.sdss.org/press-releases/astronomers-map-a-record-breaking-1-2-million-galaxies-to-study-the-properties-of-dark-energy/. Accessed: March 18th, 2019

242. NASA/WMAP Science Team – Original version: NASA; modified by Ryan Kaldari; CMB Timeline300 no WMAP.jpg; 2006

243. F. Zwicky: On the Masses of Nebulae and of Clusters of Nebulae. Astrophys. J. **86**, 217 (1937)

244. Bahcall, N.A.: Vera Rubin. Nat. **542**, 32 (2017). https://doi.org/10.1038/542032a.. Accessed: March 18th, 2019. Rubin, V.: Bright Galaxies, Dark Matters. Masters of Modern Physics. Woodbury, New York, US: Springer/AIP Press (1997). ISBN 1563962314

245. Freese, K.: Review of Observational Evidence for Dark Matter in the Universe and in upcoming searches for Dark Stars. EAS Publ. Ser. **36**, 113–126 (2009). arXiv:0812.4005 [astro-ph]. http://www-personal.umich.edu/~ktfreese/. Accessed: March 18th, 2019

246. Genzel, R. et al.: Strongly baryon-dominated disk galaxies at the peak of galaxy formation ten billion years ago. Nat. **543**, 397–401 (16 March 2017)

247. King, L.J. et al.: A complete infrared Einstein ring in the gravitational lens system B1938+666. arXiv:astro-ph/9710171

248. Photo credit: ESA/Hubble & NASA https://www.spacetelescope.org/images/potw1151a/. Accessed: March 18th, 2019; picture on https://www.spacetelescope.org/images/opo0532d/. Accessed: March 18th, 2019; Jaime Trosper, June 2015

249. Raffelt, G.G. (Max Planck Institute for Physics, Werner Heisenberg Institute, München; homepage): http://wwwth.mpp.mpg.de/members/raffelt/. Accessed: March 18th, 2019. ◇ Ryden, B.: Introduction to Cosmology. Pearson Education Limited (2013). ◇ Griest, K.: All about MACHO. http://www.slac.stanford.edu/pubs/beamline/30/1/30-1-griest.pdf. Accessed: March 18th, 2019. ◇ The MACHO Project: http://wwwmacho.anu.edu.au/. Accessed: March 18th, 2019. ◇ Peng, E.W.: The Feasibility of Obtaining Finite Source Sizes from MACHO-Type Microlensing Light Curves. Astrophys. J. **475**, 43–46 (1997)

250. Turner, M.S.: The Meaning of EROS/MACHO. arXiv: astro-ph/9310019 (1993)

251. Wittman, D.M. et al.: Detection of weak gravitational lensing distortions of distant galaxies by cosmic dark matter at large scales. Nat. **405**, 143–148 (11 May 2000)

252. Jullo, E. et al.: Cosmological Constraints from Strong Gravitational Lensing in Galaxy Clusters. arXiv:1008.4802 [astro-ph.CO] or arXiv:1008.4802v1 [astro-ph.CO]

253. Image credit: X-ray: NASA/CXC/CfA/ Markevitch, M., et al.; Gravitational Lenses: NASA/STScI; ESO WFI; Magellan/U. Arizona/ Clowe, D., et al., Optically: NASA/STScI; Magellan/U. Arizona/ D. Clowe et al.; further references under: The Matter of the Bullet Cluster: https://apod.nasa.gov/apod/ap170115.html. Accessed: March 18th, 2019

254. Kolb, R.: Nothing get repulsive. In: Cole, K.C.: The Hole in the Universe: How Scientists Peered over the Edge of Emptiness and found Everything. Mariner Books (2012). ◇ Frieman, J. (2015), The Dark Energy Survey: https://www.darkenergysurvey.org/scientistoftheweek/josh-frieman/. Accessed: March 18th, 2019. ◇ Kolb, E., Turner, M.: The Early Universe. Westview Press (1994)

255. Formaggio, J.A.: Direct neutrino mass measurements after PLANCK. Phys. Dark Universe **4**, 75–80 (2014). ◇ The Ultimate Neutrino Page, Juho Sarkamo (2005): http://cupp.oulu.fi/neutrino/index.html. Accessed: March 18th, 2019. http://cupp.oulu.fi/neutrino/nd-mass.html. Accessed: March 18th, 2019

256. Laboratory measurements and limits for neutrino properties: http://cupp.oulu.fi/neutrino/nd-mass.html. Accessed: March 18th, 2019 (Hannestad)

257. Planck 2015 results. XIII. Cosmological parameters. https://arxiv.org/pdf/1502.01589.pdf. Accessed: March 18th, 2019. ◇ Patrignani, C. et al. (Particle Data Group), Chin. Phys. C **40**, 100001 (2016). ◇ Cosmological Neutrinos, G.G. Raffelt, http://wwwth.mpp.mpg.de/members/raffelt/mypapers/200507.pdf. Accessed: March 18th, 2019

258. http://www.katrin.kit.edu/. Accessed: March 18th, 2019. Drexlin, G., Hannen, V., Mertens, S., Weinheimer, C.: Current Direct Neutrino Mass Experiments. Adv. High Energy Phys. **2013**, Article ID 293986 (2013)

259. Aker, M. et al.: An improved upper limit on the neutrino mass from a direct kinematic method by KATRIN (KArlsruhe TRItium Neutrino Experiment). arXiv:1909.06048 (2019)

260. Aprile, E. et al. (XENON Collaboration): First Dark Matter Search Results from the XENON1T Experiment. Phys. Rev. Lett. **119**, 181301 (30 October 2017) and references therein

261. Vogel, J.K.: Searching for solar axions in the eV-mass region with the CCD detector at CAST; https://e-docs.geo-leo.de/handle/11858/00-1735-0000-0001-309A-C. Accessed: March 18th, 2019. ◇ Benson, B.A. et al., Kavli Institute for Cosmological Physics at the University of Chicago: CAST, The Solar Axion Search at CERN; http://collargroup.uchicago.edu/projects/axion/index.html. Accessed: March 18th, 2019

262. CAST: CERN Axion Solar Telescope: Computational Structure Formation. http://cast.web.cern.ch/CAST. Accessed: March 18th, 2019

263. Thomas, R. et al.; The extended epoch of galaxy formation: Age dating of $\sim$ 3600 galaxies with $2 < z < 6.5$ in the VIMOS Ultra-Deep Survey. Astron. Astrophys. **602**, A35 (2017). https://doi.org/10.1051/0004-6361/201628141. Accessed: March 18th, 2019

264. Raffelt, G.G. (Max Planck Institute for Physics, Werner Heisenberg Institute, München), home page: http://wwwth.mpp.mpg.de/members/raffelt/. Accessed: March 18th, 2019; and private communication, 2018

265. Schmitz, H.A.: Survey of Evidence for Top-Down versus Bottom-Up Evolution of Structure on Various Scales. In Potter, F. (ed.): 2nd Crisis in Cosmology Conference, CCC-2; ASP Conference Series **413**, Proceedings of the conference held 7–11 September 2008, at Port Angeles, Washington, USA; San Francisco, Astronomical Society of the Pacific, 98–108 (2009). ◇ W. Guzmán: https://www.astro.ufl.edu/~guzman/ast1002/class_notes/Ch15/Ch15b.html. Accessed: March 18th, 2019

266. Amanullah, R., Lidman, C. et al.: Spectra and Light Curves of Six Type Ia Supernovae at $0.511 < z < 1.12$ and the Union2 Compilation. Astrophys. J. **716**, 712–738 (2010). https://doi.org/10.1088/0004-637X/716/1/712. arXiv:1004.1711 [astro-ph.CO] Accessed: March 18th,

2019. ◇ Wolschin, G. (Heidelberg): Conference probes the dark side of the universe. CERN Courier (Feb. 2009). ◇ Durrer, R., Maartens, R.: Dark Energy and Modified Gravity. In: Ruiz-Lapuente, P. (ed.): Dark Energy: Observational and Theoretical Approaches. Cambridge University Press, 48–91 (2010). https://doi.org/10.1017/CBO9781139193627.003. Accessed: March 18th, 2019 (2008). arXiv:0811.4132v1 [astro-ph]. ◇ See also https://cerncourier.com/conference-probes-the-dark-side-of-the-universe/. Accessed: March 18th, 2019

267. NASA Exoplanet Archive, operated by the California Institute of Technology, under contract with the National Aeronautics and Space Administration under the Exoplanet Exploration Program. http://exoplanetarchive.ipac.caltech.edu/. Accessed: March 18th, 2019. California Institute of Technology, NASA Exoplanet Archive, NASA Exoplanet Science Institute

268. http://kepler.nasa.gov/images/mws/BatalhaPNAS_F1.large.jpg. Accessed: March 18th, 2019. ◇ Exploring exoplanet populations with NASA's Kepler Mission. http://www.pnas.org/content/111/35/12647.full. Accessed: March 18th, 2019. With courtesy of S. Howell, Project Scientist cc Michele Johnson NASA Ames Research Center Moffett Field, CA 94035-1000. ◇ Batalha, N.M.: Exploring exoplanet populations with NASA's Kepler Mission. Proc. Natl. Acad. Sci. U.S.A. **11(35)**, 12647–12654 (2014). https://doi.org/10.1073/pnas.1304196111. Accessed: March 18th, 2019

269. Beaulieu, J.P. et al.: Discovery of a cool planet of 5.5 Earthmasses through gravitational microlensing. Nat. **439**, 437–440 (2006). ◇ http://planet.iap.fr. Accessed: March 18th, 2019. https://www.nature.com/articles/nature04441. Accessed: March 18th, 2019

270. Dai, X., Guerras, E.: Probing extragalactic planets using quasar microlensing. Astrophys. J. Lett. **853(2)**, L27 (2018)

271. Benneke, B. et al.: Water Vapor on the Habitable-Zone Exoplanet K2-18b. Submitted to Astron. J. (2019). ◇ https://www.sciencedaily.com/releases/2019/09/190911121950.htm. Accessed: September 15th, 2019

272. Rees, M.: Just Six Numbers: The Deep Forces That Shape the Universe. Basic Books, New York (2001)

273. Oberhummer, H., Puntigam, M., Gruber, W.: Das Universum ist eine Scheißgegend (The universe is a fucking place). Carl Hanser Verlag, Berlin (2015)

274. Park, R.L.: Superstition: Belief in the Age of Science. Princeton University Press, New Jersey (2009), p. 11. ISBN 978-0-691-13355-3

275. Stenger, V.J.: Anthropic Design: Does the Cosmos Show Evidence of Purpose? Skeptical Inquirer **23(4)**, 40–63 (July–August 1999)

276. http://en.wikipedia.org/wiki/Triple-alpha_process. Accessed: March 18th, 2019

277. Csótó, A. et al.: Fine-tuning the basic forces of nature through the triple-alpha process in red giant stars. Nucl. Phys. A **688(1–2)**, 560–562 (2001). https://arxiv.org/abs/nucl-th/0010052. Accessed: March 18th, 2019

278. http://www.fourmilab.ch/documents/OhMyGodParticle/. Accessed: September 15th, 2019

279. 'Introduction to Astronomy' (2017): https://www.physik.uni-bielefeld.de/~verbiest/Teaching/Intro_to_Astronomy/Week15_Ex.pdf. Accessed: September 15th, 2019

280. Fließbach, T.: Allgemeine Relativitätstheorie. 7th ed., Springer Spektrum, Berlin, Heidelberg (2016)

281. Primordial Alchemy – G. Steigman: Counting Relativistic Degrees of Freedom https://ned.ipac.caltech.edu/level5/March02/Steigman/Steigman1_2_1.html. Accessed: March 18th, 2019

Further Reading

Rossi, B.: High-Energy Particles, Prentice-Hall, Englewood Cliffs, N.J., 4th print (1965)

Lee, T.D., Yang, C.N.: Question of Parity Conservation in Weak Interactions. Phys. Rev. **104**, 254 (1956)

Furth, H.P.: Perils of Modern Living. Publisher: Condé Nast (USA), New York City. Originally published in the New Yorker Magazine (1956)

Wu, C.S. et al.: Experimental Test of Parity Conservation in Beta Decay. Phys. Rev. **105**, 1413 (1957)

Hagedorn, R.: Relativistic Kinematics. W. A. Benjamin Inc. Reading, Mass. (1963)

Christenson, J.H. et al.: Evidence for the 2π Decay of the K_2^0 Meson. Phys. Rev. Lett. **13**, 138 (1964)

Hayakawa, S.: Cosmic Ray Physics. Wiley-Interscience, New York (1969)

Georgi, H., Glashow, S.L.: Unity of All Elementary Particle Forces. Phys. Rev. Lett. **32**, 438 (1974)

Allkofer, O.C.: Introduction to Cosmic Radiation, Thiemig, Munich (1975)

Adams, D.J.: Cosmic X-ray Astronomy, Adam Hilger Ltd., Bristol (1980)

Silk, J.: The Big Bang—The Creation and Evolution of the Universe, Freeman, New York (1980)

Shu, F.H.: The Physical Universe: An Introduction to Astronomy, Univ. Science Books, Mill Valley, California (1982)

Allkofer, O.C., Grieder, P.K.F.: Cosmic Rays on Earth. Physics Data Series, Fachinformationszentrum Karlsruhe, Report 25–1, Karlsruhe, Germany (1984)

Hillier, R.: Gamma-Ray Astronomy. Oxford University Press, Oxford, UK (1985)

Arnett, W.D. et al.: Supernova 1987A. Ann. Rev. Astron. Astrophys. **27**, 629–700 (1989)

Bahcall, J.A.: Neutrino Astronomy, Cambridge University Press, Cambridge (1990)

Gaisser, T.K.: Cosmic Rays and Particle Physics, Cambridge University Press, Cambridge (1990)

Weinberg, S.: The First Three Minutes, BasicBooks, New York (1993)

Gehrels, N. et al.: The Compton Gamma Ray Observatory. Sci. Am., 38–45 (Dec. 1993)

Rohlf, J.W.: Modern Physics from α to Z^0. J. Wiley & Sons, New York (1994)

Biermann, P.L.: The Origin of Cosmic Rays. MPI Radioastronomie Bonn, MPIfR 605, astro-ph/9501003. Phys. Rev. D **51**, 3450 (1995)

Vignaud, D.: Solar and Supernovae Neutrinos. 4th School on Non-Accelerator Particle Astrophysics, Trieste 1995, DAPNIA/SPP 95-26, Saclay

Gaisser, T.K. et al.: Particle Astrophysics with High Energy Neutrinos, Phys. Rep. **258(3)**, 173–236 (1995)

Plunkett, S.P. et al.: A Search for Ultra-High Energy Counterparts to Gamma-Ray Bursts. Astrophys. Space Sci. **231(1–2)** 271–276 (1995). astro-ph/9508083 (1995)

© Springer Nature Switzerland AG 2020
C. Grupen, *Astroparticle Physics*, Undergraduate Texts in Physics,
https://doi.org/10.1007/978-3-030-27339-2

Milgrom, M., Usov, V.: Possible Association of Ultra-High Energy Cosmic Ray Events with Strong Gamma-Ray Bursts, Astrophys. J. **449**, 37 (1995). astro-ph/9505009 (1995)

Berezinsky, V.: High Energy Neutrino Astronomy. Gran Sasso Lab. INFN, LNGS-95/04 (1995)

Jelley, J.V., Weekes, T.C.: Ground-Based Gamma-Ray Astronomy. Sky & Telescope **Sept. 95**, 20–24 (1995)

Ferguson, K.: Prisons of Light—Black Holes. Cambridge University Press, Cambridge (1996)

Grupen, C.: Particle Detectors. Cambridge University Press, Cambridge (1996). ◇ Grupen, C., Shwartz, B.: Particle Detectors. Cambridge University Press, Cambridge, 2nd ed. (1998)

Spiro, M., Aubourg, E.: Experimental Particle Astrophysics. Int. Conf. on High Energy Physics, Warsow, Poland (1996)

Mannheim, K. et al.: The Gamma-Ray Burst Rate at High Photon Energies. Astrophys. J. **467**, 532–536 (1996). astro-ph/9605108 (1996)

Melissinos, A.C.: Lecture Notes on Particle Astrophysics. Univ. of Rochester 1995, UR-1841 (Sept. 1996)

Battistoni, G., Palamara, O.: Physics and Astrophysics with Multiple Muons, Gran Sasso Lab. INFN/AE 96/19, L'Aquila, Abruzzo, Italy (1996)

Ellis, J.: Astroparticle Physics—A Personal Outlook, Nucl. Phys. Proc. Suppl. **48** 522–544 (1996). CERN-TH/96-10. astro-ph/9602077 (1996)

Raffelt, G.G.: Astro-Particle Physics. Europhysics Conference on High-Energy Physics, August 1997, Jerusalem. hep-ph/9712548 (1997)

Halzen, F.: The Search for the Source of the Highest Energy Cosmic Rays. Univ. of Madison Preprint MAD-PH 97-990, Madison, USA. astro-ph/9704020 (1997)

Burdman, G. et al.: The Highest Energy Cosmic Rays and New Particle Physics. Phys. Lett. B **417**, 107–113 (1998). Univ. of Madison Preprint MAD-PH 97-1014, Madison, USA. hep-ph/9709399 (1997)

Lipari, P.: Cosmology, Particle Physics and High Energy Astrophysics. Conf. Proc. Frontier Objects in Astrophysics and Particle Physics; Giovannelli, F., Mannocchi, G. (eds.); Workshop, Vulcano, Italy; **57**, 595 (1997)

Raffelt, G.G.: Dark Matter: Motivation, Candidates and Searches. Proc. 1997 European School of High Energy Physics, Menstrup, Denmark. hep-ph/9712538 (1997)

De Marzo, C.N. (for the Airwatch Collaboration): Extreme Energy Cosmic Rays (EECR), Observation Capabilities of an "Airwatch from Space" Mission. Nucl. Phys. B Proc. Suppl. **70(1–3)**, 515–517 (1999). physics/9712039 (1997)

Meyer, H. (University of Wuppertal, Germany): Photons from the Universe: New Frontiers in Astronomy. hep-ph/9710362 (1997)

Guth, A.: The Inflationary Universe. Vintage, Vancouver, USA (1998)

Shiozawa, M. et al. (Super-Kamiokande collaboration): Search for Proton Decay via $p \rightarrow e^{+}\pi^{0}$ in a Large Water Cherenkov Detector. Phys. Rev. Lett. **81**, 3319 (1998)

Cowan, G.: Statistical Data Analysis, Oxford University Press, Oxford (1998)

Filippenko, A.V., Riess, A.G.: Results from the High-z Supernova Search Team. Phys. Rep. **307** 31–44 (1998). astro-ph/9807008 (1998)

Greiner, J.: Gamma-Ray Bursts: Old and New. astro-ph/9802222 (1998). Mem. Soc. Ast. It. **70** 891 (1999)

Schramm, D.N., Turner, M.S.: Big bang nucleosynthesis enters the precision era. Rev. Mod. Phys. **70**, 303 (1998)

Bock, R.K., Vasilescu, A.: The Particle Detector Briefbook. Springer, Berlin, Heidelberg (1998)

Allday, J.: Quarks, Leptons and the Big Bang. Inst. of Physics Publ., Bristol (1998)

Börner, G.: The Early Universe: Facts and Fiction. Springer, Berlin, 4th ed., 2nd corr. print (2004)

Hogan, C.J.: The Little Book of the Big Bang. Copernicus, Springer, New York (1998)

Krauss, L.M.: A New Cosmological Paradigm: The Cosmological Constant and Dark Matter. hep-ph/9807376 (1998)

Capelle, K.S. et al.: On the Detection of Ultra High Energy Neutrinos with the Auger Observatory. Astropart. Phys. **8**, 321–328 (1998). astro-ph/9801313

Farrar, G.R.: Can Ultra High Energy Cosmic Rays be Evidence for New Particle Physics? astro-ph/9801020 (1998)

Super-Kamiokande Collaboration: Measurement of a Small Atmospheric v_μ / v_e Ratio. Phys. Lett. B **433**, 9–18 (1998). hep-ex/9803006 (1998)

Turner, M.S.: Cosmology Solved? Quite Possibly! Publ. Astron. Soc. Pac. **111**, 264–273 (1999). astro-ph/9811364 (1998)

Izotov, Y.I., Thuan, T.X.: Heavy-element abundances in blue compact galaxies, Astrophys. J. **511**, 639–659 (1999)

Burles, S. et al.: Sharpening the predictions of big-bang nucleosynthesis. Phys. Rev. Lett. **82**, 4176 (1999). astro-ph/9901157 (1999)

Olive, K.A.: Primordial Big Bang Nucleosynthesis. astro-ph/9901231 (1999). NATO Advanced Study Institute: Summer School on Theoretical and Observational Cosmology, Cargese, Corsica, France, 1998. CNUM: C98-08-17.3

Veneziano, G.: Challenging the Big Bang: A Longer History of Time. CERN-Courier **March 1999**, 18

Greene, B.: The Elegant Universe. Vintage 2000, London (1999)

Grupen, C.: Astroteilchenphysik: das Universum im Licht der kosmischen Strahlung. Vieweg, Braunschweig (2000)

Hogan, C.J.: Why the universe is just so. astro-ph/9909295 (Sept. 1999). Rev. Mod. Phys. **72**, 1149–1161 (2000)

Klapdor-Kleingrothaus, H.V., Zuber, K.: Teilchenastrophysik. Teubner, Stuttgart (1997). And: Particle Astrophysics. Inst. of Physics Publ., Bristol (2000)

Rees, M.: New Perspectives in Astrophysical Cosmology. Cambridge University Press (2000)

Grieder, P.K.F.: Cosmic Rays at Earth. Elsevier Science, Amsterdam, The Netherlands (2001)

Cole, K.C.: The Hole in the Universe. The Harvest Book/Harcourt Inc., San Diego (2001)

Rees, M.: Just Six Numbers: The Deep Forces That Shape the Universe, Basic Books, New York (2001)

Bennett, C.L. et al.: First Year Wilkinson Microwave Anisotropy Probe (WMAP) Observations: Maps and Basic Results. Astrophys. J. Suppl. **148** 1 (2003). http://map.gsfc.nasa.gov. Accessed: April 1st, 2019

Steinhardt, P.J.: Quintessential Cosmology and Cosmic Acceleration. http://physics.princeton.edu/~steinh/prit4.pdf. Accessed: April 1st, 2019. And: Peebles, P.J.E., Ratra, B.: The cosmological constant and dark energy. Rev. Mod. Phys. **75**, 559 (2003)

Peacock, J.A.: Cosmological Physics. Cambridge University Press (2003)

Bilenky, S.M.: The History of Neutrino Oscillations. hep-ph/0410090 (2004)

Particle Data Group (S. Eidelman et al.): Review of Particle Physics. Phys. Lett. B **592**, 1–1109 (2004). ◇ Latest ed.: Particle Data Group (C. Patrignani et al.), Chin. Phys. C **40**, 100001 (2016)

Greene, B.: The Fabric of the Cosmos: Space, Time and the Texture of Reality. Alfred A. Knopf, New York (2004)

Bergström, L., Goobar, A.: Cosmology and Particle Astrophysics, Wiley, Chichester, 2nd ed. (2004)

Sarkar, U.: Particle and Astroparticle Physics. CRC Press, Taylor & Francis, Abingdon, UK (2007)

Weinberg, S.: Cosmology. Oxford University Press, Oxford (2008)

Stanev, T.: High Energy Cosmic Rays. Springer, Heidelberg, 2nd ed. (2009)

Perkins, D.H.: Particle Astrophysics. Oxford University Press, Oxford, UK, 2nd ed. (2009)

Giani, S., Leroy, C., Rancoita, P.G.: Cosmic Rays for Particle and Astroparticle Physics. World Scientific, Singapore 596224 (2011)

Bertolotti, M.: Celestial Messengers: Cosmic Rays: The Story of a Scientific Adventure (Astronomers' Universe). Springer, Berlin (2013)

Catling, D.C.: Astrobiology. Oxford University Press, Oxford, UK (2013)

Liddle, A.: An Introduction to Modern Cosmology. Wiley, Hoboken, New Jersey, USA, 3rd ed. (2015)

Longstaff, A.: Astrobiology: An Introduction (Astronomy and Astrophysics). Taylor & Francis, Abingdon, UK (2015)

Unsöld, A., Baschek, B.: Der Neue Kosmos. Springer Spektrum, Berlin, Heidelberg, 7th ed. (2015)

Ryden, B.: Introduction to Cosmology. Cambridge University Press, Cambridge, 2nd ed. (2016). And: Astrophysics and Cosmology, Oxford University Press, Oxford (2018)

Gaisser, T.K., Engel, R., Resconi, E.: Cosmic Rays and Particle Physics. Cambridge University Press, Cambridge, 2nd ed. (2016)

Carroll, B.W., Ostlie, D.A.: An Introduction to Modern Astrophysics. Cambridge University Press, Cambridge, 2nd ed. (2017)

De Angelis, A., Pimenta, M.J.M.: Introduction to Particle and Astroparticle Physics: Multimessenger Astronomy and its Particle Physics Foundations. Springer, Heidelberg and New York, 2nd ed. (2018)

Maggiore, M.: Gravitational Waves. Vol. 2: Astrophysics and Cosmology. Oxford University Press, Oxford (2018)

Rothery, D.A., Gilmour, I., Sephton, M.A. (eds.): An Introduction to Astrobiology. Cambridge University Press, Cambridge, UK, 3rd ed. (2018)

Web site of the ALEPH collaboration: http://alephwww.cern.ch. Accessed: April 1st, 2019

Halzen, F.: Ice Fishing for Neutrinos. AMANDA Homepage: http://icecube.berkeley.edu/amanda/ice-fishing.html. Accessed: April 1st, 2019

Gamma Ray Bursts, NASA: http://science.nasa.gov/ems/12_gammarays.html. Accessed: April 1st, 2019

Web site of the European Organisation for Nuclear Research (CERN): http://www.cern.ch. Accessed: April 1st, 2019

Web site of the Super-Kamiokande experiment: http://www-sk.icrr.u-tokyo.ac.jp/sk/index-e.html. Accessed: April 1st, 2019

Index*

A

Abell 1689, 417
Abell 754, 218
Absolute brightness, *513*
Absolute zero, *513*
Absorber crystal, 88
Absorption
   coefficient, 67
   line, neutrino, 420
Abundance, *513*
   cosmic, *521*
   deuterium, 345
   helium-4 (^{4}He), 30, 339, 345, 347, 351
   light nuclei, 340, 347, 348
   lithium-7 (^{7}Li), 348, 350
   measurement, 350
   predicted, 348
   primordial, 20
     deuterium, 348
     helium-3 (^{3}He), 348
     helium-4 (^{4}He), 30
     light nuclei, 347
Accelerated expansion, 384, 432
Acceleration
   by binaries, 112
   by pulsars, 109
   by shock waves, 102
   by sunspots, 101
   equation, 306, 306, 384, *513*
   in gravitational potentials, 112
   mechanism, 99, 99, 107, *526*
   of particles in pulsars, 450
   of particles in supernovae, 104, 450
   of protons, 167
   parameter, 309, 427, 451, *513*

shock, 116
Accelerator, *513*
   contributions to cosmic rays, 20
   cosmic, 450
   data, number of neutrino families, 354
   experiments, 449
Accretion disk, 112, 113, 215, 258, *513*
Acoustic peak, 369, 370, *514*
Active galactic nuclei, 23, 114, 180, 194, 216, 287, 450, *514*
Active galaxy, *514*
Activity, potassium-40 (^{40}K), 170
Adams, D., 401
Additional Higgs field, 335
Additional neutrino families, 352
Adiabatic expansion, 305
Advanced X-ray Astrophysics Facility, *see* AXAF
AGASA, 126, *514*
Age
   of life, 442
   of the universe, *514*
Aggregation, gravitational, 402
AGILE satellite, 195
AGN, *see* active galactic nuclei
AIGO facility, 227
Air scintillation, 77, 191, 266
Air shower
   acoustic detection, 281
   Cherenkov technique, 82, 124, 189, 191, 195, *514*
   extensive, 12, 124, 258, 262, *526*
     correlations, 273
     measurement technique, 265
   radio measurement, 275

*Underlined page numbers refer to main entries. Page numbers in italics apply to the glossary.
© Springer Nature Switzerland AG 2020
C. Grupen, *Astroparticle Physics*, Undergraduate Texts in Physics,
https://doi.org/10.1007/978-3-030-27339-2

Printed in the United States
By Bookmasters